Wireless Sensor Networks

Wireless Sensor Networks

Signal Processing and Communications Perspectives

Edited by

Ananthram Swami
Army Research Laboratory, USA

Qing Zhao
University of California at Davis, USA

Yao-Win Hong
National Tsing Hua University, Taiwan

Lang Tong
Cornell University, USA

John Wiley & Sons.

Other Wiley Editorial Offices

John Wiley & Sons Inc., 111 River Street, Hoboken, NJ 07030, USA

Jossey-Bass, 989 Market Street, San Francisco, CA 94103-1741, USA

Wiley-VCH Verlag GmbH, Boschstr. 12, D-69469 Weinheim, Germany

John Wiley & Sons Australia Ltd, 42 McDougall Street, Milton, Queensland 4064, Australia

John Wiley & Sons (Asia) Pte Ltd, 2 Clementi Loop #02-01, Jin Xing Distripark, Singapore 129809

John Wiley & Sons Canada Ltd, 6045 Freemont Blvd, Mississauga, Ontario, L5R 4J3, Canada

Wiley also publishes its books in a variety of electronic formats. Some content that appears
in print may not be available in electronic books.

Anniversary Logo Design: Richard J. Pacifico

Library of Congress Cataloging-in-Publication Data

Wireless sensor networks : signal processing and communications perspectives / edited by
A. Swami ... [et al.].
        p. cm.
  Includes bibliographical references.
  ISBN 978-0-470-03557-3 (cloth)
  1. Sensor networks. 2.   Wireless LANs. 3.   Signal processing – Digital techniques.   I.
Swami, Ananthram.
  TK7872.D48W585 2007
681′.2 – dc22

                                                                    2007021100

British Library Cataloguing in Publication Data

A catalogue record for this book is available from the British Library

ISBN: 978-0-470-03557-3 (HB)

Typeset in 10/12 Times by Laserwords Private Limited, Chennai, India
Printed and bound in Great Britain by Antony Rowe Ltd, Chippenham, Wiltshire
This book is printed on acid-free paper responsibly manufactured from sustainable forestry
in which at least two trees are planted for each one used for paper production.

Contents

Part II   Signal Processing for Sensor Networks   117

8   Distributed Learning in Wireless Sensor Networks   185
Joel B. Predd, Sanjeev R. Kulkarni, and H. Vincent Poor

9   Graphical Models and Fusion in Sensor Networks   215
Müjdat Çetin, Lei Chen, John W. Fisher III, Alexander T. Ihler,
O. Patrick Kreidl, Randolph L. Moses, Martin J. Wainwright,
Jason L. Williams, and Alan S. Willsky

List of Contributors

Müjdat Çetin
Sabanci University
Faculty of Engineering and Natural
   Sciences
Orhanli – Tuzla
34956 Istanbul
mcetin@sabanciuniv.edu

Jean-François Chamberland
Texas A&M University
Department of Electrical and Computer
   Engineering
College Station, TX 77843-3128
USA
chmbrlnd@tamu.edu

Lei Chen
Massachusetts Institute of Technology
77 Massachusetts Avenue, 32-D568
Cambridge, MA 02139
USA
lchen@mit.edu

Yunxia Chen
Department of Electrical and Computer
   Engineering
University of California, Davis
Davis, CA 95616
USA
yxchen@ece.ucdavis.edu

Anthony Ephremides
Dept. of Electrical Engineering
University of Maryland
College Park, MD 20742
USA
etony@ece.umd.edu

John W. Fisher III
Massachusetts Institute of Technology
77 Massachusetts Avenue, 32-D468
Cambridge, MA 02139
USA
fisher@csail.mit.edu

Michael Gastpar
University of California, Berkeley
Department of Electrical Engineering and
   Computer Sciences
265 Cory Hall
Berkeley, CA 94110-1770
USA
gastpar@eecs.berkeley.edu

Georgios B. Giannakis
Dept. of Electrical and Computer
   Engineering
University of Minnesota
200 Union Street SE
Minneapolis, MN 55455,
USA
georgios@umn.edu

Arvind Giridhar
Interest Rate Products Strategies, FICC
Goldman Sachs & Co
85 Broad Street, New York, NY 10004
USA
arvind.giridhar@gmail.com

Yao-Win Hong
101 Section 2 Kuang-Fu Rd.
National Tsing Hua University
EECS Building 620B

Hsinchu, 30013
Taiwan
ywhong@ee.nthu.edu.tw

Alexander T. Ihler
Toyota Technological Institute
University Press Building
1427 East 60th Street, Second Floor
Chicago, Illinois 60637
USA
ihler@tti-c.org

Pradeep Khosla
Department of Electrical and Computer
   Engineering
Carnegie Mellon University
5000 Forbes Ave
Carnegie Mellon University
Pittsburgh
PA 15213
USA
pkk@ece.cmu.edu

O. Patrick Kreidl
Massachusetts Institute of Technology
77 Massachusetts Avenue, 32-D572
Cambridge, MA 02139
USA
opk@mit.edu

Vikram Krishnamurthy
Department of Electrical and Computer
   Engineering
University of British Columbia
Vancouver, V6T 1Z4
Canada
vikramk@ece.ubc.ca

Sanjeev R. Kulkarni
Department of Electrical Engineering
Princeton University
Princeton, NJ 08540
USA
kulkarni@princeton.edu

P.R. Kumar
University of Illinois

CSL
1308 West Main Street
Urbana, IL 61801
USA
prkumar@uiuc.edu

Zhi-Quan Luo
Dept. of Electrical and Computer
   Engineering
University of Minnesota
200 Union Street SE
Minneapolis, MN 55455,
USA
luozq@umn.edu

Michael Maskery
Department of Electrical and Computer
   Engineering
University of British Columbia
Vancouver, V6T 1Z4
Canada
mikem@ece.ubc.ca

Saswat Misra
Army Research Laboratory
AMSRD-ARL-CI-CN
2800 Powder Mill Rd.
Adelphi, MD 20783
USA
sm353@cornell.edu

Randolph L. Moses
708 Dreese Laboratory
Department of Electrical and Computer
   Engineering
The Ohio State University
2015 Neil Avenue
Columbus, OH 43210
USA
moses.2@osu.edu

Rohit Negi
Department of Electrical and Computer
   Engineering
Carnegie Mellon University
5000 Forbes Ave

Carnegie Mellon University
Pittsburgh, PA 15213
USA
negi@ece.cmu.edu

Minh Hanh Ngo
Department of Electrical and Computer
   Engineering
University of British Columbia
Vancouver V6T 1Z4
Canada
minhn@ece.ubc.ca

H. Vincent Poor
Department of Electrical
   Engineering
Princeton University
Princeton, NJ 08540
USA
poor@princeton.edu

Joel B. Predd
RAND Corporation
4570 Fifth Avenue
Pittsburgh, PA 15213
USA
jpredd@rand.org

Yaron Rachlin
Accenture Technology Labs
161 North Clark Street
Chicago, IL 60601
USA
yaron.rachlin@alumni.cmu.edu

Alejandro Ribeiro
Deptartment of Electrical and Computer
   Engineering
University of Minnesota
200 Union Street SE
Minneapolis, MN 55455
USA
aribeiro@ece.umn.edu

Anna Scaglione
Cornell University

School of Electrical and Computer
   Engineering
325 Rhodes Hall
Ithaca, NY 14853
USA
anna@ece.cornell.edu

Ioannis D. Schizas
Dept. of Electrical and Computer
   Engineering
University of Minnesota
200 Union Street SE
Minneapolis, MN 55455
USA
schiz001@umn.edu

Birsen Sirkeci-Mergen
UC Berkeley
EECS Department
273 Cory Hall
University of California
Berkeley, CA 94720
USA
bs233@eecs.berkeley.edu

Ananthram Swami
PO Box 4640
Silver Spring
MD 20914-4640
USA
a.swami@ieee.org

Lang Tong
School of Electrical and Computer
   Engineering
Center for Applied Mathematics
384 Frank H.T. Rhodes Hall
Cornell University
Ithaca, NY 14853
USA
ltong@ece.cornell.edu

Pramod K. Varshney
Department of Electrical Engineering and
   Computer Science
335 Link Hall

Syracuse University
Syracuse
New York 13244
USA
varshney@syr.edu

Venugopal V. Veeravalli
ECE Department and Coordinated
    Science Lab
University of Illinois at Urbana-Champaign
1308 West Main Street
Urbana IL 61801
USA
vvv@uiuc.edu

Martin J. Wainwright
University of California at Berkeley
Department of Electrical Engineering and
    Computer Sciences
263 Cory Hall
Berkeley, CA 94720
USA
wainwrig@eecs.berkeley.edu

Jason L. Williams
Massachusetts Institute of Technology
77 Massachusetts Avenue, 32-D572

Cambridge, MA 02139
USA
jlwil@mit.edu

Alan S. Willsky
Massachusetts Institute of Technology
77 Massachusetts Avenue, 32-D582
Cambridge, MA 02139
USA
willsky@mit.edu

Jin-Jun Xiao
Dept. of Electrical and Systems
    Engineering
Washington University
One Brookings Drive
St. Louis, MO 63130
USA
xiao@ese.wustl.edu

Qing Zhao
Department of Electrical and Computer
    Engineering
University of California Davis
Davis, CA 95616
USA
qzhao@ece.ucdavis.edu

1

Introduction

Modern wireless sensor networks are made up of a large number of inexpensive devices that are networked via low power wireless communications. It is the networking capability that fundamentally differentiates a sensor network from a mere collection of sensors, by enabling cooperation, coordination, and collaboration among sensor assets. Harvesting advances in the past decade in microelectronics, sensing, analog and digital signal processing, wireless communications, and networking, wireless sensor network technology is expected to have a significant impact on our lives in the twenty-first century. Proposed applications of sensor networks include environmental monitoring, natural disaster prediction and relief, homeland security, healthcare, manufacturing, transportation, and home appliances and entertainment. Sensor networks are expected to be a crucial part in future military missions, for example, as embodied in the concepts of network centric warfare and network-enabled capability.

Wireless sensor networks differ fundamentally from general data networks such as the internet, and as such they require the adoption of a different design paradigm. Often sensor networks are application specific; they are designed and deployed for special purposes. Thus the network design must take into account the specific intended applications. More fundamentally, in the context of wireless sensor networks, the broadcast nature of the medium must be taken into account. For battery-operated sensors, energy conservation is one of the most important design parameters, since replacing batteries may be difficult or impossible in many applications. Thus sensor network designs must be optimized to extend the network lifetime. The energy and bandwidth constraints and the potential large-scale deployment pose challenges to efficient resource allocation and sensor management. A general class of approaches – cross-layer designs – has emerged to address these challenges. In addition, a rethinking of the protocol stack itself is necessary so as to overcome some of the complexities and unwanted consequences associated with cross-layer designs.

This edited book focuses on theoretical aspects of wireless sensor networks, aiming to provide signal processing and communication perspectives on the design of large-scale sensor networks. Emphasis is on the fundamental properties of large-scale sensor networks, distributed signal processing and communication algorithms, and novel cross-layer design paradigms for sensor networking.

Wireless Sensor Networks: Signal Processing and Communications Perspectives   A. Swami, Q. Zhao, Y.-W. Hong and L. Tong
© 2007 John Wiley & Sons, Ltd

The design of a sensor network requires the fusion of ideas from several disciplines. Of particular importance are the theories and techniques of distributed signal processing, recent advances in collaborative communications, and methodologies of cross-layer design.

This book elucidates key issues and challenges, and the state-of-the-art theories and techniques for the design of large-scale wireless sensor networks. For the signal processing and communications research community, the book provides ideas and illustrations of the application of classical theories and methods in an emerging field of applications. For researchers and practitioners in wireless sensor networks, this book complements existing texts with the infusion of analytical tools that will play important roles in the design of future application-specific wireless sensor networks. For students at senior and the graduate levels, this book identifies research directions and provides tutorials and bibliographies to facilitate further investigations.

The book is divided into three parts: I Fundamental Properties and Limits; II Signal Processing for Sensor Networks; and III Communications, Networking and Cross-Layer Designs.

Part I Fundamental Properties and Limits

Despite the remarkable theoretical advances in link-level communications, scientific understanding of and design methodologies for large-scale complex networks, such as wireless sensor networks, are still primitive. The variety of potential applications and sensor devices, the dynamics and unreliability of the wireless communication medium, and the stringent resource constraints all present major obstacles to a fundamental understanding of the structure, behavior, and dynamics of large-scale possibly heterogeneous sensor networks.

Part I presents representative samples of recent developments in the discovery of fundamental properties and performance limits of large-scale sensor networks. The aim is to show that despite the vast differences in applications and communication environments, there exist universal laws and performance bounds, especially in the asymptotic regime, that may lead to systematic approaches to the design of such large-scale complex networks.

Chapter 2 by Gastpar focuses on communication aspects: the rate and fidelity of transporting sensor measurements to a fusion center for data processing. Based on a digital communication architecture that separates source coding from channel coding, limits on the achievable rate-distortion regions under power constraints are presented. Compelling examples are given to illustrate the possible performance loss incurred by such a separated design.

Chapter 3 by Giridhar and Kumar addresses in-network information processing. Instead of transmitting measurements to a fusion center for processing, sensor nodes are responsible for computing a certain function of all measurements, for example, the mean or the maximum, through inter-node communications. The quantities of interest are the maximum rate at which such in-network computation can be performed and how it scales with network size. Interestingly, the scaling behavior depends not only on the communication topology of the network, but also on the properties of the function being calculated.

Chapter 4 by Negi, Rachlin, and Khosla is concerned with the fundamental relationship between the number of sensor measurements and the ability of the network to identify the state of the environment being monitored. The focus of the chapter is on detection problems where the number of possible hypotheses is large. For this problem of large-scale detection, a lower limit on the sensing capacity of sensor networks is derived that characterizes the minimum rate at which the number of sensor measurements should scale with the number of hypotheses in order to achieve the desired detection accuracy. An intriguing analogy between the sensing capacity of sensor networks and channel coding theory for communication channels points to the possibility of porting the large body of results available on communication channels to the design of large-scale sensor networks.

The last chapter of Part I by Chen and Zhao focuses on the lifetime of sensor networks to address the energy constraint. Given that the sensor network lifetime depends on network architectures, specific applications, and various parameters across the entire protocol stack, an accurate characterization of network lifetime as a function of key design parameters is notably difficult to obtain. It is shown in this chapter that there is, in fact, a simple law that governs the network lifetime for all applications (event-driven, clock-driven, or query-driven), under any network configuration (centralized, ad hoc, or hierarchical). This law of network lifetime reveals the key role of two physical layer parameters – residual energy and channel state – and a general principle for the design of upper layer network protocols.

This set of four chapters points to promising directions toward a scientific understanding of core principles and fundamental properties of large complex sensor networks. Many problems, however, remain. When is the separated design of source coding and channel coding sufficient to achieve the best scaling behavior? How can delay and energy constraints be adequately modeled within the information theoretic framework? What are the fundamental tradeoffs between communication and computation under energy and complexity constraints? These are only a few of the many challenges we face in advancing the basic science of large-scale wireless sensor networks.

Part II Signal Processing for Sensor Networks

Part II of this book focuses on signal processing problems in sensor networks. Fundamental to sensor signal processing are distributed information processing at the individual sensor nodes and the fusion of sensor measurements for global signal processing.

Distributed detection is a classical subject that attracted considerable interest in the late 1980s and early 1990s when the power of DSP and wired communications enabled the networking of distributed radar systems for target detection and tracking. Radars generate enormous amount of data, and transmitting all the measurements to a central processing location is neither feasible nor necessary. The natural research focus then was how to quantize measurements at the local sensor nodes and how to derive optimal inference algorithm at the fusion center.

While many technical issues in classical distributed detection remain in modern wireless sensor networks, several new challenges have arisen. The fading and broadcast aspects of the wireless transmission medium, the presence of interference, and constraints on energy and power demand a new design paradigm. Chapter 6 by Veeravalli and Chamberland is an introduction to distributed detection for modern wireless sensor networks. This chapter provides an informative survey of classical results and sheds new light on the interplay among quantization, sensor fusion under resource constraints, and optimal detection performance. The approach based on asymptotic statistical techniques is especially appropriate for large sensor networks.

Distributed estimation deals with statistical inference problems when the underlying phenomenon cannot be modeled by a few disjoint hypotheses; there are in general innumberable possible distributions from which sensor measurements are generated. It is thus not possible to design a sensor quantization scheme that is uniformly optimal. Chapter 7 on distributed estimation by Ribeiro, Schizas, Xiao, Giannakis and Luo provides a broad coverage of estimation problems in wireless sensor networks when sensor measurements must be quantized or compressed. Both point estimation and Bayesian setups are considered, and performance bounds provided.

Chapter 8 on distributed learning by Predd, Kulkarni, and Poor introduces learning theory and techniques for sensor networks. The focus here is on nonparametric statistical inference under bandwidth and energy constraints. The authors develop a framework for distributed learning and draw connections with classical concepts. Different network architectures and learning techniques are presented.

Chapter 9 by Çetin, Chen, Fisher, Ihler, Moses, Wainwright, Williams and Willsky introduces graphical models and fusion for sensor networks. Statistical correlations in sensor measurements have a natural graphical model representation in which the graph vertices represent the random variables and corresponding edges their statistical dependency. The study of graphical models has led to fundamental insights in coding and decoding techniques in communications. For statistical inference using wireless sensor networks, one can take the view that inference should be derived from *a posteriori* distributions (belief), and the calculation of such distributions in a distributed fashion is at the core of sensor information processing. This chapter provides an introduction to various message passing techniques and their applications in sensor self-localization, tracking, and data association problems. Energy and bandwidth constraints are once again key design parameters.

The set of four chapters in Part II have explored important aspects of signal processing in sensor networks, including detection, estimation, learning and fusion. However, many challenges remain. What is the role of quantization when nodes must code their bits to cope with fading and noisy channels, or when they must otherwise packetize the data? What is the right architecture for decentralized inference in a sensor network, keeping in mind that the sensing graph is not identical to the communications graph? How should multi-hop delays and temporal (de)correlation be modeled and handled? What is the role of collaboration and consensus in a sensor network? Given that energy and bandwidth constraints are severe, overhead in bits (e.g., in the headers, or number of messages) or Joules (e.g., energy consumed in processing, reception, and transmission) should not be ignored. Finally, while asymptotic analyses provide critical insights and design guidelines, issues related to finite networks need to be explored.

Part III Communications, Networking and Cross-Layered Designs

Conventional networking and communication protocols provide generic designs that are suitable for a large number of applications and utilize performance measures such as throughput, fairness, delay and bit-error-rate (BER) etc. as design criteria. These methods are suitable for applications such as telecommunications or computer data networks, where users act as equal individuals and transmit messages that have little relation with others. The main concern in these cases is the quality-of-service (QoS) that each user receives.

In contrast to conventional communication and data networks, sensor networks consist of users that are deployed to achieve a common goal, to sense a common event or to measure highly correlated data due to the spatial correlation of most physical phenomenon. The sensors are cooperative in nature and should work together to fulfill their application needs. In fact, two properties of sensor networks can be exploited to improve communication efficiency: the *cooperative nature* of the sensors and *application-dependent* performance measures. In Part III of this book, we gather four chapters that consider these properties in the design of physical-layer transmissions, medium access control policies, routing protocols, sensor actuation and transmission scheduling.

Cooperation can be applied to many areas of communications and networking. At the physical layer, cooperation has been realized by allowing users to relay messages and by adopting signal-combining techniques at the destination to enhance reception. Diversity and multiplexing gains can thereby be achieved by exploiting the independent fading paths attained through cooperative relaying. Local resources such as battery-energy and channel bandwidth can be shared among sensors and optimally allocated from a system-wide perspective. This differs from that in conventional networks where fairness is a critical issue and may reduce the effectiveness of cooperation. In Chapter 10 by Sirkeci-Mergen and Scaglione, a tutorial review of cooperative communication schemes is given along with novel randomized approaches that are used to reduce the system complexity and to enhance the bandwidth efficiency of cooperative methods.

The efficiency of resource utilization can be further improved if the network is designed to optimize application-dependent performance measures. Specifically, data aggregation has been proposed to reduce traffic in multi-hop sensor networks. In contrast to data networks, here data that are unreliable or have low information context can be dropped. The efficiency of data aggregation techniques is highly dependent on the specific routing algorithm. For example, in data gathering applications, sensors may compress the incoming data along with their local data before relaying to the next sensor in the multi-hop route. In this case, the compression efficiency is highly dependent on the correlation between the measurements of the sensors in neighboring hops. A discussion of cross-layer routing protocols is given in Chapter 11 by Misra, Tong and Ephremides. Emphasis is placed on distributed detection applications where the performance depends on the data gathered through the multi-hop transmission routes.

An efficient sensor network MAC protocol also plays an important role in improving the efficiency of resource utilization. Conventional MAC protocols are designed for users that have independent data to transmit and that are competing for the use of the channel. The goal is to avoid interference and collision between different users. In contrast, in a well-designed cooperative sensor network, users that access the same channel simultaneously may improve the detection performance, as opposed to causing interference or collision. A survey of sensor network MAC protocols and design concepts for cooperative MAC protocols is given in Chapter 12 by Hong and Varshney. More interestingly, the cooperative advantages are further exploited by taking into consideration the properties of the underlying application or the statistics of the sensors' measurements. It is shown that MAC efficiency can be improved by allocating the same transmission channel to users that have highly correlated messages to transmit. Examples are given for two specific sensor applications: a data gathering application and a distributed detection application.

Duty-cycling is a technique used to reduce energy consumption and extend network lifetime. Nodes may enter a sleep state when their presence is not necessary to maintain the functionality of the system, e.g., when no event occurs in the sensor's vicinity or when no message is routed through the sensor. In this case, the activation of sensors should be optimized according to the statistics of the underlying measurements or the goal of the application. Due to the large-scale deployment of sensors, no centralized control can be applied to schedule the activation period of the sensors and, therefore, decentralized methods are required. In Chapter 13 by Krishnamurthy, Maskery and Ngo, decentralized sensor activation and transmission scheduling methods are discussed from a game-theoretic point of view. The sensors are able to learn the reliability of their measurements and decide locally when they should schedule their activation and transmissions.

In Part III, the importance of cross-layer communication and networking protocols is emphasized. These theoretical studies can provide insights for sensor network design. Nevertheless, caution should be taken when designing cross-layer strategies since it may obviate the advantages of modularization and result in high system complexity. Moreover, when only partial functionalities of two modules are jointly optimized, it is not clear whether the remaining functionalities will be as effective as before. These issues should also be taken into consideration in the future design of sensor systems.

Part I

Fundamental Properties and Limits

2

Information-theoretic Bounds on Sensor Network Performance

Michael Gastpar

2.1 Introduction

Sensor networks are subject to tight communication and computation constraints. These constraints are due to size and cost limitations: many of the most interesting applications require tiny and cheap sensors. This creates an interesting interplay between energy (and device size) spent on communication versus computation, and it is unclear which of the two is the true bottleneck. Some experts show that in current circuitry, the amount of energy spent on computation at least equals if not exceeds the transmit energy. Other experts use 'Moore's law' to argue that any computational bottleneck will disappear and that we are still very far away from the ultimate limits of quantum computation.

In this chapter, we will follow the second kind of experts and focus exclusively on the constraints imposed by *communication*. The goal is to understand at a fundamental level the impact of these constraints on the overall usefulness of the sensor network. Specifically, while we do not claim that they are irrelevant, we will entirely omit computational constraints and instead assume that all involved devices can perform *arbitrary* computations at no cost. This abstraction permits us to understand the effect of communication constraints and to derive relatively compact performance bounds. It forms the centerpiece of the success of information-theoretic methods in the analysis of communication systems.

For the most part, this chapter is an overview of known techniques. The chapter is divided into four main sections. In the first section, we discuss the basic modeling features that permit an information-theoretic analysis. The second section discusses bounds for

Wireless Sensor Networks: Signal Processing and Communications Perspectives A. Swami, Q. Zhao, Y.-W. Hong and L. Tong
© 2007 John Wiley & Sons, Ltd

digital architectures, by which we mean that the sensors first apply an optimum source
coding stage, and then communicate the resulting source codeword indices across the noisy
channel without making further errors. The third section briefly illustrates the possible
penalties incurred by a digital communication architecture, and the fourth section gives
a short overview of information-theoretic techniques for general (not necessarily digital)
architectures.

Throughout the chapter, we discuss general information-theoretic bounds. We also show
how these bounds apply to a specific scenario that we refer to as the *linear Gaussian sensor
network*. This helps in making the presented results more concrete: our bounds can then
be expressed in terms of the structure of matrices, rather than as general (and somewhat
abstract) information-theoretic quantities. We also use this example to discuss some of
the 'scaling-law' implications of the information-theoretic bounds (i.e., as the number of
sensors becomes very large).

2.2 Sensor Network Models

The main goal of this chapter is to shed light on how the *communications resources* impact
the overall performance of the sensor network. The specific methods discussed here are
information-theoretic,[1] and cannot be applied to all types of sensor networks. They are of
particular interest for sensor networks that monitor an underlying physical reality over a
'long' time and need to reproduce some recurring aspects of this physical reality. In other
words, the methods discussed here will not generally lead to relevant performance bounds
for sensor networks that are designed to raise one single alarm in their lifetime.

The sensor network model studied in this chapter is shown in Figure 2.1. There is an
underlying physical phenomenon which we characterize by L variables. These could be
thought of as the degrees of freedom of the system, or, equivalently, its current state. For
the scope of this chapter, each degree of freedom will be modeled as a random process *in
discrete time*.[2] Though we also discuss the scenario where the sensors can directly observe
the state of the system, some of the most interesting considerations discussed in this chapter

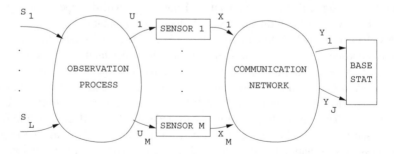

Figure 2.1 The general sensor network problem considered in this chapter.

---

[1]Sometimes also referred to as *Shannon-theoretic*, referring to the original paper of Shannon (1948), in an
attempt to distinguish them from statistical meanings of the term 'information-theoretic' (as in Fisher information).

[2]The discrete-time model is justified by arguing that the state of the system does not change very rapidly. This
may be a serious restriction for certain scenarios. The continuous-time extension is currently under investigation.

apply when the underlying degrees of freedom (or state) *cannot* be observed directly. Here, each sensor measures a (different) noisy version of a combination of all of these variables. We model this observation process in a probabilistic fashion as a conditional distribution of the observations given the state. The sensors may have the chance to cooperate to some (generally limited) extent, and there may be feedback from the base stations to each of the sensors. Based on the respective sensor readings, the inter-sensor communication, and the feedback signals, each sensor has to produce an output to be transmitted over the communication link (e.g., a wireless link). This channel is again modeled in a probabilistic fashion by a conditional distribution. The channel outputs are received by the base stations. For the information-theoretic bounds discussed in this chapter, we shall assume that the central data collection unit is ideally linked (e.g., over a backbone network) to the base stations. The goal of the data collector is to get to know, not the raw sensor readings, but the values of the underlying degrees of freedom (or state) of the physical system.

More precisely, and to fix notations, the underlying physical phenomenon is characterized by the sequence of random vectors

$$\{S[n]\}_{n\geq 0} = \{(S_1[n], S_2[n], \ldots, S_L[n])\}_{n\geq 0}, \tag{2.1}$$

where n is the time index. The arguments presented in this chapter address the case where $\{S[n]\}_{n\geq 0}$ is a sequence of independent and identically distributed (iid) random vectors. To simplify the notation in the rest of the chapter, we denote sequences as $S^N \overset{def}{=} \{S[n]\}_{n=1}^N$. We use the upper case S to denote the random variable, and the lower case s to denote its realization. The distribution of S is denoted by $P_S(s)$. To simplify notation, we will also use the shorthand $P(s)$ when the subscript is just the capitalized version of the argument in the parentheses. The random vector $S[n]$ is not directly observed by the sensors. Rather, sensor m observes a sequence $U_m^N = \{U_m[n]\}_{n=1}^N$ which depends on the physical phenomenon according to a conditional probability distribution,

$$p\left(\{u_m[n]\}_{n\geq 0}, m = 1, 2, \ldots, M \middle| \{s_\ell[n]\}_{n\geq 0}, \ell = 1, 2, \ldots, L\right) \tag{2.2}$$

For the scope of this chapter, the observation process is memoryless in the sense that the observation at time n only depends on the source outputs at time n. Hence, the observation process can be characterized by $P(u_1, \ldots, u_M | s_1, \ldots, s_L)$. Sensor m may also receive information from other sensors as well as from the destination. Denoting the totality of this information as it is available to sensor m up to time $n - 1$ by V_m^{n-1}, the signal transmitted by sensor m at time n can be expressed as

$$X_m[n] = F_m^{(n)}\left(U_m^N, V_m^{n-1}\right). \tag{2.3}$$

The transmitted signals satisfy a power, or more generally, a cost constraint of the form

$$E\left[\rho_m(X_m^N)\right] \leq \Gamma_m. \tag{2.4}$$

More generally, we may also allow constraints of the form

$$E\left[\rho(X_1^N, X_2^N, \ldots, X_M^N)\right] \leq \Gamma. \tag{2.5}$$

One example of a constraint of this kind is a *sum power constraint* on the outputs of the sensors, allowing for power allocation between them.

The final destination uses the outputs of the communication channel, $Y^N = (Y_1^N, Y_2^N, \ldots, Y_J^N)$ to construct estimates $\hat{S}^N = (\hat{S}_1^N, \hat{S}_2^N, \ldots, \hat{S}_L^N)$. The task is to design the decoder G such that $\hat{S}^N = G(Y^N)$ is as close to S^N as possible, in the sense of an appropriately chosen distortion measure $d(s^N, \hat{s}^N)$. For a fixed code, composed of the encoders $F_1, F_2, \ldots, F_M$ at the sensors and the decoder G, the achieved distortion Δ is computed as follows:

$$\Delta = E\left[d\left(S^N, \hat{S}^N\right)\right]. \tag{2.6}$$

The relevant figure of merit is therefore the trade-off between the *cost* Γ of the transmission (Eqn. (2.4)), and the achieved *distortion level* Δ (Eqn. (2.6)). The problem studied in this chapter is that of finding the optimal trade-offs (Γ, Δ), where optimal is understood in an information-theoretic sense, i.e., irrespective of delay and complexity, as $N \to \infty$.

2.2.1 The Linear Gaussian Sensor Network

For illustration purposes and to get a sense of the value of the information-theoretic techniques discussed in this chapter, we keep as our running example the paradigmatic scenario referred to as the *linear Gaussian sensor network*. This is schematically shown in Figure 2.2. Sensor m observes a linear combination of the L underlying complex-valued source signals,

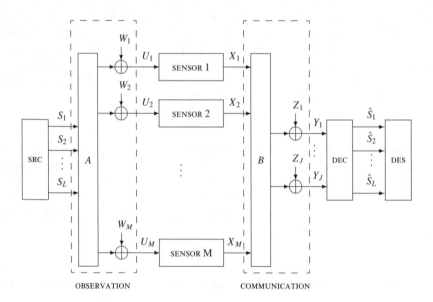

Figure 2.2 The Gaussian sensor network: A vector source S, not necessarily Gaussian, is observed M-fold through a matrix A and in additive white Gaussian noise, independently by M sensors. The M sensors communicate over a AWGN MIMO channel, characterized by the matrix B, to a base station that houses the central estimation officer. The sensors may have (generally limited) cooperation capabilities, indicated by the dotted lines in the figure.

subject to additive white Gaussian observation noise, as follows:

$$U_m[n] = W_m[n] + \sum_{\ell=1}^{L} a_{m,\ell} S_\ell[n], \tag{2.7}$$

where $a_{m,\ell}$ are appropriate complex-valued constants that are assumed to be fixed and known throughout. The observation noises $W_m[n]$ are i.i.d. (both over n and over m) circularly symmetric complex Gaussian random variables of mean zero and variance σ_W^2.

We will also use vector (and matrix) notation to express this in a more compact form as

$$\mathbf{U}[n] = A^{(M)}\mathbf{S}[n] + \mathbf{W}[n], \tag{2.8}$$

where $\mathbf{U}[n] = (U_1[n], U_2[n], \ldots, U_M[n])^T$ is a complex-valued column vector of length M, $A^{(M)}$ is a complex-valued matrix of dimensions $M \times L$, with entries $\{A^{(M)}\}_{i,j} = a_{i,j}$, $\mathbf{S}[n] = (S_1[n], S_2[n], \ldots, S_L[n])^T$ is a complex-valued column vector of length L, and $\mathbf{W}[n] = (W_1[n], W_2[n], \ldots, W_M[n])^T$ is a column vector of circularly symmetric complex Gaussian random variables of mean zero and covariance matrix $\sigma_W^2 I_M$.

The encoding task can then be described as follows: Sensor m makes N consecutive observations $\{u_m[n]\}_{n=1}^N$. In some versions of the problem, sensor m may also receive a limited amount of additional information from other sensors, and we will denote this additional information collectively by V_m. Based on the observed sequence $\{u_m[n]\}_{n=1}^N$ and the additional information V_m, sensor m decides on a codeword $\{x_m[n]\}_{n=1}^{KN}$ to be transmitted across the channel. Here, the parameter K specifies the relative (temporal) bandwidth of the communication channel with respect to the source. Concretely, one can think of K channel uses that are available for the transmission of each source sample. This induces a probability distribution over the codewords of sensor m, and we can thus think of the output of sensor m as a random vector $\{X_m[n]\}_{n=1}^{KN}$. The codewords of sensor m must be designed to satisfy an average power constraint as follows:

$$\frac{1}{KN} \sum_{n=1}^{KN} E\left[|X_m[n]|^2\right] \leq P_m. \tag{2.9}$$

Rather than enforcing such a constraint individually for each sensor, we will often allow power allocation, i.e., any choice of powers $P_1, P_2, \ldots, P_M$ satisfying

$$\sum_{m=1}^{M} P_m \leq P_{tot}(M). \tag{2.10}$$

The receiver observes these codewords across a Gaussian vector channel. More specifically, at time n, the receiver observes a vector $\{Y_j[n]\}_{j=1}^J$ with components

$$Y_j[n] = Z_j[n] + \sum_{m=1}^{M} b_{j,m} X_m[n], \tag{2.11}$$

where the channel noises $Z_j[n]$ are i.i.d. (both over n and over j) circularly symmetric complex Gaussian random variables of mean zero and variance σ_Z^2.

Again, we will find it convenient to use vector notation occasionally, and rewrite (2.11) more compactly as

$$\mathbf{Y}[n] = B^{(M)}\mathbf{X}[n] + \mathbf{Z}[n], \tag{2.12}$$

where $\mathbf{Y}[n] = (Y_1[n], Y_2[n], \ldots, Y_J[n])^T$ is a complex-valued column vector of length J, $B^{(M)}$ is a complex-valued matrix of dimensions $J \times M$, with entries $\{B^{(M)}\}_{i,j} = b_{i,j}$, $\mathbf{X}[n] = (X_1[n], X_2[n], \ldots, X_M[n])^T$ is a complex-valued column vector of length M, and $\mathbf{Z}[n] = (W_1[n], W_2[n], \ldots, W_J[n])^T$ is a column vector of circularly complex Gaussian random variables of mean zero and covariance matrix $\sigma_W^2 I_J$.

Based on the sequence of channel output vectors $\mathbf{Y}[n]$, the goal is to construct an estimate $\hat{\mathbf{S}}[n]$ of the underlying source vector sequence $\mathbf{S}[n]$. The success of such reconstruction will be assessed in terms of the resulting mean-squared error, defined as

$$D_N(M) = \frac{1}{N} \sum_{n=1}^{N} \frac{1}{L} E\left[\|\mathbf{S}[n] - \hat{\mathbf{S}}[n]\|^2\right]. \tag{2.13}$$

The bounds discussed in this chapter will be interpreted in terms of fundamental relationships between the matrices $A^{(M)}$ and $B^{(M)}$, the total power $P_{tot}(M)$, and the end-to-end distortion $D_N(M)$.

2.3 Digital Architectures

One of the crown jewels of information theory is the source/channel separation theorem, given by Shannon (1948), Section 13, and more generally in Shannon (1959), Theorem 5, stating that an overall coding problem of the kind defined in Section 2.2 can, in some cases, be split into two separate coding problems: one in which the source is compressed using R bits per source sample, and one in which R bits are communicated reliably across the channel (i.e., without making any errors). This two-stage approach will be referred to as 'digital architecture' for the purpose of this chapter. Apart from optimality, such an architecture also has very attractive modularity properties, and therefore, this approach has guided the design of virtually all contemporary communication systems.

In this section, we discuss information-theoretic bounds on the performance of digital architectures in sensor networks. While such an architecture is interesting in and of itself, it *does* entail a performance penalty in sensor networks, by contrast to the standard point-to-point communication problem. That is, there is no general guarantee that a digital architecture will have a performance close to the optimum, not even in an asymptotic sense (as the number of nodes becomes large). We discuss this issue in more detail in Section 2.4. In the final section (Section 2.5), we discuss performance bounds for general architectures.

2.3.1 Distributed Source Coding

In this section, we review mostly known bounding techniques for the distributed source coding problem. The linear Gaussian special case is illustrated in Figure 2.3, but the bounds discussed here are valid more generally.

More specifically, as in the discussion in Section 2.2, consider a set of L source sequences, denoted by $\{S_\ell[n]\}$, for $\ell = 1, 2, \ldots, L$, see Equation (2.1). There are M encoders, each of

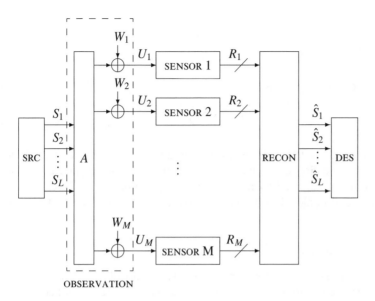

Figure 2.3 The (distributed) source coding problem associated with the linear Gaussian sensor network considered in this chapter and illustrated in Figure 2.2.

which makes a sequences of observations $\{U_m[n]\}_{n \geq 0}$, for $m = 1, 2, \ldots, M$, and we will restrict attention to a memoryless observation model, as described in the discussion following Equation (2.2). Encoder m must produce a bit sequence that appears at a rate of R_m bits per source symbol, for $m = 1, 2, \ldots, M$. The decoder ('fusion center') observes all M bit streams and is required to produce source reconstruction sequences

$$\hat{S}_\ell[n], \text{ for } \ell = 1, 2, \ldots, L, \tag{2.14}$$

such as to meet fidelity criteria:

$$D_\ell = \lim_{N \to \infty} \frac{1}{N} \sum_{n=1}^{N} E\left[d\left(S_\ell[n], \hat{S}_\ell[n] \right) \right]. \tag{2.15}$$

Hence, the performance of such a rate-distortion code can be captured in terms of a *rate-distortion vector* $(R_1, R_2, \ldots, R_M, D_1, D_2, \ldots, D_L)$ in the positive quadrant of $(M + L)$-dimensional real space. The set of all rate-distortion vectors that correspond to actual source coding schemes is called the *rate-distortion region*. For a fixed set of distortion requirements $(D_1, D_2, \ldots, D_L)$, we will denote the set of rate vectors that permit to satisfy these distortion constraints as

$$\mathcal{R}(D_1, \ldots, D_L). \tag{2.16}$$

For brevity, and in order to gain insight, we will often focus on two simple key parameters of the rate-distortion vector, rather than its entirety. Specifically, we will consider the total (sum-)rate

$$R_{tot}(D) = \sum_{m=1}^{M} R_m, \qquad (2.17)$$

and the sum of the distortion terms,

$$D = \sum_{\ell=1}^{L} D_\ell. \qquad (2.18)$$

The task is then to characterize the optimal trade-offs between R_{tot} and D. Note that this reduced characterization conceptually assumes that rate can be allocated arbitrarily between the sensors, and that all sources S_ℓ are equally important. However, none of the bounding techniques considered in this chapter fundamentally relies on such a symmetry assumption.

Direct source observation

We first consider the simplified case where the sources are observed directly, i.e., the number of sources is equal to the number of observations, $L = M$, and where there is no observation noise. Specifically, with a small loss of generality, we consider

$$U_m[n] = S_m[n]. \qquad (2.19)$$

A first lower bound results by dropping the constraint that encoding be performed in a distributed manner. For the purpose of this chapter, we will refer to this lower bounding argument as 'centralized'. Clearly, for this case, the answer is merely the standard rate-distortion function, leading to the following bound.

Theorem 2.3.1 (centralized lower bound, sum rate) *The sum rate R_{tot} required to enable the decoder to reconstruct the sources S_ℓ, $\ell = 1, 2, \ldots, L$, at sum distortion D satisfies*

$$R_{tot}(D) \geq \min I(U_1, \ldots, U_M; \hat{S}_1, \ldots, \hat{S}_L), \qquad (2.20)$$

where the minimum is over all $p(\hat{s}_1, \ldots, \hat{s}_L | u_1, \ldots, u_M)$ for which $\sum_{\ell=1}^{L} E[d_\ell(S_\ell, \hat{S}_\ell)] \leq D$.

A proof of this theorem can be found in Cover and Thomas (2006), p. 315, or Berger (1971).

Using the results of Wyner and Ziv (1976), we can refine this characterization.

Theorem 2.3.2 (centralized lower bound, rate region) *The rate vectors $(R_1, R_2, \ldots, R_M)$ required to enable the decoder to reconstruct the sources S_ℓ, $\ell = 1, 2, \ldots, L$, at distortions D_ℓ, respectively, satisfy, for each subset $S \subseteq \{1, 2, \ldots, M\}$,[3]*

$$\sum_{m \in S} R_m \geq R_S \stackrel{def}{=} \min_V I(\{U_S; V) - I(U_{S^c}; V) \qquad (2.21)$$

where the minimum is taken separately for each S over all random variables V with $p(v|u_S)$ for which there exist reconstruction functions $\hat{S}_\ell(V, U_{S^c})$ such that $E[d_\ell(S_\ell, \hat{S}_\ell)] \leq D_\ell$, for $\ell = 1, 2, \ldots, L$.

---

[3]Throughout, we will use S to denote a subset of the integers $\{1, 2, \ldots, M\}$, and S^c to denote its complement in the set $\{1, 2, \ldots, M\}$. Moreover, we will use the notation X_S to denote the set of all variables X_m with index $m \in S$.

A proof of this theorem can be found in Cover and Thomas (2006), p. 581. The function R_S is often referred to as the Wyner-Ziv rate-distortion function for encoding $\{U_m\}_{m \in S}$ for a decoder that has access to the 'side information' $\{U_m\}_{m \in S^c}$.

For the linear Gaussian scenario, we obtain

Corollary 2.3.3 (centralized lower bound, direct source observation) *The sum rate R_{tot} required to encode an L-dimensional i.i.d. Gaussian vector source with mean zero and covariance matrix $\Sigma_s = \sigma_S^2 I_L$ in such a way that reconstruction incurs an average sum distortion of at most D satisfies*

$$R_{tot}(D) \geq \sum_{\ell=1}^{L} \log \frac{\alpha_\ell}{D_\ell}, \tag{2.22}$$

where α_ℓ, $\ell = 1, 2, \ldots, L$, denote the L eigenvalues of the matrix Σ_s, and

$$D_\ell = \begin{cases} \nu, & \text{if } \nu < \alpha_\ell, \\ \alpha_\ell, & \text{otherwise.} \end{cases} \tag{2.23}$$

where $\sum_{\ell=1}^{L} D_\ell = D$.

This corollary follows straightforwardly from Theorem 2.3.1 by noting that $\hat{S}_\ell$ should be selected jointly Gaussian with S_ℓ, for $\ell = 1, 2, \ldots, L$. See Cover and Thomas (2006), p. 312.

Clearly, this bound will not generally be interesting for the sensor network scenarios considered in this chapter: the constraint that encoding be done in a distributed fashion must be expected to crucially impact the overall performance. Consequently, information-theoretic arguments have been developed to explicitly take into account this constraint. To date, these bounds tend to be rather difficult to evaluate, even for the simple Gaussian case that we use to illustrate the concepts.

The first general bound was presented by Berger (1977) and Tung (1978). It can be expressed as follows:

Theorem 2.3.4 (Berger-Tung lower bound) *The sum rate R_{tot} required to enable the decoder to reconstruct the sources S_ℓ, $\ell = 1, 2, \ldots, L$, at sum distortion D satisfies*

$$R_{tot}(D) \geq \min I(U_1, \ldots, U_L; V_1, \ldots, V_L), \tag{2.24}$$

where the minimum is over all joint distributions $p(u_1, \ldots, u_L, v_1, \ldots, v_L)$ that satisfy the Markov condition

$$p(v_m | u_1, \ldots, u_L) = p(v_m | u_m), \text{ for } m = 1, 2, \ldots, L, \tag{2.25}$$

and over all mappings $\hat{S}_\ell(V_1, \ldots, V_L)$ for which $\sum_{\ell=1}^{L} E[d_\ell(S_\ell, \hat{S}_\ell)] \leq D$.

The first major difficulty with evaluating this bound for concrete cases at hand is the selection of the auxiliary random variables. Note that a lower bound is only guaranteed if the actual *minimum* is identified. The reason why we cannot solve this for the linear Gaussian case at hand is because it is not clear to date whether it is sufficient to restrict the auxiliary random variables with conditional distributions $p(v_m|u_m)$ to be Gaussian. This question has been settled recently for the special case $L = M = 2$ in the work by Wagner et al. (2005), where it is shown that Gaussian auxiliaries are sufficient.

To get a sense of the bound of Theorem 2.3.4, let us suppose that the source vector $(S_1, S_2, \ldots, S_L)$ is Gaussian with mean zero and covariance matrix Σ_s, as in Corollary 2.3.3. We may evaluate the expression given in Equation (2.24) *assuming* Gaussian statistics for the auxiliaries. At this point, the problem becomes a relatively standard minimization problem. To make things simple, let us assume a sufficiently symmetric scenario. More specifically, we will assume that all the distributions $p(v_m|u_m)$ are equal. Then, we obtain the following expression:

$$R_{tot}(D) = \sum_{m=1}^{L} \log\left(1 + \frac{\alpha_m}{\mu}\right), \tag{2.26}$$

where α_ℓ, $\ell = 1, 2, \ldots, L$, denote the L eigenvalues of the matrix Σ_s, and

$$D = \sum_{m=1}^{M} \frac{\alpha_m \mu}{\alpha_m + \mu}. \tag{2.27}$$

It is important to recall that at this point, it is *not known* whether this is a lower bound. However, in the shape of an idle thought, we may still compare it to the 'centralized' lower bound of Corollary 2.3.3 and note that the main difference lies in what is sometimes referred to as (inverse) *water-filling* (see e.g. Cover and Thomas (2006)). Therefore, roughly speaking, as long as the total rate R_{tot} is large enough, and the spread of the eigenvalues α_ℓ, for $\ell = 1, 2, \ldots, L$, is small enough such that in Equation (2.23), we have that $D_\ell = v$, for $\ell = 1, 2, \ldots, L$, then the difference between the above and the centralized bound (Corollary 2.3.3) will be insignificant.

The second drawback of the Berger-Tung lower bound of Theorem 2.3.4 is that it cannot be shown (and should not be expected) to be tight in all but trivial cases. As a case in point, Wagner (2005) and Wagner and Anantharam (2005) have presented several examples where the bound is strictly loose.

While in this chapter, we focus on fundamental bounds on sensor network performance, the natural subsequent question is that of how tight these bounds are. That is, the question is whether there are actual coding schemes that attain or at least come close to the bounds discussed above. Most of the information-theoretic analysis of this question follows the lines of a code construction due to Slepian and Wolf (1973), extended to the rate-distortion case by Wyner and Ziv (1976), Berger (1977), Tung (1978), Omura and Housewright (1977). Recent work by Neuhoff and Pradhan (2005) has also studied these questions under very interesting constraints such as scalar quantization. A detailed discussion of this is beyond the scope of the present chapter.

Indirect source observation

The situation is subtly different for indirect (or noisy) source observations, sometimes referred to as remote source coding and as the CEO problem in the distributed setting (see Berger et al. (1996)).

First of all, it is important to point out that the centralized lower bounds of Theorems 2.3.1 and 2.3.2 apply without any changes. That is, from a centralized coding perspective, there are no information-theoretic differences between the direct and the indirect source observation problem. This was already observed in the early work on remote source coding (see Dobrushin and Tsybakov (1962) and Wolf and Ziv (1970)) and in more generality by Berger (1971), p. 78–81.

Some interesting differences between the direct and the noisy indirect source coding problem appear upon evaluation of the bounds, and hinge on the inherent balance between the ability to estimate the source through the observation process and the encoding quality.

To start the discussion, let us consider the linear Gaussian case, but without any observation noise, i.e., the scenario illustrated in Figure 2.3, but with the observation noises set to zero, $W_m = 0$, for $m = 1, 2, \ldots, M$. For the centralized encoding, it is easy to see what to do: as long as the matrix A has full (column) rank, i.e., $\text{rank}(A) = L$, one can recover the source sequences $\{S_\ell[n]\}$ from the observations, and hence, the observation process does not matter at all. More precisely, Corollary 2.3.3 applies without any changes. (If the matrix A is rank-deficient, some underlying sources cannot be estimated at all, and a simple model reduction, omitting the unobservable sources, will lead to a revised, full-rank matrix A.)

Merely for future comparisons, let us record that this implies that for the simple case $L = 1$ (and arbitrary $M \geq 1$), the centralized bound says that the necessary rate is lower bounded by the standard rate-distortion function,

$$R_{tot}(D) \geq \log \frac{\sigma_S^2}{D},$$
(2.28)

or conversely,

$$D(R_{tot}) \geq \sigma_S^2 2^{-R_{tot}}.$$
(2.29)

The next step is to explicitly include the observation noise into our considerations. As mentioned above, the centralized lower bound of Theorem 2.3.1 still applies, and the remaining problem is merely to evaluate it for the linear Gaussian sensor network. The resulting behavior can be characterized as follows:

Corollary 2.3.5 (centralized lower bound, indirect noisy observation) *The sum rate R_{tot} required to encode an L-dimensional i.i.d. Gaussian vector source with mean zero and covariance matrix $\Sigma_s = \sigma_S^2 I_L$, observed through the matrix A and Gaussian observation noise with covariance matrix Σ_w, in such a way that reconstruction incurs an average sum distortion of at most D satisfies*

$$R_{tot} \geq \sum_{\ell=1}^{L} \log \frac{\lambda_\ell}{D_\ell},$$
(2.30)

where λ_ℓ, $\ell = 1, 2, \ldots, L$, denote the L eigenvalues of the matrix

$$\Sigma_s A^H \left(A \Sigma_s A^H + \Sigma_w \right)^{-1} A \Sigma_s,$$
(2.31)

and

$$D_\ell = \begin{cases} \nu, & \text{if } \nu < \lambda_\ell, \\ \lambda_\ell, & \text{otherwise,} \end{cases}$$
(2.32)

where $\sum_{\ell=1}^{L} D_\ell = D - D_0$, where

$$D_0 = \sum_{\ell=1}^{L} (\sigma_S^2 - \lambda_\ell). \tag{2.33}$$

Specifically, for a single underlying source ($L = 1$) with symmetric conditionally independent observations (A is a column vector of M ones and $\Sigma_w = \sigma_w^2 I_M$, where I_M denotes the M-dimensional identity matrix) this can be evaluated in closed form to yield the following formula:

$$R_{tot}(D) = \log \frac{\sigma_S^2}{D} + \log \frac{M\sigma_S^2}{M\sigma_S^2 + \sigma_W^2 - \frac{\sigma_S^2}{D}\sigma_W^2}, \tag{2.34}$$

or, conversely,

$$D(R_{tot}) = \frac{\sigma_S^2}{1 + \frac{M\sigma_S^2}{\sigma_W^2}} + \frac{\sigma_S^2 2^{-R_{tot}}}{1 + \frac{\sigma_W^2}{M\sigma_S^2}}. \tag{2.35}$$

To understand the difference between the expressions (2.28) and (2.34), let us consider two different scenarios. First if the total rate R_{tot} increases *logarithmically* with the number of sensors M, then the two expressions describe the same overall behavior, namely a distortion that decreases like

$$D \sim \frac{1}{M} \tag{2.36}$$

However, if the total rate increases at a much faster pace, then the two expressions describe very different behavior. For example, suppose that the total rate increases *linearly* in M, then, for the noiseless case, the distortion in (2.28) decreases *exponentially* in M, while for the noisy case as in (2.34), it still decreases inversely proportionally. This illustrates the balance between coding accuracy and observation quality.

We want to briefly illustrate how to evaluate the centralized bound for the more general case where the underlying source sequences $\{S_\ell[n]\}$ are *not* drawn from a Gaussian distribution. This has been studied by Eswaran (2005).

Theorem 2.3.6 (centralized lower bound, non-Gaussian sources) *The sum rate R_{tot} required to encode an arbitrary scalar source S with probability density function $p_S(s)$ with differential entropy $h(S) > -\infty$, observed in additive white Gaussian noise with $a_m = 1$, for $m = 1, 2, \ldots, M$, satisfies*

$$D(R_{tot}) \geq \frac{\frac{2^{h(S)}}{2\pi e}}{1 + \frac{M\sigma_S^2}{\sigma_W^2}} + \frac{2^{h(E[S|U])}}{2\pi e} 2^{-R}, \tag{2.37}$$

where $h(\cdot)$ denotes the differential entropy function with the binary logarithm.

For a proof outline, we first note that we can consider the sum of the observations for coding since it is a sufficient statistic for the underlying source. The proof then follows from the bound in (Eswaran 2005, p. 19, Theorem 2.6), which relies on the fact that for additive noise and squared error distortion, finding the minimum mean squared error estimate at the encoder and then relying on best mean squared error quantizer for the induced distribution is an optimal coding strategy. The resulting distortion is then the minimum mean squared error of the underlying source given the observations at the encoder plus a Shannon lower bound on the distortion given a rate R for quantizing the induced source, which is the distribution of the conditional mean of S given the observation. That is,

$$D(R_{tot}) \geq E\left(S - E[S|S + \frac{1}{M}\sum_i W_i]\right)^2 + \frac{2^{h(E[S|U])}}{2\pi e}2^{-R}, \tag{2.38}$$

The lower bound for the first term can be found by exploiting the fact that the W_i are jointly Gaussian. A detailed argument was given by Eswaran (2005), p. 57, Theorem A.3.

The 'centralized' bounds studied so far are generally loose, and therefore, we will now discuss bounds that explicitly take into account the constraint that the encoding step be implemented in a distributed fashion. Specifically, the Berger-Tung bound (Theorem 2.3.4) can be straightforwardly extended to the case of noisy observations, but is again non-trivial to evaluate, and has been shown to be loose for several interesting scenarios, including a simple version of the linear Gaussian sensor network source coding problem illustrated in Figure 2.3 (namely, the version addressed in Theorem 2.3.7). The bound developed by Wagner and Anantharam (2005) also directly extends to this scenario, and in fact, appears to be a good match, a potential that remains yet to be explored. Again, however, it is a rather elaborate task to evaluate it in special cases.

Nevertheless, using a more direct and specialized approach (specialized to the Gaussian case), Oohama (1997) (with subsequent extensions due to Chen et al. (2004) and Prabhakaran et al. (2004)) found a simple and explicit bound for the case of a single underlying source ($L = 1$) with conditionally independent observations ($\Sigma_w = \sigma_w^2 I_M$, where I_M denotes the M-dimensional identity matrix).

Theorem 2.3.7 (distributed coding, indirect noisy observation ('CEO problem')) *The sum rate R_{tot} required to encode a (one-dimensional) i.i.d. Gaussian source with mean zero and variance σ_S^2, observed through the all-ones vector and independent Gaussian observation noises with variance σ_W^2, in such a way that reconstruction incurs an average sum distortion of at most D satisfies, for $D > \sigma_S \sigma_W^2/(M\sigma_S^2 + \sigma_W^2)$,*

$$R_{tot}(D) = \log\frac{\sigma_S^2}{D} + M\log\frac{M\sigma_S^2}{M\sigma_S^2 + \sigma_W^2 - \frac{\sigma_S^2}{D}\sigma_W^2}. \tag{2.39}$$

It is also interesting to compare this bound to the centralized coding bound (2.34). In order to do so, let us again think of the 'scaling-law' case, i.e., when M becomes large. Then, if the total rate increases faster than like $\log M$, Equation (2.39) and the centralized bound of Equation (2.35) both show that the distortion scaling behavior is $1/M$ (more precisely, that $\lim_{M\to\infty} D(M)/(1/M)$ is a non-zero, finite constant). However, if the total rate increases like $\log M$ or slower, then the bottleneck becomes the coding performance (rather than the estimation performance), and in this case, Equation (2.39) is significantly larger than Equation (2.35), showing a large penalty due to distributed processing. This is summarized in Table 2.1.

Table 2.1 Scaling behavior of the distortion for indirect source encoding of a single Gaussian source based on M noiseless or noisy measurements.

Total rate scaling	Centralized encoding		Distributed encoding	
	noiseless	noisy	noiseless	noisy
$\sim \log M$	$1/M$	$1/M$	$1/M$	$1/\log M$
$\sim M$	$\exp(-M)$	$1/M$	$\exp(-M)$	$1/M$

Encoding functions of source observations

Another variation on the standard source coding problem that may be of interest in sensor networks concerns the encoding of functions of source observations. Consider the coding problem illustrated in Figure 2.4. Note that some versions of this problem correspond to the noisy indirect source coding problem discussed in the previous paragraph, but not all.

Example 2.3.8 *An interesting example was presented by Körner and Marton (1979). In Figure 2.4, suppose that $M = 2$. Assume that U_1 and U_2 are correlated binary sources, and suppose that $L = 1$ and*

$$f(U_1, U_2) = U_1 \oplus U_2, \tag{2.40}$$

that is, the goal of the decoder is to recover the (point-wise) modulo-2 sum of the two source sequences. Clearly, one way to achieve this goal is to merely compress each of the sequences separately. This requires a rate R_1 corresponding to the entropy of the observation sequence U_1 and a rate R_2 corresponding to the entropy of the observation sequence U_2. The main question is whether one can get away with a smaller rate pair (R_1, R_2).

A first improvement follows as a direct consequence of the seminal work of Slepian and Wolf (1973) and gives gains whenever the two source sequences are correlated. Specifically,

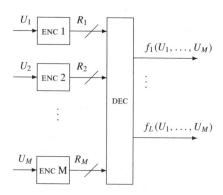

Figure 2.4 Distributed encoding of functions of the source observations.

suppose that $U_1[n]$ is simply a Bernoulli(1/2) process, and that

$$U_2[n] = U_1[n] \oplus E[n], \tag{2.41}$$

where $E[n]$ is a Bernoulli(p) process independent of $U_1[n]$, which makes $U_2[n]$ also a Bernoulli(1/2) process. Hence, each source has entropy $H(U_1) = H(U_2) = 1$, and so, a rate of $R_1 = R_2 = 1$ bits per source symbol will definitely be sufficient, but it is not necessary. In fact, for this scenario, the work of Slepian and Wolf (1973) has shown that for any rate pair satisfying[4]

$$R_1 > H(U_1|U_2) = H_b(p) \tag{2.42}$$

$$R_2 > H(U_2|U_1) = H_b(p) \tag{2.43}$$

$$R_1 + R_2 > H(U_1, U_2) = 1 + H_b(p), \tag{2.44}$$

there exists a code that permits the decoder to recover both U_1 and U_2 perfectly; a considerably lower total rate than for the naive scheme. It turns out, however, that even this rate is wasteful: after all, the decoder is only interested in the modulo-2 sum $U_1 \oplus U_2$. This problem has been resolved by Körner and Marton (1979), who proved that the rate pair

$$R_1 = R_2 = H_b(p) \tag{2.45}$$

is sufficient. To show this, they devise a random linear code construction; it turns out that unstructured random coding of the usual kind (such as the ones discussed in Cover and Thomas (2006), pp. 61–62) are not sufficient to prove this result. For the perspective of this chapter, we are more interested in the converse statement, i.e., that this is the smallest possible rate. For the case at hand, this can be answered by giving U_2 for free to encoder 1 and to the decoder. For the decoder to determine E, encoder 1 still needs to encode at a rate of $H_b(p)$. This provides a somewhat interesting and at first perhaps surprising insight: If the two sources U_1 and U_2 are independent of each other, then even if the decoder only needs the modulo-2 sum, the encoders must provide a full description of both sources separately. No 'rate savings' are available. We will come back to this example in Theorem 2.3.15.

A related but different example concerns the case where $U_1, \ldots, U_M$ are jointly Gaussian, and the function to be recovered is merely their sum,

$$f(U_1, \ldots, U_M) = \sum_{m=1}^{M} U_M. \tag{2.46}$$

For the special case of $M = 2$ and positive correlation between the sources, this problem has been shown to be equivalent to a simple modification of the CEO problem (see Theorem 2.3.7), see the work of Wagner et al. (2005).

Specifically, it can again be shown that if the sources are independent, then no gains are possible over separately describing each source, even though the decoder only needs to recover their sum. We will return to this example in the context of Section 2.4.

---

[4] $H_b(p) = -p \log p - (1 - p) \log(1 - p)$ denotes the binary entropy function.

2.3.2   Distributed Channel Coding

In Section 2.3.1, we discussed known and novel bounds on the performance of source coding techniques for sensor networks. For these to be useful, it must be possible to reliably communicate the resulting compressed source descriptions to the fusion center. This is the standard and well-known problem of (reliable) channel coding, leading to the information-theoretic notion of capacity. In this section, we discuss known and novel bounding techniques on the capacity of such noisy communication networks.

We will proceed by analogy to Section 2.3.1. That is, we will briefly discuss the different techniques, and then apply them to the considered linear Gaussian sensor network example. While it is well known that exact capacity results are rare, recent work has uncovered several instances where a more or less precise result concerning the scaling behavior of capacity, i.e., its fundamental dependence on the number of nodes, can be characterized. This development has been spearheaded by the work of Gupta and Kumar (2000) for the case of ad-hoc wireless networks. As we will see in more detail below, the standard ad-hoc network setup is not always meaningful for sensor networks, due in part to the fact that the information at the different terminals is typically not independent in sensor networks.

For the type of sensor network studied in this chapter, the goal is generally to read out the information at some collection point. Therefore, we will concentrate on the so-called multiple access channel (MAC), though most of the techniques apply more generally.

Multiple access with independent messages

In the information theoretic literature, this is considered the canonical case of multiple access: Consider a set of M encoders or channel input terminals. We will assume that each encoder has a long bit stream to send, appearing at R_m bits per channel use, for $m = 1, 2, \ldots, M$. As in Figure 2.5, when necessary, we will denote these bit streams as V_m, for $m = 1, 2, \ldots, M$. The key is that the bit streams observed by different encoders are independent of one another. Each encoder uses only its own bits in order to produce a suitable channel input signal. We assume that the channel is used in discrete time (i.e., that the total channel bandwidth is limited). Hence, there will be M channel input sequences, denoted by $\{X_m[i]\}_{i \in Z}$, for $m = 1, 2, \ldots, M$, and J corresponding channel output sequences, denoted by $\{Y_j[i]\}_{i \in Z}$, for $j = 1, 2, \ldots, J$. The decoder observes all channel outputs and needs to recover the original bit streams with an error probability that goes to zero as the coding block length is increased.

For this problem, a meaningful notion of capacity can be defined. More precisely, the capacity region is the set of all rate vector $(R_1, \ldots, R_M)$ for which this error probability requirement can be satisfied. As discussed in Section 2.2, there is also typically a cost constraint on the input sequences, i.e., not all possible input sequences are allowed. For the purpose of our discussion, this cost constraint takes the shape of an expected value, characterized by P_m, for $m = 1, 2, \ldots, M$, which may be thought of as an average power constraint. This leads to a vector of simultaneously sustainable rates and costs $(R_1, R_2, \ldots, R_M, P_1, P_2, \ldots, P_M)$. For a fixed cost vector $(P_1, P_2, \ldots, P_M)$, we will denote the corresponding capacity region by

$$\mathcal{C}(P_1, P_2, \ldots, P_M). \tag{2.47}$$

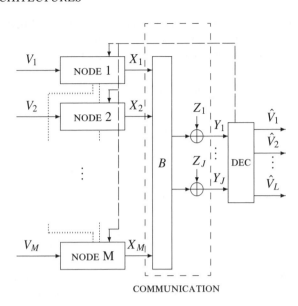

COMMUNICATION

Figure 2.5 The (distributed) channel coding problem associated with the linear Gaussian sensor network considered in this chapter and illustrated in Figure 2.2. This problem is often referred to as the *multiple access channel*, here with generally limited cooperation between the nodes and some forms of feedback.

For the standard capacity problem, this question has been resolved by Ahlswede (1971) and Liao (1972) in the shape of the following theorem.

Theorem 2.3.9 *The capacity region of the multiple access channel with independent messages and without feedback and encoder cooperation is given by the convex closure of the union over all product distributions*

$$p(x_1, x_2, \ldots, x_M) = \prod_{m=1}^{M} p_m(x_m) \qquad (2.48)$$

of the sets of rate vectors

$$\{(R_1, R_2, \ldots, R_M) : R_S \leq I(X_S; Y | X_{S^c}), \text{ for all } S \subseteq \{1, 2, \ldots, M\}\}. \qquad (2.49)$$

Multiple access with independent messages and feedback: cut-set bounds

Let us now generalize the previous considerations slightly. Specifically, we want to allow for feedback, as illustrated by the dashed arrows from the decoder to all of the encoders in Figure 2.5. Feedback has long been known to increase the capacity of channels with memory (the roots of this insight go back to Shannon (1956)). It also increases the capacity region of general networks, even of memoryless ones. This was first observed in Gaarder and Wolf (1975). The general capacity of the multiple-access channel with feedback is unknown to date. Instead, we discuss two different upper bounding techniques: In this paragraph, we discuss a well-known *cut-set* approach. In the next paragraph, we discuss a somewhat less well-known *dependence-balance* approach.

The heart of the cut-set approach is as follows: Any partial sum of the rates must satisfy

$$\sum_{m \in S} R_m = H(V_S) = H(V_S | V_{S^c})$$ (2.50)

However, as long as the following Markov chain holds:

$$(V_S, Y) \longleftrightarrow X_S[i] \longleftrightarrow Y_i$$ (2.51)

this implies that (Cover and Thomas 2006, Thm.15.10.1)

$$\sum_{m \in S} R_m \leq I(X_S; Y_{S^c} | X_{S^c}).$$ (2.52)

The important observation is that even when feedback is available, this Markov chain holds.[5] Since any code must *simultaneously* satisfy *all* such bounds, one obtains the following theorem.

Theorem 2.3.10 (max-min cut-set bound, Cover and Thomas (2006)) *The capacity region is contained within the convex closure of the union over all distributions*

$$p(x_1, x_2, \ldots, x_M)$$ (2.53)

of the sets of rate vectors

$$\{(R_1, R_2, \ldots, R_M) : R_S \leq I(X_S; Y_{S^c} | X_{S^c}), \text{ for all } S \subseteq \{1, 2, \ldots, M\}\}.$$ (2.54)

It is important to observe that in spite of the apparent similarity to Theorem 2.3.9, this theorem describes a fundamentally different behavior. Mathematically, this is reflected by the fact that in Theorem 2.3.10, the maximization is performed over all possible joint distributions $p(x_1, \ldots, x_M)$ whereas in Theorem 2.3.9, it is only over product distributions. For the linear Gaussian sensor network example, let us first consider the case $M = 2$. Then, Theorem 2.3.9 can be written as

$$R_1 \leq \log\left(1 + \frac{P_1}{\sigma_Z^2}\right)$$ (2.55)

$$R_2 \leq \log\left(1 + \frac{P_2}{\sigma_Z^2}\right)$$ (2.56)

$$R_1 + R_2 \leq \log\left(1 + \frac{P_1 + P_2}{\sigma_Z^2}\right)$$ (2.57)

---

[5]Notice, however, that the Markov chain does not hold any longer if the underlying *messages* $V_1, V_2, \ldots, V_M$ are allowed to be dependent. We discuss this issue in the sequel.

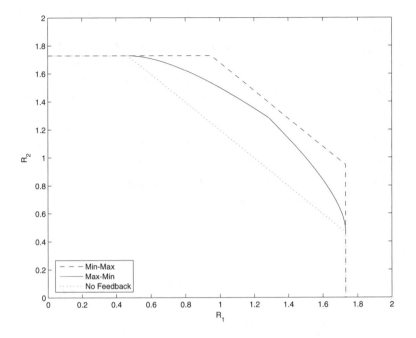

Figure 2.6 Capacity region and cut-set bounds for the Gaussian two-user MAC with $P_1 = P_2 = 10$ and unit noise variance. Rates are in bits.

By contrast, Theorem 2.3.10 can be expressed as the union over all ρ of the regions

$$R_1 \leq \log\left(1 + \frac{P_1(1 - \rho^2)}{\sigma_Z^2}\right) \tag{2.58}$$

$$R_2 \leq \log\left(1 + \frac{P_2(1 - \rho^2)}{\sigma_Z^2}\right) \tag{2.59}$$

$$R_1 + R_2 \leq \log\left(1 + \frac{P_1 + P_2 + 2\rho\sqrt{P_1 P_2}}{\sigma_Z^2}\right). \tag{2.60}$$

This is illustrated in Figure 2.6 by the region labeled 'Max-Min.' It was shown by Ozarow (1984) that the max-min cut-set bound is not only an outer bound; it is the actual feedback capacity region. Hence, in the (Gaussian) two-user case, the bound is tight. This is not true for more than two users. We briefly discuss this in the next paragraph. Finally, the region labeled 'Min-Max' is a weaker but more general outer bound, given in Theorem 2.3.12.

Multiple access with independent messages and feedback: dependence-balance bounds

The arguments discussed in the previous paragraph for the special case of the multiple-access channel have illustrated a special property: If channel inputs can somehow be made dependent, then a better performance can be achieved in general. There are of course many different ways in which such dependence can be attained. One way of obtaining a bound out of this intuition was presented by Hekstra and Willems (1989). This bound crucially relies on the fact that initially, the source information at the nodes is independent, and that dependence must be attained via a communication channel. This bound appears to be particularly useful for two-way channels and for scenarios that involve feedback (see Gastpar and Kramer (2006a)).

The following bound was originally developed for the two-way channel and for the two-user multiple-access channel by Hekstra and Willems (1989); the current shape (and extension to more than two users) was presented by Kramer and Gastpar (2006).

Theorem 2.3.11 (dependence-balance bound) *The capacity region is contained within the (convex closure of the) union over all distributions*

$$p(t, x_1, x_2, \ldots, x_M) p(y|x_1, x_2, \ldots, x_M) \tag{2.61}$$

that satisfy the conditions

$$I(X_1, \ldots, X_M; Y|T) \leq \frac{1}{K-1} \sum_{k=1}^{K} I(X_{\mathcal{W}_k^c}; Y|X_{\mathcal{W}_k}, T) \tag{2.62}$$

for any partition $\{\mathcal{W}_k\}_{k=1}^{K}$ *of the set* $\{1, 2, \ldots, M\}$ *into* K *subsets where* $K \geq 2$, *and where* $\mathcal{W}^c$ *denotes the complement of the set* $\mathcal{W}$ *in the set* $\{1, 2, \ldots, M\}$, *of the sets of rate vectors*

$$\{(R_1, R_2, \ldots, R_M) : R_{\mathcal{S}} \leq I(X_{\mathcal{S}}; Y_{\mathcal{S}^c}|X_{\mathcal{S}^c}), \text{ for all } \mathcal{S}\}. \tag{2.63}$$

For the two-way channel (and the two-user MAC), this theorem was first established by Hekstra and Willems (1989), where its centerpiece, Equation (2.62), was referred to as the dependence-balance condition. The evaluation of the theorem for concrete cases is complicated by the presence of the auxiliary random variable T. Specifically, while omitting some of the subsets $\mathcal{S}_m$ from Equation (2.62) yields a valid outer bound to the capacity region, one must maximize the expressions in the theorem over *all* choices of the auxiliary T. This problem can be illustrated by the aid of the linear Gaussian scenario considered in this chapter. To evaluate Theorem 2.3.10 for this case, one can start with an arbitrary $p(x_1, x_2, \ldots, x_M)$ and then argue that switching to a Gaussian distribution with the same second-order statistics cannot decrease any of the bounds in Equation (2.54). This argument cannot be used in conjunction with Theorem 2.3.11 since switching to Gaussian increases both sides in the dependence-balance condition (Eq. 2.62). Instead, a more elaborate argument is needed. One such argument is given by Kramer and Gastpar (2006), and another by Gastpar and Kramer (2006b). For the two-user case illustrated in Figure 2.6, the dependence-balance bound can be shown to coincide with the max-min cut-set bound. However, for M users, it can be shown that the cut-set bound and the dependence-balance bound lead to different scaling behaviors, and that the cut-set bound is loose.

Multiple access with dependent messages

For sensor networks, it may very well be the case that the messages at different nodes are *not* independent of each other (since they are typically related to one and the same underlying phenomenon), or that the transmitting terminals have some form of cooperation available. The case of dependent messages has been studied by Cover et al. (1980), though Dueck (1981) has shown that these results are not entirely general. Some extensions were given by Ahlswede and Han (1983). No conclusive or general capacity results are available. It is important to note that the max-min cut-set bound given in Theorem 2.3.10 no longer applies: the Markov chain condition of Equation (2.51) cannot be established.

More specifically, and in line with the distributed source models that we have discussed in Section 2.3.1, we can consider the situation of a discrete memoryless distributed source, that is, a sequence of independent and identically distributed discrete random vectors

$$\{(V_1[i], V_2[i], \ldots, V_M[i])\}_{i \geq 0}, \tag{2.64}$$

distributed according to a fixed and known distribution $p(v_1, v_2, \ldots, v_M)$. A natural question is: when is it feasible to communicate these sources across the given multiple-access channel? A generally non-computable ('infinite-letter') answer to this question was given by Cover et al. (1980). However, the focus of this chapter is on insightful and sufficiently simple performance upper bounds. By analogy to the discussion in Section 2.3.1, the simplest upper bounds are again the one for 'centralized' coding, for which the answers are well known. To see what we mean by this, consider Figure 2.5: The sum of all the individual rates, $R_{tot} = \sum_{m=1}^{M} R_m$, cannot be larger than the capacity of the point-to-point (MIMO) channel with input vector $(X_1, X_2, \ldots, X_M)^T$ and output vector $(Y_1, Y_2, \ldots, Y_J)^T$. This, in turn, can be considered a 'centralized coding' upper bound: it corresponds to merging all the M encoders into a single 'superencoder' that has simultaneous access to all M messages, $V_1, V_2, \ldots, V_M$. It can also be seen as a cut-set argument: We cut the network into two parts, one comprised of the nodes $1, 2, \ldots, M$, the other comprised of the base station.

More generally, one can consider arbitrary networks, and partition the nodes into two disjoint sets, S and S^c.

Theorem 2.3.12 (min-max cut-set bound) *If it is feasible to communicate the sources* $(V_1, V_2, \ldots, V_M)$ *across an M-user multiple access channel, then we must have*

$$H(V_S) \leq C_S \text{ for all } S \subseteq \{1, 2, \ldots, M\}, \tag{2.65}$$

where

$$C_S \stackrel{\text{def}}{=} \max I(X_S; Y_{S^c} | X_{S^c}), \tag{2.66}$$

where the maximum is over all joint distributions $p(x_{S,S^c})$ that satisfy the cost constraints $P_1, \ldots, P_M$.

In spite of its apparent coarseness, this bound turns out to be tight in some non-trivial cases, including, for example, some Rayleigh-fading AWGN relay channels, see Kramer et al. (2005). Generally, however, the bound must be expected to be loose. This is illustrated for the Gaussian MAC (with $M = 2$ and $J = 1$) and *independent* sources V_1 and V_2 (of entropy R_1 and R_2, respectively) in Figure 2.6 by the region labeled 'Min-Max'.

We can evaluate Theorem 2.3.12 for the linear Gaussian case. The solution to this problem is well known (see e.g. Telatar (1995)), as follows.

Corollary 2.3.13 *For the Gaussian multiple access channel with J receive antennas, characterized by a matrix $B \in C^{J \times M}$ of rank $\tilde{J}$ with singular values β_n, the maximum joint source entropy that can be supported satisfies*

$$H(\{V_m\}_{m=1}^M) \leq \sum_{n=1}^{\tilde{J}} \left(\log(\beta_n^2 \gamma)\right)^+, \tag{2.67}$$

where γ is chosen such that

$$\sum_{n=1}^{\tilde{J}} \left(\gamma - \frac{1}{\beta_n^2}\right)^+ = P_{tot}. \tag{2.68}$$

To understand the difference between Theorem 2.3.10 and Theorem 2.3.12, it suffices to consider the scenario where the messages at all terminals are identical. Then, as a matter of fact, the sum rate is exactly equal to the upper bound given by Theorem 2.3.12, which is strictly larger than the convex closure of the union taken in Theorem 2.3.10. Specifically, for the two-user Gaussian MAC with independent sources (i.e., $R_1 = H(V_1)$, $R_2 = H(V_2)$ and $R_1 + R_2 = H(V_1, V_2)$), Theorem 2.3.12 yields the following three bounds:

$$R_1 \leq \log\left(1 + \frac{P_1}{\sigma_Z^2}\right) \tag{2.69}$$

$$R_2 \leq \log\left(1 + \frac{P_2}{\sigma_Z^2}\right) \tag{2.70}$$

$$R_1 + R_2 \leq \log\left(1 + \frac{P_1 + P_2 + 2\sqrt{P_1 P_2}}{\sigma_Z^2}\right). \tag{2.71}$$

This is illustrated in Figure 2.6.

For certain cases, better bounds have been developed by Kang and Ulukus (2006).

Special cases

For a few special cases of interest, all of the above notions of capacity (with feedback, with potentially partially cooperating encoders, with dependent messages) can be determined exactly, rather than merely bounded.

Fast fading with uniform phases known only at the decoder. As a first example of a special case, consider the linear Gaussian situation illustrated in Figure 2.5. To capture the crux of the argument, it suffices to consider $J = 1$. However, contrary to our earlier considerations, we now assume that B is randomly selected, independently for each time unit i. Hence, we can think of a sequence $B[i]$, and we assume that $B[i]$ is known only to the decoder, i.e., the standard fast fading model. To deal with this situation, one can introduce an augmented channel output, given by the pair $(Y[i], B[i])$, and apply all the above theorems by substituting this pair for the channel output $Y[i]$. It is then easy to calculate that the distribution that maximizes the bounds in Theorem 2.3.12 is simply to make $X_1, X_2, \ldots, X_M$ independent Gaussian random variables. It is immediately clear that the corresponding rates are achievable, even without exploiting the feedback, any other form of cooperation, or the dependence of the underlying messages, thus establishing a capacity result.

Received-power constraints. As another variation on the problem of multiple access, suppose that instead of having separate power constraints for each user, there is one power constraint on the channel output signal. Specifically, we require that the codebooks be designed in such a way as to guarantee that

$$\frac{1}{n} \sum_{i=1}^{n} E\left[|Y[i]|^2\right] \le Q. \tag{2.72}$$

For this problem, it can again be shown that independently chosen codebooks maximize the bounds in Theorem 2.3.12, directly implying a capacity result. A detailed study of this was given by Gastpar (2007a).

2.3.3 End-to-end Performance of Digital Architectures

In this section, we briefly discuss the overall performance that can be attained by a 'digital architecture', by which we mean a scheme in which the source observations are first compressed into bit sequences (or discrete messages), and these messages are then communicated across the noisy channel network in a reliable fashion. This performance can be bounded by *combining* the arguments for distributed source coding discussed in Section 2.3.1 with the bounds for distributed channel coding considered in Section 2.3.2.

Let us first briefly discuss the scenario of a simple point-to-point communication problem (such as, for example, the setting of Figure 2.2 with $L = M = J = 1$). This problem is well understood to date (at least as long as all involved random processes are stationary and ergodic). Specifically, for this case, the rate-distortion region defined in Equation (2.16) simply becomes the standard rate-distortion function, often denoted by $R(D)$ (see e.g. Berger (1971)). Similarly, the capacity-cost region defined in Equation (2.47) simply becomes the standard capacity-cost function (or merely capacity), often denoted by $C(P)$ (see e.g. Cover and Thomas (2006, p. 263), for the Gaussian case or Csiszár and Körner (1981, p. 108), for the general discrete case).

For such a point-to-point communication problem, a digital architecture can attain any cost-distortion trade-off (P, D) that satisfies[6]

$$R(D) < C. \tag{2.73}$$

This is achievable simply by first encoding the source using $R(D)$ bits. Then, since the channel capacity is larger than $R(D)$, these bits can indeed be communicated to the destination.

In a sense more pertinent to the main points discussed in this chapter is the fact that there is no coding scheme whatsoever (not necessarily digital) that attains a cost-distortion trade-off (P, D) for which

$$R(D) > C. \tag{2.74}$$

This fact is often referred to as the source/channel separation theorem. We discuss this in the context of general bounds below in Section 2.5.

---

[6]The case $R(D) = C$ is attainable in some cases, but not in all (see e.g. Gastpar et al. (2003) for a more detailed discussion).

For networks, the analogous question is phrased in terms of the cost-distortion vectors $(P_1, \ldots, P_M, D_1, \ldots, D_L)$. Specifically, using the rate-distortion region defined in Equation (2.16) and the capacity-cost region defined in Equation (2.47), one can establish that a cost-distortion vector $(P_1, \ldots, P_M, D_1, \ldots, D_L)$ can be attained using a digital communication strategy if and only if

$$\mathcal{R} \cap \mathcal{C} \neq \emptyset. \tag{2.75}$$

By contrast to the point-to-point setting, this is a necessary condition only for digital schemes; more general schemes may attain better cost-distortion trade-offs. We discuss this in detail in Sections 2.4 and 2.5.

Again, we want to illustrate this by the aid of the linear Gaussian example.

Example 2.3.14 *Consider the linear Gaussian example shown in Figure 2.2, and assume that $L = J = 1$ and that A and B are merely vectors of all ones, i.e., $A = B^T = (1, 1, \ldots, 1)$. As we have seen in Theorem 2.3.7, for this simple case, we can express the rate of the source code (in order to attain a mean-square distortion level of D) as*

$$R_{tot}(D) = \log \frac{\sigma_S^2}{D} + M \log \frac{M\sigma_S^2}{M\sigma_S^2 + \sigma_W^2 - \frac{\sigma_S^2}{D}\sigma_W^2}. \tag{2.76}$$

Similarly, let us consider the special case $J = 1$, and where B is simply a vector of all ones. Then, we can evaluate Corollary 2.3.13 to yield

$$C(P_{tot}) \leq \log \left(1 + \frac{MP_{tot}}{\sigma_Z^2} \right). \tag{2.77}$$

Combining these two formulas, we conclude that the power-distortion pairs (P_{tot}, D) attainable by any digital architecture must satisfy the relationship

$$\frac{\sigma_S^2}{D} + \left(\frac{M\sigma_S^2}{M\sigma_S^2 + \sigma_W^2 - \frac{\sigma_S^2}{D}\sigma_W^2} \right)^M \leq 1 + \frac{MP_{tot}}{\sigma_Z^2}. \tag{2.78}$$

This implies the following lower bound on the distortion that can be attained with total power P_{tot} :

$$D \geq \frac{\sigma_S^2 \sigma_W^2}{\sigma_S^2 \log \left(1 + \frac{MP_{tot}}{\sigma_Z^2} \right) + \sigma_W^2} \tag{2.79}$$

As a second example, let us consider the simple computation problem discussed earlier.

Example 2.3.15 (distributed computation over a MAC) *As a second (toy) example, let us reconsider Example 2.3.8, i.e., the case where a function of the source observations needs to be encoded. That is, the problem is now to convey the sum across a multiple access channel. Clearly, if the capacity region of that MAC includes the rate point*

$$(R_1 = H_b(p), R_2 = H_b(p)), \tag{2.80}$$

then it is feasible to reliably communicate the sum via a digital strategy, and otherwise, it is not. We will discuss this example again from a slightly different perspective in Section 2.4 (Example 2.4.3).

2.4 The Price of Digital Architectures

There are many different reasons why digital architectures may be desirable, including modularity of system design and some degree of robustness. However, in terms of pure communication quality (which we have defined in Section 2.2 in terms of the trade-off between communication cost and benefit/distortion), a digital architecture generally comes at a penalty, which is perhaps not counterintuitive.

Shannon's so-called source/channel separation theorem comes as a miracle in this universe, proving that this penalty vanishes for the simple (stationary, ergodic) point-to-point communication link. In other words, a cost-distortion vector (P, D) is achievable if and only if

$$R(D) \le C(P). \tag{2.81}$$

The proof that no coding strategy can perform better only requires the data processing inequality (see e.g. Cover and Thomas (2006)). Specifically, one proceeds as follows. For any encoder,

$$I(S^n; \hat{S}^n) \le I(X^n; Y^n). \tag{2.82}$$

But then, the minimum of the left-hand side is the rate-distortion function, and the maximum of the right-hand side the capacity-cost function, establishing that $R(D) \le C(P)$. It should also be pointed out that if the assumptions of stationarity and ergodicity are dropped, this is no longer true. An interesting illustration of this was given by Vembu et al. (1995).

In this section, we illustrate that in general networks, there is a strict performance penalty for digital communication architectures. In Section 2.3.3, we expressed the best possible performance of a digital architecture by equating the rate (region) for the source coding to the capacity (region) of the channel. More formally, a digital architecture can attain any performance characterized by a relationship of the following kind:

$$\mathcal{R}(D) \cap \mathcal{C}(P) \ne \emptyset, \tag{2.83}$$

where $\mathcal{R}$ is the rate region required for the source coding step (to satisfy distortion constraints D), and $\mathcal{C}$ is the capacity region of the channel network (for given resource constraints P). However, it has long been known that Equation (2.83) is not a necessary condition. In other words, even if the intersection of the rate-distortion region and the capacity region is empty, there may exist a code that achieves the prescribed distortion levels D at a cost P. However, that code is *not* a digital code – that is, it cannot be understood in terms of source compression followed by reliable communication across noisy channels. Rather, it requires joint source-channel coding.

Example 2.4.1 (Cover et al. (1980)) *A classical example illustrating the fact that source/ channel separation does not hold for networks is the following: The channel is the binary adder multiple access channel, taking two binary $\{0, 1\}$ inputs and outputting their sum $\{0, 1, 2\}$. The capacity region $\mathcal{C}$ of this channel has the pentagonal shape given in Figure 2.7, see Cover and Thomas (2006, Figure 15.13), for more details. Now suppose that the two transmitting terminals each observe a binary sequence, call them S_1^n and S_2^n. The two sequences are correlated with each other such that for each time instant, the events $(S_1, S_2) = (0, 0), (0, 1),$*

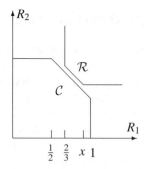

Figure 2.7 Capacity region $\mathcal{C}$ and rate-distortion region $\mathcal{R}$ do not intersect in this example.

and $(1, 1)$ *are all equally likely, and* $(1, 0)$ *does not occur. Clearly, at least* $H(S_1, S_2) = \log 3 \approx 1.58$ *bits per source sample are required. The full Slepian-Wolf rate region* $\mathcal{R}$ *is also given in Figure 2.7; the point labeled* x *is* $\log 3 - \frac{2}{3}$. *The two regions do not intersect, and hence, one is tempted to guess that these two sources cannot be transmitted across this MAC. However, this conclusion is wrong: While there is no 'digital' architecture that achieves this, there is a simple 'analog' strategy: pure uncoded transmission will always permit recoverery of both source sequences without error, due to the fact that the dependence structure of the sources is perfectly matched to the channel. This illustrates that no separation theorem applies to general networks.*

Perhaps a more relevant and slightly less contrived example of the price of digital communication in networks can be given by the aid of the linear Gaussian sensor network that was studied throughout this chapter.

Example 2.4.2 (Gastpar and Vetterli (2002, 2003, 2005)) *To illustrate the price of digital architectures by the aid of the linear Gaussian example shown in Figure 2.2, let us assume that* $L = J = 1$ *and that* A *and* B *are merely vectors of all ones, i.e.,* $A = B^T = (1, 1, \ldots, 1)$. *Suppose that the 'code' used by the sensors is given by*

$$X_m[i] = \sqrt{\frac{P_m}{\sigma_S^2 + \sigma_W^2}} U_m[i], \tag{2.84}$$

i.e., the sensor instantaneously (in each time unit i*) scale their observation in such a way as to meet their power constraint, but otherwise do not apply any further coding. It is easy to see that the optimum decoding rule is also instantaneous. More precisely, it is simply given by the conditional mean,*

$$\hat{S}[i] = E\left[S[i]|Y[i]\right], \tag{2.85}$$

which can be easily evaluated for the case where $\{S[i]\}_{i \geq 0}$ *is a sequence of i.i.d. Gaussian random variables of mean zero and variance* σ_S^2. *By noting that the received signal* $Y[i]$ *is given by*

$$Y[i] = Z[i] + \sum_{m=1}^{M} X_m[i] = Z[i] + \sqrt{\frac{M P_{tot}}{(\sigma_S^2 + \sigma_W^2)}} S[i] + \sum_{m=1}^{M} \sqrt{\frac{P_{tot}}{M(\sigma_S^2 + \sigma_W^2)}} W_m[i],$$

we find that

$$\hat{S}[i] = \frac{MP_{tot}\sigma_S^4/(\sigma_S^2 + \sigma_W^2)}{\sigma_Z^2 + MP_{tot}\sigma_S^2/(\sigma_S^2 + \sigma_W^2) + P_{tot}\sigma_W^2/(\sigma_S^2 + \sigma_W^2)} Y[i]. \tag{2.86}$$

This lets us determine the resulting distortion as

$$D = \frac{\sigma_S^2 \sigma_W^2}{M\sigma_S^2 + \sigma_W^2}\left(1 + \frac{M(\sigma_S^2 \sigma_Z^2/\sigma_W^2)}{\frac{M\sigma_S^2 + \sigma_W^2}{\sigma_S^2 + \sigma_W^2}P_{tot}(M) + \sigma_Z^2}\right). \tag{2.87}$$

This can be proved to be the optimal behavior, see Gastpar (2007b).

The most interesting aspect of this insight is not the formula given in Equation (2.87), but its comparison with Equation (2.79): In the former, the distortion scales like $1/M$, whereas in the latter, it scales like $1/\log M$. Hence, for this simple example, the digital architecture not only performs suboptimally, it entails an unbounded penalty, exponential in the number of nodes M.

It may be tempting to guess at this point that it is the dependence of the source observations that is responsible for the shortcoming of the digital communication paradigm. However, this argument is not sufficiently precise to capture the gist of the story. Interestingly, the distortion criteria play an equally important (or perhaps dual) role. We illustrate this by way of the following example.

Example 2.4.3 (Gastpar (2002)) *Let us reconsider the setup of Examples 2.3.8 and 2.3.15. However, this time, rather than merely specifying the capacity region of the multiple access channel (as in Example 2.3.15), we define the precise structure. To first make a simple point, consider a discrete memoryless multiple access channel with binary inputs X_1 and X_2 and a binary output Y given by*

$$Y = X_1 \oplus X_2, \tag{2.88}$$

where $\oplus$ denotes modulo-2 addition. It is immediately clear that if we get to use the channel once per source symbol, then we can always attain our goal: We simply map each source symbol separately onto a channel input, and the channel computes the modulo-2 sum for us. Can we also always attain this via a digital communication strategy? The best such strategy will use the code discussed in Example 2.3.8, and it will work whenever the resulting source coding rate point will come to lie in the capacity region of the channel. It is easy to verify that for the considered MAC, the capacity region satisfies

$$R_1 + R_2 \leq 1, \tag{2.89}$$

since the channel output is binary. Hence, the digital strategy will work if and only if $2H_b(p) \leq 1$, or $p < 0.11$ (or equivalently, $p > 0.89$). That is, there is a price of separation. This example was considerably generalized by Nazer and Gastpar (2005).

It is important at this point to emphasize that there are networks for which digital communication does not incur a penalty. That is, there are also classes of networks where a separation theorem of the shape of Equation (2.83) can be given, including the transmission of independent sources with respect to independent fidelity criteria across any multiple access channel, see e.g. Gastpar (2002), and the error-free transmission of discrete correlated sources across separate (parallel) channels, see Barros and Servetto (2006); Xiao and Luo (2005). We will discuss this in somewhat more detail in the next section.

2.5   Bounds on General Architectures

In the previous section, we showed that there can be an arbitrarily high price for digital architectures. In order to get a better understanding and assessment of this penalty, we need fundamental information-theoretic bounds on the performance for general architectures. There are only very few tools known today that permit to derive such bounds, which is reflected by the length of the present section. One way of providing such bounds is along the lines of cut-set (i.e., 'centralized coding') arguments. We will again concentrate on the multiple-access topology. The simplest cut-set bound takes the following shape:

Theorem 2.5.1 (cut-set bound) *Any achievable cost-distortion vector* $(P_1, P_2, \ldots, P_M, D_1, D_2, \ldots, D_L)$ *must satisfy, for all subsets* $\mathcal{S} \subseteq \{1, 2, \ldots, M\}$,

$$R_{\mathcal{S}} \leq C_{\mathcal{S}}. \tag{2.90}$$

where $R_{\mathcal{S}}$ *is defined in Equation (2.21) and* $C_{\mathcal{S}}$ *is defined in Equation (2.66).*

To prove this theorem, it suffices to merge, for each subset $\mathcal{S}$, all encoders with indices in the set $\mathcal{S}$ into one encoder, and to provide the observation streams with indices outside the set $\mathcal{S}$ directly to the fusion center. The resulting communication system is a point-to-point system with side information at the decoder, for which the optimum performance is characterized precisely by the condition $R_{\mathcal{S}} \leq C_{\mathcal{S}}$ (Gastpar 2002, Thm.1.10). In line with the general structure of this chapter, we again want to illustrate this bound for the special case of the linear Gaussian sensor network example. For a derivation, see Gastpar and Vetterli (2005).

Theorem 2.5.2 *For large enough total sensor power* P_{tot}, *the distortion that can be achieved in the Gaussian sensor network cannot be smaller than*

$$D \geq \sum_{\ell=1}^{L} \frac{\sigma_S^2 \sigma_W^2}{\alpha_\ell^2 \sigma_S^2 + \sigma_W^2} + c_1 \left(\frac{1}{c_2 + \frac{P_{tot}}{K J \sigma_Z^2} \mathcal{G}(\beta)} \right)^{K \tilde{J}/L} \tag{2.91}$$

where $\alpha_1, \alpha_2, \ldots, \alpha_L$ *are the singular values of A,* $\beta_1, \beta_2, \ldots, \beta_{\tilde{J}}$ *are the non-zero singular values of B,* σ_S^2 *is the variance of the underlying sources,* σ_W^2 *is the variance of the observation noises,* σ_Z^2 *is the variance of the noises in the communication channel,* P_{tot} *is the total sensor transmit power for the K channel uses,* $\tilde{J}$ *is the rank of the matrix B, and*

$$c_1 = 2^{\frac{1}{L} \sum_{\ell=1}^{L} \log \frac{\alpha_\ell^2 \sigma_S^4}{\alpha_\ell^2 \sigma_S^2 + \sigma_W^2}}, \tag{2.92}$$

$$c_2 = \mathcal{G}(\beta)/\mathcal{H}(\beta), \tag{2.93}$$

where the harmonic mean of the squares of the non-zero singular values of B is denoted by

$$\mathcal{H}(\beta) = \left(\frac{1}{\tilde{J}} \sum_{n=1}^{\tilde{J}} \frac{1}{\beta_n^2} \right)^{-1}, \tag{2.94}$$

and their geometric mean by

$$\mathcal{G}(\beta) = \sqrt[\tilde{J}]{\prod_{n=1}^{\tilde{J}} \beta_n^2}. \tag{2.95}$$

This theorem can be used to establish a number of scaling-law separation theorems, i.e., sensor network situations for which a digital architecture, while not provably optimal, attains the optimal performance scaling-law behavior as a function of the number of sensors. Preliminary results can be found in the work of Gastpar et al. (2006).

Theorem 2.5.1 can be tightened slightly for several special cases. One interesting bound exploits the concept of *maximal correlation*, which we introduce here along the lines of the work of Witsenhausen (1975). Consider two random variables U_1 and U_2, with joint distribution $p(u_1, u_2)$. Let $f_1(\cdot)$ and $g_2(\cdot)$ be functions satisfying $E[f_1(U_1)] = E[g_1(U_2)] = 0$ and $E[f_1^2(U_1)] = E[g_1^2(U_2)] = 1$. Then, the maximum correlation $\rho_{max}(U_1, U_2)$ is defined to be

$$\rho_{max}(U_1, U_2) = \sup_{f_1, g_1} E[f_1(U_1)g_1(U_2)]. \tag{2.96}$$

Theorem 2.5.3 (Witsenhausen (1975)) *For the two-user multiple-access channel, suppose that the source observations are sequences of independent and identically distributed pairs of random variables $\{(U_1[n], U_2[n])\}_{n \geq 0}$. If*

$$H(U_1, U_2) > \max I(X_1, X_2; Y), \tag{2.97}$$

where the maximum is over all joint distribution $p(x_1, x_2)$ for which the correlation coefficient is no larger than $\rho_{max}(U_1, U_2)$, then the sources cannot be transmitted across the multiple-access channel.

As shown by Witsenhausen (1975), the bound can be extended to encompass sequences of pairs of independent, but not necessarily identically distributed random variables. This bound has been used by Lapidoth and Tinguely (2006) and later by Gastpar (2007b). We discuss an application of this theorem in the next section. A related but different bound was found by Kang and Ulukus (2006).

Comparison of the bounds for the linear Gaussian sensor network

In order to summarize the arguments discussed in this chapter, we now compare the various bounds for the simplest case of the linear Gaussian sensor network example in its simplest case, namely with $L = J = 1$.

We saw that a digital architecture attains a distortion that satisfies

$$D \geq \frac{\sigma_S^2 \sigma_W^2}{\sigma_S^2 \log\left(1 + \frac{M P_{tot}}{\sigma_Z^2}\right) + \sigma_W^2} \qquad (2.98)$$

whereas there is a simple 'uncoded' strategy attaining

$$D = \frac{\sigma_S^2 \sigma_W^2}{M \sigma_S^2 + \sigma_W^2} \left(1 + \frac{M(\sigma_S^2 \sigma_Z^2 / \sigma_W^2)}{\frac{M \sigma_S^2 + \sigma_W^2}{\sigma_S^2 + \sigma_W^2} P_{tot} + \sigma_Z^2} \right). \qquad (2.99)$$

Theorem 2.5.3 can be extended to show that this is the smallest possible distortion, i.e., that the 'uncoded' strategy is exactly optimal for this simple special case, see Gastpar (2007b). The general cut-set bound given in Theorem 2.5.2 provides a slightly less tight lower bound, as follows:

$$D \geq \frac{\sigma_S^2 \sigma_W^2}{M \sigma_S^2 + \sigma_W^2} \left(1 + \frac{M(\sigma_S^2 \sigma_Z^2 / \sigma_W^2)}{M P_{tot} + \sigma_Z^2} \right). \qquad (2.100)$$

2.6 Concluding Remarks

In this chapter, we discussed information-theoretic bounding techniques for sensor network performance. We first considered 'digital' strategies, which are characterized by the paradigm that each sensor transforms its observations into a bit sequence (in the best possible way), and all these bit sequences are then communicated without incurring further errors, using sufficiently long codes. We gave an overview of existing and novel techniques to bound the performance of any overall strategy that falls into this class. For (stationary, ergodic) point-to-point communication, these strategies are as good as the best strategies. This fact is known as the source/channel separation theorem. As we briefly discussed, there is no such theorem for communication problems of the sensor network type, and the extent of this lack is non-trivial: the performance deficiency of the best digital strategy with respect to the best general strategy can be exponential in the number of sensor nodes.

More generally, the area of multi-terminal information theory is full of open problems. While not all of them should be expected to be of key relevance to the sensor network problem, some that we believe to be of interest include:

- *Beyond cut-set bounds.* One of the most fundamental problems concerns the derivation of performance upper bounds that do not rely on simplistic cut-set arguments, but rather permit us to directly address the fact that coding must be distributed. Such bounds are needed both to get a better sense of the performance of digital schemes as well as to get better bounds for general architectures. Theorems 2.3.11 (the dependence-balance bound) and 2.5.3 (Witsenhausen's approach) are examples of such arguments.

- *Scaling-law source/channel separation theorems.* For certain sensor network topologies, while 'digital' schemes may not be exactly optimal, they are (almost) optimal in the limit as the network becomes large. Preliminary results can be found in the work of Gastpar et al. (2006).

- *Partial orderings.* Perhaps a more active formulation of the previous point is the need for partial orderings, that is, arguments that permit comparison of different sensor network problems in terms of their potential and difficulty. One promising approach is the graph-based perspective advocated and explored by Pradhan et al. (2004).

- *Beyond digital code constructions.* Finally, a problem that we omitted entirely from this chapter is the development of actual code constructions. Based on our bounds, we believe that this is a rich area, reaching considerably beyond the digital codes that are presently available. In preliminary work, we have found some non-digital code constructions for the problem of communicating functions. Details were given by Nazer and Gastpar (2005, 2006).

Acknowledgements

The author gratefully acknowledges stimulating discussions with K. Eswaran, B. Nazer, and A. D. Sarwate (University of California, Berkeley), as well as with P. L. Dragotti (Imperial College, London), G. Kramer (Bell Labs, Alcatel-Lucent), and M. Vetterli (EPFL/UC Berkeley). This work was supported in part by the National Science Foundation under awards CNS-0326503 and CCF-0347298 (CAREER).

Bibliography

Ahlswede R 1971 Multi-way communication channels. *Proc IEEE Int Symp Info Theory*, Tsahkadsor, Armenian S.S.R.

Ahlswede R and Han TS 1983 On source coding with side information via a multiple-access channel and related problems in multi-user information theory. *IEEE Transactions on Information Theory* **IT–29**(3), 396–412.

Barros J and Servetto SD 2006 Network information flow with correlated sources. *IEEE Trans. Inform. Theory* **52**(1), 155–170.

Berger T 1971 *Rate Distortion Theory: A Mathematical Basis for Data Compression.* Prentice-Hall, Englewood Cliffs, NJ.

Berger T 1977 Multiterminal source coding. Lectures presented at CISM summer school on the Information Theory Approach to Communications. Princeton University Press, Princeton, NJ. pp. 171–231.

Berger T, Zhang Z and Viswanathan H 1996 The CEO problem. *IEEE Transactions on Information Theory* **IT–42**, 887–902.

Chen J, Zhang X, Berger T and Wicker S 2004 An upper bound on the sum-rate distortion function and its corresponding rate allocation schemes for the CEO problem. *IEEE Journal on Selected Areas in Communications.* **22**, 977–987.

Cover TM and Thomas JA 2006 *Elements of Information Theory.* 2nd edn. Wiley, New York.

Cover TM, El Gamal AA and Salehi M 1980 Multiple access channels with arbitrarily correlated sources. *IEEE Transactions on Information Theory* **26**(6), 648–657.

Csiszár I and Körner J 1981 *Information Theory: Coding Theory for Discrete Memoryless Systems.* Academic Press, New York.

Dobrushin RL and Tsybakov BS 1962 Information transmission with additional noise. *IRE Transactions on Information Theory* **IT–18**, S293–S304.

Dueck G 1981 A note on the multiple access channel with correlated sources. *IEEE Transactions on Information Theory* **27**(2), 232–235.

Eswaran K 2005 Remote source coding and AWGN CEO problems, Master's thesis, University of California Berkeley, CA.

Gaarder NT and Wolf JK 1975 The capacity region of a multiple-access discrete memoryless channel can increase with feedback. *IEEE Transactions on Information Theory* **IT–21**, 100–102.

Gastpar M 2002 To code or not to code, PhD thesis, Ecole Polytechnique Fédérale (EPFL), Lausanne, Switzerland.

Gastpar M 2007a On capacity under receive and spatial spectrum-sharing constraints. *IEEE Transactions on Information Theory* **53**(2), 471–487.

Gastpar M 2007b Uncoded transmission is exactly optimal for a simple Gaussian sensor network. *In Proc. 2007 Information Theory and Applications Workshop*, San Diego, CA.

Gastpar M and Kramer G 2006a On cooperation via noisy feedback. *In Proc International Zurich Seminar*, Zurich, Switzerland.

Gastpar M and Kramer G 2006b On noisy feedback for interference channels. *In Proc 40th Asilomar Conference on Signals, Systems, and Computers*, Asilomar, CA.

Gastpar M and Vetterli M 2002 On the capacity of wireless networks: The relay case. *In Proc IEEE Infocom 2002*, New York.

Gastpar M and Vetterli M 2003 Source-channel communication in sensor networks. In *2nd International Workshop on Information Processing in Sensor Networks (IPSN'03)* (ed. Guibas LJ and Zhao F) Lecture Notes in Computer Science, Springer, New York, pp. 162–177.

Gastpar M and Vetterli M 2005 Power, spatio-temporal bandwidth, and distortion in large sensor networks. *IEEE Journal on Selected Areas in Communications (Special Issue on Self-Organizing Distributive Collaborative Sensor Networks)* **23**(4), 745–754.

Gastpar M, Rimoldi B and Vetterli M 2003 To code, or not to code: Lossy source-channel communication revisited. *IEEE Transactions on Information Theory* **49**(5), 1147–1158.

Gastpar M, Vetterli M and Dragotti PL 2006 Sensing reality and communicating bits: A dangerous liaison. *IEEE Signal Processing Magazine* **23**(4), 70–83.

Gupta P and Kumar PR 2000 The capacity of wireless networks. *IEEE Transactions on Information Theory* **IT–46**(2), 388–404.

Hekstra AP and Willems FMJ 1989 Dependence balance bounds for single-output two-way channels. *IEEE Transactions on Information Theory* **IT–35**(1), 44–53.

Housewright KB 1977 Source coding studies for multiterminal systems, PhD thesis, University of California, Los Angeles, CA.

Kang W and Ulukus S 2006 An outer bound for multiple access channels with correlated sources. *In Proc. 40th Conf. on Information Sciences and Systems (CISS)*, Princeton, NJ.

Körner J and Marton K 1979 How to encode the modulo-two sum of binary sources. *IEEE Transactions on Information Theory* **IT–25**(2), 219–221.

Kramer G and Gastpar M 2006 Dependence balance and the Gaussian multiaccess channel with feedback. *Proc 2006 IEEE Information Theory Workshop*, pp. 198–202, Punta del Este, Uruguay.

Kramer G, Gastpar M and Gupta P 2005 Cooperative strategies and capacity theorems for relay networks. *IEEE Transactions on Information Theory* **51**(9), 3037–3063.

Lapidoth A and Tinguely S 2006 Sending a bi-variate Gaussian source over a Gaussian MAC. *In Proc IEEE Int Symp Info Theory*, Seattle, WA.

Liao H 1972 A coding theorem for multiple access communications. *In Proc IEEE Int Symp Info Theory*, Asilomar, CA.

Nazer B and Gastpar M 2005 Reliable computation over multiple access channels. *In Proc 43rd Annual Allerton Conference on Communication, Control, and Computing*, Monticello, IL.

Nazer B and Gastpar M 2006 Computing over multiple-access channels with connections to wireless network coding. *In Proceedings of the 2006 International Symposium on Information Theory (ISIT 2006)*, Seattle, WA.

Neuhoff D and Pradhan S 2005 An upper bound to the rate of ideal distributed lossy source coding of densely sampled data. *In Proc IEEE Int Conf Acoustics Speech Sig.* Toulouse, France.

Oohama Y 1997 Gaussian multiterminal source coding. *IEEE Transactions on Information Theory* IT–43(6), 1912–1923.

Ozarow LH 1984 The capacity of the white Gaussian multiple access channel with feedback. *IEEE Transactions on Information Theory* IT–30(4), 623–629.

Prabhakaran V, Tse D and Ramchandran K 2004 Rate region of the quadratic Gaussian CEO problem. *In Proc IEEE Int Symp Info Theory*, Chicago, IL.

Pradhan SS, Choi S and Ramchandran K 2004 An achievable rate region for multiple access channels with correlated messages. *In Proc IEEE Int Symp Info Theory*, Chicago, IL.

Shannon CE 1948 A mathematical theory of communication. *Bell Sys. Tech. Journal* 27, 379–423, 623–656.

Shannon CE 1956 The zero-error capacity of a noisy channel. *IRE Transactions on Information Theory* IT–2, 8–19.

Shannon CE 1959 Coding theorems for a discrete source with a fidelity criterion. *In IRE Nat Conv Rec*, pp. 142–163.

Slepian D and Wolf J 1973 Noiseless coding of correlated information sources. *IEEE Trans. Inform. Theory* 19, 471–480.

Telatar IE 1995 Capacity of multi-antenna Gaussian channels. *Bell Labs Technical Memorandum.* Also published in *European Transactions on Telecommunications*, 10(6): 585-596, Nov.–Dec. 1999.

Tung SY 1978 Multiterminal source coding. PhD thesis, Cornell University, Ithaca, NY. See also *IEEE Trans. Info. Theory*, November 1978.

Vembu S, Verdú S and Steinberg Y 1995 The source-channel separation theorem revisited. *IEEE Transactions on Information Theory* IT–41(1), 44–54.

Wagner A 2005 Methods of offline distributed detection: interacting particle models and information-theoretic limits. PhD thesis, University of California, Berkeley, CA.

Wagner A, Tavildar S and Viswanath P 2005 The rate region of the quadratic Gaussian two-terminal source-coding problem. *http://arXiv:cs.IT/0510095.*

Wagner AB and Anantharam V 2005 An improved outer bound for the multiterminal source coding problem. *In Proc IEEE Int Symp Info Theory*, pp. 1406–1410, Adelaide, Australia.

Witsenhausen HS 1975 On sequences of pairs of dependent random variables. *SIAM J. Appl. Math.* 28(1), 100–113.

Wolf JK and Ziv J 1970 Transmission of noisy information to a noisy receiver with minimum distortion. *IEEE Transactions on Information Theory* IT–16, 406–411.

Wyner A and Ziv J 1976 The rate-distortion function for source coding with side information at the decoder. *IEEE Trans. Inform. Theory* 22, 1–10.

Xiao JJ and Luo ZQ 2005 Multiterminal source-channel communication under orthogonal multiple access. *In Proc 43rd Annual Allerton Conference on Communication, Control, and Computing*, Monticello, IL.

3

In-Network Information Processing in Wireless Sensor Networks

Arvind Giridhar and P. R. Kumar

3.1 Introduction

In recent years there has been a great deal of interest in wireless sensor networks. These are formed by nodes that can sense their environment, perform some computations on data that they have either sensed or received from other nodes, and can communicate over the wireless medium with other nodes. Because of the underlying wireless communication fabric, one possibility would be to view them as wireless networks with sensors at nodes replacing files as sources of data. On the other hand, because these wireless sensor networks are designed to output some 'answer', such as the mean temperature over all the sensors, another possibility would be to view them simply as computers.

They are actually combinations of both. There are some features that distinguish wireless sensor networks from both data communication networks as well as traditional computers. In traditional data networks, intermediate nodes do not alter the payload of packets they are forwarding. They are only allowed to read and modify the headers for packets, which essentially contain information such as the origin and destination of packets. In contrast, in sensor networks, a node may create a 'fused' packet that contains only the maximum temperature measurement that it has received from its neighbors and forward that while discarding all the packets that it has received from its neighbors. Thus, the

Wireless Sensor Networks: Signal Processing and Communications Perspectives A. Swami, Q. Zhao, Y.-W. Hong and L. Tong
© 2007 John Wiley & Sons, Ltd

nodes process the data *in-network*. So it is more appropriate to consider wireless sensor networks as also having the functionality traditionally associated with 'computers' rather than networks.

However, nodes in a wireless sensor network can only collaborate over the wireless medium. This medium is distinguished by it being a shared medium and an unreliable medium at that. These aspects are typically best analyzed from a communication networking point of view.

Thus, we are motivated to investigate what sort of a theory is best developed to understand how in-network processing must be performed in wireless sensor networks. This is the motivation for this chapter. We provide an account of some results of interest in this emerging area.

Our exposition will be in a staged manner by progressively introducing distinguishing aspects of wireless sensor networks, and at the same time showing what sorts of theoretical techniques are used to obtain answers to the questions that are posed. In the first model studied, we bring in the idea of nodes possessing different portions of the information, and exchanging messages with each other in order to compute a function of their joint data. In the second model studied, we introduce the issue of spatial reuse in wireless networks, the possibility of speedup of computational throughput by block computation, the idea of geographical modeling of wireless sensor networks either as collocated networks or random multi-hop wireless networks, and last but not least the study of classes of functions with respect to how efficiently they can be computed. In the third model, we will bring in the fact that the communications in a wireless network are themselves noisy, and the notion of computing an answer that is correct with a certain guaranteed high probability. Finally, we raise the issue of correlated information at nodes, and show how information theory can model sophisticated modalities of computation, which, if understood, would provide fundamental limits on what is achievable in wireless sensor networks.

Our exposition begins with some results from the field of communication complexity, where nodes exchange bits until some given function of nodal information is resolved. Then we turn to a theoretical framework that combines the aspects of viewing a wireless network as a computational fabric, takes into explicit account the shared medium nature of wireless communication, and also incorporates the lessons learnt from information theory, such as the fact that block computation can lead to throughput efficiencies. After that we turn to a model where communication is unreliable, but nevertheless the goal is to compute a given function with low probability of error. Finally, we sketch an information theoretic agenda for the problem of in-network information processing in wireless sensor networks. A theory of in-network information processing is still very much in its infancy and much remains to be done, as we note toward the end of the chapter.

Wireless sensor networks are intended to be deployed for several types of applications. An example is environmental monitoring, where each node may be endowed with a temperature sensor, and the goal may be to monitor temperature over the domain. Another example may be to monitor the maximum temperature over the domain, which may, in turn, be used to flag an alarm when it exceeds a certain critical threshold value.

An interesting feature of such wireless sensor networks is that information collected at the nodes of the network is processed within the network itself. A distinguished node in the sensor network may be dubbed a 'collector node', and it is the goal of the information processing within the network to deliver the desired statistic, say, the mean temperature or the maximum temperature, to this collector node, from where this desired information may be filtered out of the network. Thus the collector node may be regarded as an 'output unit'. The 'input units' are the measurements taken at the sensing nodes in the network. Everything else that comes in between, that is, the entire network itself, is the information processing system. Wireless sensor networks thus combine three aspects – sensing by nodes, computational capabilities at nodes, and wireless information transfer between nodes.

What distinguishes wireless sensor networks from mere 'communication networks' or 'data networks' is the feature that information can be processed at the nodes of the network. Thus, a node may elect to add the value of the temperatures contained in two packets it has received from two of its neighboring nodes, and pass on this sum to a third neighbor. Or, in an alarm network, it could even elect to ignore a packet it has received because it knows of an even higher temperature elsewhere in the network. This feature of information processing at nodes distinguishes wireless sensor networks from traditional data networks where nodes are only allowed to process or modify packet headers. Concerning the actual payload of the packet. they are only allowed to forward a packet without altering its content.

We should note that the functionality of nodes to process packets is also a subject of much recent interest in the emerging area called network coding. For the particular problem of multicast, network coding is in fact emerging as an attractive choice (Wu, Chou and Kung 2005). In sensor networks the scope of what nodes are allowed is very broad, since each node can simply be viewed as a 'computer' in its own right.

Another aspect of wireless sensor networks that distinguishes them is the medium that is used for communication – wireless. The wireless medium is a shared medium, and is further unreliable, susceptible as it is to interference and fading. These bring in several interesting aspects related to how the wireless medium is to be exploited for information processing.

Combining the above two characteristics, one can envision a wireless sensor network as a distributed computational system where communication is by wireless – a sort of 'Maxwellian computer'. This raises the issue of how such a system ought to be exploited to deliver the functionality that is sought. That is the subject of this chapter – how should information collected at the nodes of a wireless sensor network be processed in-network? This is very much a topic that is still evolving and much work remains to be done.

The outline of this chapter is as follows. We begin in Section 3.2 by studying a simple scenario where two nodes exchange 'bits' with the goal of computing a function whose two arguments are separately measured at the two nodes. If the goal is to compute the function in minimum time, it can be regarded as a problem in determining communication complexity of distributed computation, and was introduced by Yao (1979).

Next, in Section 3.3, we turn to a model (Giridhar and Kumar 2005) that introduces four additional distinguishing features of wireless sensor networks. One is the *spatial reusability* aspect of wireless networks. That is, the communication resource – the spectrum – can be simultaneously utilized at two distant locations in the network, for different message transmissions. We will consider a simple geometric model where transmissions simply create 'interference footprints' within which other receptions are impossible, as in Gupta and Kumar (2000). The second aspect that is introduced in this section is that, unlike simply considering a sensor network as only being required to perform a 'one-shot computation' of a function at just one given set of nodal measurements, we will suppose that computations must be repeated by the network with differing sets of nodal measurements, taken, say daily. That is, the network continues to monitor the environment every day. This continuing nature of the problem raises the issue of whether *block computation* can lead to greater efficiencies, as was pointed out and exploited by Shannon for communication systems. To put it another way, we are motivated whether there is an extension of block communication, called, say, 'block computation', that can provide 'computational throughput' speedups, just as block communication provides for the problem of communication. We combine both these issues in specific *geographical models* of wireless sensor networks, the third aspect introduced in this section. One specific class of model we will study is *random multi-hop wireless networks*, where nodes are uniformly located in a disk, with nodes choosing a common range sufficient for network connectivity, and communicating by multi-hop relaying of packets. The fourth aspect introduced in this section is the study of how the wireless sensor network can perform the computation of *symmetric functions*, which remain unaltered even if the sensor measurements at the nodes are permuted, and certain subclasses of them. This class includes many functions of interest such as 'mean' or 'maximum' alluded to above, and indeed several statistical functions.

Finally, in Section 3.4, we turn to another class of models in which yet another new ingredient is introduced – noisy communication (Gallager 1988; Rajagopalan and Schulman 1994; Schulman 1996; Kushilevitz and Mansour 1998). We will consider a broadcast network, i.e., collocated network, and present results on two specific functions.

In Section 3.5, we conclude with some comments on the difficulties confronting us in an information theoretic approach to in-network processing in sensor networks.

We note that the focus in this chapter is only on the communication cost of the process of distributed computation. We will not specifically address issues related to the computational cost.

3.2   Communication Complexity Model

Consider the following problem where there are two nodes, each having a different piece of information, say, a sensor measurement, and they desire to compute a function that depends on both pieces of information. Let us denote the two nodes, as is traditional, by Alice and Bob. Alice has access to a variable x, while Bob has access to a variable y. There is a function f, and both Alice and Bob want to determine the value of $f(x, y)$. Alice and Bob can communicate with each other, let us say over a link of 1 bit per time slot. The issue we will address is the minimum time required for Alice and Bob to determine the value of $f(x, y)$. More formally, we seek the minimum time over all protocols, of the maximum time to compute $f(x, y)$ over all possible inputs (x, y). We will suppose that x takes values in a finite set X, whose cardinality we denote by $|X|$. Similarly, we suppose that $y \in Y$ and $f(x, y) \in Z$. We will also use $|\text{Range}(f)|$ to denote the cardinality of the range, i.e., the set of values taken by $f(x, y)$ for $x \in X$ and $y \in Y$.

It should be noted that the specification of the deterministic protocol to compute $f(x, y)$ must include which node should transmit on each slot, as a causal function of all previous transmissions, as well as what that node should then transmit, which can be based not only on previous transmissions but also on its own variable.

Let us start with the simplest protocol. Alice sends Bob the value of x, which takes $\log |X|$ slots, and then Bob sends back Alice the value $f(x, y)$, which takes $\log |\text{Range}(f)|$ slots. This protocol takes a total of $\log |X| + \log |\text{Range}(f)|$ slots. Similarly, $\log |Y| + \log |\text{Range}(f)|$ is also feasible, which leads to the minimum of the two aforesaid expressions being an upper bound on computation time.

Conversely, it can be shown that $\log |\text{Range}(f)|$ is a lower bound. To see this, note that to encode the set of possible values of f takes $\log |\text{Range}(f)|$ bits. If there were a protocol for which the number of transmitted bits were less, then the mapping between function values and strings of transmitted bits could not be one-to-one. Therefore, there would have to be two inputs (x_1, y_1) and (x_2, y_2) for which the transmitted bits are exactly the same, but yet $f(x_1, y_1) \neq f(x_2, y_1)$. Note now that either $f(x_1, y_1) \neq f(x_1, y_2)$ or $f(x_1, y_2) \neq f(x_2, y_2)$, (for if both were equalities, one would have $f(x_1, y_1) = f(x_2, y_2)$). Suppose that $f(x_1, y_1) \neq f(x_1, y_2)$. Then the protocol would transmit the same bits for (x_1, y_2) as it would for either (x_1, y_1) or (x_2, y_2). But then Alice cannot resolve between $f(x_1, y_1)$ and $f(x_1, y_2)$, because she sees the same set of transmitted bits from Bob in each case and has the same input in each case as well.

A conceptually powerful representation of a protocol is as a binary tree. Each internal node v corresponds to either Alice or Bob, and as such is assigned either a function $a_v : X \rightarrow \{0, 1\}$, or $b_v : Y \rightarrow \{0, 1\}$. Making a sequence of transmissions corresponds to traversing down the binary tree. Each node represents the current state after all the previous transmissions, which both nodes are aware of. At each node, Alice or Bob will transmit depending on which function is assigned to that node, and the value of the transmission is the function value given the particular input held by Alice or Bob. The transmitted value then determines which of the two child nodes to pick (0 corresponds to left, 1 to right), after which a single time step has passed and the new node state is now known.

The protocol terminates when a leaf node is reached. Each leaf node corresponds to a particular function value (of the original function of interest). Thus, if both nodes know the structure of the protocol tree prior to commencing transmissions, and know the node state throughout the duration of the protocol, both will know the function value once the leaf node is reached. Some thought will reveal that this is indeed the most general notion of a deterministic protocol between two agents.

The communication complexity of the function in terms of the protocol tree is then the depth of the tree, or the maximum distance of any leaf node from the root, which is also the worst case number of time steps taken by the protocol.

One can also envision the process of computation through a matrix, as in Figure 3.1.

Alice\Bob	y_1	y_2	y_3	y_4
x_1	0	0	0	0
x_2	0	0	0	1
x_3	0	0	0	1
x_4	1	0	0	1

Figure 3.1 A protocol envisioned as a matrix.

The values of $f(x, y)$ are indicated. The rows correspond to the possible values of the x variable, while the columns correspond to the possible values of the y variable. Suppose Bob goes first, and transmits 'right' in the first transmission. This indicates that his input is either y_3 or y_4. Suppose Alice then sends 'bottom', indicating that her variable is either x_2, x_3, or x_4. If now Bob replies with 'right' again, then this shows that the function value is 1. Thus the protocol breaks the matrix into sub-matrices which are 'monochrome', and the communication complexity in the maximum number of transmissions until a monochrome sub-matrix is reached.

We can show that the logarithm (to base two) of the rank of the matrix $[f(x, y)]$ is a lower bound, as follows. It is sufficient to note that after each transmission the *maximum* rank of the resulting sub-matrices can be at most reduced by a factor of 2. Since the final rank of the monochrome sub-matrix reached is 1, it follows that $\log \mathrm{Rank}[f(x, y)]$ is a lower bound on communication complexity.

Yet another technique to obtain a lower bound is to use what is called a 'fooling set' technique. Suppose $f(x_1, y_1) = f(x_2, y_2) = \cdots = f(x_m, y_m)$ but that for any (x_1, y_1) and (x_j, y_j) with $1 \le i, j \le m$, either $f(x_i, y_j) \ne f(x_i, y_i)$ or $f(x_j, y_i) \ne f(x_i, y_i)$. Then similarly to the above techniques one can show that $\log m$ is a lower bound on communication complexity.

We now provide some examples of functions and their communication complexity.

(i) Let $X = Y$. Consider the equality testing function:

$$f(x, y) = 1 \text{ if } x = y,$$
$$= 0 \text{ if } x \ne y.$$

Then the fooling set method provides a lower bound of $\log |X|$. In fact it can be shown that its communication complexity is $1 + \log |X|$.

(ii) Again, suppose $X = Y = \{1, 2, \ldots, n\}$, and consider the '$x$ greater than y' function:

$$f(x, y) = 1 \text{ if } x > y,$$
$$= 0 \text{ otherwise .}$$

Its communication complexity is $1 + \log n$.

(iii) Let $X = Y = \{1, 2, \ldots, n\}$, and consider the averaging function whose value is equal to the average of the two subsets x and y. Its communication complexity is $O(\log n)$.

(iv) For the same sets X and Y as above, the median of the sets x and y also has communication complexity $O(\log n)$.

There are some features of wireless sensor networks that are not captured by the formulation of communication complexity. For example, the notion of spatial reuse is not modeled. Communication is also assumed to be noise-free. Also any possible correlations between measurements at the different nodes are not exploited. A continuum version of the problem is studied by Luo and Tsitsiklis (1994).

3.3 Computing Functions over Wireless Networks: Spatial Reuse and Block Computation

We now consider a model of a wireless network which incorporates the notion of spatial reuse in communications, thus making it more relevant to wireless sensor networks. We will also suppose computations are to be repeatedly performed with new sets of measurements. As in the spirit of information theory we will allow for block computation to see if that can lead to greater computational throughput.

3.3.1 Geographical Models of Wireless Communication Networks

First we begin with a geographical model of the wireless communication network. We consider a domain, say, a disk of unit area, and suppose that there are n nodes located in the disk. We will suppose that all nodes choose a common range r for all their transmissions. To model the interference aspect of the wireless medium we will suppose that whenever a node i transmits, it creates an interference footprint which is a disk of radius $(1 + \Delta)r$ centered at the transmitter. For a node j to receive a transmission from node i, we will require that its distance to node i be no more than r, while its distance to every other concurrent transmitter is at least $(1 + \Delta)r$. That is, it must be within reception range r of its own transmitter, and outside the interference footprint of every other concurrent transmitter. This model has been considered in Gupta and Kumar (2000) for the study of wireless networks, and is a simplification of more general models allowing for reception based on the magnitude of the signal-to-interference plus noise ratio (Xue and Kumar 2006). We will suppose that all transmissions have a data rate of 1 bit per second.

We will consider three models of wireless networks:

1. **Collocated Networks.** In these networks we will suppose that the n nodes are arbitrarily located within the unit disk. We will also suppose that the range of each transmission r is greater than the diameter of this disk, so that every node's transmission reaches every other node in one hop; hence the name 'collocated network'; see Figure 3.2. Thus there is no need for multi-hop delivery of packets. Given the model of interference, it also follows that at any given time there can only be one active transmitter. In this model there is therefore no exploitation of the notion of spatial reuse.

2. **Random Multi-Hop Wireless Networks.** Here we will suppose that the n nodes are randomly located, uniformly and independently in the disk of unit area; see Figure 3.3. We will suppose that the common range of all transmissions by all nodes is

$$r_n = (1 + \epsilon)\sqrt{\frac{\log n}{\pi n}}, \tag{3.1}$$

for some $\epsilon > 0$. The reason for this choice of range is that if the range is only $r_n = \sqrt{\frac{\log n}{\pi n}}$, then it turns out that as $n \to \infty$ the probability that the network is

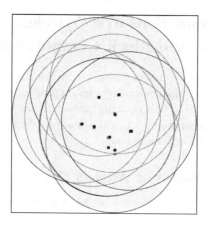

Figure 3.2 A collocated wireless network.

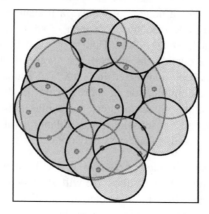

Figure 3.3 A random multi-hop wireless network.

connected converges to 0. By 'connected' we refer to the random graph that is formed by including a link between any two nodes which are at a distance no more than r_n apart. On the other hand, for any $\epsilon > 0$, the range (3.1) is asymptotically sufficient in that the probability that the random graph is connected converges to 1 as $n \to \infty$. Another consequence of the choice of range as in Eq. (3.1) is that with probability approaching 1 as $n \to \infty$, it follows that every node in the graph has $\Theta(\log n)$ neighbors, i.e., $\Theta(\log n)$ nodes are within range $r(n)$. (Note that we say that $f(n) = \Theta(g(n))$ if $f(n) = O(g(n))$ as well as $g(n) = O(f(n))$).

3. **Regular Networks.** Consider a network where there are n nodes located at coordinates (i, j) with $1 \le i, j \le \sqrt{n}$; see Figure 3.4. We will also suppose that the range of all nodes is $r(n) = \sqrt{2}$ so that every node not on the boundary has eight neighbors. This too is a connected network, though with a bounded neighborhood, in contrast to the random, multi-hop wireless network model above.

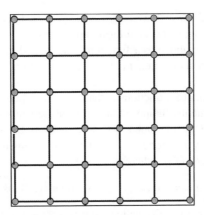

Figure 3.4 A regular planar wireless network.

3.3.2 Block Computation and Computational Throughput

In contrast to the one-shot computation in Section 3.2, we will allow for block computation. Such a strategy is motivated by information theory (Cover and Thomas 1991) where it is critical to attaining the capacity of a channel.

Let $[x_1(i), x_2(i), \ldots, x_n(i)]$ be the vector of measurements taken by the n nodes at epoch i. Given a function $f(\cdot)$; it is desired that the nodes in the network collaborate in computing the function $f(x_1(i), x_2(i), \ldots, x_n(i))$, and make it available at some given distinguished node in the network, designated as a collector node. Instead of performing a one-shot computation of the function value of $f(x_1(i), x_2(i), \ldots, x_n(i))$, we consider the case where N sets of such measurements $[x_1(1), x_2(1), \ldots, x_n(1)]$, $[x_1(2), x_2(2), \ldots, x_n(2)]$, $\ldots$, $[x_1(N), x_2(N), \ldots, x_n(N)]$ are available, and the network computes the N values $f(x_1(1), x_2(1), \ldots, x_n(1))$, $f(x_1(2), x_2(2), \ldots, x_n(2))$, $\ldots$, $f(x_1(N), x_2(N), \ldots, x_n(N))$. Let us suppose that using some protocol or strategy $\mathcal{P}$, this computation is conducted in an amount of time $T_{\mathcal{P}}(N)$. Then $R_{\mathcal{P}}(N) := \frac{N}{T_{\mathcal{P}}(N)}$ is the rate at which this computation has been performed for this block size N. We are interested in $R_{\max} := \sup_{\mathcal{P}} \sup_{N \geq 1} R_{\mathcal{P}}(N)$, and call this the computational throughput of the network. On occasion we may refer to this as $R_{\max}(f)$, when it is desired to highlight that this is the throughput for the particular function f of interest. To be more precise we actually need to define a sequence of functions f_n, since the domain varies as the number of nodes varies, but we will omit this for brevity; full details can be found in Giridhar and Kumar (2005).

3.3.3 Symmetric Functions and Types

Our goal is to determine $R_{\max}(f)$ for various functions f of interest. Two functions of particular interest are the *mean*,

$$f_{mean}(x_1, \ldots, x_n) = \frac{x_1 + x_2 + \cdots + x_n}{n},$$

and the *max*,

$$f_{\max}(x_1, x_2, \ldots, x_n) = \max_{1 \leq i \leq n} x_i.$$

These, together with functions such as f_{median}, and f_{mode}, defined in the obvious ways, are of interest in several environmental monitoring applications.

One characteristic of the above four functions, and indeed many statistical functions, is that they are symmetric functions, i.e., $f(x) = f(\pi x)$ where π is any permutation applied to x.

A key property of symmetric functions is that they are dependent only on the frequency histogram of the measurements. Suppose that the measurements taken by the sensor nodes are from a finite subset of D real numbers, which we will call the alphabet, as in information theory. Then the frequency histogram of a set of measurements $x = (x_1, x_2, \ldots, x_n)$ is $\tau(x) := (\tau_1(x), \tau_2(x), \ldots, \tau_D(x))$, where

$$\tau_k(x) := \text{number of } x_i\text{'s in } x = (x_1, x_2, \ldots, x_n) \text{ which are the } k\text{-th letter in the alphabet.}$$

This frequency histogram $\tau(x)$ is also called the type vector in information theory. We note that the number of possible type vectors is $O(n^D)$. Hence, a type vector can be represented by $O(\log n)$ bits.

Since a symmetric function is dependent on x only through its type vector $\tau(x)$, the maximally difficult symmetric function to compute is just the type vector function. A lower bound on the computational rate at which any other symmetric function can be computed is provided by the computational rate for the type vector function.

3.3.4   The Collocated Network

Let us consider the problem of computation for a collocated network. The first issue is to make precise what a protocol or strategy entails. We begin by noting that a protocol does need to specify when a particular node should or should not transmit. We will further constrain our protocol to be one which does not result in any collisions. That is, every transmission should be successful, which in the special case of a collocated network simply means that at most one node should be allowed to transmit at any given time. It should be noted that the collision-free schedule will have to be arrived at in a causal distributed manner by the nodes, with each node determining whether it should transmit based only on prior transmissions it has heard. The content of what it then transmits can additionally depend on its private information. On the very first transmission, the choice of which node transmits will have to be predetermined. Recursively at the m-th transmission, the decision on which node will transmit is based only on previous broadcasts since that is the only information common to all nodes, while the content of the broadcast is a function of the previous broadcasts as well as the private information possessed by the node. We formulate this as a definition.

Collision-free protocols

We will assume for simplicity that each transmission is binary, either 0 or 1. A collision-free protocol (CFP) $\mathcal{P}$ for the collocated network consists of the following:

1. Let us denote by $\phi_m(\cdot)$ the function that determines which node transmits at time m. Each such ϕ_m is a mapping $\phi_m : \{0, 1\}^{m-1} \longrightarrow \{1, 2, \ldots, n\}$, $2 \leq m \leq T_{\mathcal{P}}$, and $\phi_1 \in \{1, 2, \ldots, n\}$. The argument of ϕ_m is to be interpreted as the history of past broadcasts that is heard by all the nodes.

2. Let us denote by ψ_m the function that determines *what* is broadcast on the m-th broadcast. Each such ψ_m is a function $\psi_m : \mathcal{X}^N \times \{0, 1\}^{m-1} \longrightarrow \{0, 1\}, 1 \leq m \leq T_\mathcal{P}$. Let $\underline{X}_i$ denote the block of N measurements $[x_1(1), x_1(2), \ldots, x_1(N)]$ of node i. The mth transmission Z_m is (recursively) defined as follows: $Z_1 := \psi_1(\underline{X}_{\phi_1})$, $Z_m := \psi_m(\underline{X}_i, Z_{m-1}, \ldots, Z_1)$, for $1 < m \leq T(\mathcal{P})$, where $i = \phi_m(Z_{m-1}, Z_{m-2}, \ldots, Z_1)$.

3. Finally we need to specify how the final value of the function is arrived at. This is given by a decoding function $\xi : \{0, 1\}^{T_\mathcal{P}} \longrightarrow \mathcal{Y}^N$, such that $\underline{f}(\overline{X}) = \xi(Z_1, Z_2, \ldots, Z_{T_\mathcal{P}})$.

The functions $\phi_m(\cdot)$ and $\psi_m(\cdot)$ are the analog of a *codebook*. They are fixed a-priori, and are known to all nodes.

The node designated to transmit at time m is $\phi_m(Z_{m-1}, Z_{m-2}, \ldots, Z_1)$. Since it is based only on the common broadcasts heard by all nodes, every node knows if it is the one that should or should not transmit, and since only one node is designated as a transmitter, the transmissions are collision-free. The medium access problem is thus resolved in a distributed but collision-free fashion. Concerning the content of the transmission, it can depend on (1) what the sensor itself 'knows,' which knowledge is comprised of its own data vector, and (2) all the previous transmissions that it has heard. We note also that since each node knows who the transmitter is, 'packet headers' specifying who is the transmitter are not required.

This definition of the class of protocols generalizes the notion of communication protocols in communication complexity (Kushilevitz and Nisan 1997).

We note that, as an alternative model, one could conceivably allow for collisions. Then, however, the issue arises of whether collisions can be discerned as such by all the nodes. If they can indeed be discerned as collisions, then collisions also carry information, reminiscent of classical models of the ALOHA type (Bertsekas and Gallager 1987). Since we are considering a 'packet capture' model of communication, we have pursued the alternative model of computation achieved only through collision-free packets.

3.3.5 Subclasses of Symmetric Functions: Type-sensitive and Type-threshold

Having defined the class of allowable protocols, we now turn to the functions themselves. Defined below are two disjoint subclasses of symmetric functions. The interest in these two function classes arises from two facts. First, they include several functions of interest, such as *mean* and *max*, as we will see below. In fact, they include most statistical functions of interest. Second, it turns out that the order of the maximum rate can be characterized uniformly over each subclass, in both collocated and random multi-hop wireless networks. It should be noted, however, that these subclasses together do not comprise all possible symmetric functions.

Type-sensitive functions

Consider a symmetric function f. We say that it is a *type-sensitive* function if there exists some $0 < \gamma < 1$, and an integer n', such that for $n \geq n'$, and any $j \leq n - \lceil \gamma n \rceil$, given any subset $\{x_1, x_2, \ldots, x_j\}$, there are two subsets of values $\{y_{j+1}, y_{j+2}, \ldots, y_n\}$ and $\{z_{j+1}, z_{j+2}, \ldots, z_n\}$, such that

$$f(x_1, \ldots, x_j, z_{j+1}, \ldots z_n) \neq f(x_1, \ldots, x_j, y_{j+1}, \ldots y_n). \qquad (3.2)$$

It is trivial to observe that if the above is true for $j = n - \lceil \gamma n \rceil$, then it also holds for all lower values of j as well.

Essentially, this definition asserts that if a large enough fraction of the measurements are unknown, then the function cannot really be determined. Thus these are functions that one would intuitively expect to be difficult to determine, and in fact we will quantify this below.

The other class of functions that we consider is the following.

Type-threshold functions

Recall that we use the notation $\tau(x)$ to denote the type vector of a measurement vector x. Again, consider a symmetric function f. We say that f is a type-threshold function if there exists a vector θ of D non-negative entries, which we shall call a threshold vector, such that $f(x) = f(y)$ whenever $min(\tau(x), \theta) = min(\tau(y), \theta)$. Above, the *min* operation applied to vectors denotes the vector of element-wise minima.

In contrast to the class of type-sensitive functions, it can be seen that the value of a type-threshold function can be determined from a fixed number of known arguments. These two subclasses are clearly disjoint. Indeed if f is a type-threshold function, then as $n \to \infty$ the fraction $\frac{\sum_i \theta_i}{n} \to 0$. Hence it cannot be a type-sensitive function.

Now we list several examples of functions that fall within these classes.

1. The *mode* of a vector of measurements x is the value which occurs the most frequently in the set of measurements. Clearly if more than half the x_is are unknown, the mode is undetermined. Hence it is a type-sensitive function.

2. The *mean* of the entries of a vector x is also a type-sensitive function. Even if we require the mean to be computed to only within some finite precision, it is still a type-sensitive function. The median and the standard deviation are also type-sensitive functions.

3. The *max* function, i.e., the maximum among the entries x_i's, is also a type-threshold function. In fact, its threshold vector is simply $[1, 1, \ldots, 1]$. Similarly, the *min* function and the *range* function ($max_i \, x_i - min_i \, x_i$) are also type-threshold functions.

4. The kth *largest value* among the x_i's is a type-threshold function.

5. The function that computes the *mean of the k largest values* of a vector is also a type-threshold function. Its threshold vector is the vector of constant elements $[k, k, \ldots, k]$.

6. Consider the function that determines whether there is some particular letter in the alphabet that is indeed present in the set of measurements taken by the nodes at an epoch. This is captured by an indicator function $I(x_i = \alpha, \, for \, some \, i)$. It is clearly a type-threshold function,. In fact its threshold vector is simply $[0, 0, \ldots, 0, 1, 0 \ldots 0]$, where the 1 is in the position corresponding to the letter α in the alphabet.

To see that there are nevertheless functions that do not lie in either of these classes, one may consider the function $f(x)$ which is 1 whenever the number of entries that are 1 in the vector x is greater than $\lceil \sqrt{n} \rceil$, and 0 otherwise. No matter how large a constant we choose, knowing a constant (i.e., independent of n) number of values cannot fix the value of the function. Thus it cannot be a type-threshold function. It cannot also be a type-sensitive function since there is no $c < 1$ such that $cn > n - \lceil \sqrt{n} \rceil$ for arbitrarily large n.

3.3.6 Results on Maximum Throughput in Collocated Networks

Now we turn to the quantification of the difficulty in computing certain functions over wireless sensor networks. The metric we focus on is that of 'computational throughput'. Our results aim at characterizing the order of growth (or reduction) of this computational throughput. That is, we eschew pre-constants, and aim for the scaling law of the order of growth for this metric.

In this section we will focus on collocated networks. What we provide below are sharp characterizations, order-wise, of the maximum computational throughput for type-sensitive functions and type-threshold functions in collocated networks.

Theorem 3.3.1   *1. For the problem of computing a type-sensitive function over a collocated network, the maximum computational throughput is $\Theta(\frac{1}{n})$.*

2. For the problem of computing a type-threshold function over a collocated network, the maximum computational throughput is $\Theta(\frac{1}{\log n})$.

In order to understand the import of these results, it is worth considering an extreme case: the data downloading problem. This is the problem of delivering all the measurements from all the sensors to the collector node. It can be shown that the computational throughput for this data downloading problem is in fact $\Theta(\frac{1}{n})$. This sets result (1) above in context. It shows that, at least order-wise, computing a type-sensitive function in a collocated network is maximally difficult. Since the mean, mode and median are exemplars of this class, they too are all seen to be maximally difficult order-wise. This result for type-sensitive functions is clearly pessimistic. On the other hand, it is not very surprising, since the very core of the definition of a type-sensitive function indicates that $\Omega(n)$ amount of data must be communicated for each set of network-wide readings, for the function to be computable. It should also be noted that block coding in this case does not provide any improvement in order.

In contrast, what is surprising is that an exponential improvement is possible for type-threshold functions. In this case, it is block coding that does the trick, since, for a one-shot communication scenario, it is easily shown that $\Omega(n)$ time slots are required.

We now provide a proof of achievability of this latter result for the special case of computing the *max* function $f(x_1, x_2, \ldots, x_n) = max\{x_i : 1 \le i \le n\}$ with an alphabet $\mathcal{X} = \{0, 1\}$ of size two. Our proof is constructive and is based on constructing a sequence of collision-free protocols $S_{l(n+1),n}$, for $l = 1, 2, 3 \ldots$, which asymptotically achieve the rate $\Omega(\frac{1}{\log n})$. Let us denote the block-length by $N = l(n + 1)$. Let us denote the block of measurements at node i by $\underline{X}_i$. Also, recalling that the measurements, say, 'alarm values', are binary 0 or 1, suppose that the number of 1s in this block $\underline{X}_i$ of measurements at node i is N_i. It is convenient to consider the set of time instants $S_i := \{1 \le j \le N : X_i(j) = $

$1, X_k(j) = 0$ *for all* $k < i$}. For node 1, S_1 is the set of time instants at which it attains the max alarm value of 1. For node 2, S_2 is the set of time instants at which it attains the max alarm value of 1, but node 1 does not. For node 3, S_2 is the set of time instants at which it attains the max alarm value of 1, but nodes 1 and 2 do not, and so on. Clearly if the union of all the S_is could be communicated to the collector node, then it would know the set of all times at which the max alarm value was 1, and (in this binary case) the complement of this set would be the set of times at which the max alarm value is 0.

Now we address the issue of how the nodes can individually communicate their S_is to the collector node. The first point to note is that communication itself is particularly easy in the collocated network since all a node needs to do is broadcast at a time at which no other node broadcasts. Also, it should be noted that such broadcasts are heard by *all* the nodes in the network. The second point to address is, how can a node i know what S_i is? The point is that the definition of S_i needs knowledge of S_k for $k < i$. In fact, we turn this requirement into an advantage and the basis of an algorithm: The nodes simply broadcast their Ss in order, beginning with node 1, then node 2, then node 3, etc. Node i can then listen to the previously broadcast $S_1, S_2, \ldots, S_{i-1}$, and determine its own S_i. Moreover, after the last node n has broadcast its S_n, all nodes, which specifically includes the collector node, know all the S_is, and in particular $\bigcup_{1 \le i \le n} S_i$. The S_is are disjoint, and $\bigcup_i S_i = \{j : f(X_1(j), X_2(j), \ldots, X_n(j)) = 1\}$. Thus the collector node obtains the max alarm value at all time instants. Therefore, communicating the sets $S_1, S_2, \ldots, S_n$ to the collector node suffices to reconstruct the max function.

Now we proceed to calculate how efficiently all this can be done. Denote by $\overline{N_i} := |S_i|$, the number of epochs whose identity node i needs for its broadcast. Clearly $\overline{N_i} \le N_i$. The $\overline{N_i}$'s count the number of 1s in the ith vector, in positions that are all 0s in the previous $k < i$ vectors. In a sense, these are the only 'new' 1s as far as the max function is concerned.

As outlined above, the protocol uses n broadcasts, with node i transmitting at stage i. In the ith stage, S_j for $j < i$ is known to all nodes, and the ith node can compute $\overline{N_i}$. Communicating its value takes $\log N$ bits. In addition, the list of times of these N_i entries must also be encoded. The length of this encoding is easy to compute. Each S_i is one of $\binom{N - \sum_{j<i} \overline{N_j}}{\overline{N_i}}$ possibilities. Also, by this stage, $\overline{N_i}$ as well as the previous S_js are known to all nodes, whether for decoding or encoding. Node i can accordingly encode the identity of the set S_i in $\log \binom{N - \sum_{j<i} \overline{N_j}}{\overline{N_i}}$ bits. The question of how to parse successive broadcasts is resolved in an iterative manner as the N_is are revealed. Thus all nodes know when to begin their own broadcasts. Similarly the collector node knows when to begin its decoding. To determine the total computational time we simply add up all these packet lengths. The total number of bits required is

$$T(\mathcal{P}) = n \log N + \sum_i \log \binom{N - \sum_{j<i} \overline{N_j}}{\overline{N_i}}. \tag{3.3}$$

This expression in Eq. (3.3) can in turn be bounded as follows. The quantity $\prod_i \times \left(\binom{N - \sum_{j<i} \overline{N_j}}{\overline{N_i}} \right)$ is the multinomial coefficient $\binom{N}{\overline{N_1}, \overline{N_2}, \ldots, \overline{N_n}, \overline{N'}}$, where $\overline{N'} = N - \sum_i^n \overline{N_i}$. Now we exploit the fact that a multinomial coefficient attains its maximum

when all the $\overline{N}_i$s and $\overline{N}'$ are equal to $N/(n+1) = l$. Hence,

$$\left(\begin{array}{c} N \\ \overline{N}_1, \overline{N}_2, \ldots, \overline{N}_n, \overline{N}' \end{array} \right) \leq \left(\begin{array}{c} N \\ \frac{N}{n+1}, \frac{N}{n+1}, \ldots, \frac{N}{n+1} \end{array} \right)$$

$$= \prod_{0 \leq i \leq n} \left(\begin{array}{c} (n-i+1)l \\ l \end{array} \right) < \left(\frac{l(n+1)}{l} \right)^{n+1}.$$

Now we invoke the simple combinatorial inequality $\left(\begin{array}{c} n \\ k \end{array} \right) < (\frac{ne}{k})^k$. This yields

$$\left(\left(\begin{array}{c} l(n+1) \\ l \end{array} \right) \right)^{n+1} < \left(\frac{l(n+1)e}{l} \right)^{l(n+1)} = ((n+1)e)^{l(n+1)}.$$

Hence the total time to communicate the block of N maxima, $T(\mathcal{P})$, satisfies

$$T(\mathcal{P}) < n \log l(n+1) + l(n+1) \log (n+1)e.$$

Now for $n > 10$, $n^2 > e(n+1)$. This shows that

$$R_{max} = \limsup_{N \to \infty} \sup_{\mathcal{S}_{N,n} \in CFP} \frac{N}{T(\mathcal{P})}$$

$$\geq \limsup_{l \to \infty} \frac{l(n+1)}{T(\mathcal{P})} \geq \frac{1}{2 \log n}.$$

The central importance of allowing block coding in the above encoding scheme should be noted.

The next question that arises is whether this is indeed a sharp bound order-wise. By using fooling-set type arguments it can be shown that it is indeed so.

Thus we conclude that the maximum computational rate for computing the *max* function is $\Theta \left(\frac{1}{\log n} \right)$.

Let us denote the reciprocal of the computational rate as the computational cycle-time. Then we see that the computation cycle-time is exponentially larger for the mean in comparison to the max.

3.3.7 Multi-Hop Networks: The Random Planar Network

Consider now the random planar network on a unit area disk (Figure 3.3). Further suppose that every node has the same transmission range $r(n)$. The range $r(n)$ essentially determines the degree of each node. The following lemma (Gupta and Kumar 2000) shows how to so choose $r(n)$ so that the resulting graph $G^{(n)}$ is connected, and shows that the order of the degree $d(G^{(n)})$ that results is $O(\log n)$ with high probability.

Lemma 3.3.2 *For random planar networks, if range $r(n) \geq \sqrt{\frac{2 \log n}{n}}$, then*

$$\lim_{n \to \infty} Pr[G^{(n)} \text{ is connected}] = 1,$$

and

$$\lim_{n \to \infty} Pr[d(G^{(n)}) \leq c \log n] = 1,$$

for some c > 0.

In collocated networks, there did not seem to be any apparent advantage in considering symmetric functions, since for the fairly large subclass of type-sensitive functions, the maximum rate is of the same order as that of communicating all measurements. Although the quantity of information, i.e., the log of the range of the function in bits, is logarithmic in the size of the network, as opposed to linear in the case of all measurements, this could not be exploited.

However, in a multi-hop network such as the random planar network, the possibility of spatial reuse in transmitting, and of performing in-network aggregation in relaying data, means that a significant compression can be achieved for all symmetric functions.

The key idea is that in the random planar network, the histogram or type can be communicated to the collector at a rate $O(\frac{1}{nr^2(n)})$, which specializes to $O(\frac{1}{\log n})$ for a suitable chosen $r(n)$. This is exponentially better that what is possible in the collocated network case, as well as exponentially better than the achievable rate of communicating all the data in the random network case. This further means that all symmetric functions can be computed at least such a rate, because symmetric functions are completely determined by knowledge of the frequency histogram.

The protocol to achieve such a rate can be constructed by dividing the unit disk into appropriate size cells, and forming on the basis of this division a tree rooted at the collector node; see Figure 3.5. Together with an appropriately defined schedule and use of pipelining, it can be shown that such a rate can be achieved. A general version of such a protocol can be applied to a more general class of graphs and functions. This is described in (Giridhar and Kumar 2005).

For type-sensitive functions, the above rate is order optimal. The optimality proof uses the result in the previous section for collocated networks.

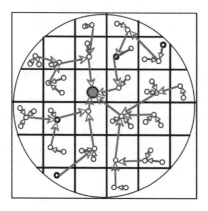

Figure 3.5 Forming a tree rooted at the collector.

For type-threshold functions, similar upper and lower bounds on order can be derived, by suitably extending the results of the previous section, and by constructing a tree-based achievability protocol. The results are summarized in Theorem 3.3.3.

Theorem 3.3.3 *Consider a wireless network with n nodes uniformly and independently located in a unit area disk on the plane. Suppose that all nodes employ a common range $r(n)$ which is chosen to be large enough so that the network is connected.*

1. *Suppose $f(\cdot)$ is a type-sensitive function. Then there exist constants $c_2 > c_1 > 0$, such that*

$$\lim_{n \to \infty} Pr\left[R_{max}(f) \geq \frac{j}{nr^2(n)} \right] = \begin{cases} 1 & \text{if } j \leq c_1 \\ 0 & \text{if } j \geq c_2 \end{cases}. \tag{3.4}$$

Then there exist constants $c_4 > c_3 > 0$, such that

$$\lim_{n \to \infty} Pr\left[R_{max} \geq \frac{j}{\log(nr^2(n))} \right] = \begin{cases} 1 & \text{if } j \leq c_3 \\ 0 & \text{if } j \geq c_4 \end{cases}. \tag{3.5}$$

By choosing $r(n) = \sqrt{\frac{2\log n}{n}}$, the results specialize to $\Theta(\frac{1}{\log n})$ for type-sensitive functions and $\Theta(\frac{1}{\log \log n})$ for type-threshold functions, respectively. The proofs can be found in Giridhar and Kumar (2005).

3.3.8 Other Acyclic Networks

Let us now consider the class of networks with bounded node degrees. For concreteness, we can consider regular networks, such as linear regular arrays and two dimensional grids (Figure 3.4), in which each node can transmit to any of its nearest neighbors. A simple protocol to compute symmetric functions in such networks is to communicate the histogram to the collector. This protocol has an associated rate of $O(\frac{1}{\log n})$.

For the subclass of type-threshold functions, one can do even better. It is fairly easy to show that the maximum rate is $O(1)$. Consider the subclass of type-sensitive functions. Is histogram communication order optimal for such functions? The answer is no. Consider the following type-sensitive function:

$$f(x_1, \ldots, x_n) = \begin{cases} 0 & \text{if } x_1 + \ldots x_n \text{ is even} \\ 1 & \text{if } x_1 + \ldots x_n \text{ is odd}. \end{cases} \tag{3.6}$$

It is easily verified that such a function is type-sensitive. It is equally easy to construct a protocol which has $O(1)$ rate: In the linear case, for example, each node receives the parity of the partial sum of previous nodes, and forwards the parity of the partial sum including its own measurement.

For some type-sensitive functions, however, such as the sum function, it is clear that $O(\frac{1}{\log n})$ rate is optimal, since the inverse of this is the logarithm of the size of the function range. Thus, the characterization of maximum rate for the class of type-sensitive functions does not hold for the linear array, and more generally for constant degree networks.

The general problem for these networks, which is as yet unresolved, is the problem of determining maximum rate (or its order) for symmetric functions which are not type-threshold, and which do not have large ranges. Two examples of functions for which the answer is not known in general are the median and mode.

The problem is somewhat more tractable for acyclic graphs. For the linear array, and more generally for tree graphs, the maximum rate for the median and mode can be determined: The rates for both are $O(\frac{1}{\log n})$. Upper bounding the performance of any protocol is, as is typical, the difficult part of such problems.

One feature to note about the above results is that they are scaling results in the network size, whereas communication complexity results are often order results in terms of alphabet size. The former are thus more applicable when the network size is considerably larger than alphabet size. In a scaling sense, we see that the computational throughput for type-threshold functions is exponentially higher than for type-sensitive functions, in both the collocated network as well as the random multi-hop network. Also, multi-hop networks allow for a far greater degree of in-network compression, and consequently allow a higher computational throughput than the collocated network.

There are some drawbacks with the model and results described above. First, the model does not incorporate node or link failures. It is not clear if these results easily extend to such unreliable scenarios. Second, there may be considerable overhead required for nodes to know what computational operations they must carry out; the 'roles' of individual nodes depend on their locations and thus must be dynamically assigned.

A further drawback of the model is that it does not take into account correlations in the source measurements, which could be exploited if the requirement of exact computation with probability 1 were relaxed.

3.4   Wireless Networks with Noisy Communications: Reliable Computation in a Collocated Broadcast Network

A natural generalization of the above models is the incorporation of noise into the channel model. While the results in the previous section could be utilized by implementing error correction, along with the already present block computation, and thereby constructing protocols which are of the same order as their deterministic counterparts, the framework of the previous chapter does not directly translate to a network with noisy links with one-shot computation.

We present an example of such a problem. This problem was posed by El Gamal, and subsequently studied by (Gallager 1988). We consider a network where all nodes are within range of each other, i.e., a collocated network. This can also be called a broadcast network. We will suppose that it consists of n nodes. Suppose for simplicity that each of the n nodes has a measurement which is just one bit, 0 or 1.

Now we describe the model for communication in this broadcast network. At any given time only one node can broadcast. The message to be broadcast can either be 0 or 1. We will suppose that every other node receives this broadcast; however, each such reception can be erroneously received. In particular we will model the channel from the transmitter to each node as a separate binary symmetric channel. By this we mean that there is a probability $\epsilon > 0$ with which a 1 is received as 0, and vice-versa. We will also suppose the channels to different nodes are independent, that is, each node's reception is independent.

3.4.1 The Sum of the Parity of the Measurements

Now we turn to the function to be computed. Let us consider the *parity function* defined by:

$$f_{parity}(x_1, x_2, \ldots, x_n) = 1 \text{ if the number of } x_i\text{'s that are 1 is even,}$$

$$= 0 \text{ otherwise.}$$

We note that parity is a particularly sensitive function of its arguments since even one node's message erroneously received can lead to a flip in the parity of the receptions. We will address this problem as a one-shot computation problem.

If all receptions are noiseless, clearly n transmissions are necessary and sufficient to compute the parity function exactly. This is therefore a lower bound even for the noisy case.

Due to the noise in the channel, any information transfer is liable to be erroneous with some positive probability. Exact computation with probability one is therefore impossible. Necessitated by this, we relax the requirement of exact computation. We will consider a relaxation of the criterion to require computation of the correct answer to within some probability of error.

The problem can now be posed as follows: Minimize the total number of transmitted bits which will guarantee that the parity will be known to within a desired probability of error.

The key to the solution is to exploit the fact that even though the information obtainable on any single transmission by any one receiver is liable to be wrong with significant error probability, when we look at the totality of information that becomes available at all the nodes, it can be used for error correction.

Since any broadcast is received by n nodes, and these receptions are independent, by pooling their receptions, they can collectively make a good estimate of the single bit of information involved in a transmission. There is, however, the issue that even pooling the bits is a noisy process.

One simple solution is to use some error correction code, of which an example is repetition coding – simply repeat the same transmission several times, say k times. This, however, results in a total of kn transmissions. After these kn transmissions, a receiver could make a maximum likelihood estimate of the bit. How large must k be in this approach? An elementary computation shows that for a fixed small probability of error, the number k must grow like $\log n$. Note that this is the number of transmissions needed to communicate the information at each node. Since there are a total of n nodes, it follows that this scheme requires a total of $\Theta(n \log n)$ transmissions. But this approach does not utilize the broadcast nature of the receptions at all. The question is whether one exploit that feature to reduce the communication cost of computing the parity function.

The above problem was addressed by (Gallager 1988). Let δ denote the stipulated maximum probability of error allowed in the final answer. Gallager constructed an innovative protocol that requires only $O(n \log \log \frac{n}{\delta})$ transmissions to guarantee a probability of error in the computation of parity of less than δ.

We now provide a an outline of this scheme. First we divide the nodes into several subsets, each consisting of $\Theta(\log n)$ nodes. The protocol itself consists of two phases:

1. In this first phase, every node repeatedly transmits its measurement (here just 0 or 1), a total of k times. Next, in each of the $\Theta\left(\frac{n}{\log n}\right)$ subsets, each node makes an estimate of the parity of the sum of the bits in its own subset.

2. In a second phase, every node transmits its estimate of the parity of the sum. This is done only once by each node. After all nodes have so transmitted, the collector simply makes a maximum likelihood estimate.

A calculation shows that if one chooses $k = \Theta(\log \log \frac{n}{\delta})$, then this is sufficient to guarantee a probability of error less than δ.

In fact, one can even download all the data to the collector node. By suitably modifying this scheme, it can be shown that using a scheme with a number of transmissions per node that is of the same order $k = \Theta(\log \log \frac{n}{\delta})$, in fact all the nodal bit values can be communicated to the collector node with high probability. It is still an open question whether this order in n is sharp, i.e., whether $\Omega(n \log \log n)$ transmissions is also necessary for this purpose.

3.4.2   Threshold Functions

Another class of functions, threshold functions (which should be distinguished from the class of type-threshold functions considered earlier), has also been addressed for this same model of a noisy broadcast network by Kushilevitz and Mansour (1998). These are defined as functions whose value is completely determined by whether the number of 1s in the network exceeds some threshold. The difference between this class of threshold functions and type-threshold functions is that in the latter class the threshold needs to be fixed and independent of n. Examples of threshold functions include the AND, OR, and $majority$. Kushilevitz and Mansour (1998) have provided the construction of a protocol that computes a threshold function in time $O(n)$. It can be shown, (see Giridhar and Kumar 2005), that even if all transmissions were noiseless, the network would still need n transmissions. Hence this scheme is order optimal.

We note that in both the scheme of Gallager as well as the scheme of Kushilevitz and Mansour, the order of transmissions is fixed a-priori, and is not affected by the contents of the transmissions. This is in contrast to several schemes for the noise-free model. For example, the computation of the median in the two-party communication complexity problem (Kushilevitz and Nisan 1997), and the max computation in the broadcast network, both are not so oblivious to order.

3.5   Towards an Information Theoretic Formulation

To truly understand the fundamental limitations on what can be computed in a wireless sensor network requires an information theoretic treatment. This is as true for wireless sensor networks as it is true for plain wireless networking. However, the difficulties involved in wireless sensor networks are considerably more than in just wireless networking. In this section we illustrate how one can model wireless sensor networks at the information theoretic level, illustrate some known results, and outline the significant challenges that still face us in the development of an information theoretic foundation for wireless sensor networks.

Information theory allows modeling of the medium. It also allows modeling one feature that we have not alluded to so far – the presence of spatial correlations in measurements.

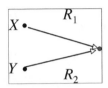

Figure 3.6 The Slepian-Wolf problem.

Taking these into account and developing algorithms that exploit such correlations can lead to the elimination of redundancies and greater efficiencies. Finally, information theory can also be used to model the distributed aspects of the problem, including source or channel coding. More broadly one can aim for a comprehensive formulation of the distributed computation problem. We should note though that while we can formulate such problems, we are still very far off from any comprehensive solutions.

We begin by providing an outline of some of the known results in this area. We will start with the problem of distributed compression, or what is called source coding. The simplest setting to consider is two sensors that each have a separate time-sequence of measurements, Suppose that for each time instant, the two measurements at the sensors are correlated. This correlation between the two measurements is captured by a joint probability distribution that is known to both nodes. We will, however, suppose that the measurements are independent in the time dimension. We will also suppose for simplicity that the joint distribution does not change with time. Fixing the time instant and suppressing it in the notation, let us denote the two measurements by the two random variables X and Y, with joint distribution $p(x, y)$.

Let us suppose that each of the two sensors is connected to a collector node by a communication link; see Figure 3.6. We will suppose that the two communication links are noiseless, each with a throughput rate that we will denote by R_1 and R_2, respectively. The collector node wishes to determine the values of both X and Y. This can thus be regarded as a data collection problem, however simplified by the assumption of 'wired links' for each node to the collector node.

Let us suppose that the entropy rate in the sequence of measurements $\{X_1, X_2, \ldots, X_t, \ldots\}$ taken at node 1 is $H(X)$. Then it is well known that a link of rate $R_1 \geq H(X)$ suffices to communicate the sequence reliably to the collector node. Similarly, if the entropy rate of the sequence of measurements $\{Y_1, Y_2, \ldots, Y_t, \ldots\}$ taken at node 2 is $H(Y)$, then a link of rate $R_2 \geq H(Y)$ suffices for it to communicate its measurements to the collector.

However, we can do better. Since the two measurements are correlated, there is redundancy in the two measurements. Hence if the two nodes each had access to the other node's measurements, then they could cooperatively compress the data and transmit it over the links. Thus if node 1 has communicated its information via a link of rate $R_1 \geq H(X)$, then node 2 only needs to send its 'new information'. The entropy rate of this new information is the conditional entropy $H(Y|X)$. Thus all it needs is a link of rate $R_2 \geq H(Y|X)$. However, this line of reasoning presumes that nodes 1 and 2 can share information, or at least that node 2 has access to node 1's information. A symmetric argument shows that if node 2 has a link of rate $R_2 \geq H(Y)$, then if nodes 1 and 2 could cooperate, then node 1 only needs a link of rate $R_1 \geq H(X|Y)$. By a convexity argument, and using the fact that the joint entropy $H(X, Y)$ can be written as $H(X, Y) = H(X) + H(Y|X) = H(Y) + H(X|Y)$, it can be deduced that any *rate vector* $R = (R_1, R_2)$ satisfying:

$$R_1 \geq H(X|Y),$$

$$R_2 \geq H(Y|X),$$

$$R_1 + R_2 \geq H(X, Y),$$

suffices when the two nodes are cooperating.

The question we wish to address is: What rate vectors suffice when the two nodes cannot cooperate, i.e., when the two nodes are 'distributed'? This is the problem of distributed data compression. More formally, we wish to determine the rate region in the space of two-dimensional vector rates, at which the sources can be individually compressed and sent to the collector, and for which there are encoding and decoding schemes such that the collector can reconstruct the two sources with an arbitrarily small probability of error. This problem was solved in the classic paper of Slepian and Wolf (1973). The surprising result proved by Slepian and Wolf is that the same rate region suffices as noted above for the cooperative compression problem. Thus distributed source compression can be performed even without node cooperation with the same efficiency as it could be in the case of node cooperation!

Now we examine what the further features are that need to be introduced to more fully understand in-network information processing in wireless sensor networks. First, one may be willing to tolerate some specified level of distortion in reconstructing the two sources. This requires the specification of a distortion measure ρ (assumed to be the same for X and Y just for simplicity), and the requirement that if $\hat{X}$ and $\hat{Y}$ are the two estimates, then the distortion between X and $\hat{X}$, and also between Y and $\hat{Y}$ satisfy $E[\rho(X, \hat{X})] \leq d$, $E[\rho(Y, \hat{Y})] \leq d'$. This gives rise to a rate vs. distortion curve which could serve as a bandwidth versus QoS curve. This problem is open.

One special case of this problem can be formulated as follows. Suppose that one of the sources is available as side information to the collector, and only the other source is to be determined. This problem was solved by (Wyner and Ziv 1976) (Figure 3.7).

One can treat the problem of function computation, say of a function $f(X, Y)$ of the two sources, by formulating a distortion criterion that models the error in determining the value of this function. Orlitsky and Roche (1995) have addressed the problem with the side information Y. They determined the required rate of the link in terms of a measure defined on the two random variables and a certain graph defined by the function $f(\cdot)$, called the conditional graph entropy.

Generalizing in the direction of the communication model, one can bring in wireless aspects where the medium itself is shared, and the two (or more sources) are not regarded as

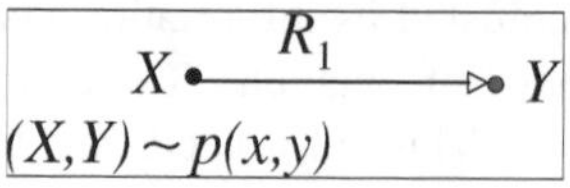

$$
\begin{array}{c}
\boxed{
\begin{array}{c}
X \bullet \xrightarrow{\quad R_1 \quad} \bullet Y \\
\underline{(X, Y) \sim p(x, y)}
\end{array}
}
\end{array}
$$

Figure 3.7 The Wyner-Ziv side information problem.

having separate 'links'. Recently, progress has been made on information theoretic scaling laws for wireless networks; (see Xie and Kumar 2004).

Combining the more general wireless network model, correlated sources, fidelity criteria, and distributed processing, will bring us closer to an information theoretic foundation for wireless sensor networks. However, it would still require addressing aspects that to date have not been captured as ideally by information theory as one would like, even for communication problems – latency and energy aspects.

All this remains as a final frontier.

3.6 Conclusion

Since wireless sensor networks are often application specific, may be intended for unattended operation over long periods of time, and may even involve large numbers of nodes, there is great interest in designing their operations to be as efficient as possible, so that batteries need to be replaced only infrequently, low cost nodes with low communication bandwidths can be deployed, or very little memory or processing capability is needed. One can draw an analogy with the area of wireless networking, where also there are similar pressures to maximally exploit the capabilities of the wireless communication medium. Such pressures in that area have resulted in investigation of cross-layer approaches (Kawadia and Kumar 2005). The goal of such cross-layer designs is to somehow optimally merge the functionalities of various layers of the communication protocol stack so that one eliminates the inefficiencies involved in layering. We believe that in the area of wireless sensor networks, similar pressures will necessitate a joint study of communication, computation, and inference.

As we have seen in this chapter, wireless networks combine interesting aspects of communication networks as well as distributed computation. We have seen how the notion of block coding, central to communication system design for increasing throughput, can also be exploited for wireless sensor networks to enhance computational throughput. Indeed it can lead to significant order of magnitude increases. Similarly we have seen how exploiting the shared wireless medium, for example, by spatial reuse, can lead to exponential decreases in computational cycle-time. We have also seen how the intrinsic nature of a function itself can allow for enhanced computational throughput, an idea that originates in the theory of computational complexity. For example, the max function allows exponential speed-up in computational cycle-time in comparison to the mean function. We have also seen how even for a particularly sensitive function such as parity, one can still exploit an unreliable wireless medium. All these developments suggest that a theory of in-network processing of information in wireless sensor network, that is rich in ideas, can be very useful for efficiency improvement of wireless sensor networks.

This theoretical endeavor is very much in its infancy. Results on the tradeoffs between latency, energy consumption or lifetime, and computational throughput, are needed. Also, one would like to exploit better models of the physical environment producing the measurements, for example, spatial correlation or band-limitedness.

We hope that these motivations and possibilities will galvanize further research in this emerging area.

Acknowledgments

This material is based upon work partially supported by DARPA under the Contract N66001-06-C-2021, DARPA/AFOSR under Contract No. F49620-02-1-0325, AFOSR under Contract No. F49620-02-1-0217, USARO under Contract No. DAAD19-01010-465, NSF under Contract Nos. NSF CNS 05-19535 and CCR-0325716.

Bibliography

Bertsekas D and Gallager R 1987 *Data Networks*. Prentice-Hall, Englewood Cliffs, NJ.

Cover T and Thomas J 1991 *Elements of Information Theory*. Wiley, New York.

Dolev D and Feder T 1989 Multiparty communication complexity. *In the 30th Annual Symposium on Foundations of Computer Science (FOCS-89)*, pp. 428–435.

Gallager R 1988 Finding parity in a simple broadcast network. *IEEE Transactions on Information Theory* **34**(2), 176–180.

Giridhar A and Kumar PR 2005 Computing and communicating functions over sensor networks. *IEEE Journal on Selected Areas of Communication* **23**(4), 755–764.

Gupta P and Kumar PR 2000 The capacity of wireless networks. *IEEE Transactions on Information Theory* **46**(2), 388–404.

Kawadia V and Kumar PR 2005 A cautionary perspective on cross layer design. *IEEE Wireless Communication Magazine*, *12*(1), pp. 3–11, February.

Kushilevitz E and Mansour Y 1998 Computation in noisy radio networks. *SODA '98: Proceedings of the Ninth Annual ACM-SIAM Symposium on Discrete Algorithms*, pp. 236–243, Philadelphia, PA.

Kushilevitz E and Nisan N 1997 *Communication Complexity*. Cambridge University Press, Cambridge, UK.

Luo Zhi-Quan and Tsitsiklis John 1994 Data fusion with minimal communication. *IEEE Transactions on Information Theory*, *40*(5), pp. 1551–1563, September.

Orlitsky A and Roche JR 1995 Coding for computing. *IEEE Symposium on Foundations of Computer Science*, pp. 502–511.

Rajagopalan S and Schulman L 1994 A coding theorem for distributed computation. In *STOC '94: Proceedings of the Twenty-Sixth Annual ACM Symposium on Theory of Computing*, pp. 790–799, New York.

Schulman LJ 1996 Coding for interactive communication. *IEEE Transactions on Information Theory* **42**(6), 1745–1756.

Slepian D and Wolf J 1973 Noiseless coding of correlated information sources. *IEEE Transactions on Information Theory* **19**(4), 471–480.

Tiwari P 1987 Lower bounds on communication complexity in distributed computer networks. *J. ACM* **34**(4), 921–938.

Wu Y and Chou PA, and Kung SY 2005 Minimum-energy multicast in mobile ad hoc networks using network coding. *IEEE Trans. Comm.* **53**(11), 1906–1918.

Wyner A and Ziv J 1976 The rate distortion function for source coding with side information at the receiver. *IEEE Transactions on Information Theory* **IT-22**, 1–11.

Xie L-L and Kumar PR 2004 A network information theory for wireless communication: scaling laws and optimal operation. *IEEE Transactions on Information Theory*, **50**(5), 748–767, May.

Xue F and Kumar PR 2006 *Scaling Laws for Ad Hoc Wireless Networks: An Information Theoretic Approach*. NOW Publishers, Delft, The Netherlands, 2006.

Yao AC 1979 Some complexity questions related to distributive computing(preliminary report) *STOC '79: Proceedings of the Eleventh Annual ACM Symposium on Theory of Computing*, pp. 209–213, New York.

4

The Sensing Capacity of Sensor Networks

Rohit Negi, Yaron Rachlin, and Pradeep Khosla

4.1 Introduction

The essential function of a sensor network is to identify the state of an environment using multiple sensors. Other aspects of sensor network design, such as the communication, power, and computational resources, are evaluated by their ability to facilitate this function. In classic literature, there are two distinct flavors to this problem – an 'estimation problem' where 'state' is continuous, and a 'detection problem' where 'state' is captured by a finite set of hypotheses, such as a binary hypothesis testing problem. In this chapter, we are interested in a structured large-scale detection problem that appears in several useful situations. In our problem, 'state' belongs to an exponentially large set. However, the structure of the set allows us to demonstrate a fundamental information theoretic relationship between the number of sensor measurements and the ability of a sensor network to identify the state of the environment to within a desired accuracy.

In a sensor network, each state of the environment results in a set of sensor measurements. The correspondence between the state and the sensor measurements can be thought of as a code, with the sensor network playing the part of an encoder. We illustrate this analogy and the models used in the remainder of this chapter by considering the following distributed sensing applications.

Wireless Sensor Networks: Signal Processing and Communications Perspectives   A. Swami, Q. Zhao, Y.-W. Hong and L. Tong
© 2007 John Wiley & Sons, Ltd

4.1.1   Large-Scale Detection Applications

Large-scale detection applications appear in several useful situations. For example, consider the problem of robotic mapping. Thrun (2002) provides a useful survey of this active area of research. In a mapping application, a group of robots collects sensor measurements in order to obtain a map of a given area, which can then be used to perform tasks such as navigation. Elfes (1989) introduced occupancy grids, a widely used approach to robotic mapping. The occupancy grid approach models an area as a grid, where each grid block takes on values from a finite set. In most mapping applications, a grid block set to '1' corresponds to an obstacle (therefore, non-traversable) while a grid block set to '0' indicates free space (therefore traversable). A group of robots collects a sequence of noisy sensor measurements, such as sonar range measurements. The state of the environment is encoded into these measurements. These sensor measurements are then used to identify the location of obstacles in the area, so as to allow navigation or other robotic tasks. While mapping has been successfully implemented in practice, many basic theoretical questions remain unaddressed. How many sensor measurements must the robots collect to obtain an accurate map of the area? How does this number vary with sensor field of view and resolution? Is it more effective to use a large number of cheap wide-angle sensors or a small number of expensive narrow-angle sensors?

Consider another example involving a network of cameras. Yang et al. (2003) demonstrated a system of multiple cameras that counts the number of people in a room while Hoover and Olsen (1999) used multiple cameras to localize motion inside a room. In both cases, the area under observation can modeled as a grid. In the motion example, each grid block takes on two values, indicating motion or absence of motion in that grid block. Each camera only observes a subset of the room. More importantly, each image is only a two-dimensional representation of the three-dimensional space, and therefore, inferring the motion map requires fusing observations from multiple cameras. In this example, the cameras encode the true motion map of the room (i.e., state) into two-dimensional images. While algorithms are available for such applications, there is no theoretical analysis of performance, which could guide the design of such a network.

A third application is the use of chemical sensor arrays to detect complex combinations of chemicals in a sample. Burl et al. (2002) discuss the design of such arrays. A complex substance can be modeled as a mixture of multiple constituent chemicals at various discrete concentrations. A chemical sensor array consists of several sensors, each of which is designed to be semi-selective. This means that each sensor in the array reacts to only a subset of the constituent chemicals. The chemical sensor array encodes the complex chemical being sensed (i.e., state) into a vector of measurements. The chemical sensors which comprise the array are based on different technologies and possess diverse characteristics. Again, a theoretical analysis of performance would shed light on the design of such chemical sensor arrays.

As a final example of large-scale detection problems, consider the problem of detection and classification of targets over a large geographical area, as illustrated in Figure 4.1. Consider a seismic sensor-based approach to this problem, as demonstrated in Li et al. (2002), Tian and Qi (2002). For such applications, the environment can be modeled as a discrete binary field, where each entry represents the presence or absence of a target at a grid

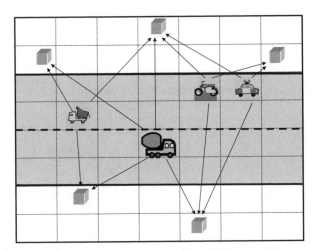

Figure 4.1 Seismic sensor network with sensors (gray cubes) sensing vibrations from multiple vehicles.

point. Seismic sensors are deployed throughout the geographical area. Each seismic sensor produces an output, in response to the vibrations caused by moving targets in its proximity. However, such sensors cannot distinguish between a few close targets and a larger number of distant targets. Therefore, measurements from multiple seismic sensors must be fused in order to obtain an accurate map of the environment. In this example, the network of seismic sensors encodes the locations and class of targets in the field.

All these examples have the following common feature. They are detection problems, with a large number of possible hypotheses (states). However, the state can be expressed as either a discrete vector or a discrete field. Each sensor in the network reacts to only a portion of this vector or field. The function of the sensor network, then, is to infer the state by fusing the noisy measurements. In this chapter, we will model and analyze the performance of sensor networks in such large-scale detection problems by utilizing a key insight – that such problems bear a striking similarity to the problem of communicating over a noisy channel.

4.1.2 Sensor Network as an Encoder

The examples discussed in Section 4.1.1 motivate the following abstract model, shown in Figure 4.2. In this model, the state of the environment is represented by a discrete 'target vector' v or field. An ideal sensor network would react to the state to produce the vector of noiseless sensor outputs x. However, x is corrupted by sensor and other noise so that we only observe the noisy sensor output vector y. A detection algorithm processes y to produce a guess $\hat{v}$ of the state. In the subsequent discussion, we use 'state' and 'target vector' interchangeably. We call each entry of the target vector a 'target position'. In the case of binary v, each target position is a bit, denoting the presence or absence of a target at that position.

Figure 4.2 Sensor network model.

Figure 4.3 Communication channel model.

Notice that this sensor network model bears a strong similarity to the classical communication channel model in Figure 4.3. The state v is the 'message' being transmitted through a noisy channel. The sensor network plays the role of a channel encoder, producing codewords x. The detection algorithm is the channel decoder, which guesses the message. The fundamental limit of a communication channel is characterized by the celebrated results in Shannon (1948) on channel capacities. In a communication channel, the maximum data rate that allows communication with arbitrarily low error probability is called the 'channel capacity'. Moura et al. (2002) proposed the idea of a 'sensing capacity' in order to find the minimum number of sensors required to detect the state of multiple targets. The question of whether there exists a positive sensing capacity, and therefore the practical value of the sensing capacity, remained open. The sensing capacity as defined in Moura et al. (2002) is zero, and therefore lacks practical value. However, the idea of a sensing capacity motivated the theoretical work described in this chapter. Rachlin et al. (2004) introduced a definition of sensing capacity that allowed for detection of multiple targets to within a tolerable error. Using this definition, Rachlin (2007) proves the existence of a strictly positive sensing capacity for a number of sensor network models. These sensing capacity results bound the minimum number of sensors for which detection error arbitrarily close to zero is achievable. The sensing capacity differs significantly from the classical Shannon channel capacity because of fundamental differences between the two models.

The most important difference between a sensor network channel model and a communication channel model arises due to the constrained encoding of sensor networks. In communications, a set of messages can be encoded in an arbitrary manner, and so, a message can be de-coupled from its codeword representation. As a result, similar messages can be distinguished with high accuracy by choosing sufficiently different codewords. In contrast, a sensor network couples a state and its codeword representation. Codewords produced by a sensor network must respect the constraints of the sensing mechanism, and so cannot be completely arbitrary. Further, given a fixed set of sensors and sensor locations, it is clear that similar states (vs that differ in only a few elements) are likely to produce similar codewords x. Therefore, it is not possible to distinguish between two very similar states to an arbitrary accuracy. Whereas the similarities of the sensor network model and the communications channel model motivate the application of the large body of insights in the latter to the sensor network problem, the differences between them caution us to apply such insights carefully, and to understand the impact of these differences on the theoretical properties of sensor networks.

4.1.3 Information Theory Context

Before we delve more deeply into the theory of sensing capacity, we put this work into context by providing a brief review of other results on the information theoretic analysis of sensor networks.

Recent years have seen a large number of papers applying ideas of information theory to obtain performance limits on sensor networks. An idea that has been heavily studied in the context of sensor networks is distributed source coding. Slepian and Wolf (1973) and Wyner and Ziv (1976) provide limits on the compression of separately encoded correlated sources. Pradhan et al. (2002) apply these results to sensor networks. Xiong et al. (2004) provide an overview of this area of research. This work focuses on compressing correlated sensor observations to reduce the communication bandwidth required. The distributed nature of the compression is the object of analysis in that work. In contrast, we focus directly on the limits of detecting the underlying state of the environment using noisy sensor observations. The notion of sensing capacity characterizes the limits that sensing (e.g. sensor type, range, and noise) imposes on the attainable accuracy. We do not examine the compression of sensor observations, or the resources required to communicate sensor observations to a point in the network. Instead, we focus on the limits of detection accuracy assuming complete availability of noisy sensor observations. Thus, our large-scale detection problem is quite unlike a distributed source coding problem. An easy way to distinguish between the two is to consider the case where the sensor network has infinite communication and computation resources. In that case, the distributed source coding problem becomes irrelevant, since each sensor can communicate all its observations in their entirety to a computer, which can then perform centralized compression. However, even in this scenario, there will exist fundamental limits for our large-scale detection problem.

The work presented in this chapter is most closely related to work on the limits of estimation or detection accuracy in sensor networks. Varshney (1997) describes a large body of work in distributed detection which focuses on hypothesis testing problems where the number of hypotheses is small. Chamberland and Veeravalli (2003) and Chamberland and Veeravalli (2004) extend this work to consider a decentralized binary detection problem with noisy communication links to obtain error exponents. D'Costa et al. (2004) analyze the performance of various classification schemes in classifying a Gaussian source in a sensor network, which is an m-ary hypothesis testing problem where the number of hypotheses is small. Kotecha et al. (2005) analyze the performance suboptimal classification schemes for classifying a fixed number of targets. While in this work the number of hypotheses is exponential in the number of targets, the fundamental limits of sensing for a large number of targets, and therefore an exponentially large number of hypotheses, are not considered. Chakrabarty et al. (2001) consider the problem of sensor placement in detecting a single or few targets in a grid. This problem is similar to large-scale detection problems studied in this chapter, though due to the restrictions on the number of targets, the number of hypotheses is comparatively small. A coding-based approach was used to propose specific sensor configurations, and to propose bounds on the minimum number of sensors required for discrimination using this structured approach. Sensors were noiseless, and of limited type, and no notion of sensing capacity was considered. In contrast to existing work on detection and classification in sensor networks, we focus on fundamental performance limits for large-scale detection problems.

Another set of results examines the limits of estimating a continuous field using sensors which obtain point samples. Scaglione and Servetto (2002) study the relationship of transport capacity and the rate distortion function of a continuous random processes. Nowak et al. (2004) study the estimation of an inhomogeneous random field using sensor that collect noisy point samples. Other work on the problem of estimating a continuous random field includes Marco et al. (2003), Ishwar et al. (2003), Bajwa et al. (2005), Kumar et al. (2004). Working on a different sensor network application, Gastpar and Vetterli (2005) consider the estimation of continuous parameters of a set of underlying random processes through a noisy communications channel. Unlike the estimation problems discussed in these papers, this chapter considers the problem of large-scale detection.

4.2   Sensing Capacity of Sensor Networks

The sensing capacity limits the number of sensor measurements required to achieve a desired detection accuracy. In this section, we define and limit the sensing capacity for a specific sensor network model that allows arbitrary sensor connections. Extensions to other sensor network models are discussed in Section 4.3. We emphasize that the sensing capacity is a quantity that describes the fundamental limits of a sensor network model, rather than a particular sensor network.

4.2.1   Sensor Network Model with Arbitrary Connections

We describe a sensor network model with arbitrary connections which was introduced in Rachlin et al. (2004). An example of this model is shown in Figure 4.4. The graphical nature of the arbitrary connections model is inspired by a general graphical model for

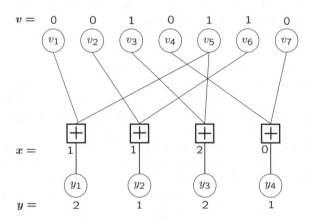

Figure 4.4   A sensor network model with $k = 7$, $n = 4$, $c = 2$.

sensor networks introduced in Moura et al. (2003), the first publication of which we are aware that introduced the idea of modeling sensor networks as a graphical model. The state of the environment is modeled as a k-bit binary target vector v. Each entry of the vector, which we call 'target position', may represent the presence or absence of a target in some spatial region, or may have other interpretations, such as the presence or absence of a chemical. The possible target vectors are denoted v_i, $i \in \{1, \ldots, 2^k\}$. We say that 'a certain v has occurred' if that vector represents the true state of the environment.

Sensor network definition

We define a sensor network as a bipartite graph showing the connections of n identical sensors to k target positions. In the sensor network model that we describe in this section, we assume that each of the c connections of each sensor can be made arbitrarily to any of the k target positions (allowing replacement). Thus, sensor ℓ is connected to (i.e., senses) up to c out of the k target positions, with some positions possibly being sensed more than once. We refer to such sensors as having a range c. Ideally, each sensor produces a value $x \in \mathcal{X}$ that is an arbitrary function of the targets which it senses, $x_\ell = \Psi(v_{\ell t_1}, \ldots, v_{\ell t_c})$. We call the ideal output vector corresponding to target vector v_i as x_i. However, we assume that the ideal output of a sensor x is corrupted by noise to produce the observed output $y \in \mathcal{Y}$. We assume that the conditional p.m.f. $P_{Y|X}(y|x)$ determines the sensor output. Since the sensors are identical, $P_{Y|X}$ is the same for all sensors. Further, we assume that the noise is independent in the sensors, so that the 'sensor output vector' y relates to the ideal output x as $P_{Y|X}(y|x) = \prod_{\ell=1}^{n} P_{Y|X}(y_\ell|x_\ell)$. Observing the output y, a detector $\hat{v} = g(y)$ must determine which of the 2^k target vectors v_i has actually occurred.

Comment: The arbitrary connections model for sensor networks is particularly easy to analyze and provides useful insight into the large-scale detection problem. However, it is also an accurate model for sensing situations where the target vectors do not have a 'spatial interpretation'. For example, such a model can represent a chemical sensor array, where the target positions represent different chemicals. A network of sensors placed on various hosts to monitor a computer network such as the internet is another such situation. An interesting application of this model is to the problem of testing a large set of individuals for a particular disease which has a low incidence rate. i.e., instead of testing each individual separately, a test ('sensor') is made on the combined samples of a subset of the individuals.

Example: Figure 4.4 shows the target vector $v = (0, 0, 1, 0, 1, 1, 0)$ indicating 4 targets present among the 7 positions. There are four sensors, each of which senses $c = 2$ target positions. The ideal sensor output is equal to the number of targets present in the target positions that it observes, $x_\ell = \sum_{u=1}^{c} v_{\ell t_u}$, so that $x \in \mathcal{X} = \{0, 1, \ldots, c\}$. Such a model can describe a seismic sensor which can sense the intensity of vibration to detect the number of targets. A more refined model may allow a weighted sum, for example. Given the sensor network configuration, represented as the connections between sensors and target positions in the graph, the ideal output (codeword) $x = (1, 1, 2, 0)$ is associated with the given v. Unlike in classical codes in communications, where the codeword associated with a message can be chosen arbitrarily, the codeword of a sensor network is computed as a function of the state of the environment. Consider the target vector $v' = (0, 0, 1, 0, 1, 1, 1)$, which differs

Table 4.1 Noise model $P_{Y|X}$ with noise probability p.

$P_{Y|X}$	$Y = 0$	$Y = 1$	$Y = 2$
$X = 0$	$1 - p$	p	0
$X = 1$	$\frac{p}{2}$	$1 - p$	$\frac{p}{2}$
$X = 2$	0	p	$1 - p$

from v in only one target position. The codeword associated with v' is $x' = (1, 1, 2, 1)$, which differs from x in only one position. This demonstrates that unlike in communications, where codewords can be chosen arbitrarily, the codewords corresponding to two states of the environment are dependent due to the fixed sensor network configuration. This dependence suggests that similar target vectors will result in similar codewords, which may be easily confused at the detector. Thus, it is important to allow for some distortion D in decoding in sensor networks, if the error probability is to converge to zero. A sample noise model $P_{Y|X}(y|x)$ is shown in Table 4.1. The codeword in Figure 4.4 is corrupted by noise so that $y = (2, 1, 2, 1)$. In this case, given y, v' has a higher likelihood than v. Therefore, due to codeword dependence, similar states of the environment can be difficult to distinguish using noisy sensors.

It was argued that the detector must allow for some distortion D due to the dependence between sensor codewords. The distortion is the fraction of target positions that are misclassified by the detector. Denoting $d_H(v_i, v_j)$ as the Hamming distance between two target vectors, we define the tolerable distortion region of v_i as $\mathcal{D}_{v_i} = \{j : \frac{1}{k} d_H(v_i, v_j) < D\}$. Then, given that v_i occurred and given a fixed sensor network s, the probability of error is $P_{e,i,s} = \Pr[\text{error}|v_i, s, x_i, y] = \Pr[g(y) \notin \mathcal{D}_{v_i}|i, s, x_i, y]$. The mean probability of error is $P_{e,s} = \frac{1}{2^k} \sum_i P_{e,i,s}$. The rate R of a sensor network is defined as the ratio of target positions being sensed to the number of sensor measurements, $R = \frac{k}{n}$. The sensing capacity of the arbitrary connections sensor network model, $C(D)$, is defined as the maximum rate R such that below this rate there exists a sequence of sensor networks (with increasing k, n) with a mean probability of error approaching zero, i.e., $P_{e,s} \to 0$ as $n \to \infty$ at a fixed rate R.

It is not clear a priori that $C(D)$ is non-zero. The main contribution of this section is to show that $C(D)$ can be non-zero in a variety of situations, i.e., that it is possible to obtain arbitrarily low error probabilities, as long as the number of sensors grows proportionally to the number of target positions, with the correct proportionality factor. The sensing capacity does not provide a guarantee for the error probability of a particular sensor deployment. However, at rates below the sensing capacity, for large k and n, a random sensor deployment can achieve arbitrarily low probability of error with high probability.

4.2.2 Random Coding and Method of Types

How many sensor measurements are necessary to distinguish among all target vectors to within a desired accuracy? In order to answer this question, we use Shannon's random coding idea. In communications, the random coding idea is used to prove the existence of non-zero channel capacities. This is accomplished by analyzing the average probability of error of an ensemble of randomly generated codes. Analogously, instead of analyzing the probability of error of a fixed sensor network s, we analyze the average probability of error of an ensemble of randomly generated sensor networks. Using this approach, we limit the rate, such that below this rate there exist sensor networks that can asymptotically achieve the desired detection accuracy.

We generate sensor networks randomly in the following manner. Since the model in this section allows an arbitrary choice (with replacement) of target position for each sensor connection, we define a probabilistic model where each of the c connections of each sensor independently picks one of the k target positions with equal probability. In particular, we allow more than one connection of a given sensor to pick the same target position. Thus, a given sensor network s has a certain probability of occurrence. We denote the expectation of the probability $P_{e,i,s}$ over all sensor networks, given that v_i occurred, as $P_{e,i} \doteq E_S[P_{e,i,S}]$. We denote the expected mean error probability as $P_e \doteq \frac{1}{2^k} \sum_i P_{e,i} = E_S[P_{e,S}]$. Then, Theorem 4.2.1 provides a lower bound $C_{LB}(D)$ for the sensing capacity $C(D)$ of the sensor network model with arbitrary connections.

The proof of Theorem 4.2.1 relies on the method of types Csiszar (1998), and its statement requires an explanation of types and joint types. The ideal (noiseless) output vector of sensor outputs x depends on the sensor network s, and on the target vector v that occurs. Suppose that the target vector v_i occurs. Since each sensor makes each of its c connections independently and uniformly over the k target positions, the distribution of its ideal output x_i depends only on the type $\gamma = (\gamma_0, \gamma_1)$ of v_i. The type of v_i is the fraction of 0s and 1s in v_i, denoted γ_0 and γ_1 respectively. It follows that $P_{X_i}(x_i) = P^{\gamma,n}(x_i) = \prod_{\ell=1}^n P^\gamma(x_{i\ell})$ for all v_i of the same type γ.

Due to the fact that a single sensor network encodes all target vectors, the codewords associated with different target vectors are dependent, unlike the case for coding in communication. The joint probability of two codewords P_{X_i,X_j} depends on the joint type of the corresponding target vectors v_i, v_j. The joint type is $\lambda = (\lambda_{00}, \lambda_{01}, \lambda_{10}, \lambda_{11})$. Here, λ_{01} is the fraction of positions in v_i, v_j where v_i has bit '0' while v_j has bit '1'. Similarly, we define $\lambda_{00}, \lambda_{10}, \lambda_{11}$. Again, due to the independent and uniform sensor connections, P_{X_i,X_j} depends only on the joint type λ, i.e., $P_{X_i,X_j}(x_i, x_j) = P^{\lambda,n}(x_i, x_j) = \prod_{\ell=1}^n P^\lambda(x_{i\ell}, x_{j\ell})$ for all v_i, v_j of the same joint type λ. Note that the joint type λ also specifies the type γ of v_i as $\gamma_0 = \lambda_{00} + \lambda_{01}$, $\gamma_1 = \lambda_{10} + \lambda_{11}$.

Example: We continue the earlier example based on the sensor network in Figure 4.4. We now assume that each of the $c = 2$ connections of each sensor can be made independently and uniformly over the target positions. Table 4.2 lists the types of four vectors v_j, and their joint type with $v_i = 0010110$. Given a target vector v_i, a sensor will output '2' only if both of its connections connect to positions with a '1'. For a vector of type γ, this occurs with probability $(\gamma_1)^2$. Table 4.3 describes the p.m.f. for the other sensor values, given that a vector of type γ occurred. Given two target vectors v_i, v_j of joint type λ, a sensor will output '0' for both target vectors only if both its connections are connected to target positions that have a '0' bit in both these target vectors. This happens with probability $(\lambda_{00})^2$. Table 4.4 lists the joint p.m.f. $P_{X_iX_j}(x_i, x_j) = P^\lambda(x_i, x_j)$ for all output pairs x_i, x_j corresponding to joint type λ.

Table 4.2 Joint types λ for four pairs of target vectors.

v_j	γ of v_j	λ of v_j with $v_i = 0010110$
0010110	$\left(\frac{4}{7}, \frac{3}{7}\right)$	$\left(\frac{4}{7}, 0, 0, \frac{3}{7}\right)$
0000110	$\left(\frac{5}{7}, \frac{2}{7}\right)$	$\left(\frac{4}{7}, 0, \frac{1}{7}, \frac{2}{7}\right)$
1000011	$\left(\frac{4}{7}, \frac{3}{7}\right)$	$\left(\frac{2}{7}, \frac{2}{7}, \frac{2}{7}, \frac{1}{7}\right)$
0000000	$\left(1, 0\right)$	$\left(\frac{5}{7}, 0, \frac{3}{7}, 0\right)$

Table 4.3 Distribution of X_i in terms of the type γ of v_i when $c = 2$.

X_i	$X_i = 0$	$X_i = 1$	$X_i = 2$
P_{X_i}	$(\gamma_0)^2$	$2\gamma_0\gamma_1$	$(\gamma_1)^2$

Table 4.4 Joint distribution of X_j and X_i in terms of the joint type λ of v_i, v_j when $c = 2$.

$P_{X_iX_j}$	$X_j = 0$	$X_j = 1$	$X_j = 2$
$X_i = 0$	$(\lambda_{00})^2$	$2\lambda_{00}\lambda_{01}$	$(\lambda_{01})^2$
$X_i = 1$	$2\lambda_{00}\lambda_{10}$	$2(\lambda_{10}\lambda_{01} + \lambda_{00}\lambda_{11})$	$2\lambda_{01}\lambda_{11}$
$X_i = 2$	$(\lambda_{10})^2$	$2\lambda_{10}\lambda_{11}$	$(\lambda_{11})^2$

4.2.3 Sensing Capacity Theorem

We specify two probability distributions required to state the Sensing Capacity Achievability Theorem. The first is the joint distribution of the ideal output x_i when v_i occurs and the noise corrupted output y caused by it. i.e., $P_{X_iY}(x_i, y) = \prod_{\ell=1}^{n} P_{X_iY}(x_{i\ell}, y_\ell) = \prod_{\ell=1}^{n} P_{X_i}(x_{i\ell})P_{Y|X}(y_\ell|x_{i\ell})$. The second distribution is the joint distribution of the ideal output x_i corresponding to v_i and the noise corrupted output y generated by the occurrence of a *different* target vector v_j. We can write this joint distribution as $Q_{X_iY}^{(j)}(x_i, y) = \prod_{\ell=1}^{n} Q_{X_iY}^{(j)}(x_{i\ell}, y_\ell)$, where $Q_{X_iY}^{(j)}(x_{i\ell}, y_\ell) = \sum_{a \in \mathcal{X}} P_{X_iX_j}(x_{i\ell}, x_j = a)P_{Y|X}(y_\ell|x_j = a)$. Notice that the joint distributions over the sensors factor into a product because the sensor connections are independent and the noise in the sensors is independent. Also, note that although Y was produced by X_j, there is dependence between X_i and Y because of the dependence of X_i and X_j. As argued earlier, P_{X_i} and P_{X_i,X_j} can be computed using the type γ and joint type λ respectively. Thus, we write

$$P_{X_iY}(x_i, y) = \prod_{\ell=1}^{n} P_{X_iY}^{\gamma}(x_{i\ell}, y_\ell)$$

$$\doteq P_{X_iY}^{\gamma}(x_i, y)$$

$$Q_{X_i Y}^{(j)}(x_i, y) = \prod_{\ell=1}^{n} Q_{X_i Y}^{\lambda}(x_{i\ell}, y_\ell)$$

$$\dot{=} Q_{X_i Y}^{\lambda}(x_i, y)$$

where $P_{X_i Y}^{\gamma}(x_i, y) = P^{\gamma}(x_i) P_{Y|X}(y|x_i)$ and $Q_{X_i Y}^{\lambda}(x_i, y) = \sum_{a \in \mathcal{X}} P^{\lambda}(x_i, x_j = a)$ $P_{Y|X}(y|x_j = a)$. We are now ready to state the main theorem of this section.

Theorem 4.2.1 (Sensing Capacity Achievability Theorem – Arbitrary Connections Model) *Denoting $D(P \| Q)$ as Kullback-Leibler distance and $H(P)$ as entropy (as defined in Cover and Thomas (1991)), the sensing capacity at distortion D is bounded as,*

$$C(D) \geq C_{LB}(D) = \min_{\substack{\lambda \\ \lambda_{01} + \lambda_{10} \geq D \\ \lambda_{00} + \lambda_{01} = \gamma_0 \\ \lambda_{10} + \lambda_{11} = \gamma_1}} \frac{D\left(P_{X_i Y}^{\gamma} \| Q_{X_i Y}^{\lambda}\right)}{H(\lambda) - H(\gamma)} \tag{4.1}$$

where $\gamma = (0.5, 0.5)$ while $\lambda = (\lambda_{00}, \lambda_{01}, \lambda_{10}, \lambda_{11})$ is an arbitrary probability mass function.

We note that the bound above is positive because it is the ratio of two positive quantities. The numerator is a Kullback-Leibler distance and the denominator is the difference of the entropy of a joint distribution and the entropy of a marginal of that distribution. This bound differs significantly from Shannon's famous channel capacity result for a communication channel. The most striking difference is that the bound depends not on a mutual information, but on a Kullback-Leibler distance. This difference is interesting because of the frequent use of mutual information as a sensor selection metric (e.g., Manyika and Durrant-Whyte (1994)). Theorem 4.2.1 shows that the probability of error in large-scale detection problems is related to a Kullback-Leibler distance. A relationship to mutual information is not obviously apparent. Intuitively, mutual information characterizes distance between codewords that are uniformly spread out over the space of possible codewords. Such a distribution is feasible in a communication channel because each message can be mapped to any arbitrary codeword within the allowable space. Figure 4.5 illustrates this fact. Figure 4.5 depicts the codewords associated with messages (target vectors) with a varying number of ones and zeros, as indicated in the legend. The codewords corresponding to the messages are independent and identically distributed (i.i.d.) as in channel capacity proofs. All the codewords are then projected along two randomly chosen orthogonal basis vectors, to allow a visual depiction. As expected, Figure 4.5 shows no discernible relationship between a message and its corresponding codeword. In contrast, the codewords in a sensor network are strongly dependent on the specific state (target vector). This fact is demonstrated in Figure 4.6. Figure 4.6 depicts the sensor network codewords associated with target vectors with a varying number of ones and zeros (i.e., type γ). All target vectors are encoded using a single randomly generated sensor network with arbitrary connections. The sensors output a count of the number of ones in the target positions which they observe. The codewords are then projected along the same pair of orthogonal vectors used to generate Figure 4.5. Figure 4.6 demonstrates a strong dependence between the type of a target vector and its codeword distribution. Thus unlike in communication, the codewords are not identically distributed. Further, the figure also demonstrates that similar target vectors yield

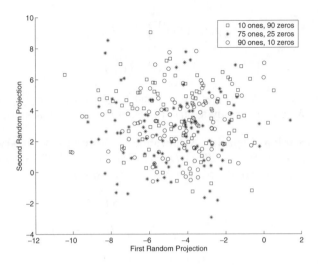

Figure 4.5 Illustration of codeword geometry of communication channel code.

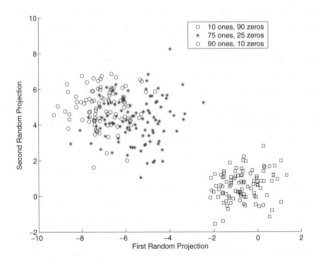

Figure 4.6 Illustration of codeword geometry of sensor network code.

similar codewords. More generally, the codewords obtained by the random sensor network are dependent, unlike the case in Shannon's random coding method for communication channels.

Another way in which the bound differs from the classical channel capacity result is due to the denominator in the bound. This denominator accounts for the non-identical codeword distribution of sensor networks. Based on the sensing model, as demonstrated in Figure 4.6, it is possible to have codewords that are 'clustered'. Finally, another difference with channel capacity is the fact that the bound is a minimum over some set of distributions. This occurs because in sensor networks, not all codewords contribute equally to the probability of an error. In fact, target vectors are more likely to be incorrectly decoded into target vectors that are close to them. In summary, if the codewords of the sensor network were independently and identically distributed, the bound shown in Theorem 4.2.1 would reduce to Shannon's classic channel capacity result based on mutual information $I(X; Y)$. Thus, one of the primary differences between the sensing capacity and channel capacity is the constrained encoding in sensor networks.

We outline the proof of Theorem 4.2.1. A complete proof can be found in Rachlin (2007). The proof broadly follows the proof of channel capacity provided by Gallager (1968), by analyzing pair-wise error probabilities, averaged over randomly generated sensor networks. However, it differs from Gallager (1968) in several important ways. The most important difference arises due to the non i.i.d codeword distribution induced by a random 'encoder' (i.e., sensor network). We use the method of types to group the exponential number of pair-wise error probability terms into a polynomial number of terms, so as to upper bound the probability of error P_e.

Proof Outline. For a fixed sensor network s there is a fixed and known correspondence between target vectors v_j and codewords x_js. We assume a Maximum-Likelihood (ML) decoder, $g_{ML}(y) = \arg \max_j P_{Y|X}(y|x_j)$. For this decoder, we consider the expected mean probability of error $P_e = \frac{1}{2^k} \sum_i P_{e,i}$, where $P_{e,i} = E_S[P_{e,i,S}]$. $P_{e,i,s}$ is the probability of error for a fixed sensor network s given that target vector v_i has occurred. $P_{e,i}$ is obtained by taking the expectation over all randomly generated sensor networks. Since the sensor network is randomly generated, the codewords are random. Denoting the set of random codewords as $C = \{X_1, \ldots, X_{2^k}\}$ and assuming a uniform distribution over the target vectors (due to the mean probability P_e), we can write,

$$P_e = E_{VYC}[P_{e,V,Y,C}] \tag{4.2}$$

where $P_{e,V,Y,C} = \Pr(g(Y) \notin D_V | V, Y, C)$. Using the fact that $P_{e|V,Y,C}$ is a probability, we can bound P_e as follows,

$$P_e \leq E_{VYC}\left[\sum_w \Pr(g(Y) \in S_w | V, Y, C)^\rho\right]$$

where $\rho \in [0, 1]$ and $\{S_1, S_2, \ldots\}$ is any partition of the complement of D_V (denoted $\overline{D_V}$). Using the union bound to upper bound $\Pr(g(y) \in S_w | i, y, C)$ in terms of pairwise error probabilities, we obtain the bound:

$$P_e \leq E_{VYC}\left[\sum_w \left(\sum_{j \in S_w} \Pr(g(Y) = v_j | V, Y, C)\right)^\rho\right] \tag{4.3}$$

The probability term $\Pr(g(Y) = v_j | V, Y, C)$ is a pairwise error term, depending only on the X_i associated with V and X_j associated with v_j. Using this fact, the concavity of x^ρ for $\rho \in [0, 1]$, and Jensen's inequality, we obtain the following bound on (4.3):

$$P_e \leq E_{VYX_i}\left[\sum_w \left(\sum_{j \in S_w} E_{X_j|X_i}\left[\Pr(g(Y) = j | V, Y, X_i, X_j)\right]\right)^\rho\right] \tag{4.4}$$

Using the ML decoder, the probability term above equals one only if v_j has the highest likelihood, and is zero otherwise. This allows us to bound the probability term, resulting in the following:

$$P_e \leq E_{VYX_i} \left[\sum_w \left(\sum_{j \in S_w} E_{X_j|X_i} \left[\left(\frac{P_{Y|X}(Y|X_j)}{P_{Y|X}(Y|X_i)} \right)^{\frac{1}{1+\rho}} \right] \right)^\rho \right] \tag{4.5}$$

The bound (4.5) has an exponential number of pairwise error terms. However, all the probability distributions can be equivalently specified by the type γ and joint type λ instead of specific i, j pairs. Since there are only a polynomial (in k) number of types, this allows us to group the terms into a polynomial number of sets, one for each joint type. For this purpose, we choose each S_w to be a distinct joint type λ and let w enumerate the set $S_\gamma(D)$ of all the λ that are the joint type of V and any $v_j \in \overline{D_V}$. Then Eq. (4.5) can be written as,

$$P_e \leq \sum_\gamma 2^{-kD(\gamma \| (0.5, 0.5))} \sum_{x_i \in \mathcal{X}^n} \sum_{y \in \mathcal{Y}^n} P^\gamma(x_i) P_{Y|X}(y|x_i)$$

$$\cdot \sum_{\lambda \in S_\gamma(D)} \left(2^{k(H(\lambda) - H(\gamma))} \sum_{x_j \in \mathcal{X}^n} P^\lambda(x_j|x_i) \left(\frac{P_{Y|X}(y|x_j)}{P_{Y|X}(y|x_i)} \right)^{\frac{1}{1+\rho}} \right)^\rho$$

The exponential term within the bracket is simply a bound on the size of set S_w, while the exponential term outside the bracket is the probability of type γ. We can write $S_\gamma(D)$ in terms of λ as below.

$$S_\gamma(D) = \{ \lambda : \lambda_{01} + \lambda_{10} \geq D, \ \gamma_0 = \lambda_{00} + \lambda_{01}, \ \gamma_1 = \lambda_{10} + \lambda_{11} \} \tag{4.6}$$

Since each sensor forms independent connections and has independent noise, the joint p.m.f.s factor into a product. Using the fact that the number of types is polynomial in k, we get the bound,

$$P_e \leq 2^{-nE_r(R,D) + o(\log(n))} \tag{4.7}$$

where $o(\log(n))$ grows logarithmically. Here, we define:

$$E_r(R, D) = \min_\gamma \min_{\lambda \in S_\gamma(D)} \max_{0 \leq \rho \leq 1} E(\rho, \lambda) - \rho R(H(\lambda) - H(\gamma)) + D(\gamma \| (0.5, 0.5)) \tag{4.8}$$

$$E(\rho, \lambda) = -\log \left(\sum_{a_i \in \mathcal{X}} \sum_{b \in \mathcal{Y}} P^{\gamma_i}(a_i) P_{Y|X}(b|a_i)^{\frac{1}{1+\rho}} \left(\sum_{a_j \in \mathcal{X}} P^\lambda(a_j|a_i) P_{Y|X}(b|a_j)^{\frac{1}{1+\rho}} \right)^\rho \right)$$

The average error probability $P_e \to 0$ as $n \to \infty$ if $E_r(R, D) > 0$. This is possible when R is bounded as follows:

$$R < \min_{\substack{\lambda \\ \lambda_{01} + \lambda_{10} \geq D \\ \lambda_{00} + \lambda_{01} = \gamma_0 \\ \lambda_{10} + \lambda_{11} = \gamma_1}} \frac{D\left(P^\gamma_{X_i Y} \| Q^\lambda_{X_i Y} \right)}{H(\lambda) - H(\gamma)} \tag{4.9}$$

where $\gamma = (0.5, 0.5)$. Therefore, the right-hand side of Eq. (4.9) is a lower bound on $C(D)$.

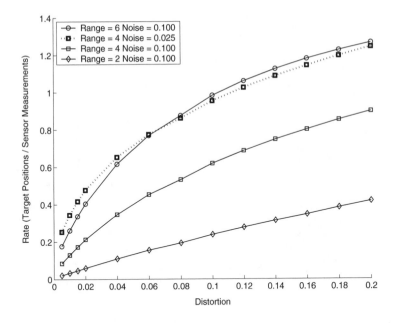

Figure 4.7 $C_{LB}(D)$ for sensors of varying noise levels and range.

4.2.4 Illustration of Sensing Capacity Bound

We continue the previous example based on the sensor network in Figure 4.4. We compute the sensing capacity bound $C_{LB}(D)$ in Theorem (4.2.1) for this network. However, we experiment with various values of noise level and sensor range c. The computed bounds are shown in Figure 4.7. In Figure 4.7, 'Noise = p' indicates that for a sensor, $P(Y \neq X) = p$, with $\mathcal{Y} = \mathcal{X}$ assumed. The probability p is assumed equally distributed over the two (one in the case of $x = 0$ and $x = c$) closest values to x. In all cases, $C_{LB}(D)$ approaches 0 as D approaches 0. This occurs because similar target vectors have similar codewords due to dependence in the codeword distribution, and therefore, more sensor measurements are required to differentiate among them. The bounds for sensors of varying range and noise levels reveal tradeoffs among different sensor classes. Some tradeoffs agree with intuition. For example, sensing capacity increases with decreasing noise levels. However, other tradeoffs are not as obvious. For example, compare the bound for sensors of range $c = 4$ and noise 0.025 with sensors of range $c = 6$ and noise 0.10. Neither sensor is clearly better than the other, and the preference for one over another depends on the desired distortion. For distortion below $D = 0.06$, the shorter range but lower noise sensor results in a higher sensing capacity than sensors of longer range but higher noise. The reverse is true for distortions greater than $D = 0.06$. Thus, the bound presented in (4.1) expresses a complex tradeoff between sensor noise, sensor range, and the desired detection accuracy.

It may be interesting to observe the dependence of the sensing capacity bound on the joint type λ. Figure 4.8 is a contour plot of the ratio in (4.1) that is minimized over λ. Since λ is constrained by $\gamma = \{0.5, 0.5\}$, it has two free variables $\lambda_{01}, \lambda_{10}$. The segment $\lambda_{01} + \lambda_{10}$ is the distortion resulting from mistaking vectors of joint type λ as one another. Notice

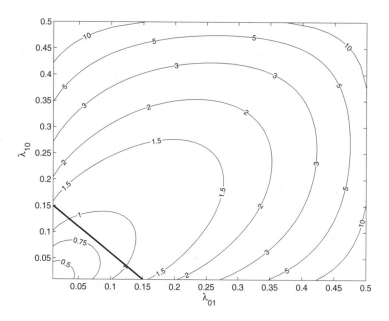

Figure 4.8 Computing $C_{LB}(D)$ as a function of λ. The diagonal segment corresponds to distortion bound $D = 0.15$.

that the ratio decreases as we move towards the origin. This is the region of low distortion, which can only be achieved at low rates. However, due to the distortion bound D (the diagonal black distortion segment) which is allowed to the decoder, only the region above this segment needs to be considered by the minimization in (4.1), resulting in non-zero sensing capacity.

Using the loopy belief propagation algorithm in Pearl (1988), we empirically examined sensor network performance as a function of rate. We generated sensor networks of various rates by setting the number of targets, and varying the number of sensors. We chose the number of connections to be $c = 4$, the distortion to be $D = 0.1$, and the noise level to be 0.1. The capacity value C_{LB} for the model used in this experiment is 0.62. We empirically evaluated the error rate averaged over a set of randomly generated target vectors and sensor networks. Figure 4.9 shows that the error probability decreases as the rate is reduced. Interestingly, the reduction becomes sharper as k, n are increased (for the given rate), which supports the 'phase transition effect' indicated by the sensing capacity theorem. This phase transition occurs close to the computed C_{LB} for this network, although the error probability is still significant below C_{LB}. We conjecture that this occurs because belief propagation is suboptimal for graphs with cycles, such as the graphs generated using our sensor network model.

4.3   Extensions to Other Sensor Network Models

Section 4.2 introduced a sensor network model where each sensor is allowed to make arbitrary connections to the target vector. An analysis of this model culminated in Theorem 4.2.1,

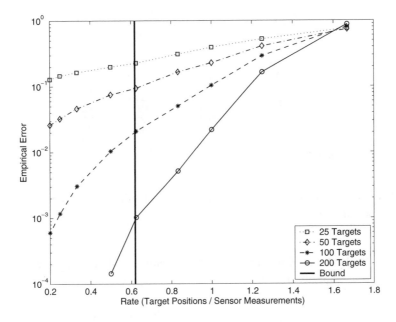

Figure 4.9 Average empirical error rate of loopy belief propagation-based detection for varying rates, and the corresponding lower bound.

which provided insight into the large-scale detection problem. In several situations, more complex sensor network models may be called for, such as those based on localized sensing functions or more complex target models. This section describes extensions of the sensing capacity concept to such complex models. We begin with a few straightforward refinements of the arbitrary connection model. The first refinement considers non-binary target vectors. Binary target vectors indicate the presence or absence of targets at the spatial positions. A target vector over a general finite alphabet may indicate, in addition to the presence of targets, the class of a target. Alternatively, the entries of non-binary vectors can indicate levels of intensity or concentration. Assuming a target vector over alphabet $\mathcal{A}$, we can define types and joint types over $\mathcal{A}$, and apply the same analysis as before to obtain the sensing capacity bound below:

$$C(D) \geq C_{LB}(D) = \min_{\substack{\lambda \\ \sum_{a \neq b} \lambda_{ab} \geq D \\ \sum_{b} \lambda_{ab} = \gamma_a}} \frac{D\left(P^{\gamma}_{X_i Y} \| Q^{\lambda}_{X_i Y}\right)}{H(\lambda) - H(\gamma)} \tag{4.10}$$

where $\gamma = \left(\gamma_a = \frac{1}{|\mathcal{A}|}, a \in \mathcal{A}\right)$, while $\lambda = (\lambda_{ab},\ a, b \in \mathcal{A})$ is an arbitrary probability mass function.

The second refinement allows the following a priori distribution over target vectors (whereas Section 4.2 did not assume an a priori distribution.) Assume that each target position is generated i.i.d. with probability P_V over the alphabet $\mathcal{A}$. This may model the fact that targets are sparsely present. The previous analysis can be extended to a Maximum-

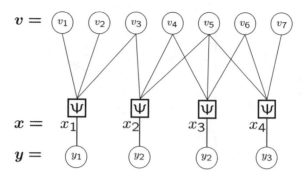

Figure 4.10 A one-dimensional sensor network model with spatially contiguous connections and $k = 7, n = 4, c = 3$.

a-Posteriori (MAP) detector, instead of the ML detector considered earlier, resulting in the following sensing capacity bound:

$$C(D) \geq C_{LB}(D) = \min_{\substack{\lambda \\ \sum_{a \neq b} \lambda_{ab} \geq D \\ \sum_b \lambda_{ab} = \gamma_{ia}}} \frac{D\left(P_{X_iY}^{\gamma_i} \| Q_{X_iY}^{\lambda}\right)}{H(\lambda) - H(\gamma_j) - D(\gamma_j \| P_V)} \tag{4.11}$$

where $\gamma_i = P_V$, $\lambda = (\lambda_{ab}, \ a, b \in \mathcal{A})$ is an arbitrary probability mass function and γ_j is the marginal of λ calculated as $\gamma_{jb} = \sum_a \lambda_{ab}$.

A third extension accounts for heterogenous sensors, where each class of sensor possibly has a different range c, noise model $P_{Y|X}$, and/or sensing function Ψ. Let the sensor of class l be used with a given relative frequency α_l. For such a model, the sensing capacity bound is as follows:

$$C(D) \geq C_{LB}(D) = \min_{\substack{\lambda \\ \sum_{a \neq b} \lambda_{ab} \geq D \\ \sum_b \lambda_{ab} = \gamma_{ia}}} \frac{\sum_l \alpha_l D\left(P_{X_iY}^{\gamma_i,l} \| Q_{X_iY}^{\lambda,l}\right)}{H(\lambda) - H(\gamma_j) - D(\gamma_j \| P_V)} \tag{4.12}$$

where $\gamma_i = P_V$, $\lambda = (\lambda_{ab}, \ a, b \in \mathcal{A})$ is an arbitrary probability mass function and γ_j is the marginal of λ calculated as $\gamma_{jb} = \sum_a \lambda_{ab}$.

4.3.1 Models with Localized Sensing

The sensor network model with arbitrary connections depicted in Figure 4.4 does not capture the geometrical properties of many classes of sensors. For example, a seismic sensor receives vibrations from nearby targets only. A camera can only image a contiguous portion of space.

In Rachlin et al. (2005) and Rachlin (2007), we analyzed the sensing capacity of a one-dimensional sensor network model that captures contiguity in sensor connections. In this work, we assumed a one-dimensional target array, such as positions along the perimeter of

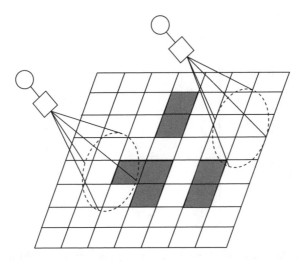

Figure 4.11 A two-dimensional target grid with spatially contiguous sensor observations. The state of the environment is modeled as a Markov random field.

a wall. An example of this model is shown in Figure 4.10. In Figure 4.10, sensors observe a contiguous set of three target positions and output an arbitrary function Ψ of the target positions they observe. For example, a sensor could output a weighted sum of the targets they observe, such as in the case of infrared temperature sensors.

The analysis of the contiguous connections model in one dimension requires the application of 'c-order types'. The c-order type of a target vector is the fraction of occurrences of each possible sub-string of length c in the target vector. For example, for a binary target vector and $c = 2$, the c-order type is defined as $\gamma = (\gamma_{00}, \gamma_{01}, \gamma_{10}, \gamma_{11})$. Here, γ_{ab} is the fraction of occurrences of the string 'ab' in the target vector. The primary difference in analyzing the sensing capacity of the contiguous connections model arises due to the use of c-order types instead of types, which requires different counting arguments. The basic method of proving the sensing capacity bound, however, is similar to the case of the arbitrary connections model.

In Rachlin (2007), we extend our analysis of contiguous connections models to two-dimensional target and sensor configurations. In this case, the set of target positions can be represented by a matrix instead of a target vector. An example of such a network is shown in Figure 4.11. Such a model can model applications such as surveillance of an area by seismic or camera sensors. The analysis of this model requires the use of two-dimensional types. These types enumerate the fraction of occurrences of all possible two dimensional patterns of a given size, similar to the c-order type in the one-dimensional case.

4.3.2 Target Models

Section 4.2 which introduced sensing capacity did not assume any prior distribution on the target vectors. The extensions to this model that considered an a priori distribution modeled the target positions as i.i.d. with probability distribution P_V. In Rachlin (2007),

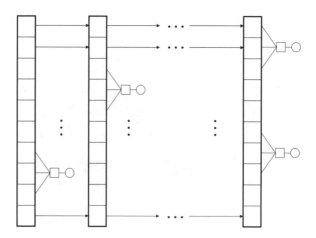

Figure 4.12 Sensor network model for a sequence of dependent target vectors.

we considered extensions to target models that allow for correlation between the target positions. In that work, the state of the environment is modeled as a two-dimensional Markov random field. Such a model can be used to capture spatial structure, such as target clustering, as would occur in a surveillance application with groups of people. The analysis of the Markov random field model requires the introduction of field-types which specify the probability of a field. These depend on the target distribution and are distinct from the types and c-order types which depend on sensor connections.

Many large-scale detection applications require a sensor network (Figure 4.12) to monitor an environment whose state varies over time. Examples include pollution, traffic, and agricultural monitoring, and surveillance. In Rachlin et al. (2006c), we analyzed the sensing capacity of a sensor network that observes a sequence of dependent states of the environment. Each target position was assumed to independently evolve as a Markov chain over time. The analysis of this model requires the use of temporal-types, which specify the probability of a sequence of target vectors, based on the Markov chain.

4.4 Conclusion

The results presented in this chapter provide limits on the performance of sensor networks for large-scale detection problems. The notion of sensing capacity indicates that it should be possible to obtain arbitrarily good performance (to within the distortion constraint) so long as the number of sensors used is proportional to the scale of the sensing problem (the number of target positions), with the correct proportionality factor. As importantly, this chapter demonstrates a close connection between sensor networks and communication channels. It is tantalizing to consider that one could transpose the large body of results available on communication channels, into the sensor network setting. For example, channel coding theory contains a large number of results that are used to build practical communication systems. Can we build on our analogy to communication channels to bring insights from coding theory into the design of sensor networks? As a first step, in Rachlin et al. (2006a,b) and Rachlin (2007) we proposed extending ideas from convolutional coding to sensor networks. We demonstrated that a version of sequential decoding (which is a low complexity decoding heuristic for convolutional codes) can be applied to detection in sensor networks, as an alternative to the complex belief propagation algorithm. Our empirical results indicate that above a certain number of sensor measurements, the sequential decoding algorithm achieves accurate decoding with bounded computations per bit (target position). This empirical result suggests the existence of a 'computational cut-off rate', similar to one that exists for channel codes.

We believe that the current state of the art on the theory of sensing capacity merely scratches the surface of a large set of problems on large-scale detection. A host of interesting and useful directions can be followed to explore this field. The most obvious direction is to strengthen the theory by considering alternative settings of the problem, tightening the sensing capacity bounds, and proving a converse to sensing capacity. Another direction is to explore the connection between sensor networks and communication channels, including the exploitation of existing channel codes to design sensor networks (or provide heuristics for such design). A third direction involves questions about sensor selection and sensor design for practical applications, guided by the notion of sensing capacity. In the authors' experience, the technical specifications of off-the-shelf commercial sensors are not directly amenable to performance analysis in a large-scale detection setting. For example, the published specification of an infrared temperature sensor characterizes it assuming a single target in its field of view. In a large-scale detection setting, where the sensor may possibly see multiple targets, the specification does not provide a model for sensor behavior. Hopefully, engineering ideas about sensor design, selection, and specification can be re-examined based on the insights obtained from the theory of sensing capacity.

Bibliography

Bajwa W, Sayeed A and Nowak R 2005 Matched source-channel communication for field estimation in wireless sensor networks. In *Proc. Fourth Int. Symp. on Information Processing in Sensor Networks*.

Burl M, Sisk B, Vaid T and Lewis N 2002 Classification performance of carbon black-polymer composite vapor detector arrays as a function of array size and detector composition. *Sensors and Actuators B* **87**, 130–149.

Chakrabarty K, Iyengar SS, Qi H and Cho E 2001 Coding theory framework for target location in distributed sensor networks. In *Proc. Int. Conf. on Inform. Technology: Coding and Computing*.

Chamberland J and Veeravalli V 2003 Decentralized detection in sensor networks. *IEEE Transactions on Signal Processing* **51**(2), 407–416.

Chamberland J and Veeravalli V 2004 Asymptotic results for decentralized detection in power constrained wireless sensor networks. *IEEE JSAC Special Issue on Wireless Sensor Networks* **22**(6), 1007–1015.

Cover TM and Thomas JA 1991 *Elements of Information Theory*. Wiley-Interscience, New York.

Csiszar I 1998 The method of types. *IEEE Trans. Inform. Theory*.

D'Costa A, Ramachandran V and Sayeed A 2004 Distributed classification of Gaussian space-time sources in wireless sensor networks. *IEEE J. Selected Areas in Communications (special issue on Fundamental Performance Limits of Wireless Sensor Networks)*, pp. 1026–1036.

Elfes A 1989 Occupancy grids: a probabilistic framework for mobile robot perception and navigation. PhD thesis, Electrical and Computer Eng. Dept., Carnegie Mellon University.

Gallager R 1968 *Information Theory and Reliable Communications*. Wiley, Chichester.

Gastpar M and Vetterli M 2005 Power, spatio-temporal bandwidth, and distortion in large sensor networks. *IEEE Journal on Selected Areas in Communications* **23**(4), 745–754.

Hoover A and Olsen B 1999 A real-time occupancy map from multiple video streams. In *Proc. Int. Conf. on Robotics and Automation*.

Ishwar P, Kumar A and Ramachandran K 2003 Distributed sampling for dense sensor networks: A bit-conservation principle. *Information Processing in Sensor Networks*.

Kotecha J, Ramachandran V and Sayeed A 2005 Distributed multi-target classification in wireless sensor networks. *IEEE JSAC Special Issue on Self-Organizing Distributed Collaborative Sensor Networks*.

Kumar A, Ishwar P and Ramchandran K 2004 On distributed sampling of smooth non-bandlimited fields. *Int. Symp. on Information Processing in Sensor Networks*.

Li D, Wong K, Hu Y and Sayeed A 2002 Detection, classification and tracking of targets in distributed sensor networks. *IEEE Signal Processing Magazine* pp. 17–29.

Manyika J and Durrant-Whyte H 1994 *Data Fusion and Sensor Management: A Decentralized Information-Theoretic Approach*. Prentice Hall, Englewood Cliffs, NY.

Marco D, Duarte-Melo E, Liu M and Neuhoff D 2003 On the many-to-one transport capacity of dense wireless sensor networks and the compressibility of its data. In *Information Processing in Sensor Networks*, pp. 1–16.

Moura J, Liu J and Kleiner M 2003 Intelligent sensor fusion: A graphical model approach. *IEEE Int. Conf. on Sig. Proc.*

Moura J, Negi R and Pueschel M 2002 Distributed sensing and processing: a graphical model approach. DARPA ACMP Integrated Sensing and Processing Workshop, Annapolis, MD.

Nowak R, Mitra U and Willett R 2004 Estimating inhomogeneous fields using wireless sensor networks. *IEEE Journal on Selected Areas in Communications* **22**(6), 999–1006.

Pearl J 1988 *Probabilistic Reasoning in Intelligent Systems: Networks of Plausible Inference*. Morgan Kaufmann, San Francisco.

Pradhan S, Kusuma J and Ramachandran K 2002 Distributed compression in a dense microsensor network. *IEEE Signal Processing Magazine* **19**, 51–60.

Rachlin Y 2007 On the interdependence of sensing, accuracy, and complexity in large-scale detection applications. PhD thesis, Carnegie Mellon University.

Rachlin Y, Narayanaswamy B, Negi R, Dolan J and Khosla P 2006a Increasing sensor measurements to reduce detection complexity in large-scale detection applications. In *Proc. Military Communications Conference*.

Rachlin Y, Negi R and Khosla P 2004 Sensing capacity for target detection. In *Proc. IEEE Inform. Theory Wksp.*

Rachlin Y, Negi R and Khosla P 2005 Sensing capacity for discrete sensor network applications. In *Proc. Fourth Int. Symp. on Information Processing in Sensor Networks*.

Rachlin Y, Negi R and Khosla P 2006b On the interdependence of sensing and estimation complexity in sensor networks. In *Proc. Fifth Int. Conf. on Information Processing in Sensor Networks*.

Rachlin Y, Negi R and Khosla P 2006c Temporal sensing capacity. *Proc. Allerton Conference on Communication, Control, and Computing*.

Scaglione A and Servetto SD 2002 On the interdependence of routing and data compression in multi-hop sensor networks. In *Proc. 8th ACM Int. Conference on Mobile Computing and Networking*.

Shannon C 1948 A mathematical theory of communication. *Bell System Technical Journal* **27**, 379–423, 623–656.

Slepian D and Wolf J 1973 Noiseless coding of correlated information sources. *IEEE Trans. Inform. Theory* **19**, 471–480.

Thrun S 2002 Robotic mapping: A survey. In *Exploring Artificial Intelligence in the New Millenium.* (ed. Lakemeyer G and Nebel B) Morgan Kaufmann, San Francisco.

Tian Y and Qi H 2002 Target detection and classification using seismic signal processing in unattended ground sensor systems. *International Conference on Acoustics Speech and Signal Processing (ICASSP)*, vol. 4.

Varshney P 1997 *Distributed Detection and Data Fusion.* Springer-Verlag, New York.

Wyner A and Ziv J 1976 Noiseless coding of correlated information sources. *IEEE Trans. Inform. Theory* **22**, 1–10.

Xiong Z, Liveris A and Cheng S 2004 Distributed source coding for sensor networks. *IEEE Signal Processing Magazine* **21**, 80–94.

Yang D, Gonzalez-Banos H and Guibas L 2003 Counting people in crowds with a real-time network of image sensors. In *Proc. Int. Conf. on Computer Vision.*

5

Law of Sensor Network Lifetime and Its Applications

Yunxia Chen and Qing Zhao

5.1 Introduction

The performance measure of network lifetime is particularly relevant to sensor networks where battery-powered, dispensable sensors are deployed to collectively perform a certain task. For a communication network, which is generally designed to support individual users, network lifetime is subject to interpretation; a communication network may be considered dead by one user while continuing to provide required quality of service (QoS) for others. In contrast, a sensor network is not deployed for individual nodes, but for a specific collaborative task at the network level. The lifetime of a sensor network thus has an unambiguous definition: it is the time span from the deployment to the instant when the network can no longer perform the task.

Network lifetime is crucial to large-scale sensor networks since in many applications, it is undesirable or infeasible to replace or recharge sensors once the network is deployed. Much has been said about maximizing sensor network lifetime. The lack of an accurate characterization of network lifetime as a function of key design parameters, however, presents a fundamental impediment to optimal protocol design. Given that network lifetime depends on network architectures, specific applications, and various parameters across the entire protocol stack, existing techniques tend to rely on either a specific network setup or the use of upper bounds on lifetime. As such, it is difficult to develop a general design principle.

Wireless Sensor Networks: Signal Processing and Communications Perspectives   A. Swami, Q. Zhao, Y.-W. Hong and L. Tong
© 2007 John Wiley & Sons, Ltd

In this chapter, we show that there is, in fact, a simple law that governs network lifetime for all applications, under any network configuration (Chen and Zhao 2005). This law of lifetime not only identifies two key physical layer parameters that affect network lifetime, but also reveals a general design principle for lifetime maximization. An example of applying this law of lifetime to MAC design is provided. Specifically, we first obtain the limiting performance on network lifetime achieved by centralized scheduling that optimally exploits the two key physical layer parameters. We then demonstrate that, by applying the general design principle revealed by the law of lifetime, we can obtain a simple distributed scheduling protocol that asymptotically achieves the limiting performance defined by centralized scheduling. A brief overview of existing analytical work on network lifetime is also provided to give a more complete picture on recent advances in this area.

5.2 Law of Network Lifetime and General Design Principle

In this section, we present the law of lifetime, which provides an exact characterization of network lifetime under a general network setting. Based on this law of lifetime, we obtain a general principle for any lifetime-maximizing design.

5.2.1 Network Characteristics and Lifetime Definition

Major network characteristics that affect network lifetime include network architecture, energy consumption model of sensor nodes, and data collection mode and lifetime definition determined by the underlying application. Below, we take a closer look at these network characteristics to identify factors affecting network lifetime.

Sensor network architecture

Three types of sensor network architecture have been considered in the literature: flat ad hoc, hierarchical, and SEnsor Network with Mobile Access (SENMA). Under the flat ad hoc architecture, sensors collaboratively relay their data to access points (a.k.a. base stations or sinks). In hierarchical networks, sensors are organized into clusters, and cluster heads (a.k.a. relay nodes) are responsible for collecting and aggregating data from sensors and then reporting to access points (APs). In SENMA, sensors communicate directly with mobile APs moving around the sensor field. Transmissions from sensors to APs are typically multi-hop in the flat ad hoc networks, single-hop in SENMA, single-hop within clusters and multi-hop between clusters in the hierarchical networks.

Data collection initiation

According to network applications, the data collection process in a sensor network can be initiated by the internal clock of sensors, the event of interest, or the demand of the end-user. In clock-driven networks, sensors collect and transmit data at pre-determined time intervals. In event-driven or demand-driven networks, data collections are triggered by an event of interest or a request from APs.

Energy consumption model

The energy consumption model characterizes the sources of energy consumption in a network. According to the rate of energy expenditure, we classify the network energy consumption into two general categories: the continuous energy consumption and the reporting energy consumption. The continuous energy consumption is the minimum energy needed to sustain the network during its lifetime without data collection. It includes, for example, the battery leakage and the energy consumed in sleeping, sensing, and signal processing. The reporting energy consumption is the additional energy consumed in a data collection process. It depends on the channel model as well as the network architecture and protocols. In particular, the reporting energy consumption includes the energy consumed in transmission, reception, and possibly channel acquisition.

Lifetime definition

Network lifetime is the time span from the deployment to the instant when the network is considered nonfunctional. When a network should be considered nonfunctional is, however, application-specific. It can be, for example, the instant when a certain fraction of sensors die, loss of coverage occurs (i.e., a certain portion of the desired area can no longer be monitored by any sensor), loss of connectivity occurs (i.e., sensors can no longer communicate with APs), or the detection probability drops below a certain threshold.

5.2.2 Law of Lifetime

The above discussion demonstrates the variety of network parameters that may affect network lifetime. We show below that, behind the vast differences in communication environment and network configuration and application, there exists a simple law governing network lifetime as well as a general design principle applicable to various network aspects.

Theorem 5.2.1 (Law of network Lifetime) *For a sensor network with total non-rechargeable initial energy E_0, the expected network lifetime $\mathbb{E}[\mathcal{L}]$, measured as the average amount of time until the network is considered nonfunctional, is given by*

$$\mathbb{E}[\mathcal{L}] = \frac{E_0 - \mathbb{E}[E_w]}{P_c + \lambda \mathbb{E}[E_r]}, \tag{5.1}$$

where P_c is the constant continuous power consumption of all sensors in the network, $\mathbb{E}[E_w]$ is the expected wasted energy (i.e., the total unused energy in the network when it dies), λ is the expected sensor reporting rate defined as the number of data collections per unit time, and $\mathbb{E}[E_r]$ is the expected reporting energy consumed by all sensors in a randomly chosen data collection.

Proof. See (Chen and Zhao 2005) for details.

The law of lifetime given in Theorem 5.2.1 is proven based on the strong law of large numbers. It holds for all network applications under a general setting: arbitrary network architecture, arbitrary channel and radio models, and arbitrary definition of lifetime. Inspection of Eq. (5.1) reveals that reducing the expected reporting energy $\mathbb{E}[E_r]$ and the expected wasted energy $\mathbb{E}[E_w]$ leads to a prolonged network lifetime. This observation helps us identify two key physical layer parameters that affect network lifetime: channel state and residual energy. Specifically, to reduce $\mathbb{E}[E_r]$, a lifetime-maximizing protocol should exploit the channel state information (CSI) to prioritize sensors with better channels. On the other hand, to reduce $\mathbb{E}[E_w]$, a lifetime-maximizing protocol should exploit the residual energy information (REI) to favor sensors with more residual energies and thus balance the energy consumption across the network. Since channel states are independent of residual energies (the sensor with the better channel may have less residual energy), a lifetime-maximizing protocol needs to strike a balance between these two often conflicting objectives.

5.2.3 A General Design Principle For Lifetime Maximization

To obtain the optimal tradeoff between CSI and REI, we again resort to the law of lifetime. Consider first the expected reporting energy consumption $\mathbb{E}[E_r]$ in a randomly chosen data collection. As shown by Chen and Zhao (2005), $\mathbb{E}[E_r]$ can be obtained by averaging the expected reporting energy $\mathbb{E}[E_r(k)]$ consumed in the kth data collection over the randomly chosen data collection index K:

$$\mathbb{E}[E_r] = \mathbb{E}_K\{\mathbb{E}[E_r(K)]\}, \tag{5.2}$$

where $\mathbb{E}_K\{\cdot\}$ denotes the expectation over the randomly chosen data collection index K. Note that the probability mass function $\Pr\{K = k\}$ decreases with the data collection index k (Chen and Zhao 2005). This observation leads to the conclusion that the energy consumed at the early stage of network lifetime carries more weight. Thus, reducing the reporting energy consumption $\mathbb{E}[E_r(k)]$ in the kth data collection is crucial when k is small (i.e., when the network is young). On the other hand, the wasted energy E_w only depends on sensor residual energies when the network dies. Hence, maintaining small dispersion of sensor residual energies is only crucial when the network is approaching the end of its lifetime.

The above discussion suggests that a lifetime-maximizing protocol should be adaptive with respect to the age of the network. Specifically, lifetime-maximizing protocols should be more opportunistic by favoring sensors with better channels (focusing on reducing $\mathbb{E}[E_r]$) when the network is young and more conservative by favoring sensors with more residual energies (focusing on reducing $\mathbb{E}[E_w]$) when the network is old. We see here a connection between extending network lifetime and the retirement-planning strategy. When we are young, we can afford to be more aggressive, putting retirement savings to relatively more risky investments. As we age, we become more conservative. Since the law of lifetime holds under a general network setting, this general principle can be used to guide the design of various lifetime-maximizing protocols including MAC, routing as well as network configuration.

5.3 Fundamental Performance Limit: A Stochastic Shortest Path Framework

We now present an example of applying the law of lifetime to MAC design. As shown in Figure 5.1, we consider sensor networks with mobile APs which can be UAVs, UGVs, and

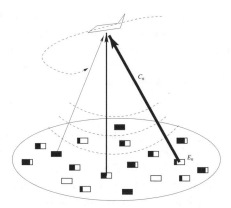

Figure 5.1 Sensor network with mobile access point. C_n: channel gain; E_n: residual energy.

robotics. Due to node redundancy and spatial correlation among sensor measurements, it is often sufficient to retrieve data from a fraction of sensors. Transmission scheduling is thus a key issue in sensor networks: which set of sensors should be chosen to transmit their measurements so that the network can carry out its task for the longest period of time.

In this section, we explore the fundamental performance limit of sensor transmission scheduling within the stochastic shortest path (SSP) framework. We show that the limiting performance on network lifetime is defined by the optimal policy of an SSP problem. We further show that the rich structure of the sensor scheduling problem leads to a polynomial-time solution to this SSP when the network is dense.

5.3.1 Problem Statement

Network model

Consider a network of N sensors. In each data collection initiated by the AP, N_0 ($1 \leq N_0 \leq N$) out of N sensors are chosen to transmit their measurements directly to the AP through a fading channel. The number N_0 of sensors required to transmit is determined by the underlying application and the QoS requirement of the network. For simplicity, we assume $N_0 = 1$. Extensions to $N_0 > 1$ are discussed in (Chen and Zhao 2007; Chen et al. 2007).

We assume that sensor measurement is encoded in a packet with fixed size. The channels between the AP and the sensors follow the block fading model with block length equal to the transmission time of one packet. That is, channel gains are independently and identically distributed (i.i.d.) across data collections (but not necessarily i.i.d. across sensors).

Energy model

We assume that sensors can adjust their transmission power according to the channel condition to ensure successful reception at the AP. Let W_n denote the energy required for sensor n to successfully transmit its packet to the AP in a data collection. Due to the presence of small-scale fading, the required transmission energy W_n is a random variable determined by the current channel state associated with sensor n. Since channel gains are i.i.d. across data collections, the required transmission energies $\mathbf{W}$ are i.i.d. across data collections. In general, the better the channel, the lower the required transmission energy. In practice, sensors can only transmit at a finite number L of power levels due to hardware and power limitations. Hence, the transmission energy W_n has realizations restricted to a finite set $\mathcal{W}$

$$\mathcal{W} \triangleq \{\varepsilon_1, \varepsilon_2, \ldots, \varepsilon_L\}, \quad 0 < \varepsilon_1 < \ldots < \varepsilon_L < \infty, \tag{5.3}$$

where ε_k is the energy consumed by a sensor in transmitting at the kth power level in a data collection.

Assume that each sensor is powered by a non-rechargeable battery with initial energy $\mathcal{E}_0$. Let E_n denote the residual energy of sensor n at the beginning of a data collection. Depending on the channel conditions and the sensor selections in previous data collections, the residual energy E_n of a sensor is a random variable. Since the transmission energy W_n is restricted to the finite set $\mathcal{W}$, the residual energy E_n takes values from a finite set $\mathcal{E}$:

$$\mathcal{E} \triangleq \left\{ e : e = \mathcal{E}_0 - \sum_{k=1}^{L} \alpha_k \varepsilon_k \geq 0 \text{ for some } \alpha_k \in \mathbb{Z} \text{ and } \alpha_k \geq 0 \right\}. \tag{5.4}$$

Lifetime definition

According to the required transmission energy W_n and the residual energy E_n at the beginning of a data collection, sensor n can be in one of the following states: active, inactive, and dead. Sensor n is considered active if it has enough energy for transmission in the current data collection, i.e., $E_n \geq W_n$. Sensor n is considered dead if its residual energy E_n drops below the minimum required transmission energy ε_1. In other words, it does not have enough energy for transmission under any channel condition. If sensor n has residual energy higher than ε_1 but insufficient for the current transmission ($\varepsilon_1 \leq E_n < W_n$), then it is considered inactive in the current data collection.

In each data collection, an active sensor is scheduled for transmission. If there is no active sensor in the network, this data collection is considered invalid. We define network lifetime $\mathcal{L}$ as the number of data collections until the number of dead sensors in the network reaches a certain threshold N_T ($1 \leq N_T \leq N$). We also assume that the network lifetime terminates when an invalid data collection occurs. This condition on lifetime allows us to ignore the tail portion of the network lifetime when sensors only have enough energy for exceptionally good channel states. In this case, data collection may suffer from large delay. All results presented below, however, can be extended straightforwardly without posing this condition on network lifetime.

5.3.2   SSP Formulation

We show that the problem of dynamically choosing which sensor to communicate with the AP for maximum network lifetime can be formulated as an SSP problem, a special class of Markov decision process (MDP) with non-discounted rewards and inevitable terminating states.

Network state space

The law of lifetime reveals the key role of residual energy and channel state in the design of upper layer protocols. We thus characterize the network state by the network energy profile $\mathbf{E} \triangleq (E_1, \ldots, E_N)$ and the sensor transmission energy requirement $\mathbf{W} \triangleq (W_1, \ldots, W_N)$. The state space $\mathcal{S}$ is given by

$$\mathcal{S} \triangleq \{(\mathbf{e}, \mathbf{w}) : \mathbf{e} \triangleq [e_1, \ldots, e_N] \in \mathcal{E}^N, \mathbf{w} \triangleq [w_1, \ldots, w_N] \in \mathcal{W}^N\}. \tag{5.5}$$

The size of the state space grows exponentially with network size N: $|\mathcal{S}| = M^N L^N$ where $M = |\mathcal{E}|$ and $L = |\mathcal{W}|$ denote, respectively, the number of possible residual energies and power levels.

The network enters a terminating state when its lifetime expires. According to the network lifetime definition, we define the set of terminating states $\mathcal{S}_t \subset \mathcal{S}$ as

$$\mathcal{S}_t \triangleq \{(\mathbf{e}, \mathbf{w}) : |\{n : e_n < \varepsilon_1\}| \geq N_T \text{ or } e_n < w_n, \forall n\}, \tag{5.6}$$

where $|\{n : e_n < \varepsilon_1\}| \geq N_T$ indicates that the number of dead sensors in the network reaches threshold N_T, and $e_n < w_n, \forall n$, indicates that there is no active sensor in the network, i.e., an invalid data collection occurs.

Action space

At the beginning of each data collection, an active sensor is chosen, based on the current network state $(\mathbf{e}, \mathbf{w})$, to communicate with the AP. The action space can thus be defined as

$$\mathcal{A}(\mathbf{e}, \mathbf{w}) \triangleq \{n : n \in \{1, \ldots, N\}, e_n \geq w_n\}. \tag{5.7}$$

Controlled Markovian dynamics

At the end of each data collection, the network transits to a new state according to the sensor selection and the channel states. Let $P_n(\mathbf{e}', \mathbf{w}' | \mathbf{e}, \mathbf{w})$ denote the probability that the network state transits from $(\mathbf{e}, \mathbf{w})$ to $(\mathbf{e}', \mathbf{w}')$ after sensor n is chosen. Since terminating states are absorbing, we have $P_n(\mathbf{e}, \mathbf{w} | \mathbf{e}, \mathbf{w}) = 1$ if $(\mathbf{e}, \mathbf{w}) \in \mathcal{S}_t$. In each data collection, only the residual energy of the scheduled sensor n changes according to its transmission energy while other sensors' residual energies remain the same. Hence, the transition probability after sensor n is chosen can be written as

$$P_n(\mathbf{e}', \mathbf{w}' | \mathbf{e}, \mathbf{w}) = p(\mathbf{w}') 1_{[\mathbf{e}' = \mathbf{e} - \mathbf{I}_n w_n]}, \quad \text{if } (\mathbf{e}, \mathbf{w}) \in \mathcal{S} \backslash \mathcal{S}_t, \tag{5.8}$$

where $\mathbf{I}_n$ is a $1 \times N$ unit vector whose nth element is 1, and $p(\mathbf{w}') = \Pr\{\mathbf{W} = \mathbf{w}'\}$ is the probability mass function (PMF) of the sensor transmission energy requirement $\mathbf{W}$, which is determined by the channel fading statistics for a given set $\mathcal{W}$ of transmission energy levels.

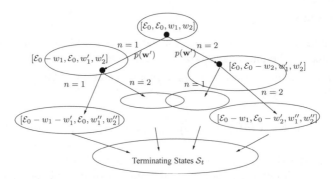

Figure 5.2 Stochastic shortest path formulation of sensor scheduling: $N = 2$.

Example: Figure 5.2 illustrates a network comprising of two sensors ($N = 2$). The network state transits from $(\mathcal{E}_0, \mathcal{E}_0, w_1, w_2)$ to $(\mathcal{E}_0 - w_1, \mathcal{E}_0, w_1', w_2')$ with probability $p(\mathbf{w}')$ if sensor $n = 1$ is chosen in the first data collection; it transits to state $(\mathcal{E}_0, \mathcal{E}_0 - w_2, w_1', w_2')$ with probability $p(\mathbf{w}')$ if sensor $n = 2$ is chosen. The bottom ellipse in Figure 5.2 indicates the set $\mathcal{S}_t$ of terminating states. The fact that sensor batteries have finite initial energy and that each transmission consumes non-zero energy implies that the network always reaches a terminating state in a finite but random time. The inevitable termination makes the sensor scheduling problem an instance of an SSP problem whose design objective is to maximize the total expected reward before a terminating state is reached.

Transmission reward

Maximizing the expected network lifetime is equivalent to assigning a unit reward to each data collection until the network dies after which no reward is earned. Accordingly, given network state $(\mathbf{e}, \mathbf{w})$, we define the instantaneous reward obtained in this data collection as

$$R(\mathbf{e}, \mathbf{w}) \triangleq 1_{[(\mathbf{e}, \mathbf{w}) \in \mathcal{S} \backslash \mathcal{S}_t]}, \tag{5.9}$$

where $1_{[x]}$ is the indicator function. Hence, the total reward accumulated until the network reaches a terminating state in $\mathcal{S}_t$ represents network lifetime.

SSP formulation

We have formulated the sensor transmission scheduling problem as an SSP. A transmission scheduling protocol is thus a policy π of this SSP. A policy π is given by a sequence of functions $\pi = \{\mu_1, \mu_2, \dots, \}$ where $\mu_k : \mathcal{S} \to \{1, \dots, N\}$ specifies the sensor chosen in the kth data collection. If μ_k is identical for all k, π is a stationary policy.

5.3.3   Fundamental Performance Limit on Network Lifetime

Let $V^* : \mathcal{S} \to \mathbb{R}$ denote the value function, where $V^*(\mathbf{e}, \mathbf{w})$ represents the maximum expected network lifetime (i.e., the maximum total expected reward of the SSP problem) starting from state $(\mathbf{e}, \mathbf{w})$. It has been shown that the value function is the unique solution to Bellman's optimality equation (Bertsekas 1995):

$$V^*(\mathbf{e}, \mathbf{w}) = R(\mathbf{e}, \mathbf{w}) + \max_{n \in \mathcal{A}(\mathbf{e}, \mathbf{w})} \left\{ \sum_{(\mathbf{e}', \mathbf{w}') \in \mathcal{S}} P_n(\mathbf{e}', \mathbf{w}' | \mathbf{e}, \mathbf{w}) V^*(\mathbf{e}', \mathbf{w}') \right\}. \tag{5.10}$$

Since terminating states are absorbing states with a zero reward, the maximum expected lifetime starting from a terminating state is zero, i.e., $V^*(\mathbf{e}, \mathbf{w}) = 0$ if $(\mathbf{e}, \mathbf{w}) \in \mathcal{S}_t$.

The fundamental performance limit on network lifetime can be obtained as

$$\mathbb{E}[\mathcal{L}^{opt}] = \sum_{\mathbf{w} \in \mathcal{W}} p(\mathbf{w}) V^*(\mathcal{E}_0 \mathbf{I}, \mathbf{w}), \tag{5.11}$$

where $\mathcal{E}_0$ is the initial energy of each sensor, and $\mathbf{I}$ is a $1 \times N$ vector of all ones. A stationary optimal transmission scheduling policy that achieves this limiting performance $\mathbb{E}[\mathcal{L}^{opt}]$ is given by

$$\mu(\mathbf{e}, \mathbf{w}) = \arg \max_{n \in \mathcal{A}(\mathbf{e}, \mathbf{w})} \left\{ \sum_{(\mathbf{e}', \mathbf{w}') \in \mathcal{S}} P_n(\mathbf{e}', \mathbf{w}' | \mathbf{e}, \mathbf{w}) V^*(\mathbf{e}', \mathbf{w}') \right\}. \tag{5.12}$$

Clearly, if we obtain the maximum expected lifetime $V^*(\mathbf{e}, \mathbf{w})$ for all network states, both the limiting performance $\mathbb{E}[\mathcal{L}^{opt}]$ and the optimal policy can be readily computed. The optimal design of transmission scheduling thus hinges on an efficient computation of the value function given in (5.10).

The value iteration algorithm is a widely used iterative procedure to solve the optimality equation (Bertsekas 1995, p.303). Specifically, we initialize the value iteration algorithm at $V_0(\mathbf{e}, \mathbf{w}) = 0$ for all $(\mathbf{e}, \mathbf{w}) \in \mathcal{S}$. In the kth iteration, we calculate value function $V_k(\mathbf{e}, \mathbf{w})$ for all non-terminating network states $(\mathbf{e}, \mathbf{w}) \in \mathcal{S} \backslash \mathcal{S}_t$ as

$$V_k(\mathbf{e}, \mathbf{w}) = R(\mathbf{e}, \mathbf{w}) + \max_{n \in \mathcal{A}(\mathbf{e}, \mathbf{w})} \left\{ \sum_{(\mathbf{e}', \mathbf{w}') \in \mathcal{S}} P_n(\mathbf{e}', \mathbf{w}' | \mathbf{e}, \mathbf{w}) V_{k-1}(\mathbf{e}', \mathbf{w}') \right\}. \tag{5.13}$$

It has been shown in (Bertsekas 1995) that the value iteration algorithm always converges, i.e.,

$$V^*(\mathbf{e}, \mathbf{w}) = \lim_{k \to \infty} V_k(\mathbf{e}, \mathbf{w}). \tag{5.14}$$

Unfortunately, it generally requires an infinite number of iterations to converge (Bertsekas 1995). Furthermore, in each iteration, the computational complexity (measured as the number of multiplications) is quadratic in the number $|\mathcal{S} \backslash \mathcal{S}_t|$ of non-terminating states and linear in the number $|\mathcal{A}(\mathbf{e}, \mathbf{w})|$ of actions (as can be seen from (5.13)). Hence, the complexity of computing the value function V^* is on the order of $N(LM)^{2N}$ per iteration, which increases exponentially with network size N.

5.3.4 Computing the Limiting Performance with Polynomial Complexity in Network Size

By exploiting the underlying structure of the sensor scheduling problem, we can significantly reduce the computational complexity of the value iteration algorithm. We first show that for the scheduling problem, the value iteration algorithm converges in one iteration. We then reduce the computational complexity of this iteration from exponential to polynomial in network size when the network is dense.

Acyclic transition graph

Due to the fact that the total residual energy in the network decreases after each data collection, we can obtain the value function in one iteration by calculating (5.10) in an increasing order of the network energy.

Proposition 5.3.1 *For any transmission scheduling policy, the transition graph of the under-lying Markov chain is acyclic. As a consequence, the value function V^* can be obtained in one iteration.*

Proof. See (Chen et al. 2007).

Sparse transition matrix

Next, we focus on reducing the computational complexity of Bellman's optimality equation (5.10) by reducing the size of the state space. We note from (5.8) that the transition matrix is sparse. Substituting (5.8) into (5.10), we obtain the optimality equation as

$$V^*(\mathbf{e}, \mathbf{w}) = R(\mathbf{e}, \mathbf{w}) + \max_{n \in \mathcal{A}(\mathbf{e}, \mathbf{w})} \left\{ \sum_{\mathbf{w}'} p(\mathbf{w}') V^*(\mathbf{e} - \mathbf{I}_n w_n, \mathbf{w}') \right\}. \tag{5.15}$$

Note that the summation in the curly parenthesis of (5.15) is taken over $|\mathcal{W}|^N = L^N$ transmission energy requirements while that of (5.10) is taken over $|\mathcal{S}| = (LM)^N$ network states. Hence, the computational complexity of the value function is reduced from $\mathcal{O}(N(LM)^{2N})$ to $\mathcal{O}(NL^{2N}M^N)$.

Uncontrollable non-correlated channel states

Recall that a network state $(\mathbf{e}, \mathbf{w})$ consists of two components: the network energy profile $\mathbf{e}$ and the required transmission energy $\mathbf{w}$. The transition of $\mathbf{e}$ is affected by the chosen action, but the transition of $\mathbf{w}$ is not since $\mathbf{w}$ is determined solely by channel fading statistics. Define a new value function $\hat{V} : \mathcal{E}^N \to \mathbb{R}$, where $\hat{V}(\mathbf{e})$ represents the maximum expected lifetime starting from network energy profile $\mathbf{e}$:

$$\hat{V}(\mathbf{e}) = \sum_{\mathbf{w}} p(\mathbf{w}) V^*(\mathbf{e}, \mathbf{w}). \tag{5.16}$$

Averaging (5.15) over all possible transmission energies $\mathbf{W}$, we obtain a modified Bellman's optimality equation for $\hat{V}$ as

$$\hat{V}(\mathbf{e}) = \sum_{\mathbf{w}} p(\mathbf{w}) \left\{ R(\mathbf{e}, \mathbf{w}) + \max_{n \in \mathcal{A}(\mathbf{e}, \mathbf{w})} \hat{V}(\mathbf{e} - \mathbf{I}_n w_n) \right\}. \tag{5.17}$$

Since the new value function $\hat{V}$ is executed over the space of $\mathbf{e}$, its computational complexity is $\mathcal{O}(NL^N M^N)$.

In terms of the new value function $\hat{V}$, the limiting performance is given by

$$\mathbb{E}[\mathcal{L}^{opt}] = \hat{V}(\mathcal{E}_0 \mathbf{I}), \tag{5.18}$$

and the corresponding optimal scheduling policy can be calculated as

$$\mu(\mathbf{e}, \mathbf{w}) = \arg \max_{n \in \mathcal{A}(\mathbf{e}, \mathbf{w})} \hat{V}(\mathbf{e} - \mathbf{I}_n w_n). \tag{5.19}$$

Invariance to sensor permutation

The above results hold for any channel distribution. We can further simplify the calculation of the maximum expected lifetime when the channel distribution satisfies certain conditions.

Suppose that the joint distribution $p(\mathbf{w})$ of the transmission energy requirements $\mathbf{W}$ is invariant to sensor permutations, i.e., $p(\mathbf{w}) = p(\tilde{\mathbf{w}})$ if $\tilde{\mathbf{w}}$ is a permutation of $\mathbf{w}$. Note that this condition is satisfied when the channel fading is i.i.d. across sensors. From (5.16), we can show that $\hat{V}(\mathbf{e}) = \hat{V}(\tilde{\mathbf{e}})$ if $\tilde{\mathbf{e}}$ is also a permutation of $\mathbf{e}$, or equivalently $\tilde{\mathbf{e}}$ and $\mathbf{e}$ have the same pattern. Hence, we only need to compute the value function $\hat{V}$ for different patterns of network residual energy profile $\mathbf{e}$ rather than all possible $\mathbf{e}$. Since the number of $\mathbf{e}$ patterns is polynomial $\mathcal{O}(N^{M-1})$ in network size, we reduce the complexity of computing the maximum expected lifetime from $\mathcal{O}(N L^N M^N)$ as in (5.17) to $\mathcal{O}(N^M L^N)$ with respect to the network size N.

Spatial aggregation

In dense networks, closely-spaced sensors may experience approximately the same channel fading. According to sensor locations, we can classify sensors into $\tilde{N} \ll N$ spatial clusters such that the transmission energy requirement is identical for all sensors within the same spatial cluster. As a consequence, the number of all possible transmission energy requirements is reduced from L^N to $L^{\tilde{N}}$. Assuming that the distribution of transmission energy requirements $\mathbf{W}$ is invariant to cluster permutations, we find that the computational complexity of the maximum expected lifetime can be reduced to $\mathcal{O}(N^M L^{\tilde{N}}) = O(N^M)$ in network size N if the number $\tilde{N}$ of clusters is independent of the network size N, which holds in dense networks deployed over fixed geographic areas.

Summary of complexity reduction

In Table 5.1, we summarize the computational complexity of the maximum expected lifetime. By exploiting the special structures of the sensor transmission scheduling problem, we show that the value iteration algorithm converges in one iteration. For dense networks deployed over fixed geographic areas, we reduce the computational complexity from exponential to polynomial in network size.

5.4 Distributed Asymptotically Optimal Transmission Scheduling

A direct implementation of the optimal scheduling policy (5.19) obtained from the SSP formulation requires global information of channel state and residual energy, resulting in large

Table 5.1 Computational complexity of the limiting lifetime performance.

Special structure	Iterations	Complexity per Iteration in N
General SSP problem	∞	$\mathcal{O}(N(LM)^{2N})$
Acyclic transition graph	1	$\mathcal{O}(N(LM)^{2N})$
Sparse transition matrix	1	$\mathcal{O}(NL^{2N}M^N)$
Uncontrollable non-correlated channel state	1	$\mathcal{O}(NL^NM^N)$
Invariance to sensor permutation	1	$\mathcal{O}(N^ML^N)$
Spatial aggregation	1	$\mathcal{O}(N^M)$

N: number of sensors, L: number of power levels, M: number of possible residual energies.

implementation overhead. In this section, we develop distributed scheduling algorithms that exploit local CSI and REI to reduce implementation overhead while retaining the benefit of cross-layer optimization. The basic idea is to allow each sensor to determine, based on its own channel state and residual energy, whether to transmit in each data collection.

5.4.1 Dynamic Protocol for Lifetime Maximization

To formulate the design of distributed transmission scheduling protocols, we introduce the concept of energy-efficiency index (Chen and Zhao 2007). At the beginning of a data collection, every sensor n is assigned with an energy-efficiency index γ_n, which is a function of its required transmission energy W_n and residual energy E_n:

$$\gamma_n = g(W_n, E_n), \tag{5.20}$$

where g is a real-valued function. The active sensor with the largest energy-efficiency index is then scheduled for transmission using the distributed opportunistic carrier sensing scheme (see Section 5.4.4 for details). The distributed transmission scheduling design is thus reduced to the design of the energy-efficiency index. For example, if $\gamma_n = -W_n$, we have a pure opportunistic protocol, which enables the active sensor with the least required transmission energy (i.e., the best channel). Similarly, $\gamma_n = E_n$ leads to a pure conservative protocol, which schedules the active sensor with the most residual energy.

We point out that it is possible to have a time-varying definition of the energy-efficiency index, i.e., $\gamma_n = g_k(W_n, E_n)$ where k denotes the kth data collection. Here, we focus on time-invariant function g for its ease of implementation. We will show that protocols defined by a time-invariant energy-efficient index can still be dynamic with respect to the age of the network.

Following the general design principle derived from the law of lifetime, we propose a dynamic transmission scheduling protocol that adaptively trades off CSI with REI according to the age of the network. Referred to as DPLM, the proposed protocol selects the active sensor whose current channel state demands the least portion of its residual energy for the transmission. The energy-efficiency index of DPLM is defined as

$$\gamma_n = \frac{E_n}{W_n}. \tag{5.21}$$

That is, the sensor that is able to transmit the most number of times under the current channel condition is scheduled for transmission in this data collection. Note that if the sensor with the largest energy-efficiency index defined in (5.21) is inactive, i.e., $\max_n\{\gamma_n\} < 1$, then no sensor in the network is active, resulting in an invalid data collection.

5.4.2 Dynamic Nature of DPLM

Before investigating the properties of DPLM in a general setting, let us first consider a simple example to gain some intuitions into the dynamic nature of DPLM. Consider a network with two sensors. Suppose that the network energy profile at the beginning of a data collection is given by $\mathbf{E} = (e_1, e_2)$. Without loss of generality, we assume that $e_1 > e_2$. The absolute dispersiveness between sensor residual energies is given by $\Delta = e_1 - e_2$. It can be readily shown from (5.21) that sensor 2, the one with less energy, is selected when

$$\gamma_2 > \gamma_1 \quad \Rightarrow \quad \frac{W_1 - W_2}{W_1} > \frac{\Delta}{e_1}. \tag{5.22}$$

Note that the required transmission energy W_n decreases as the channel condition improves. Hence, for a given difference Δ in the energy profile, the relative improvement in channel condition required for selecting the sensor that has less residual energy decreases with e_1, which can be considered as a measure of the network age since the total network energy is given by $2e_1 - \Delta$. Consider the following two extreme cases. When e_1 approaches infinity, we have $\lim_{e_1 \to \infty} \frac{\Delta}{e_1} = 0$ and the condition (5.22) reduces to $W_2 < W_1$. That is, when there is plenty of energy in the network (the network is young), DPLM acts like the pure opportunistic protocol by selecting the sensor with the best channel (i.e., the least required transmission energy). On the other hand, when e_1 approaches zero (the network is old), we have $\lim_{e_1 \to 0} \frac{\Delta}{e_1} = \infty$ and condition (5.22) holds with probability 0. DPLM puts more weight on the REI by selecting the sensor with the most residual energy (specifically, sensor 1). This dynamic nature of DPLM is analytically characterized in Property 1.

Property 1 (Dynamic Nature of DPLM) *DPLM dynamically trades off CSI with REI according to the network age measured by the total energy $\sum_{n=1}^{N} E_n$ in the network. Specifically, let $I^* = \arg\max_n \left\{ \frac{E_n}{W_n} \right\}$ denote the index of the sensor with the largest energy-efficiency index defined for DPLM. Given the network residual energy profile $\mathbf{E} = \mathbf{e}$ at the beginning of a data collection, we have, $\forall \epsilon > 0$,*

$$\Pr\{W_{I^*} = W_{\min} \mid \mathbf{E} = \mathbf{e}\} \leq \Pr\{W_{I^*} = W_{\min} \mid \mathbf{E} = \mathbf{e} + \epsilon\}, \tag{5.23a}$$

$$\Pr\{e_{I^*} = e_{\max} \mid \mathbf{E} = \mathbf{e}\} \geq \Pr\{e_{I^*} = e_{\max} \mid \mathbf{E} = \mathbf{e} + \epsilon\}, \tag{5.23b}$$

where $\Pr\{W_{I^*} = W_{\min} \mid \mathbf{E} = \mathbf{e}\}$ and $\Pr\{e_{I^*} = e_{\max} \mid \mathbf{E} = \mathbf{e}\}$ denote the conditional probabilities that sensor I^* has the least required transmission energy $W_{\min} \overset{\Delta}{=} \min\{W_n\}_{n=1}^{N}$ and the most residual energy $e_{\max} \overset{\Delta}{=} \max\{e_n\}_{n=1}^{N}$, respectively.

Proof. See (Chen and Zhao 2007) for details.

Property 1 shows that the probability of choosing the sensor with the best channel increases while the probability of choosing the sensor with the most residual energy decreases with the total energy in the network. In other words, when the network is young, DPLM is more likely to choose the sensor with the best channel to reduce the reporting energy. When the network grows old, DPLM becomes more conservative in order to reduce the wasted energy when the network dies.

5.4.3 Asymptotic Optimality of DPLM

It has been shown that the optimal transmission scheduling policy under the unconstrained formulation is the pure opportunistic scheme which enables the sensor with the best channel realization to transmit (Knopp and Humblet 1995; Zhao and Tong 2005). One would expect that the optimal transmission scheduling policy under the constrained formulation approaches the pure opportunistic scheme when the constraint on the initial energy becomes less restrictive, i.e., $\mathcal{E}_0 \to \infty$. In Property 2, we prove this statement and characterize the maximum rate at which the network lifetime increases with the initial energy $\mathcal{E}_0$. We then show in Property 3 that DPLM is asymptotically optimal. Specifically, in the asymptotic regime, DPLM approaches the pure opportunistic scheme and its relative performance loss as compared to the limiting performance diminishes.

Property 2 (Asymptotic Behavior of the Optimal Transmission Scheduling Protocol)
Assume that the channel states are i.i.d. across data collections and across sensors.

P2.1 *Under the unconstrained formulation, the expected reporting energy consumption of the pure opportunistic scheme in a data collection is given by*

$$\mathcal{E}_{\min} = \mathbb{E}\left[W_{\min}\right] \tag{5.24}$$

where $W_{\min} \overset{\Delta}{=} \min\{W_n\}_{n=1}^{N}$ is the minimum of N i.i.d. random variable $\{W_n\}_{n=1}^{N}$.

The optimal transmission scheduling policy in terms of network lifetime approaches the pure opportunistic scheme as the initial energy goes to infinity. Specifically, the asymptotic expected reporting energy consumption $\mathbb{E}[E_r^{opt}]$ of the optimal policy in a randomly chosen data collection is given by

$$\lim_{\mathcal{E}_0 \to \infty} \mathbb{E}[E_r^{opt}] = \mathcal{E}_{\min}. \tag{5.25}$$

P2.2 *The asymptotic rate at which the limiting performance $\mathbb{E}[\mathcal{L}^{opt}]$ increases with respect to the sensor initial energy $\mathcal{E}_0$ is given by*

$$\lim_{\mathcal{E}_0 \to \infty} \frac{\mathbb{E}[\mathcal{L}^{opt}]}{\mathcal{E}_0} = \frac{N}{\mathcal{E}_{\min}}. \tag{5.26}$$

Proof. See (Chen and Zhao 2007) for details.

The limiting performance $\mathbb{E}[\mathcal{L}^{opt}]$ provides a benchmark for all transmission scheduling policies (including centralized schemes). As seen from (5.12), to achieve the performance limit for a finite initial energy $\mathcal{E}_0$, we need global rather than local CSI and REI in each data collection, resulting in large implementation overhead. As shown in Property 3 and Section 5.4.4, DPLM provides a distributed solution to approaching the performance limit in the asymptotic regime.

Property 3 (Asymptotic Optimality of DPLM) *Assume that the channel states are i.i.d. across data collections and across sensors. In the asymptotic regime ($\mathcal{E}_0 \to \infty$),*

P3.1 *DPLM approaches the pure opportunistic scheme. Specifically, the expected reporting energy consumption* $\mathbb{E}[E_r^{DPLM}]$ *of DPLM in a randomly chosen data collection approaches* $\mathcal{E}_{\min}$ *as given in* (5.24):

$$\lim_{\mathcal{E}_0 \to \infty} \mathbb{E}[E_r^{DPLM}] = \mathcal{E}_{\min}. \tag{5.27}$$

P3.2 *DPLM is asymptotically optimal. Specifically, the relative performance loss of DPLM as compared to the limiting performance* $\mathbb{E}[L^{opt}]$ *diminishes with the initial energy:*

$$\lim_{\mathcal{E}_0 \to \infty} \frac{\mathbb{E}[\mathcal{L}^{opt}] - \mathbb{E}[\mathcal{L}^{DPLM}]}{\mathbb{E}[\mathcal{L}^{opt}]} = 0, \tag{5.28}$$

where $\mathbb{E}[\mathcal{L}^{DPLM}]$ *denotes the lifetime achieved by DPLM.*

Proof. See (Chen and Zhao 2007) for details.

We point out that other designs (e.g., the pure opportunistic scheme) of energy-efficiency index γ may also achieve asymptotic optimality. As shown in Figure 5.9, however, DPLM approaches the limiting performance for small initial energy $\mathcal{E}_0$.

5.4.4 Distributed Implementation

We briefly comment on the distributed implementation of DPLM using the opportunistic carrier sensing scheme first proposed in (Zhao and Tong 2003). The basic idea is to incorporate the local information (i.e., the energy-efficiency index) of each sensor into the backoff strategy of carrier sensing. This opportunistic carrier sensing scheme provides a distributed solution to the general problem of finding the global maximum or minimum.

At the beginning of each data collection, the AP broadcasts a beacon signal to activate and synchronize all sensors in the area of current interest. Upon receiving the beacon signal, each sensor estimates its channel state (using the beacon signal) and determines the required transmission energy W_n. Based on W_n and its residual energy E_n, each sensor calculates the predefined energy-efficiency index γ_n. Then, every active sensor maps its energy-efficiency index γ_n to a backoff time τ_n (see Figure 5.3) according to a predetermined common

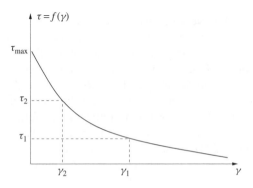

Figure 5.3 Opportunistic carrier sensing.

function $f(\gamma)$ and listens to the channel. An active sensor will transmit with its chosen backoff delay τ_n if and only if no one transmits before its backoff time expires. If $f(\gamma)$ is chosen to be a strictly decreasing function of the energy-efficiency index γ as shown in Figure 5.3, this opportunistic carrier sensing will ensure that the active sensor with the largest energy-efficiency index $\max\{\gamma_n\}_{n=1}^N$ seizes the channel.

5.4.5 Simulation Studies

We compare the performance of the proposed DPLM with the following four distributed transmission scheduling schemes: (1) the layered approach, which assumes that sensors are indistinguishable at the physical layer and randomly chooses an active sensor in each data collection; (2) the pure opportunistic protocol, which exploits solely CSI by choosing the active sensor with the best channel; (3) the pure conservative protocol, which uses only REI by choosing the active sensor with the most residual energy; and (4) the max-min protocol, a heuristic design proposed in (Chen and Zhao 2005), which jointly exploits CSI and REI by choosing the active sensor with the most residual energy after its transmission, i.e., it uses energy-efficiency index $\gamma_n = E_n - W_n$.

We assume perfect carrier sensing and ignore the energy consumed in carrier sensing, which is common to all protocols considered here. Normalizing the energy quantities by the received signal energy required to achieve the targeted SNR at the AP, we model the required transmission energy W_n as

$$W_n = \mathcal{E}_c + \mathcal{E}_{es} + \frac{1}{C_n}, \qquad (5.29)$$

where $\mathcal{E}_c = 0.01$ is the transmitter circuitry consumption, $\mathcal{E}_{es} = 0.001$ is the energy required for a sensor to estimate its channel realization, and C_n is the square of the fading amplitude. We assume that the channel state C_n follows an exponential distribution (i.e., Rayleigh fading) with normalized mean $\mathbb{E}[C_n] = 1$. The channel states are i.i.d. across data collections and across sensors. We also assume that sensors can transmit at continuous power levels. A sensor with residual energy e_n is considered dead if it does not have enough energy for transmission in 99.995% of the time, i.e., $\Pr\{e_n < \mathcal{E}_c + \mathcal{E}_{es} + \frac{1}{C_n}\} \geq 99.995\%$. The lifetime is defined as the number of data collections until any sensor in the network dies.

Impact of network size

We first study the expected network lifetime $\mathbb{E}[\mathcal{L}]$ as a function of the number N of sensors. As shown in Figure 5.4, the network lifetime $\mathbb{E}[\mathcal{L}]$ increases with N, but the rate at which $\mathbb{E}[\mathcal{L}]$ increases saturates. As expected, the layered approach, which ignores diversities at the physical layers of sensors, performs the worst. Transmission scheduling protocols exploiting CSI (such as the pure opportunistic scheme, the max-min scheme, and DPLM) outperform those without CSI (such as the layered approach and the pure conservative scheme). By jointly exploiting CSI and REI, the max-min protocol outperforms the pure opportunistic protocol when the number of sensors is large. It is, however, static to the network age as shown in (Chen and Zhao 2007). Adaptive to the network age, DPLM achieves the best performance, and its performance gain increases with the number of sensors.

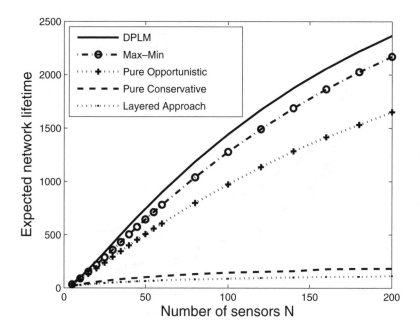

Figure 5.4 The expected network lifetime $\mathbb{E}[\mathcal{L}]$. $\mathcal{E}_0 = 5$.

In Figure 5.5, we investigate the expected energy $\mathbb{E}[W_{I*}]$ consumed by the chosen sensor I^* of different transmission scheduling protocols exploiting CSI and compare them with the asymptotic lower bound $\mathcal{E}_{\min}$ given in (5.24). Due to multiuser diversity (Knopp and Humblet 1995), the expected sensor energy consumption $\mathbb{E}[W_{I*}]$ in a randomly chosen data collection decreases with the number N of sensors. Not surprisingly, the pure opportunistic protocol, solely focusing on minimizing the transmission energy, performs the best in terms of $\mathbb{E}[W_{I*}]$ and achieves $\mathcal{E}_{\min}$ even when the initial energy $\mathcal{E}_0$ is small. As the initial energy $\mathcal{E}_0$ increases, the expected reporting energy $\mathbb{E}[W_{I*}^{\text{DPLM}}]$ of DPLM decreases and quickly approaches $\mathcal{E}_{\min}$, confirming P3.1. A small $\mathcal{E}_0$ that allows a sensor to transmit, on the average, only 10 times in its lifetime seems to be sufficient to bring $\mathbb{E}[W_{I*}^{\text{DPLM}}]$ close to $\mathcal{E}_{\min}$. We point out that the expected energy consumption of the pure opportunistic protocol under the constrained formulation may be larger than $\mathcal{E}_{\min}$ especially when N is small. This is because when the sensor with the best channel is inactive, the pure opportunistic protocol will have to choose an active sensor with a worse channel realization. DPLM, by balancing the energy consumption among sensors and thus enlarging the set of active sensors, can even outperform the pure opportunistic scheme in $\mathbb{E}[W_{I*}]$ when N is small. Compared to the pure opportunistic approach and DPLM, the max-min protocol performs the worst in terms of reporting energy consumption.

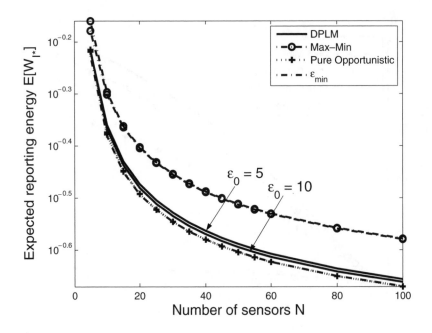

Figure 5.5 The expected energy $\mathbb{E}[W_{I^*}]$ consumed by the chosen sensor I^* in a randomly chosen data collection. $\mathcal{E}_0 = 5, 10$.

Figure 5.6 investigates the expected wasted energy $\mathbb{E}[E_w]$ of different transmission scheduling protocols. As the number N of sensors increases, $\mathbb{E}[E_w]$ of all protocols increases. The max-min protocol and DPLM offer significant reduction in the expected wasted energy as compared with the pure opportunistic and the pure conservative protocols. As the initial energy $\mathcal{E}_0$ increases, the expected wasted energies $\mathbb{E}[E_w]$ of the max-min protocol and DPLM remain almost the same while those of the pure opportunistic and the pure conservative protocols increase significantly. Combining Figures 5.5 and 5.6, we see that DPLM achieves the best balance between reducing reporting energy $\mathbb{E}[W_{I^*}]$ and reducing $\mathbb{E}[E_w]$; it consumes nearly minimum energy consumption $\mathcal{E}_{\min}$ per data collection without sacrificing $\mathbb{E}[E_w]$. The reason behind this desired property is the dynamic nature of DPLM as illustrated below.

Dynamic nature of DPLM

Figure 5.7 shows the expected dynamic range $\bar{\delta} = \mathbb{E}[\max\{E_n\}_{n=1}^N - \min\{E_n\}_{n=1}^N]$ of the network energy profile during the network lifetime. Since different transmission scheduling protocols may achieve different network lifetime, we normalize the data collection index by the expected network lifetime of the protocol. The expected dynamic range of the pure opportunistic scheme grows large toward the end of the network lifetime, resulting in its poor performance in terms of wasted energy $\mathbb{E}[E_w]$ as shown in Figure 5.6. The

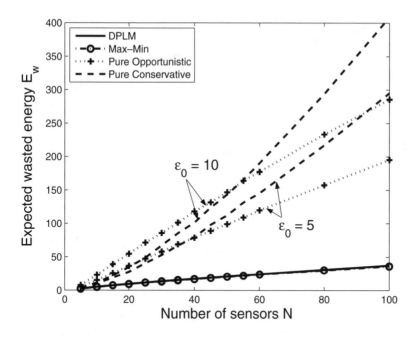

Figure 5.6 The expected wasted energy $\mathbb{E}[E_w]$. $\mathcal{E}_0 = 5, 10$.

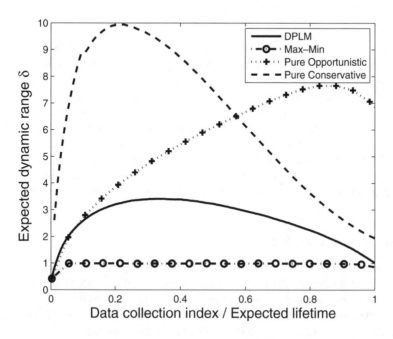

Figure 5.7 The expected dynamic range $\overline{\delta} = \mathbb{E}[\max\{E_n\}_{n=1}^N - \min\{E_n\}_{n=1}^N]$ of the network energy profile. $N = 10$, $\mathcal{E}_0 = 20$.

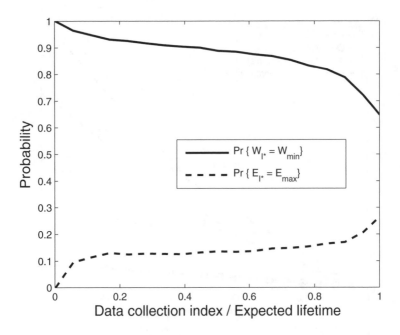

Figure 5.8 The probability that DPLM chooses the sensor with the least required transmission energy $W_{\min} = \min\{W_n\}_{n=1}^{N}$ and the probability that it chooses the sensor with the most residual energy $E_{\max} = \max\{E_n\}_{n=1}^{N}$. $N = 10$, $\mathcal{E}_0 = 20$.

dynamic range of the max-min protocol remains constant during the whole network lifetime. Adaptive to the network age, DPLM allows large variation in sensors' residual energies at the early stage of the lifetime (when reducing the transmission energy is more crucial) and brings down the dynamic range to as low as that of the max-min protocol toward the end of the lifetime (when balancing energy consumption among sensors becomes crucial). This explains how DPLM achieves nearly minimum transmission energy $\mathcal{E}_{\min}$ without sacrificing performance in wasted energy $\mathbb{E}[E_w]$.

Figure 5.8 further demonstrates the dynamic nature of DPLM. As the age of the network increases, the probability that DPLM selects the sensor with the best channel realization decreases while the probability of choosing the sensor with the most residual energy increases.

Asymptotic optimality of DPLM

The asymptotic optimality of DPLM in terms of using CSI was demonstrated in Figure 5.5. In Figure 5.9, we investigate the relative performance loss of the DPLM and the max-min protocols as compared to the asymptotic upper bound on network lifetime: $\frac{N\mathcal{E}_0}{\mathcal{E}_{\min}}$. We can see that as the initial energy $\mathcal{E}_0$ increases, the relative performance loss of DPLM approaches 0, which confirms P3.2. Moreover, its convergence rate is fast. For example, when the initial energy is $\mathcal{E}_0 = 10$, i.e., a sensor can transmit, on average, 10 times during its lifetime,

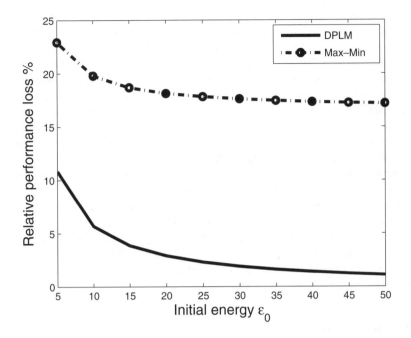

Figure 5.9 The asymptotic optimality of DPLM in network lifetime. $N = 50$.

the relative performance loss is as low as 6%. The max-min protocol, however, is not asymptotically optimal due to its static behavior during the network lifetime.

5.5 A Brief Overview of Network Lifetime Analysis

While numerous protocols and network deployment strategies have been proposed for network lifetime maximization (Chang and Tassiulas 2004; Chen et al. 2006a; Esseghir et al. 2005; Hou et al. 2003; Maric and Yates 2005; Mhatre et al. 2005; Pan et al. 2005; Park and Sahni 2006), analytical studies on network lifetime are few. As various elements affect network lifetime, including network architectures, applications as well protocols, exact lifetime characterization is notoriously difficult.

Upper bounds on lifetime have been obtained for a range of network setups. The most widely-studied network architecture is the flat ad hoc model where sensors collect data within their sensing ranges and then collaboratively relay the measurements to the AP. Without the knowledge of sensor and AP locations, Bhardwaj et al. (2001) and Hu and Li (2004) derive upper bounds on network lifetime based on the assumption that all sensor measurements are relayed via an optimal number of hops to the AP. While simple, these upper bounds can be loose since network topology has not been taken into account. Sparkled by the work of (Chang and Tassiulas 1999, 2000), network flow techniques have been used to bound the lifetime of networks with fixed and known topology (Bhardwaj and Chandrakasan 2002; Chang and Tassiulas 2004; Duarte-Melo et al. 2003; Giridhar and Kumar 2005; Kalpaki et al. 2002; Kansal et al. 2005). Meanwhile, a variety of other methods have been proposed to evaluate or bound network lifetime. For example, the reliability theory is applied to derive the lifetime distribution for both square-grid and hex-grid networks (Jain and Liang 2005). Based on stochastic dominance, lifetime upper bounds are derived in (Rai and Mahapatra 2005) for both linear and planar networks. Assuming Poisson distribution of sensor locations, Zhang and Hou (2004) bound network lifetime based on the theory of coverage processes. By calculating the lifetime of the set of sensors that can communicate directly to the AP, Zhu and Papavassiliou (2003) propose an iterative procedure to evaluate network lifetime. In (Blough and Santi 2002), the relation between lifetime and node density is explored when the network employs a cell-based geography-informed energy conservation scheme described in (Xu et al. 2001).

As compared to the flat ad hoc architecture, lifetime analysis of hierarchical and SENMA networks is scarce. Under the hierarchical architecture, network lifetime is estimated or bounded by optimally allocating energy to sensors (Duarte-Melo and Liu 2002) or placing the AP (Pan et al. 2005). Under the SENMA architecture where sensors communicate directly with the AP, network lifetime is analyzed within the SSP framework and obtained via the value iteration algorithm (Chen et al. 2006b, 2007). A lifetime upper bound is derived in (Arnon 2005) for SENMA networks employing code division multiple access (CDMA).

5.6 Conclusion

In this chapter, we have focused on the law of network lifetime and its application in MAC design. We have demonstrated that the optimal MAC design under an energy constraint should be based upon a physical layer model that captures diversities among sensors and be adaptive to the age of the network. We have shown that protocols dynamically exploiting dependencies between the MAC and the physical layers offer improved performance in network lifetime.

While the law of lifetime presented in this chapter applies to a wide range of network settings, it does not take into account of rechargeable sensors. When sensors can harvest energy from the environment or be recharged by the mobile APs, the fundamental law that governs network lifetime may change, so are the design principles. It is an interesting direction to establish the fundamental properties of rechargeable sensor networks and investigate the impact of the battery recharging rate on network lifetime.

Acknowledgement

This work was supported in part by the National Science Foundation under Grants CNS-0627090 and ECS-0622200, and by the Army Research Laboratory CTA on Communication and Networks under Grant DAAD19-01-2-0011.

Bibliography

Arnon S 2005 Deriving an upper bound on the average operation time of a wireless sensor network. *IEEE Communications Letters* **9**(2), 154–156.

Bertsekas DP 1995 *Dynamic Programming and Optimal Control,* vol. 1. Athena Scientific, Belmont, Massachusetts.

Bhardwaj M and Chandrakasan AP 2002 Bounding the lifetime of sensor networks via optimal role assignments. In *Proceedings of Annual Joint Conference of the IEEE Computer and Communications Societies (INFOCOM),* pp. 1587–1596, New York City.

Bhardwaj M, Garnett T and Chandrakasan AP 2001 Upper bounds on the lifetime of sensor networks. In *Proceedings of IEEE International Conference on Communications (ICC),* pp. 785–790, Helsinki, Finland.

Blough DM and Santi P 2002 Investigating upper bounds on network lifetime extension for cell-based energy conservation techniques in stationary ad hoc networks. In *Proceedings of ACM Annual International Conference on Mobile Computing and Networking (MOBICOM),* pp. 183–192, Atlanta, Georgia, USA.

Chang JH and Tassiulas L 1999 Routing for maximum system lifetime in wireless ad-hoc networks. In *Proceedings of the 37th Annual Allerton Conference on Communication, Control, and Computing,* Monticello, IL.

Chang JH and Tassiulas L 2000 Energy conserving routing in wireless ad-hoc networks. In *Proceedings of Annual Joint Conference of the IEEE Computer and Communications Societies (INFOCOM),* pp. 22–31, Tel Aviv, Israel.

Chang JH and Tassiulas L 2004 Maximum lifetime routing in wireless sensor networks. *IEEE/ACM Trans. on Networking* **12**(4), 609–619.

Chen Y and Zhao Q 2005 On the lifetime of wireless sensor networks. *IEEE Communications Letters* **9**(11), 976–978.

Chen Y and Zhao Q 2007 An integrated approach to energy-aware medium access for wireless sensor networks. *IEEE Transactions on Signal Processing,* vol. 55, no. 7, pp. 3429–3444, July 2007".

Chen Y, Chuah CN and Zhao Q 2006a Network configuration for optimal utilization efficiency of wireless sensor networks. *Elsevier Ad Hoc Networks.* doi:10.1016/j.adhoc.2006.09.001.

Chen Y, Zhao Q, Krishnamurthy V and Djonin D 2006b Transmission scheduling for sensor network lifetime maximization: A shortest path bandit formulation. In *Proceedings of IEEE International Conference on Acoustics, Speech, and Signal Processing (ICASSP),* pp. 145–148, Toulouse, France.

Chen Y, Zhao Q, Krishnamurthy V and Djonin D 2007 Transmission scheduling for optimizing sensor network lifetime: A stochastic shortest path approach. *IEEE Transactions on Signal Processing,* vol. 55, no. 5, pp. 2294–2309, May 2007".

Duarte-Melo E and Liu M 2002 Analysis of energy consumption and lifetime of heterogeneous wireless sensor networks. In *Proceedings of IEEE Global Telecommunications Conference (GLOBECOM),* vol. 1, pp. 21–25, Taipei, Taiwan.

Duarte-Melo EJ, Liu M and Misra A 2003 Lifetime bounds, optimal node distributions and flow patterns for wireless sensor networks. Technical Report TR 346, University of Michigan. www.eecs.umich.edu/techreports/systems/cspl/cspl-346.ps.gz.

Esseghir M, Bouabdallah N and Pujolle G 2005 Sensor placement for maximizing wireless sensor network lifetime. In *Proceedings of IEEE 62nd Vehicular Technology Conference (VTC-Fall),* vol. 4, pp. 2347–2351, Dallas, Texas.

Giridhar A and Kumar PR 2005 Maximizing the functional lifetime of sensor networks. In *Proceedings of Fourth International Symposium on Information Processing in Sensor Networks (IPSN),* pp. 5–12, Los Angeles, California.

Hou YT, Shi Y and Pan J 2003 A lifetime-aware single-session flow routing algorithm for energy-constrained wireless sensor networks. In *Proceedings of IEEE Military Communications Conference (MILCOM),* vol. 1, pp. 603–608, Boston, Massachusetts.

Hu Z and Li B 2004 On the fundamental capacity and lifetime limits of energy-constrained wireless sensor networks. In *Proceedings of IEEE Real-Time and Embedded Technology and Applications Symposium (RTAS),* pp. 2–9, Toronto, Canada.

Jain E and Liang Q 2005 Sensor placement and lifetime of wireless sensor networks: theory and performance analysis. In *Proceedings of IEEE Global Telecommunications Conference (GLOBECOM)*, vol. 1, pp. 173–177, St. Louis, Missouri.

Kalpaki K, Dasgupta K and Namjoshi P 2002 Maximum lifetime data gathering and aggregation in wireless sensor networks. Technical Report TR CS-02-12, University of Maryland, Baltimore County, Baltimore, Maryland.

Kansal A, Ramamoorthy A, Srivastava M and Pottie G 2005 On sensor network lifetime and data distortion. In *Proceedings of International Symposium on Information Theory*, pp. 6–10, Adelaide, Australia.

Knopp R and Humblet PA 1995 Information capacity and power control in single cell multi-user communications. In *Proceedings of International Conference on Communications (ICC)*, vol. 1, pp. 331–335, Seattle, WA.

Maric I and Yates RD 2005 Cooperative multicast for maximum network lifetime. *IEEE Journal on Selected Areas in Communications* **23**, 127–135.

Mhatre VP, Rosenberg C, Kofman D, Mazumdar R and Shroff N 2005 A minimum cost heterogeneous sensor network with a lifetime constraint. *IEEE Transactions on Mobile Computing* **4**, 4–15.

Pan J, Cai L, Hou YT, Shi Y and Shen SX 2005 Optimal base-station locations in two-tiered wireless sensor networks. *IEEE Transactions on Mobile Computing* **4**(5), 458–473.

Park J and Sahni S 2006 An online heuristic for maximum lifetime routing in wireless sensor networks. *IEEE Transactions on Computers* **55**(8), 1048–1056.

Rai V and Mahapatra RN 2005 Lifetime modeling of a sensor network. In *Proceedings of Design, Automation and Test in Europe (DATE)*, vol. 1, pp. 202–203, Munich, Germany.

Xu Y, Heidemann J and Estrin D 2001 Geography-informed energy conservation for ad hoc routing. In *Proceedings of ACM/IEEE International Conference on Mobile Computing and Networking (MobiCom)*, pp. 70–84, Rome, Italy.

Zhang H and Hou J 2004 On deriving the upper bound of α-lifetime for large sensor networks. In *Proceedings of ACM International Symposium on Mobile Ad Hoc Networking and Computing (MobiHoc)*, pp. 121–132, Roppongi, Japan.

Zhao Q and Tong L 2003 Quality-of-service specific information retrieval for densely deployed sensor network. In *Proceedings of IEEE Military Communications Conference (MILCOM)*, vol. 1, pp. 591–596, Boston, Massachusetts.

Zhao Q and Tong L 2005 Opportunistic carrier sensing for energy-efficient information retrieval in sensor networks. *EURASIP Journal on Wireless Communications and Networking* **2**, 231–241.

Zhu J and Papavassiliou S 2003 On the energy-efficient organization and the lifetime of multi-hop sensor networks. *IEEE Communications Letters* **7**, 537–539.

Part II

Signal Processing for Sensor Networks

6

Detection in Sensor Networks

Venugopal V. Veeravalli and Jean-François Chamberland

Detection is potentially a prominent application for emerging sensor network technologies. It often serves as the initial goal of a sensing system. Indeed, the presence of an object has to be ascertained before a sensor network can estimate attributes such as position and velocity. For systems observing infrequent events, detection may be the prevalent function of the network. Furthermore, in some applications such as surveillance, the detection of an intruder is the sole purpose of the sensor system. In the setting where local sensors pre-process observations before transmitting data to a fusion center, the corresponding decision-making problem is termed decentralized detection.

Decentralized detection with fusion was an active research field during the 1980s and early 1990s, following the seminal work of Tenney and Sandell (1981). The application driver for this research was distributed radar. That is, it was assumed that a set of radars observing the same event were positioned at various locations and their decisions needed to be fused at a command center. The high cost of data transfers at the time prompted system designers to quantize and compress data locally before information was relayed to the fusion center, hence the decentralized aspect of the problem. The goal was to design the sensor nodes and the fusion center to detect the event as accurately as possible, subject to an alphabet-size constraint on the messages transmitted by each sensor node. The reader is referred to Tsitsiklis (1993), Viswanathan and Varshney (1997), Blum et al. (1997), and to the references contained therein for a survey of the early work in this field.

More recently, decentralized detection has found applications in sensor networks. Wireless sensor nodes are typically subject to stringent resource constraints. To design an efficient system for detection in sensor networks, it is imperative to understand the interplay between data compression, resource allocation, and overall performance in distributed sensor systems. Classical results on inference problems, and on decentralized detection in

Wireless Sensor Networks: Signal Processing and Communications Perspectives   A. Swami, Q. Zhao, Y.-W. Hong and L. Tong
© 2007 John Wiley & Sons, Ltd

particular, can be leveraged and extended to gain insight into the efficient design of sensor networks. These results form a basis for much of the recent work on detection in sensor networks.

6.1 Centralized Detection

We begin this survey with a review of detection theory, by considering the centralized detection problem where all the sensor observations are available without distortion at the fusion center. In the Bayesian problem formulation, the probability of error at the fusion center is to be minimized; whereas in the Neyman-Pearson problem formulation, the probability of miss (type II error) is to be minimized, subject to a constraint on the probability of false alarm (type I error) (Kay 1998; Poor 1998; Trees 2001). In most of this chapter, we will consider Bayesian detection with the understanding that parallel developments are possible for the Neyman-Pearson formulation. The following example of centralized Bayesian detection, which illustrates some of the basic concepts, will be useful later in the chapter when we consider the decentralized setting.

Example 6.1.1 *Consider a detection problem where the fusion center must distinguish between two hypotheses, H_0 and H_1, based on L observations. Each observation consists of one of two possible signals, s_0 or s_1, corrupted by additive noise*

$$Y_\ell = s_j + N_\ell, \quad \ell = 1, \ldots, L. \tag{6.1}$$

The observation noise $\{N_\ell\}$ is assumed to be a sequence of independent and identically distributed (i.i.d.) Gaussian components with zero-mean and variance σ^2. This implies that the observed process $\{Y_\ell\}$, conditioned on the true hypothesis, is a sequence of i.i.d. Gaussian random variables. We can write the observations in vector form as

$$\underline{Y} = \underline{S}_j + \underline{N}, \tag{6.2}$$

where $\underline{Y} = (Y_1, \ldots, Y_L)^T$ is the aggregate information available for decision-making,

$$\underline{S}_j = (s_j, \ldots, s_j)^T = s_j \underline{1}^T \tag{6.3}$$

is the observed signal vector, and $\underline{N} = (N_1, \ldots, N_L)^T$ represents additive noise. Note that the observation vector $\underline{Y}$ is jointly Gaussian under either hypothesis. Since the two possible signal vectors are known, the covariance of $\underline{Y}$ is independent of the true hypothesis. The optimal procedure for deciding between the two hypotheses is a threshold rule on the log-likelihood ratio of the observation vector (Poor 1998). For known signals in Gaussian noise, an optimal detector can equivalently be implemented as a threshold test on the statistics

$$T(\underline{Y}) = \frac{1}{L} \sum_{\ell=1}^{L} Y_\ell = \frac{\underline{1}^T \underline{Y}}{L}. \tag{6.4}$$

The function $T(\underline{Y})$ is itself a Gaussian random variable. Its probability distribution is therefore completely determined by its mean and variance. For $\underline{S}_1 = -\underline{S}_0 = m\underline{1}$, we get

$$\mathrm{E}[T(\underline{Y})] = \pm m \tag{6.5}$$

where the leading sign is positive under hypothesis H_1 and negative under H_0. The variance of $T(\underline{Y})$ is given by

$$\mathrm{Var}(T(\underline{Y})) = \frac{\sigma^2}{L}. \tag{6.6}$$

If we further assume that the two hypotheses are equally likely, then the optimal decision threshold at the fusion center is $\tau = 0$. The performance of this threshold test on $T(\underline{Y})$ is characterized by the probability of error $P_e = Q(m\sqrt{L}/\sigma)$, where $Q(\cdot)$ is the complementary Gaussian cumulative distribution function

$$Q(x) = \int_x^\infty \frac{1}{\sqrt{2\pi}} e^{-\frac{\xi^2}{2}} d\xi. \tag{6.7}$$

In the Neyman-Pearson problem formulation, the optimal detection threshold is implicitly given by the equation $Q((m + \tau)\sqrt{L}/\sigma) = \epsilon$ where ϵ is the constraint on the type I error probability. The corresponding type II error probability is equal to $Q((m - \tau)\sqrt{L}/\sigma)$.

6.2 The Classical Decentralized Detection Framework

In the classical decentralized detection problem, a set of dispersed sensor nodes receives information about the state of nature H. Based on its observation, sensor node ℓ selects one of D_ℓ possible messages and sends it to the fusion center via a dedicated channel. Upon reception of the data, the fusion center produces an estimate of the state of nature by selecting one of the possible hypotheses. Evidently, a distributed sensor system in which every sensor node transmits a partial summary of its own observation to the fusion center is suboptimal compared to a centralized system in which the fusion center has access to the observations from all the sensors without distortion. Nonetheless, factors such as cost, spectral bandwidth limitations, and complexity may justify the use of compression algorithms at the nodes. Besides, in systems with a large number of sensors, unprocessed information could flood and overwhelm the fusion center, and a centralized implementation of the optimal detection rule may simply be unfeasible. A generic decentralized detection setting is illustrated in Figure 6.1.

Resource constraints in the classical framework are captured by fixing the number of sensor nodes and by further imposing a finite-alphabet constraint on the output of each sensor. This implicitly limits the amount of data available at the fusion center. The quantity of information provided to the fusion center by a network of L sensors, each sending one of D_ℓ possible messages, does not exceed

$$\sum_{\ell=1}^{L} \lceil \log_2 (D_\ell) \rceil \tag{6.8}$$

bits per channel use. Perfect reception of the sensor outputs is typically assumed at the fusion center. For applications such as decentralized detection and estimation, judicious signal processing at the nodes may enhance the overall performance of the sensor network. Nevertheless, it is important to recognize that once the structure of the information supplied by each sensor node is fixed, the fusion center faces a standard problem of statistical inference (Chair and Varshney 1988; Tsitsiklis 1993). As such, a likelihood-ratio test on

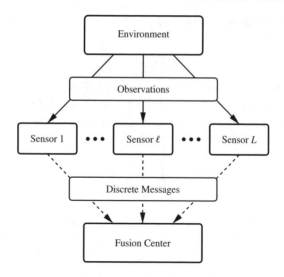

Figure 6.1 Abstract representation of the classical decentralized detection framework.

the received data will minimize the probability of error at the fusion center for a binary hypothesis testing problem, and a minimum mean-square estimator will minimize the mean-square error for an estimation problem. The crux of a standard decentralized inference problem is to determine what type of information each sensor should send to the fusion center.

Example 6.2.1 *Consider a distributed sensor network similar to the problem introduced in Example 6.1.1. This time, suppose that the observations* $\{Y_\ell\}$ *are only available at some remote sensor locations, and assume that each sensor must quantize its own observation to a single bit ($D_\ell = 2$ admissible messages). One possible quantization rule for the sensors is given by* $\gamma(y) = \mathbb{1}_{[0,\infty)}(y)$ *where* $\mathbb{1}(\cdot)$ *represents the indicator function. Accordingly, the information reaching the fusion center is of the form* $U_\ell = \gamma(Y_\ell)$. *When the two hypotheses* H_0 *and* H_1 *are equally likely, an optimal decision procedure at the fusion center for this special case is a majority rule on the received variables (Chair and Varshney 1988);* H_0 *is selected if*

$$\sum_{\ell=1}^{L} U_\ell < \frac{L}{2} \tag{6.9}$$

and H_1 *is picked otherwise. The probability of error at the fusion center is then given by*

$$P_e = \frac{1}{2} \sum_{k=0}^{\lceil L/2 \rceil} p^{L-k}(1-p)^k + \frac{1}{2} \sum_{k=0}^{\lfloor L/2 \rfloor} p^{L-k}(1-p)^k \tag{6.10}$$

where $p = Q(m/\sigma)$.

We note that the probability of error at the fusion center for the decentralized system is larger than the probability of error for the centralized system introduced in Example 6.1.1.

Furthermore, we emphasize that while the optimality of the decision rule at the fusion center is evident, it is more difficult to qualify the suitability of the local quantization rule at the sensors.

A celebrated accomplishment in decentralized detection for binary hypothesis testing is the demonstration that, for the classical framework, likelihood-ratio tests at the sensor nodes are optimal when the observations are conditionally independent, given each hypothesis (Tsitsiklis 1993). This property drastically reduces the search space for an optimal collection of local quantizers, and although the resulting problem is not necessarily easy, it is amenable to analysis in many contexts. The significance of this result is exemplified by the fact that the majority of the research on decentralized detection assumes that the observations are conditionally independent and identically distributed. In general, it is reasonable to assume conditional independence across sensor nodes if inaccuracies at the sensors are responsible for the noisy observations. However, if the observed process is stochastic in nature or if the sensors are subject to external noise, this assumption may fail. Without the conditional independence assumption, the problem of finding the optimal solution to the decentralized detection problem is computationally intractable (Tsitsiklis and Athans 1985).

Even under a conditional independence assumption, finding optimal quantization levels at the sensor nodes remains, in most cases, a difficult problem (Tsitsiklis and Athans 1985). This optimization problem is known to be tractable only under restrictive assumptions regarding the observation space and the topology of the underlying network. The solution does not scale well with the number of sensors except in some special cases, and it is not robust with respect to priors on the observation statistics.

A popular heuristic method to design decentralized detection systems is to apply a person-by-person optimization (PBPO) technique (Varshney 1996; Viswanathan and Varshney 1997). In this technique the decision rules are optimized one sensor at a time, while keeping the transmission maps of the remaining sensors fixed. The index of the sensor node being optimized is changed with every step. The overall performance at the fusion center is guaranteed to improve (or, at least, to not worsen) with every iteration of the PBPO algorithm. Specifically, in a Bayesian setting, the probability of error at the fusion center will be a monotone decreasing function of the number of PBPO iterations. Unfortunately, this algorithm does not necessarily lead to a globally optimal solution, and may only lead to a local minimum of the solution space. Other notable heuristics applicable to the design a decentralized detection system include the saddle-point approximation method (Aldosari and Moura 2007), and techniques based on empirical risk minimization and marginalized kernels (Nguyen et al. 2005). In contrast to the majority of the work on decentralized detection, the kernel method addresses system design for situations where only a collection of empirical samples is available; the joint distributions of the sensor observations conditioned on the possible hypotheses need not be known.

For wireless sensor networks with a small number of nodes, intuition regarding an optimal solution may be misleading. Consider a scenario where observations at the sensor nodes are conditionally independent and identically distributed. The symmetry in the problem suggests that the decision rules at the sensors should be identical, and indeed identical local decision rules are frequently assumed in many situations. However, counterexamples for which nonidentical decision rules are optimal have been identified (Blum and Kassam 1992; Cherikh and Kantor 1992; Tsitsiklis 1993; Zhu et al. 2000). Interestingly, identical decision rule are optimal for binary hypothesis testing in the asymptotic regime where the number of active sensors increases to infinity (Tsitsiklis 1988).

6.2.1 Asymptotic Regime

In view of the anticipated size of future sensor networks, we present an asymptotic regime where the number of sensors L and, possibly, the area covered by these sensors tend to infinity. For any reasonable collection of transmission strategies, the probability of error at the fusion center goes to zero exponentially fast as L grows unbounded. It is then adequate to compare collections of strategies based on their exponential rate of convergence to zero,

$$\lim_{L \to \infty} \frac{\log P_e (\mathcal{G}_L)}{L}. \tag{6.11}$$

The limiting value of (6.11) is sometimes referred to as the error exponent. Throughout, we use $\mathcal{G}_L$ as a convenient notation for a system configuration that contains L sensors. The following theorem describes how the class of transmission strategies with identical sensor nodes is optimal in terms of error exponent. Let Γ denote the collection of all the local decision rules that can be implemented at the sensors. Also, let $\mathbf{G}$ be the set of all finite subsets of Γ. For $G \in \mathbf{G}$, we use $G^{\mathbb{N}}$ to represent the set of system configurations of the form $\mathcal{G} = (\gamma_1, \ldots, \gamma_L)$, where L is an integer and every local decision rule γ_ℓ is an element of G. In other words, $G^{\mathbb{N}}$ is the set of all strategies with a finite number of sensor nodes where every node uses a local transmission map γ_ℓ contained in G. Likewise, let $\{\gamma\}^{\mathbb{N}}$ denote the set of strategies for which all the sensor nodes employ the same local decision rule γ.

Theorem 6.2.2 *Suppose that the observations are conditionally independent and identically distributed. Then using identical local decision rules for all the sensor nodes is asymptotically optimal;*

$$\inf_{G \in \mathbf{G}} \lim_{L \to \infty} \min_{\mathcal{G}_L \in G^{\mathbb{N}}} \frac{\log P_e (\mathcal{G}_L)}{L} = \inf_{\gamma \in \Gamma} \lim_{L \to \infty} \min_{\mathcal{G}_L \in \{\gamma\}^{\mathbb{N}}} \frac{\log P_e (\mathcal{G}_L)}{L}. \tag{6.12}$$

This theorem was originally proved by Tsitsiklis (1988) through an application of the Shannon-Gallager-Berlekamp lower bound (Gallager 1968). An alternative derivation can be obtained using the encompassing framework of large deviations (Chamberland and Veer-avalli 2004a; Chen and Papamarcou 1993). Asymptotic regimes applied to decentralized detection are convenient because they capture the dominating behavior of large systems. This leads to valuable insights into the problem structure and its solution. Design guidelines based on large deviations are expected to be applicable to all sufficiently large systems, including large wireless sensor networks. In practice, these guidelines are often found to provide good solutions, even for somewhat small systems.

6.3 Decentralized Detection in Wireless Sensor Networks

The classical decentralized detection framework has limited application to modern wireless sensor networks, as it does not adequately take into account important features of sensor

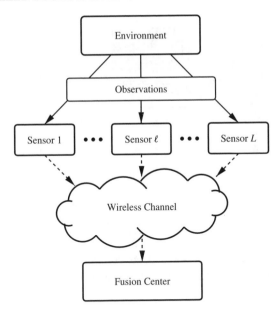

Figure 6.2 Abstract representation of an alternative decentralized detection framework.

technology and of the wireless links between the sensors and the fusion center. In particular, finite-alphabet restrictions on the sensor outputs do not capture the resource constraints of cost, spectral bandwidth and energy adequately for efficient design. Furthermore, the assumption that sensor messages are received reliably at the fusion center ignores the link variability intrinsic to wireless communications. In addition, the emphasis of the research on the classical problem has been on optimal solutions rather than scalable ones.

Reevaluating the original assumptions of the classical decentralized detection framework is an instrumental step in deriving valuable guidelines for the efficient design of sensor networks. Many recent developments in the field have been obtained by studying the classical problem while incorporating more realistic system assumptions in the problem definition. The motivation underlying many of these new research initiatives is the envisioned success of future wireless sensor networks. As such, an alternative theoretical framework tailored to decentralized detection over sensor networks is starting to emerge, as depicted in Figure 6.2. In the next section, we present interesting developments in the evolution of the decentralized detection problem formulation. We also identify important properties that seem to transcend most variations of the basic problem definition. To better understand good design strategies for distributed sensing, we present the principal constituents of a typical wireless sensor network along with their functions.

6.3.1 Sensor Nodes

Sensor nodes vary in cost and functionality. Yet their architectures and modes of operation are similar enough to draw useful conclusions about the constitution of a generic wireless sensor (Raghunathan et al. 2002). A generic node comprises four subsystems: a sensing unit,

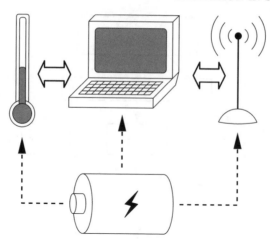

Figure 6.3 A generic wireless sensor node is composed of four subsystems: a sensing unit, a microprocessor, a communication unit, and a power supply.

a microprocessor, a communication unit, and a power supply (see Figure 6.3). The sensing unit links the sensor node to the physical world, whereas the microprocessor is responsible for controlling the sensor and processing its measurements. The microprocessor can be active or sleeping. In general, a more powerful microprocessor dissipates more power. Thus, the choice of a microprocessor should be dictated by the performance requirements of the intended application scenario, choosing the smallest microprocessor that fulfills these requirements. The communication unit, which enables the sensor node to exchange information with the fusion center and other nodes, has four distinct modes of operation: transmit, receive, idle, and sleeping. The detailed operation of the communication unit is somewhat involved. Its power consumption in transmit mode, for example, may depend on the data rate, the modulation scheme employed, and the transmission distance. Fortunately, the power consumption characteristics of the communication unit can be reduced to a few important considerations, which are discussed in more detail in Section 6.5.4.

6.3.2 Network Architectures

Network architectures for distributed sensor systems come in many different flavors. Carefully deployed systems usually form a tree, i.e., a network where nodes form a connected graph with no cycles (Tsitsiklis 1993). In a tree configuration, the information propagates from the sensor nodes to the fusion center in a straightforward manner, following a unique deterministic path. Communication overheads are therefore kept to a minimum. The parallel architecture, a subclass of the tree category where each node communicates directly with the fusion center, has received much attention in the decentralized detection literature. It is the preferred configuration to study the impact of quantization in decentralized detection as it provides a simple paradigm, easily amenable to analysis and simulations.

A distributed sensor system can also assume the form of a self-configuring wireless sensor network. In such systems, nodes are positioned randomly in an environment and then cooperate with one another to produce a dynamic communication infrastructure for the resulting network. The price paid for the greater flexibility of self-configuring networks is a much more involved communication mechanism, with substantial overheads. Communication aspects of self-configuring networks include topology management, node identification, and the choice of routing policies. In self-configuring wireless networks, nodes successively assume the roles of sensors, relays, and routers. The choice of a strategy for multi-hop routing between a source and its destination depends on the ultimate goal of the design. A reasonable abstraction for decentralized detection over sensor networks is one where the sensors local to an event of interest are used for sensing, and they transmit their information using a single hop or multiple hops to a fusion center. The other sensors in the system may be used as relays or routers. The fusion center is then responsible for final decision-making and further relaying of the information across the network if necessary.

6.3.3 Data Processing

Distributed sensing induces a natural tradeoff between performance, communication, and complexity. Combining information from neighboring nodes via in-network signal processing can improve reliability and reduce the amount of traffic on the network. On the other hand, the exchange of additional information could potentially yield better decisions. A simple technique to exchange information in the context of decentralized detection is proposed by Swaszek and Willett (1995). The authors explore the use of feedback, successive retesting, and rebroadcasting of the updated decisions as a means of reaching a consensus among sensors (also see Section 6.5.2). Two modes of operation are discussed: a fast mode where a decision is reached rapidly, and an optimum decision scheme that may require several rounds of information sharing before a consensus is reached. Both schemes lead to a unanimous decision at the nodes: the sensors never agree to disagree. These two schemes illustrate well the natural tradeoff between resource consumption and system performance, as the more intricate scheme performs better.

Under a different setting, in-network signal processing is studied by D'Costa et al. (2004). In their work, observations are assumed to possess a local correlating structure that extends only to a limited area. As such, the sensor network can be partitioned into disjoint spatial coherence regions over which the signals remain strongly correlated. The observations from different regions are assumed to be approximately conditionally independent. The resulting partitioning imposes a structure on the optimal decision rule that is naturally suited to the communication constraints of the network. Information is exchanged locally to improve the reliability of the measurements, while compressed data is exchanged among coherence regions. Under mild conditions, the probability of error of the proposed classification scheme is found to decay exponentially to zero as the number of independent node measurements increases to infinity.

6.4 Wireless Sensor Networks

A significant departure from the classical decentralized detection framework comes from the realization that wireless sensors transmit information over a shared medium, the common wireless spectrum. In a wireless environment, proximate nodes will be transmitting information over a multiple-access channel or, more generally, through a wireless network infrastructure. A problem formulation that better accounts for the physical resource constraints imposed on the system is needed for accurate performance evaluation. As discussed above, sensor nodes are often subject to very stringent power requirements. A limited spectral bandwidth and a bound on the total cost of the system may further exacerbate the design process. A flexible and adequate solution to distributed sensing should account for these important factors.

It is possible to extend the results of Theorem 6.2.2 to the case where system resources rather than the number of sensors constitute the fundamental design limitation. Let A be a global resource budget for a wireless sensor network. For instance, A may represent a sum-rate constraint, a total power requirement, a bound on system cost, or a combination thereof. We denote an admissible strategy for the total constraint A by $\mathcal{G}_A$. As before, using identical sensor nodes becomes optimal as the system grows larger (Chamberland and Veeravalli 2004a).

Theorem 6.4.1 *Suppose that the observations are conditionally independent and identically distributed. Then using identical transmission mappings for all the sensor nodes is asymptotically optimal,*

$$\inf_{G\in\mathbf{G}} \lim_{A\to\infty} \min_{\mathcal{G}_A\in G^{\mathbb{N}}} \frac{\log P_e(\mathcal{G}_A)}{A} = \inf_{\gamma\in\Gamma} \lim_{A\to\infty} \min_{\mathcal{G}_A\in\{\gamma\}^{\mathbb{N}}} \frac{\log P_e(\mathcal{G}_A)}{A}. \tag{6.13}$$

A necessary condition for this result to hold is that the number of sensor nodes must tend to infinity as the actual resource budget grows without bound. This is usually the case, as the amount of information provided by a single observation is bounded and, consequently, the amount of physical resources devoted to the corresponding sensor should also be finite. This theorem provides an extension to Theorem 6.2.2 and to the work by Tsitsiklis (1988). In the current formulation, the resource budget rather than the number of sensor nodes forms the fundamental constraint on the sensor system. Moreover, γ need not be a finite-valued function in this setting, and the communication channels between the sensor nodes and the fusion center need not be noiseless. The optimality of wireless sensor networks with identical sensor nodes is encouraging. Such networks are easily implementable, amenable to analysis, and provide robustness to the system through redundancy.

Asymptotic analyses based on error exponents also have the added benefit of decoupling the optimization across the sensors because the sensor mappings can be designed according to a local metric. For example, consider a Bayesian problem formulation where the probability of error at the fusion center is to be minimized. For wireless sensor networks with a large resource budget and conditionally i.i.d. observations, prospective sensor types should be compared according to their normalized Chernoff information

$$-\frac{1}{a(\gamma)} \min_{\lambda\in[0,1]} \left\{ \log E_{\mathcal{Q}_{0,\gamma}} \left(\frac{d\mathcal{Q}_{1,\gamma}}{d\mathcal{Q}_{0,\gamma}} \right)^{\lambda} \right\}, \tag{6.14}$$

where $a(\gamma)$ is the expected amount of system resources consumed by a node of type γ, and $\mathcal{Q}_{j,\gamma}$ is the induced probability measure on the received information $U_\ell = \gamma(Y_\ell)$ at the fusion center under hypothesis H_j. The normalized Chernoff information captures the tradeoff between resource consumption and information rendering in large sensor networks. Intuitively, allocating a larger amount of resources per node implies receiving detailed information from each node at the fusion center. On the other hand, for a fixed budget A, a reduction in resource consumption per node allows the system to contain more active sensors. The normalized Chernoff information describes in mathematical terms how this tradeoff takes place: Chernoff information divided by consumed resources. For example, doubling the Chernoff information provided by each sensor node results in the same gain in overall performance as reducing the resource consumption per node by half and doubling the number of nodes.

We can extend the preceding results to the Neyman-Pearson variant of the detection problem with little effort. In the latter problem formulation, the prior probabilities on H_0 and H_1 are unknown. The function $a(\gamma_\ell)$ then denotes the amount of resources consumed by sensor node ℓ under hypothesis H_0, and the global resource budget A is a constraint on the behavior of the system under hypothesis H_0. Again, one can show that using identical transmission mappings is asymptotically optimal as A tends to infinity. For $\epsilon \in (0, 1)$, let $\beta^\epsilon(\mathcal{G})$ represent the infimum type II error probability among all the decision tests such that the type I error probability $\alpha(\mathcal{G})$ is less than ϵ.

Theorem 6.4.2 *Using identical transmission mappings for all the sensor nodes is asymptotically optimal*

$$\inf_{G \in \mathbf{G}} \liminf_{A \to \infty} \min_{\mathcal{G}_A \in G^{\mathbb{N}}} \frac{\log \beta^\epsilon(\mathcal{G}_A)}{A} = \inf_{\gamma \in \Gamma} \liminf_{A \to \infty} \min_{\mathcal{G}_A \in \{\gamma\}^{\mathbb{N}}} \frac{\log \beta^\epsilon(\mathcal{G}_A)}{A}. \qquad (6.15)$$

In this case, the normalized relative entropy

$$\frac{1}{a(\gamma)} D\left(\mathcal{Q}_{0,\gamma} \| \mathcal{Q}_{1,\gamma}\right) \qquad (6.16)$$

plays the role of the normalized Chernoff information. Indeed, in the Neyman-Pearson framework, prospective sensor types for a sensor network with a large resource constraint should be compared according to the normalized relative entropy.

When the observations are not conditionally i.i.d., the normalized Chernoff information (or relative entropy) can no longer be shown to be the right metric for optimizing the sensor mappings. However, even in this case, the asymptotic results described in Theorems 6.4.1 and 6.4.2 can be used as some justification in order to decouple the optimization across sensors and choose each sensor mapping to maximize the normalized Chernoff information (for the Bayesian problem) or relative entropy (for the Neyman-Pearson problem). In this context, it is important to distinguish between the asymptotic results in the Bayesian and Neyman-Pearson formulations, in that in the latter formulation, the normalized relative entropy can be shown to be the right metric for optimizing the sensor mappings as long as the observations are conditionally independent and there are a large number of sensors of each type. The minimization over λ in (6.14) does not allow for a similar generalization in the Bayesian setting.

6.4.1 Detection under Capacity Constraint

The information theoretic capacity of a multiple-access channel is governed by its bandwidth, the signal power, and the noise power spectral density. More generally, the admissible rate-region of a practical system with a simple encoding scheme may depend on the bandwidth, the signal power, the noise density, and the maximum tolerable bit-error rate at the output of the decoder. Specifying these quantities is equivalent to fixing the sum-rate of the corresponding multiple-access channel. A natural initial approach to the capacity-constrained problem is to overlook the specifics of these physical parameters and to constrain the sum-capacity of the multiple-access channel available to the sensors. Specifically, a multiple-access channel may only be able to carry R bits of information per channel use. Thus, the new design problem becomes selecting L and D_ℓ to optimize system performance at the fusion center, subject to the capacity constraint

$$\sum_{\ell=1}^{L} \lceil \log_2 (D_\ell) \rceil \leq R. \tag{6.17}$$

For the time being, we ignore communication errors in the transmitted bits. Upon reception of the data, the fusion center makes a tentative decision about the state of nature H.

In the framework of Theorem 6.4.1, we have $A = R$ and $a(\gamma) = \lceil \log_2(D_\gamma) \rceil$. We know that using identical transmission functions for all the sensor nodes is asymptotically optimal. Moreover, a discrete transmission mapping γ^* is an optimal function if it maximizes the normalized Chernoff information,

$$\gamma^* = \arg\max_{\gamma} -\frac{1}{\lceil \log_2 (D_\gamma) \rceil} \min_{\lambda \in [0,1]} \left\{ \log \mathrm{E}_{\mathcal{Q}_{0,\gamma}} \left(\frac{d\mathcal{Q}_{1,\gamma}}{d\mathcal{Q}_{0,\gamma}} \right)^\lambda \right\}. \tag{6.18}$$

As an immediate corollary to this result, it can be shown that binary sensors are optimal if there exists a binary quantization function γ_b whose Chernoff information exceeds half of the information contained in an unquantized observation (Chamberland and Veeravalli 2003b, 2004a).

Corollary 6.4.3 *Let $\mathcal{P}_j$ be the probability measure of the observation Y_ℓ under hypothesis H_j, and assume that $\mathcal{P}_0$ and $\mathcal{P}_1$ are mutually absolutely continuous. If there exists a binary transmission mapping γ_b such that*

$$-\min_{\lambda \in [0,1]} \left\{ \log \mathrm{E}_{\mathcal{Q}_{0,\gamma_b}} \left(\frac{d\mathcal{Q}_{1,\gamma_b}}{d\mathcal{Q}_{0,\gamma_b}} \right)^\lambda \right\} \geq -\frac{1}{2} \min_{\lambda \in [0,1]} \left\{ \log \mathrm{E}_{\mathcal{P}_0} \left(\frac{d\mathcal{P}_1}{d\mathcal{P}_0} \right)^\lambda \right\}, \tag{6.19}$$

then having identical sensor nodes, each sending one bit of information, is asymptotically optimal.

Corollary 6.4.3 is not too surprising in itself. It asserts that if the contribution of the first bit of quantized data to the Chernoff information exceeds half of the Chernoff information offered by an unquantized observation, then using binary sensors is optimal. However, the significance of this result is that the requirements of the corollary are fulfilled for important classes of observation models (Chamberland and Veeravalli 2003b). In particular, binary sensor nodes are optimal for the problem of detecting deterministic signals in Gaussian noise, and for the problem of detecting fluctuating signals in Gaussian noise using a square-law detector. In these situations, having $L = R$ identical binary sensor nodes is asymptotically optimal. That is, the gain offered by having more sensor nodes outperforms the benefits of getting detailed information from each sensor.

This attribute can be generalized to a very important property that seems to be valid for a wide array of detection problems. In most detection settings, including the ones specified above, the number of bits necessary to capture most of the information contained in one observation appears to be very small. In other words, for detection purposes, the information contained in an observation is found in the first few bits of compressed data (Aldosari and Moura 2004; Lexa et al. 2004; Willett and Swaszek 1995). Several additional studies point to the fact that most of the information provided by an observation can be compressed to a few bits (Chen et al. 2004; Luo and Tsitsiklis 1994; Xiao and Luo 2005). The performance loss due to quantization decays very rapidly as the number of quantization levels increases. As such, message compression only plays a limited role in overall system performance. This property greatly simplifies quantizer design and system deployment. A second property worth mentioning at this point is that, for conditionally independent and identically distributed observations, the diversity obtained by using multiple sensors more than offsets the performance degradation associated with receiving only coarse data from each sensor (Chamberland and Veeravalli 2003b, 2004a).

6.4.2 Wireless Channel Considerations

Most of the early results on decentralized detection assume that each sensor node produces a finite-valued function of its observation, which is conveyed reliably to the fusion center. In a wireless system, this latter assumption of reliable transmission may fail as information is transmitted over noisy channels (Chen and Willett 2005; Duman and Salehi 1998). This limitation is made worse by the fact that most detection problems are subject to stringent delay constraints, thereby preventing the use of powerful error-correcting codes at the physical layer. Many recent research initiatives on decentralized detection consist in incorporating the effects of the wireless environment on the transmission of messages between the sensors and the fusion center. Unfortunately, to quantify the role of the wireless medium in sensor networks, very specific system assumptions must be made and the elegance and generality of the classical framework are somewhat lost. It is nonetheless possible to identify universal guidelines from such an analysis.

Example 6.4.4 *Again, consider a distributed sensor network akin to the one introduced in Example 6.1.1. However, suppose that the observations $\{Y_\ell\}$ are only available at the sensor nodes. Furthermore, assume that data must be transmitted over parallel wireless communication channels. The fusion center receives degraded information U_ℓ from sensor ℓ that is given by*

$$U_\ell = \gamma_\ell(Y_\ell) + W_\ell, \tag{6.20}$$

where W_ℓ is additive Gaussian noise with distribution $\mathcal{N}(0, \sigma_w^2)$. We study the simple situation where the additive noise is independent and identically distributed across sensor nodes. The hypothesis testing problem consists of deciding based on the received sequence $\{U_\ell\}$ whether the law generating $\{Y_\ell\}$ is $\mathcal{P}_0$ corresponding to hypothesis H_0, or $\mathcal{P}_1$ corresponding to hypothesis H_1. We focus on the specific detection problem where the wireless sensor network is subject to a total power constraint. That is, the expected consumed power summed across all the sensor nodes may not exceed a given constraint A,

$$\sum_{\ell=1}^{L} a(\gamma_\ell) \leq A \tag{6.21}$$

where $a(\gamma_\ell) > 0$ represents the expected power consumed by sensor node ℓ. This problem falls in the general framework of Theorem 6.4.1. Identical sensors are therefore optimal and system performance is maximized by using the normalized Chernoff information as the design criterion for the individual sensors.

For the purpose of illustration, we study the class of nodes where each unit retransmits an amplified version of its own observation. In this setup, a sensor node acts as an analog relay amplifier with a transmission mapping of the form $\gamma^{(s)}(y) = sy$. The transmission map $\gamma^{(s)}$ induces the following probability laws at the fusion center,

$$Q_{0,\gamma^{(s)}} \sim \mathcal{N}\left(-sm, s^2\sigma^2 + \sigma_w^2\right) \tag{6.22}$$

$$Q_{1,\gamma^{(s)}} \sim \mathcal{N}\left(sm, s^2\sigma^2 + \sigma_w^2\right). \tag{6.23}$$

The associated radiated power per node is given by

$$a\left(\gamma^{(s)}\right) = \mathrm{E}\left[s^2 y^2\right] = s^2 m^2 + s^2\sigma^2, \tag{6.24}$$

where the expectation is taken over the random variables Y and H. We can express the corresponding normalized Chernoff information as

$$-\frac{1}{a\left(\gamma^{(s)}\right)} \log\left(\int_{-\infty}^{\infty} \sqrt{Q_{0,\gamma^{(s)}}(u)Q_{1,\gamma^{(s)}}(u)}\,du\right) = \frac{m^2}{2\left(m^2 + \sigma^2\right)\left(s^2\sigma^2 + \sigma_w^2\right)}. \tag{6.25}$$

The normalized Chernoff information is a monotone decreasing function of the radiated power. Moreover, for any transmission mapping of the form $\gamma^{(s)}(y) = sy$, the asymptotic decay rate in the error probability is bounded above by

$$-\lim_{A\to\infty} \frac{\log P_e\left(\mathcal{G}_A\right)}{A} \leq \frac{m^2}{2\sigma_w^2\left(m^2 + \sigma^2\right)}. \tag{6.26}$$

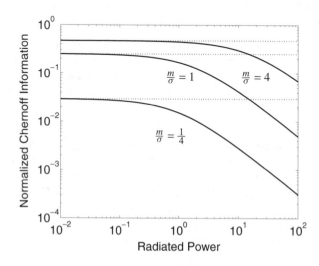

Figure 6.4 Normalized Chernoff information for analog relay amplifiers.

Table 6.1 System parameters for wireless sensor network.

Parameter Value	Description
$m^2/\sigma^2 \in \{1/16, 1, 16\}$	Signal-to-noise ratio at sensor
$a(\gamma) \in [10^{-2}, 10^2]$	Power radiated by sensor node
$\sigma_w^2 = 1$	Variance of communication noise

Figure 6.4 plots the normalized Chernoff information along with the corresponding upper bound for the transmission mapping $\gamma^{(s)}(y) = sy$ and the system parameters of Table 6.1.

Interestingly, although the problem definition of Example 6.4.4 constitutes a significant departure from the classical decentralized detection framework, a similar phenomenon is observed. Overall performance is optimized when the system uses as many independent sensors as possible, giving each sensor a minimum amount of system resources. A similar conclusion can be reached for a class of sensor nodes where each node compresses its own observation to a one-bit summary message.

Example 6.4.5 *We revisit the problem introduced in Example 6.4.4. This time, however, we adopt a different class of sensor nodes. We study the collection of nodes where each unit computes and sends a one-bit summary of its own observation to the fusion center. Binary sensors of this type are very common in the decentralized detection literature. We assume that the decision rule $\gamma^{(b)}$ employed by the nodes is a binary threshold function of the form*

$$\gamma^{(b)}(y) = \begin{cases} b & : y \geq 0 \\ -b & : y < 0 \end{cases}, \tag{6.27}$$

where $b > 0$. This binary decision rule produces the following probability measures at the fusion center,

$$\mathcal{Q}_{0,\gamma^{(b)}}(u) = \frac{1}{\sqrt{2\pi\sigma_w^2}} Q\left(\frac{m}{\sigma}\right) \exp\left(-\frac{(u-b)^2}{2\sigma_w^2}\right) \\ + \frac{1}{\sqrt{2\pi\sigma_w^2}} Q\left(-\frac{m}{\sigma}\right) \exp\left(-\frac{(u+b)^2}{2\sigma_w^2}\right) \tag{6.28}$$

$$\mathcal{Q}_{1,\gamma^{(b)}}(u) = \mathcal{Q}_{0,\gamma^{(b)}}(-u). \tag{6.29}$$

We note that the radiated power per sensor node is again independent of the prior probabilities on H_0 and H_1. It is given by $a\left(\gamma^{(b)}\right) = b^2$. The normalized Chernoff information can be computed as

$$-\frac{1}{b^2} \log\left(\int_{-\infty}^{\infty} \sqrt{\mathcal{Q}_{0,\gamma^{(b)}}(u)\mathcal{Q}_{1,\gamma^{(b)}}(u)} \, du\right). \tag{6.30}$$

Although (6.30) does not admit a closed form expression, it can easily be computed numerically. It is also possible to derive an upper bound for the Chernoff information of (6.30). First, we note from Figure 6.5 that the normalized Chernoff information is monotone decreasing in b. It follows that the normalized Chernoff information corresponding to transmission mapping $\gamma^{(b)}$ is bounded by its limiting value as b approaches zero,

$$\lim_{b\downarrow 0} -\frac{1}{b^2} \log\left(\int_{-\infty}^{\infty} \sqrt{\mathcal{Q}_{0,\gamma^{(b)}}(u)\mathcal{Q}_{1,\gamma^{(b)}}(u)} \, du\right) = \frac{1}{2\sigma_w^2}\left(Q\left(-\frac{m}{\sigma}\right) - Q\left(\frac{m}{\sigma}\right)\right)^2. \tag{6.31}$$

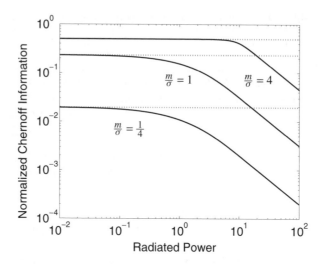

Figure 6.5 Normalized Chernoff information corresponding to wireless sensor nodes with binary local decision rule.

Figure 6.5 shows the normalized Chernoff information along with the corresponding upper bound for the transmission mapping of (6.27) and the system parameters of Table 6.1.

Once again, the tradeoff between the number of sensors and the amount of resources allocated to each sensor seems to favor large networks composed of many nodes. It is instructive to compare the transmission mappings introduced in the previous two examples. The analog sensor nodes perform better at low observation signal-to-noise ratio, whereas binary sensors are advantageous above a certain threshold signal-to-noise ratio (Chamberland and Veeravalli 2004a). This precludes an early dismissal of analog sensor nodes in favor of the more studied digital nodes. Indeed, for some detection applications, wireless sensor nodes with continuous transmission mappings may outperform sensor nodes with finite-valued transmission mappings. The use of a restricted class of sensors in Example 6.4.4 and Example 6.4.5 underscores the difficulty of finding an optimal transmission mapping for the sensors. Identifying the best possible transmission map involves a non-convex optimization problem over a space of measurable functions. Such problems are, in general, very difficult to solve. Restricting our attention to analog relay amplifiers or binary decision rules drastically reduces the solution space. An optimal design is therefore easily obtained. The more difficult problem of finding a transmission map that maximizes the normalized Chernoff information without additional constraints remains unsolved.

6.4.3 Correlated Observations

When sensor nodes are densely packed in a finite area, their observations are likely to become increasingly correlated. While the popular assumption that observations at the sensors are conditionally independent is convenient for analysis, it does not necessarily hold for arbitrary sensor systems. For instance, whenever sensor nodes lie in close proximity to one another, we expect their observations to become strongly correlated. Different approaches have been employed to study the latter problem. Willett et al. (2000) present a thorough analysis for the binary quantization of a pair of dependent Gaussian random variables. Their findings indicate that even in this simple setting, an optimal detector may exhibit very complicated behavior. Kam et al. (1992) examine the structure of an optimal fusion rule for the more encompassing scenario where multiple binary sensors observe conditionally dependent random variables. Blum and Kassam (1992) investigate the structure of an optimal detector when faced with weak signals and dependent observations. They also consider decentralized detection for dependent observations under a constant false-alarm rate criterion (Blum and Kassam 1995). Chen and Ansari (1998) propose an adaptive fusion algorithm for an environment where the observations and local decisions are dependent from one sensor to another. This adaptive approach requires the knowledge of only a few system parameters. Additional studies explore the effects of correlation on the performance of distributed detection systems (Aalo and Viswanathan 1989; Drakopoulos and Lee 1991). Blum (1996) provides a discussion of locally optimum detectors for correlated observations based on ranks. The numerical results contained in this work suggest that distributed detection schemes based on ranks and signs are less sensitive to the exact noise statistics when compared to optimum schemes based directly on the observations. This list is not intended to be exhaustive, but it offers an overview of previous work on decentralized detection with conditionally dependent observations.

Although conditional independence is a widely used assumption in the literature, it is likely to fail for dense networks. The theory of large deviations can be employed to assess the performance of wireless sensor systems exposed to correlated observations (Chamberland and Veeravalli 2006; Shalaby and Papamarcou 1994). In particular, the Gärtner-Ellis theorem and similar results from large-deviation theory have been successfully employed to assess the asymptotic performance of large, one-dimensional systems (Benitz and Bucklew 1990; Bercu et al. 1997; Bryc and Dembo 1997). For differentiating between known signals in Gaussian noise, overall performance improves with sensor density; whereas for the detection of a Gaussian signal embedded in Gaussian noise, a finite sensor density is optimal (Chamberland and Veeravalli 2006).

Example 6.4.6 *Consider the detection problem where observations become increasingly correlated as sensor nodes are placed in close proximity. Mathematically, we adopt the exact same system as in Example 6.4.4, except that the observation noise sequence $\{N_\ell\}$ is equivalent to the sampling of a 1-dimensional Gauss-Markov stochastic process. The covariance function of the observation noise is given by*

$$E[N_k N_\ell] = \sigma^2 \rho^{d(k,\ell)} \tag{6.32}$$

where $d(k, \ell)$ is the distance between sensors k and ℓ. When the sensors are equally spaced at a distance d, the best possible error exponent becomes (Chamberland and Veeravalli 2006)

$$-\lim_{A\to\infty} \frac{\log P_e(\mathcal{G}_A)}{A} \leq \frac{m^2}{2(m^2+\sigma^2)} \frac{(1-\rho^d)}{\sigma_w^2(1-\rho^d) + s^2\sigma^2(1+\rho^d)}. \tag{6.33}$$

Correlation degrades overall performance. Still, it is interesting to note that performance improves with node density. In particular, although correlation and observation signal-to-noise ratio affect overall performance, they do not necessarily change the way the sensor network should be designed. Specifically, systems with many low-power nodes will perform well for the detection of deterministic signals in Gaussian noise (Chamberland and Veer-avalli 2004b, 2006). There are situations where performance does not necessarily improves with node density. In the scenario where sensor nodes attempt to detect the presence of a stochastic signal in Gaussian noise, performance increases with sensor density only up to a certain point. Beyond this threshold, the performance starts to decay.

6.4.4 Attenuation and Fading

The performance of a sensor network depends on the nature of the wireless environment available to the sensor nodes. Wireless channels are generally prone to attenuation and fading. If sensor nodes are to be scattered around somewhat randomly, it is conceivable that their respective communication channels will feature different mean path gains, with certain nodes possibly having much better connections than others. Furthermore, changes in the environment, interference, and motion of the sensors can produce time-variations in the instantaneous quality of the wireless channels. It is then of interest to quantify the impact of fading on the performance of distributed sensor systems. At least two distinct cases are conceivable corresponding, respectively, to the situations where the channel state information is or is not available at the sensor nodes.

Having channel state information at the sensors permits the use of adaptive transmission mappings where a sensor decides what type of information to send based on the current quality of the channel. If the wireless channel is unreliable, then most of the available resources should be devoted to transmitting critical information. On the other hand, when the channel is in a good state, the sensor node can potentially use a more encompassing transmission map. Chen et al. (2004) modify the classical decentralized detection problem by incorporating a fading channel between each sensor and the fusion center. They derive a likelihood-ratio-based fusion rule for fixed local decision devices. This optimum fusion rule requires perfect knowledge of the local decision performance indices and the state of the communication channels over which messages are sent. Alternative fusion schemes that do not require as much side information are also proposed. A decision rule based on maximum-ratio combining and a two-stage approach inspired by the Chair-Varshney decision rule are analyzed. These concepts are further researched by Niu et al. (2006) for the scenario where instantaneous channel state information is not available at the fusion center. Acquiring channel state information may be too costly for a resource-constrained sensor network. It may also be impossible to accurately estimate the quality of a fast-changing channel. In their work, Niu et al. propose a fusion rule that only requires knowledge of the channel statistics. At low signal-to-noise ratio, the proposed fusion rule reduces to a statistic in the form of an equal-gain combiner; whereas at high signal-to-noise ratios, the proposed rule is equivalent to the Chair-Varshney decision procedure.

In decentralized problems, the most significant bit of a quantized observation seems to carry most of the information for the purpose of decision making. As such, it should be given more protection against noise and errors. This observation is supported by the fact that sending a one-bit message outperforms schemes where two bits of information are transmitted when the communication signal-to-noise ratio is low (Chamberland and Veeravalli 2004c). Conversely, at high signal-to-noise ratios, multiple bits of quantized data can be transferred to the fusion center.

Example 6.4.7 *Consider the simple scenario where sensor node ℓ has access to observation*

$$Y_\ell = s_j + N_\ell. \tag{6.34}$$

The observation noise $\{N_\ell\}$ is again a sequence of i.i.d. Gaussian components with zero-mean and variance σ^2. The signal s_j is equal to $-m$ under hypothesis H_0, and to m under hypothesis H_1. The sensor nodes transmit their information over wireless channels and the fusion center receives a noisy version of the data sent by the nodes,

$$U_\ell = \Theta_\ell \gamma_\ell(Y_\ell, \Theta_\ell) + W_\ell, \tag{6.35}$$

where W_ℓ is additive Gaussian noise with $\mathcal{N}\left(0, \sigma_w^2\right)$. The wireless connection Θ_ℓ is subject to Rayleigh fading:

$$f_\Theta(\theta) = 2\theta e^{-\theta^2}, \quad \theta \geq 0. \tag{6.36}$$

Based on the received data, the fusion center must choose one of the two possible hypotheses.

Since wireless sensor nodes are typically powered by small batteries, the performance and viability of sensor networks rely strongly on energy conservation. As such, a sensor node may be forced to transmit at very low power, thereby operating at levels where communication errors are non-negligible. At such levels, the choice of a signaling scheme may have a significant impact on the performance of the system. For illustrative purposes, we assume that the total energy budget per transmission is fixed and equal to a constant E. We consider the specific case where the observation at each sensor node is quantized to two bits. These bits are sent directly to the fusion center over a Rayleigh fading channel. Depending on the specific realization of its channel gain, sensor node ℓ decides how much energy should be allocated to the most significant bit and how much energy should be given to the second bit.

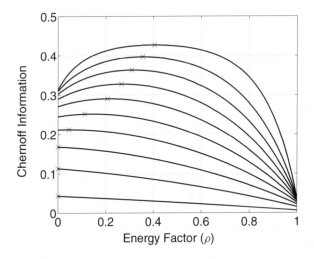

Figure 6.6 Chernoff information as a function of energy allocation for various channel gains. On the figure, the different functions correspond to increasingly favorable channel gains.

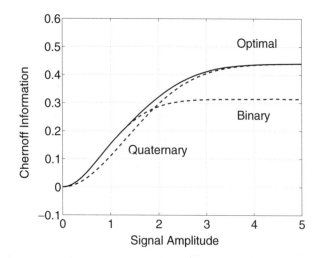

Figure 6.7 Chernoff information for binary signaling, quaternary signaling, and for the adaptive scheme where energy allocation is based on the state of the channel.

Figure 6.6 shows the Chernoff information as a function of energy allocation for different channel gains. The most significant bit has energy $E(1 - \rho)$, whereas the second bit has energy $E\rho$. As seen in Figure 6.6, the optimal energy allocation varies with channel gain. At low signal-to-noise ratio, most of the energy is given to the most significant bit; while at higher signal-to-noise ratio, energy is split between the two bits. The optimal operating points are marked with an $\times$. Figure 6.7 shows the Chernoff information as a function of received signal amplitude for a binary signaling scheme, a quaternary signaling scheme with uniform bit-energy, and the optimal allocation scheme of Figure 6.6. We gather from Figure 6.7 that using a fixed signaling scheme may result in a performance loss. For instance, a binary signaling scheme impairs performance when the signal-to-noise ratio is high. Correspondingly, quaternary signaling with uniform bit-energy underperforms at low signal-to-noise ratios. Channel state information at the sensor nodes increases overall performance by allowing for the adaptation of the signaling schemes of the individual sensor nodes based on the fading levels of their respective communication channels.

For encoded systems, this requirement entails using error-correcting codes with unequal bit protection. Numerical results suggest that Rayleigh fading only degrades performance slightly in well-designed systems. The quality of the random variables observed at the sensor nodes has a much greater impact on the probability of error at the fusion center than fading. The situation where the observations at the nodes or the gains of the wireless channels are not identically distributed across nodes is more difficult to address. Symmetry is lost and each node can then tailor its transmission function based on the distributions of its own observations and channel profile.

6.5 New Paradigms

Recently, researchers have started to explore new paradigms for detection over wireless sensor networks. These alternate points of view offer vastly different solutions to the problem of distributed sensing. The disparity among the proposed solutions can be explained in part by the perceived operation of future wireless sensor networks. Some researchers envision sensor networks to be produced as application-specific systems, giving the designer more freedom on how to best use resources. Under this assumption, every component of the network can be engineered anew. In particular, the communication infrastructure of the system and the modulation scheme employed by the sensors may be selected without constraints. Yet others believe that sensor networks will be subject to standard protocols and specifications, effectively imposing a rigid structure on the system. The exchange of information over future wireless networks may very well be governed by specifications similar to the Internet protocol suite (TCP/IP) or the Wi-Fi standard (IEEE 802.11). While more restrictive, the latter philosophy insures the inter-operability of heterogeneous network components and it allows for mass production and cost reduction. These aspects are key elements of the future success of wireless sensor technology according to many experts. The TinyOS community leads a commendable effort to create and maintain resources for generic wireless sensor nodes. TinyOS is an embedded operating system that is designed to incorporate rapid innovation and to operate within the severe constraints inherent to wireless sensor technology. It is intended to run on many academic and proprietary sensor node implementations. This may eventually be a catalyst for a dominating standard.

6.5.1 Constructive Interference

Adopting the application-specific viewpoint, Mergen and Tong (2005b) have proposed communication schemes for decentralized detection where nodes take advantage of the physical layer to transmit information efficiently and reliably. Their communication paradigms exploit the intrinsic broadcast nature of the wireless medium. When complete channel state information is available at the sensor nodes, it is possible for signals originating from various nodes to interfere constructively at the receiver through beam-forming. Recall that for binary detection with conditionally independent observations, an optimal decision-rule at the fusion center is a threshold test on the sum of individual likelihood functions. In a wireless environment, the superposition of multiple signals is equivalent to adding their amplitudes. This property can therefore be employed to sum the local likelihood ratios produced by individual sensors through the wireless channel. The fusion center can then make a final decision by applying a threshold test on the amplitude of the received aggregate signal. In this model, the wireless medium is used both to communicate information to the fusion center and to add signals coherently. This greatly reduces the spectral bandwidth requirement for the system. Physical-layer schemes are found to be asymptotically optimal as the number of sensors increases, provided that the channels from the sensors to the fusion center are statistically identical.

In related research initiatives, Mergen and Tong (2005a, 2006) propose communication schemes in which sensor nodes transmit according to the type of their observations. This strategy can be applied to decentralized parameter estimation and decentralized detection alike. The type-based multiple-access schemes lead to significant gains in performance when compared to the conventional architecture allocating orthogonal channels to the sensors. A similar concept is studied by Liu and Sayeed (2004). They consider a source-channel mapping in which every node uses the same encoder. Again, the proposed technique exploits the shared structure of the multiple-access channel to transmit data to the fusion center. The information from the sensors is collectively embedded in the conditional mean of the received signals. Based on the level of the received signal, the fusion center is able to make a decision. The authors show that the error exponent associated with a type-based multiple-access approach in Bayesian hypothesis testing coincides with the error exponent for the centralized setting where the fusion center has direct access to the observations. These observations are related to the pioneering work of Ahlswede and Csiszar (1986), where hypothesis testing based on the type of a sequence of observations is shown possible for vanishing average transmission rates. The broadcast nature of the wireless channel is exploited in a similar fashion by Hong and Scaglione (2005); Hong et al. (2005). In their work, the authors take advantage of the additive nature of a wireless multiple-access channel to address the problems of synchronization, cooperative broadcast, and decision making in sensor networks.

Under suitable channel conditions, constructive interference techniques over multiple-access channels provide an interesting solution to the problems of decentralized detection and decentralized parameter estimation. However, certain technical issues such as synchronization need to be addressed before such techniques can be exploited effectively. Furthermore, authentication and encryption cannot readily be applied to transmissions while using constructive interference.

6.5.2 Message Passing

A second paradigm that may reduce the need for spectral bandwidth is based on local message passing. In the message-passing approach, there is no designated fusion center and the goal is for all the sensors involved in the decision process to reach a consensus about the state of their environment. Every sensor possesses the same prior probability distribution about the true hypothesis and they share a common objective. They update their tentative decision whenever they make a new observation or when they receive additional information from a neighboring wireless sensor. Upon computing a new tentative decision, to a randomly selected subset of neighbors. Sensors can exchange messages in a synchronous or asynchronous manner until consensus is reached. The design problem is to find communication protocols for communication between the sensors that results in an agreement in a reasonable time. This should be achieved while respecting the constraints imposed on the communication structure and on the system resources. Conditions for asymptotic convergence of the decision sequence made by each sensor and for asymptotic agreement among all the wireless nodes are of interest.

This line of work is heavily influenced by the pioneering work of Borkar and Varaiya (1982), and the companion work by Tsitsiklis and Athans (1984) on distributed estimation. To facilitate analysis, the sensor network is modeled as a graph that represents the connectivity of various nodes. The reach of a wireless sensor is primarily limited by the power of its antenna. As such, sensors are often assumed to be linked only to proximate neighbors.

Pearl's belief propagation algorithm (Pearl 1988) is studied by Alanyali and Saligrama as a possible mechanism for the local exchange of informative data (Szymanski and Yener 2004). In message-passing, data generally take the form of a node's conditional marginal probability distribution over the possible hypotheses. Alanyali and Saligrama identify conditions under which sensors reach a consensus. They also discuss circumstances that force this consensus to be equivalent to the decision of a centralized maximum a posteriori detector. Advantages of the message-passing paradigm include a simple communication infrastructure, scalability, robustness to sensor failures, and a possible efficient use of the limited system resources.

6.5.3 Cross-Layer Considerations

In the previous section, we presented local message-passing as a way to mitigate the effects of path loss and fading in wireless communications. Another way is for the nodes to exploit a multi-hop communication scheme where data packets are relayed from sensor to sensor until they reach their respective destinations. Although a multi-hop strategy necessitates more transmissions, the non-linear attenuation intrinsic to wireless channels insures overall savings (Chamberland 2005; Kawadia and Kumar 2005).

If the data generated by the sensor nodes are to be conveyed over a multi-hop packet network, a few key observations are in order. In the context of decentralized detection, several studies point to the fact that most of the information provided by an observation can be compressed to a very few bits (Chamberland and Veeravalli 2004a; Chen et al. 2004; Luo 2005; Luo and Tsitsiklis 1994). Accordingly, the performance loss due to quantization decays rapidly as the number of information bits per transmission increases. Data packets carrying sensor information can then be assumed to contain only a few bits without much loss of generality. The exact number of bits per packet is unlikely to be a significant factor in energy consumption in view of the operations that take place at the onset of a wireless connection (Rappaport 2001; Tong et al. 2004), and also taking into consideration the size of a typical packet header (Stevens 1993). A similar phenomenon can be observed in other real-time applications such as voice-over IP, interactive games, and instant messaging. The payload of a packet in these situations is nearly of the same size or even smaller than its header. It is therefore safe to assume that once a communication link is established between two sensor nodes, the information content of an observation can be transferred essentially unaltered. This characteristic leads to an all-or-nothing model for data transmission akin to the one put forth by Rago et al. (1996), which is discussed in more detail in the following section.

6.5.4 Energy Savings via Censoring and Sleeping

As seen in Section 6.3, a generic sensor node comprises four subsystems: a sensing unit, a microprocessor, a communication unit, and a power supply. Once the components of a sensor node are fixed, the only way to reduce the average power consumption at the node is to shut off some of its units periodically. Assuming that the sensing unit is coupled to the microprocessor and that the operation of the communication unit is contingent on the microprocessor being active, a wireless sensor node has three broad modes of operation. It can be active, with all of its units powered up. Alternatively, it can be in mute mode with its communication unit off, effectively isolating itself temporarily from the rest of the

Table 6.2   Modes of operation.

Modes of operation	Microprocessor	Communication unit
Active	on	on
Mute	on	off
Sleeping	off	off

network. Finally, it can be sleeping with all of its units shut. The three modes of operation of a sensor node are summarized in Table 6.2. In most sensor networks, substantial energy savings may be achieved by having nodes communicate with the rest of the network only when necessary. For example, it may be best for a node to avoid sending data when the information content of a transmission is small (Appadwedula et al. 2005; Jiang and Chen 2005; Rago et al. 1996; Schurgers et al. 2002). While censoring sensors is a straightforward scheme to save energy, a less intuitive one consists in shutting off the sensor node completely whenever the information content of its next few observations is likely to be small.

Sensor nodes can take advantage of past observations and a priori knowledge about the stochastic processes they are monitoring to save energy and enhance performance. A small hit in performance can result in considerable energy savings for a decentralized detection system. For example, a minimal increase in expected detection delay can more than double the expected lifetime of the sensor node (Chamberland and Veeravalli 2003a). This result provides support for control policies in which wireless sensor nodes enter long sleep intervals whenever the information content of the next few observations is likely to be small. Conceptually, the sensor node uses a priori knowledge about the process it is monitoring together with its current and past observations to reduce energy consumption. More specifically, when the event of interest becomes very unlikely, sensor nodes can afford to go to sleep for an extended period of time, thus saving energy. On the other hand, when in a critical situation, sensor nodes must stay awake.

6.6   Extensions and Generalizations

The detection problems described thus far are static problems in which the sensors receive either a single observation or a single block of observations and a binary decision needs to be made at the fusion center. Many extensions and generalizations of this formulation are possible.

The case of M-ary detection (with $M > 2$) is of interest in many applications. For example, we may be interested in not simply detecting the presence of an object but classifying it into one of several categories. Extensions to M-ary detection of the classical decentralized detection problem are discussed by Sadjadi (1986); Tsitsiklis (1988). More recent work in the context of modern sensor networks is described in, e.g., D'Costa et al. (2004); Wang et al. (2005).

In the dynamic setting, each sensor receives a sequence of successive observations and the detection system has the option of stopping at any time and make a final decision, or to continue taking observations. The simplest problem in this setting is that of decentralized binary sequential detection. A decentralized version of binary sequential detection, where sensors make final decisions (linked through a common cost function) at different stopping times is studied in Teneketzis and Ho (1987); Veeravalli et al. (1994b). *Ad hoc* fusion of sequential decisions made at the sensors is considered by Hussain (1994). A more general formulation of the fusion problem was introduced by Hashemi and Rhodes (1989), and a complete solution to this problem was given in Veeravalli (1999); Veeravalli et al. (1994a).

A different binary sequential decision-making problem that first arose in quality control applications is the change detection problem. Here the distribution of the observations changes abruptly at some unknown time, and the goal is to detect the change 'as soon as possible' after its occurrence, subject to constraints on the false alarm probability. A decentralized formulation of the change detection problem is considered by Crow and Schwartz (1996); Teneketzis and Varaiya (1984) with the sensors implementing individual change detection procedures. *Ad hoc* schemes for fault detection with multiple observers are proposed by Wang and Schwartz (1994). A general formulation of decentralized change detection with a fusion center making the final decision about the change is given in Veeravalli (1999, 2001).

The design of optimal decision rules for decentralized detection problems is based on the assumption that the probability distributions of the sensor observations (under each hypothesis) are known. In many applications, however, the distributions of the sensor observations are only specified as belonging to classes which are referred to as uncertainty classes. The problem here is to design decision rules that are robust with respect to uncertainties in the distributions. A common approach for such a design is the minimax approach where the goal is to minimize the worst-case performance over the uncertainty classes. Extensions of the minimax robust detection problem to the decentralized setting are discussed in Geraniotis and Chau (1990); Veeravalli et al. (1994a). Alternatives to robust detection when partial information is available about the distributions, include composite testing based on generalized likelihood ratios, locally optimal testing for weak signals, and nonparametric detection (Poor 1998). For a discussion of some of these approaches and their relationship to censoring tests, see Appadwedula et al. (2007).

6.7   Conclusion

Detection problems provide a productive starting point for the study of more general statistical inference problems in sensor networks. In this chapter we reviewed the classical framework for decentralized detection and argued that while this framework provides a useful basis for developing a theory for detection in sensor networks, it has serious limitations. In particular, the classical framework does not adequately take into account important features of sensor technology and of the communication link between the sensors and the fusion center. We discussed an alternative framework for detection in sensor networks that has emerged over the last few years. Several design and optimization strategies may be gleaned from the alternative framework, including:

- The jointly optimum solution for the sensor mappings and fusion rule is complicated and does not scale well with the number of sensors. The focus should therefore be on good suboptimum solutions that are robust and scalable.

- In the regime of large numbers of sensors, softening the optimization metric to maximize error exponents rather than minimizing error probabilities can lead to scalable, tractable solutions.

- A key to obtaining scalable solutions is a decoupling of the optimization problem across the sensors, i.e., the sensor mappings are chosen to optimize local metrics.

- The number (density) of sensors should be considered a system design parameter that needs to be optimized before deployment. This is particularly important when the sensor observations are conditionally correlated.

- The modes of operation of sensor (sleeping, censoring, etc.) should be fully exploited to minimize resource consumption while meeting application performance criteria.

- The communication protocols within the network should be designed with due consideration to the detection application, e.g., some bits are more important than others.

Finally, while much progress has been made towards the understanding of detection problems in sensor networks using the emerging framework described in this chapter, many interesting questions remain, including:

- How do we obtain observation statistics? How do we design adaptive and robust strategies that can work even when such statistics are incomplete or partially known?

- How should the sensor outputs be communicated in the network? Is it necessary to convert the outputs to bits or packets?

- What is the role of error control coding applied to the sensor outputs? What is the tradeoff between using additional bits to protect sensor outputs versus transmitting more information about the observation?

- How much do we gain by allowing the sensors to communicate with each other in the fusion configuration?

- What is the right architecture for the network in the context of detection applications? Decentralized with fusion or distributed?

Bibliography

Aalo V and Viswanathan R 1989 On distributed detection with correlated sensors: two examples. *IEEE Transactions on Aerospace and Electronic Systems* **25**(3), 414–421.

Ahlswede R and Csiszar I 1986 Hypothesis testing with communication constraints. *IEEE Transactions on Information Theory* **32**(4), 533–542.

Aldosari SA and Moura JMF 2004 Detection in decentralized sensor networks. *International Conference on Acoustics, Speech, and Signal Processing*, vol. 2, pp. 277–280, IEEE.

Aldosari SA and Moura JMF 2007 Detection in sensor networks: The saddlepoint approximation. *IEEE Transactions on Signal Processing* **55**(1), 327–340.

Appadwedula S, Veeravalli VV and Jones DL 2005 Energy-efficient detection in sensor networks. *IEEE Journal on Selected Areas in Communications* **23**(4), 693–702.

Appadwedula S, Veeravalli VV and Jones DL 2007 Decentralized detection with censoring sensors. *IEEE Transactions on Signal Processing*. Submitted.

Benitz GR and Bucklew JA 1990 Large deviation rate calculations for nonlinear detectors in Gaussian noise. *IEEE Transactions on Information Theory* **36**(2), 358–371.

Bercu B, Gamboa F and Rouault A 1997 Large deviations for quadratic forms of stationary Gaussian processes. *Stochastic Processes and Their Applications* **71**, 75–90.

Blum RS 1996 Locally optimum distributed detection of correlated random signals based on ranks. *IEEE Transactions on Information Theory* **42**(3), 931–942.

Blum RS and Kassam SA 1992 Optimum distributed detection of weak signals in dependent sensors. *IEEE Transactions on Information Theory* **38**(3), 1066–1079.

Blum RS and Kassam SA 1995 Distributed cell-averaging CFAR detection in dependent sensors. *IEEE Transactions on Information Theory* **41**(2), 513–518.

Blum RS, Kassam SA and Poor HV 1997 Distributed detection with multiple sensors: Part II – advanced topics. *Proceedings of the IEEE* **85**(1), 64–79.

Borkar V and Varaiya P 1982 Asymptotic agreement in distributed estimation. *IEEE Transactions on Automatic Control* **27**(3), 650–655.

Bryc W and Dembo A 1997 Large deviations for quadratic functionals of Gaussian processes. *Journal of Theoretical Probability* **10**(2), 307–332.

Chair Z and Varshney PK 1988 Distributed Bayesian hypothesis testing with distributed data fusion. *IEEE Transactions on Systems, Man and Cybernetics* **18**(5), 695–699.

Chamberland JF 2005 A first look at the hidden cost of random node placement in wireless sensor networks. *Workshop on Statistical Signal Processing* IEEE.

Chamberland JF and Veeravalli VV 2003a The art of sleeping in wireless sensing systems *Workshop on Statistical Signal Processing*, pp. 17–20, IEEE.

Chamberland JF and Veeravalli VV 2003b Decentralized detection in sensor networks. *IEEE Transactions on Signal Processing* **51**(2), 407–416.

Chamberland JF and Veeravalli VV 2004a Asymptotic results for decentralized detection in power constrained wireless sensor networks. *IEEE Journal on Selected Areas in Communications* **22**(6), 1007–1015.

Chamberland JF and Veeravalli VV 2004b Decentralized detection in wireless sensor systems with dependent observations. *International Conference on Computing, Communications and Control Technologies*, vol. 6, pp. 171–175, IIIS.

Chamberland JF and Veeravalli VV 2004c The impact of fading on decentralized detection in power constrained wireless sensor networks. *International Conference on Acoustics, Speech, and Signal Processing*, vol. 3, pp. 837–840, IEEE.

Chamberland JF and Veeravalli VV 2006 How dense should a sensor network be for detection with correlated observations? *IEEE Transactions on Information Theory* **52**(11), 5099–5106.

Chen B and Willett PK 2005 On the optimality of the likelihood-ratio test for local sensor decision rules in the presence of nonideal channels. *IEEE Transactions on Information Theory* **51**(2), 693–699.

Chen B, Jiang R, Kasetkasem T and Varshney PK 2004 Channel aware decision fusion in wireless sensor networks. *IEEE Transactions on Signal Processing* **52**(12), 3454–3458.

Chen JG and Ansari N 1998 Adaptive fusion of correlated local decisions. *IEEE Transactions on Systems, Man and Cybernetics, Part C* **28**(2), 276–281.

Chen PN and Papamarcou A 1993 New asymptotic results in parallel distributed detection. *IEEE Transactions on Information Theory* **39**(6), 1847–1863.

Cherikh M and Kantor PB 1992 Counterexamples in distributed detection. *IEEE Transactions on Information Theory* **38**(1), 162–165.

Crow RW and Schwartz SC 1996 Quickest detection for sequential decentralized decision systems. *IEEE Transactions on Aerospace and Electronic Systems* **32**(1), 267–83.

D'Costa A, Ramachandran V and Sayeed AM 2004 Distributed classification of Gaussian space-time sources in wireless sensor networks. *IEEE Journal on Selected Areas in Communications* **22**(6), 1026–1036.

Dembo A and Zeitouni O 1998 *Large deviations techniques and applications: Stochastic Modelling and Applied Probability.* 2nd edn. Springer, Berlin.

Drakopoulos E and Lee CC 1991 Optimum multisensor fusion of correlated local decisions. *IEEE Transactions on Aerospace and Electronic Systems* **27**(4), 593–606.

Duman TM and Salehi M 1998 Decentralized detection over multiple-access channels. *IEEE Transactions on Aerospace and Electronic Systems* **34**(2), 469–476.

Gallager RG 1968 *Information Theory and Reliable Communication.* Wiley, Chichestes.

Geraniotis E and Chau YA 1990 Robust data fusion for multisensor detection systems. *IEEE Transactions on Information Theory* **IT-36**(6), 1265–1279.

Hashemi HR and Rhodes IB 1989 Decentralized sequential detection. *IEEE Transactions on Information Theory* **35**(3), 509–520.

Hong YW and Scaglione A 2005 A scalable synchronization protocol for large scale sensor networks and its applications. *IEEE Journal on Selected Areas in Communications* **23**(5), 1085–1099.

Hong YW, Scaglione A and Varshney PK 2005 A communication architecture for reaching consensus in decision for a large network. *Workshop on Statistical Signal Processing,* IEEE.

Hussain AM 1994 Multisensor distributed sequential detection. *IEEE Transactions on Aerospace and Electronic Systems* **30**(3), 698–708.

Jiang R and Chen B 2005 Fusion of censored decisions in wireless sensor networks. *IEEE Transactions on Wireless Communications* **4**(6), 2668–2673.

Kam M, Zhu Q and Gray WS 1992 Optimal data fusion of correlated local decisions in multiple sensor detection systems. *IEEE Transactions on Aerospace and Electronic Systems* **28**(3), 916–120.

Kawadia V and Kumar PR 2005 Principles and protocols for power control in wireless ad hoc networks. *IEEE Journal on Selected Areas in Communications* **23**(1), 76–88.

Kay SM 1998 *Fundamentals of Statistical Signal Processing, Volume 2: Detection Theory.* Prentice Hall, Englewood Cliffs, NJ.

Lexa MA, Rozell CJ, Sinanovic S and Johnson DH 2004 To cooperate or not to cooperate: detection strategies in sensor networks. *International Conference on Acoustics, Speech, and Signal Processing,* vol. 3, pp. 841–844, IEEE.

Liu K and Sayeed AM 2004 Asymptotically optimal decentralized type-based detection in wireless sensor networks. *International Conference on Acoustics, Speech, and Signal Processing,* vol. 3, pp. 873–876, IEEE.

Luo ZQ 2005 An isotropic universal decentralized estimation scheme for a bandwidth constrained ad hoc sensor network. *IEEE Journal on Selected Areas in Communications* **23**(4), 735–744.

Luo ZQ and Tsitsiklis JN 1994 Data fusion with minimal communication. *IEEE Transactions on Information Theory* **40**(5), 1551–1563.

Mergen G and Tong L 2005a Asymptotic detection performance of type-based multiple access in sensor networks. *Workshop on Signal Processing Advances in Wireless Communications,* pp. 1018–1022, IEEE.

Mergen G and Tong L 2005b Sensor-fusion center communication over multiaccess fading channels. *International Conference on Acoustics, Speech, and Signal Processing,* vol. 4, pp. 841–844, IEEE.

Mergen G and Tong L 2006 Type based estimation over multiaccess channels. *IEEE Transactions on Signal Processing* **54**(2), 613–626.

Nguyen X, Wainwright MJ and Jordan MI 2005 Nonparametric decentralized detection using kernel methods. *IEEE Transactions on Signal Processing* **53**(11), 4053–4066.

Niu R, Chen B and Varshney PK 2006 Fusion of decisions transmitted over Rayleigh fading channels in wireless sensor networks. *IEEE Transactions on Signal Processing* **54**(3), 1018–1027.

Pearl J 1988 *Probabilistic Reasoning in Intelligent Systems: Networks of Plausible Inference*. Morgan Kaufmann.

Poor HV 1998 *An Introduction to Signal Detection and Estimation*. Springer Texts in Electrical Engineering, 2nd edn. Springer, New York.

Raghunathan V, Schurgers C, Park S and Srivastava MB 2002 Energy-aware wireless microsensor networks. *IEEE Signal Processing Magazine* **19**(2), 40–50.

Rago C, Willett P and Bar-Shalom Y 1996 Censoring sensors: a low-communication-rate scheme for distributed detection. *IEEE Transactions on Aerospace and Electronic Systems* **32**(2), 554–568.

Rappaport TS 2001 *Wireless Communications: Principles and Practice* 2nd edn. Prentice Hall, Englewood Cliffs, NJ.

Sadjadi F 1986 Hypothesis testing in a distributed environment. *IEEE Transactions on Aerospace and Electronic Systems* **22**(2), 134–137.

Schurgers C, Tsiatsis V, Ganeriwal S and Srivastava M 2002 Optimizing sensor networks in the energy-latency-density design space. *IEEE Transactions on Mobile Computing* **1**(1), 70–80.

Shalaby MH and Papamarcou A 1994 Error exponents for distributed detection of Markov sources. *IEEE Transactions on Information Theory* **40**(2), 397–408.

Stevens WR 1993 *The Protocols (TCP/IP Illustrated, Volume 1)*. Addison-Wesley Professional, Reading, MA.

Swaszek P and Willett P 1995 Parley as an approach to distributed detection. *IEEE Transactions on Aerospace and Electronic Systems* **31**(1), 447–457.

Szymanski BK and Yener B 2004 *Advances in Pervasive Computing and Networking* Springer, New York. pp. 119–136.

Teneketzis D and Ho YC 1987 The decentralized Wald problem. *Information and Computation* **73**(1), 23–44.

Teneketzis D and Varaiya P 1984 The decentralized quickest detection problem. *IEEE Transactions on Automatic Control* **29**(7), 641–644.

Tenney RR and Sandell NR 1981 Detection with distributed sensors. *IEEE Transactions on Aerospace and Electronic Systems* **17**(4), 501–510.

Tong L, Sadler BM and Dong M 2004 Pilot-assisted wireless transmissions: general model, design criteria, and signal processing. *IEEE Signal Processing Magazine* **21**(6), 12–25.

Trees HLV 2001 *Detection, Estimation, and Modulation Theory, Part I* reprint edn. Wiley-Interscience, New York.

Tsitsiklis JN 1988 Decentralized detection by a large number of sensors. *Mathematics of Control, Signals, and Systems* **1**(2), 167–182.

Tsitsiklis JN 1993 Decentralized detection. *Advances in Statistical Signal Processing* **2**, 297–344.

Tsitsiklis JN and Athans M 1984 Convergence and asymptotic agreement in distributed decision problems. *IEEE Transactions on Automatic Control* **29**(1), 42–50.

Tsitsiklis JN and Athans M 1985 On the complexity of decentralized decision making and detection problems. *IEEE Transactions on Automatic Control* **30**(5), 440–446.

Varshney PK 1996 *Distributed Detection and Data Fusion*. Springer, Berlin.

Veeravalli VV 1999 Sequential decision fusion: Theory and applications. *J. Franklin Inst.* **336**(2), 301–322.

Veeravalli VV 2001 Decentralized quickest change detection. *IEEE Transactions on Information Theory* **47**(4), 1657–1665.

Veeravalli VV, Başar T and Poor HV 1994a Minimax robust decentralized detection. *IEEE Transactions on Information Theory* **40**(1), 35–40.

Veeravalli VV, Başar T and Poor HV 1994b Decentralized sequential detection with sensors performing sequential tests. *Math. Contr. Signals Syst.* **7**(4), 292–305.

Viswanathan R and Varshney PK 1997 Distributed detection with multiple sensors: Part I – fundamentals. *Proceedings of the IEEE* **85**(1), 54–63.

Wang C and Schwartz M 1994 Fault detection with multiple observers. *IEEE/ACM Trans. Network.* **1**(1), 48–55.

Wang TY, Han YS, Varshey PK and Chen PN 2005 Distributed fault-tolerant classification in wireless sensor networks. *IEEE Journal on Selected Areas in Communications* **23**(4), 724–734.

Willett P and Swaszek PF 1995 On the performance degradation from one-bit quantized detection. *IEEE Transactions on Information Theory* **41**(6), 1997–2003.

Willett P, Swaszek PF and Blum RS 2000 The good, bad and ugly: distributed detection of a known signal in dependent Gaussian noise. *IEEE Transactions on Signal Processing* **48**(12), 3266–3279.

Xiao JJ and Luo ZQ 2005 Universal decentralized detection in a bandwidth-constrained sensor network. *IEEE Transactions on Signal Processing* **53**(8), 2617–2624.

Zhu Y, Blum RS, Luo ZQ and Wong KM 2000 Unexpected properties and optimum-distributed sensor detectors for dependent observation cases. *IEEE Transactions on Automatic Control* **45**(1), 62–72.

7

Distributed Estimation under Bandwidth and Energy Constraints[1]

Alejandro Ribeiro, Ioannis D. Schizas, Jin-Jun Xiao,
Georgios B. Giannakis and Zhi-Quan Luo

In parameter estimation problems a sequence of observations $\{\mathbf{x}(n)\}_{n=1}^{N}$ is used to estimate a random or deterministic parameter of interest $\mathbf{s}$. Optimal estimation exploits the statistical dependence between $\mathbf{x}(n)$ and $\mathbf{s}$ that is described either by the joint probability distribution function (pdf) $p(\mathbf{x}(n), \mathbf{s})$ when $\mathbf{s}$ is assumed random; or by a family of observation pdfs $p(\mathbf{x}(n); \mathbf{s})$ parameterized by $\mathbf{s}$ when $\mathbf{s}$ is assumed deterministic. The optimal estimator function producing an estimate $\hat{\mathbf{s}}$ for a given set of observations $\{\mathbf{x}(n)\}_{n=1}^{N}$ is different for random and deterministic parameters. It also depends on the joint pdf $p(\mathbf{x}(n), \mathbf{s})$ (or family of pdfs $p(\mathbf{x}(n); \mathbf{s})$) and the degree of knowledge about them; i.e., whether they are known, dependent on some other (nuisance) parameters, or completely unknown (Kay 1993).

The distributed nature of a WSN implies that observations are collected at different sensors and consequently it dictates that between collection and estimation a communication is present. If bandwidth and power were unlimited, the $\mathbf{x}(n)$ observations could be conveyed with arbitrary accuracy and, intuitively, no major impact would be expected. However, bandwidth and power *are* limited, and the seemingly innocuous communication stage turns out to have a significant impact on the design of optimal estimators and their performance

[1] © 2006 IEEE. Reprinted, with permission, from Distributed Learning in Wireless Sensor Networks by Joel B. Predd, Sanjeev R. Kulkarni and H. Vincent Poor, IEEE Signal Processing Magazine, Vol 23(4), 56–69pp, July 2006.

assessed by the estimator variance. On the one hand, if digital communications are to be employed, individual observations have to be quantized, transforming the estimation problem into that of estimating s using a set of quantized observations – certainly different from estimating s using the original analog-amplitude observations. On the other hand, since components of the (vector) observation $\mathbf{x}(n)$ are typically correlated, bandwidth and power constraints can be effected by transmitting vectors $\mathbf{y}(n)$ with smaller dimensionality than that of $\mathbf{x}(n)$.

As the discussion in the previous paragraph suggests, the distributed nature of observations coupled with stringent bandwidth and power constraints so that estimation in WSNs requires: (i) a means of combining local sensor observations is in order to reduce their dimensionality while keeping the estimation MSE as small as possible; (ii) quantization of the combined observations prior to digital transmission; and (iii) construction of estimators based on the quantized digital messages. While addressing these issues jointly is challenging, the present chapter describes recent advances pertaining to all these three requirements.

7.1   Distributed Quantization-Estimation

Consider a WSN consisting of N sensors deployed to estimate a scalar deterministic parameter s. The nth sensor observes a noisy version of s given by

$$x(n) = s + w(n), \qquad n \in [0, N-1], \tag{7.1}$$

where $w(n)$ denotes zero-mean noise with pdf $p_w(w)$, that is known possibly up to a finite number of unknown parameters. We further assume that $w(n_1)$ is independent of $w(n_2)$ for $n_1 \neq n_2$; i.e., noise variables are independent across sensors.

Due to bandwidth limitations, the observations $x(n)$ have to be quantized and estimation of s can only be based on these quantized values. We will henceforth think of quantization as the construction of a set of indicator variables

$$b_k(n) = \mathbf{1}\{x(n) \in B_k(n)\}, \qquad k \in [1, K], \tag{7.2}$$

taking the value 1 when $x(n)$ belongs to the region $B_k(n) \subset \mathbf{R}^M$, and 0 otherwise. Estimation of s will rely on this set of *binary* random variables $\{b_k(n), k \in [1, K]\}_{n=0}^{N-1}$. The latter are Bernoulli distributed with parameters $q_k(n)$ satisfying

$$q_k(n) := \Pr\{b_k(n) = 1\} = \Pr\{x(n) \in B_k(n)\}. \tag{7.3}$$

In the ensuing sections, we will present the Cramér-Rao Lower Bound (CRLB) to benchmark the variance of all unbiased estimators $\hat{s}$ constructed using the binary observations $\{b_k(n), k \in [1, K]\}_{n=0}^{N-1}$. We will further show that it is possible to find maximum likelihood estimators (MLEs) that (at least asymptotically) are known to achieve the CRLB. Finally, we will reveal that the CRLB based on $\{b_k(n), k \in [1, K]\}_{n=0}^{N-1}$ can come surprisingly close to the clairvoyant CRLB based on $\{x(n)\}_{n=0}^{N-1}$ in certain applications of practical interest.

7.2   Maximum Likelihood Estimation

Let us start by assuming that $p_w(w)$ is known and let $F_w(u) := \int_u^\infty p_w(w)\, dw$ denote the complementary cumulative distribution function (CCDF) of the noise. With the pdf known, it suffices to rely on a single region $B_1(n)$ in (7.2) to generate a single bit $b_1(n)$ per sensor, using a threshold τ_c common to all N sensors: $B_1(n) := B_c = (\tau_c, \infty)$, $\forall n$. Based on these binary observations, $b_1(n) := \mathbf{1}\{x(n) \in (\tau_c, \infty)\}$ received from all N sensors, the fusion center (FC) seeks estimates of s.

An expression for the MLE of s follows readily from the following argument. Using (7.3), we can express the Bernoulli parameter as

$$q_1 = \int_{\tau_c - s}^{\infty} p_w(w)dw = F_w(\tau_c - s). \tag{7.4}$$

On the other hand, it is well known that the MLE of q_1 is given by $\hat{q}_1 = N^{-1}\sum_{n=0}^{N-1} b_1(n)$ (Kay 1993, p. 200). These two facts combined with the invariance property of MLE (Kay 1993, p. 173), readily yield the MLE of s as (Ribeiro and Giannakis 2006a):

$$\hat{s} = \tau_c - F_w^{-1}\left(\frac{1}{N}\sum_{n=0}^{N-1} b_1(n)\right). \tag{7.5}$$

It can be further shown that the CRLB on the variance of any unbiased estimator $\hat{s}$ based on $\{b_1(n)\}_{n=0}^{N-1}$ is (Ribeiro and Giannakis 2006a)

$$\text{var}(\hat{s}) \geq \frac{1}{N}\frac{F_w(\tau_c - s)[1 - F_w(\tau_c - s)]}{p_w^2(\tau_c - s)} := B(s). \tag{7.6}$$

If the noise is Gaussian and we define the σ-*distance* between the threshold τ_c and the (unknown) parameter s as $\Delta_c := (\tau_c - s)/\sigma$, then (7.6) reduces to

$$B(s) = \frac{\sigma^2}{N}\frac{2\pi Q(\Delta_c)[1 - Q(\Delta_c)]}{e^{-\Delta_c^2}} := \frac{\sigma^2}{N}D(\Delta_c), \tag{7.7}$$

with $Q(u) := (1/\sqrt{2\pi})\int_u^{\infty} e^{-w^2/2} \, dw$ denoting the Gaussian tail probability function (Figure 7.1).

The bound $B(s)$ is the variance of $\bar{x} := N^{-1}\sum_{n=0}^{N-1} x(n)$, scaled by the factor $D(\Delta_c)$ – recall that $\text{var}(\bar{x}) = \sigma^2/N$ (Kay 1993, p.31). Optimizing $B(s)$ with respect to Δ_c, yields the optimum at $\Delta_c = 0$ and the minimum CRLB as

$$B_{\min} = \frac{\pi}{2}\frac{\sigma^2}{N}. \tag{7.8}$$

Eq. (7.8) reveals something unexpected: relying on a single bit per $x(n)$, the estimator in (7.5) incurs a minimal (just a $\pi/2$ factor) increase in its variance relative to the clairvoyant $\bar{x}$ which relies on the unquantized data $x(n)$. But this minimal loss in performance corresponds to the ideal choice $\Delta_c = 0$, which implies $\tau_c = s$ and requires perfect knowledge of the unknown s for selecting the quantization threshold τ_c. How do we select τ_c and how much do we lose when the unknown s lies anywhere in $(-\infty, \infty)$, or when s lies in $[S_1, S_2]$, with S_1, S_2 finite and known a priori? Intuition suggests selecting the threshold as close as possible to the unknown parameter s. This can be realized with an iterative estimator $\hat{s}^{(i)}$, which can be formed as in (7.5), using $\tau_c^{(i)} = \hat{s}^{(i-1)}$, the parameter estimate from the previous $(i - 1)^{st}$ iteration.

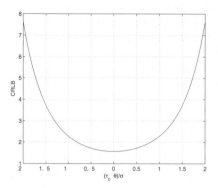

Figure 7.1 CRLB in (7.7) for the estimation of s in (7.1) when the noise is Gaussian. The minimum CRLB is $B_{\min} = (\pi/2)\sigma^2/N$, just $\pi/2$ times larger than the variance of the sample mean estimator. The increase is exponential in $\Delta_c := (\tau_c - s)/\sigma$, though.

But in the batch formulation considered herein, selecting τ_c is challenging; and a closer look at $B(s)$ in (7.6) will confirm that the loss can be huge if $\tau_c - s \gg 0$. Indeed, as $\tau_c - s \to \infty$ the denominator in (7.6) goes to zero faster than its numerator, since F_w is the integral of the non-negative pdf p_w; and thus, $B(s) \to \infty$ as $\tau_c - s \to \infty$. The implication of the latter is twofold: (i) since it shows up in the CRLB, the potentially high variance of estimators based on quantized observations is inherent to the possibly severe bandwidth limitations of the problem itself and is not unique to a particular estimator; ii) for any choice of τ_c, the fundamental performance limits in (7.6) are dictated by the end points $\tau_c - S_1$ and $\tau_c - S_2$ when s is confined to the interval $[S_1, S_2]$. On the other hand, how successful the τ_c selection is depends on the dynamic range $|S_1 - S_2|$ which makes sense because the latter affects the error incurred when quantizing $x(n)$ to $b_1(n)$. Notice that in such joint quantization-estimation problems one faces two sources of error: quantization and noise. To account for both, the proper figure of merit for estimators based on binary observations is what we will term quantization signal-to-noise ratio (Q-SNR):

$$\gamma := \frac{|S_1 - S_2|^2}{\sigma^2};\qquad(7.9)$$

Notice that contrary to common wisdom, the smaller Q-SNR is, the easier it becomes to select τ_c judiciously. Furthermore, the variance increase in (7.6) relative to the variance of the clairvoyant $\bar{x}$ is smaller, for a given σ. This is because as the Q-SNR increases the problem becomes more difficult in general, but the rate at which the variance increases is smaller for the CRLB in (7.6) than for $\text{var}(\bar{x}) = \sigma^2/N$.

7.2.1 Known Noise pdf with Unknown Variance

Perhaps more common than a perfectly known pdf is the case when the noise pdf is known except for its variance $\text{E}[w^2(n)] = \sigma^2$. Introducing the standardized variable $v(n) :=$

$w(n)/\sigma$ we write the signal model as

$$x(n) = s + \sigma v(n). \tag{7.10}$$

Let $p_v(v)$ and $F_v(v) := \int_v^\infty p_v(u)du$ denote the known pdf and CCDF of $v(n)$. Note that according to its definition, $v(n)$ has zero mean, $E[v^2(n)] = 1$, and the pdfs of v and w are related by $p_w(w) = (1/\sigma)p_v(w/\sigma)$. Note also that all two parameter pdfs can be standardized likewise.

To estimate s when σ is also unknown while keeping the bandwidth constraint to 1 bit per sensor, we divide the sensors in two groups each using a different region (i.e., threshold) to define the binary observations:

$$B_1(n) := \begin{cases} (\tau_1, \infty) := B_1, & \text{for } n = 0, \dots, (N/2) - 1 \\ (\tau_2, \infty) := B_2, & \text{for } n = (N/2), \dots, N. \end{cases} \tag{7.11}$$

That is, the first $N/2$ sensors quantize their observations using the threshold τ_1, while the remaining $N/2$ sensors rely on the threshold τ_2. Without loss of generality, we assume $\tau_2 > \tau_1$.

The Bernoulli parameters of the resultant binary observations can be expressed as [c.f. (7.4)]:

$$q_1(n) := \begin{cases} F_v\left[\frac{\tau_1 - s}{\sigma}\right] := q_1 & \text{for } n = 0, \dots, (N/2) - 1, \\ F_v\left[\frac{\tau_2 - s}{\sigma}\right] := q_2 & \text{for } n = (N/2), \dots, N. \end{cases} \tag{7.12}$$

Given the noise independence across sensors, the MLEs of q_1, q_2 can be found, respectively, as

$$\hat{q}_1 = \frac{2}{N} \sum_{n=0}^{N/2-1} b_1(n), \qquad \hat{q}_2 = \frac{2}{N} \sum_{n=N/2}^{N-1} b_1(n). \tag{7.13}$$

Mimicking (7.5), we can invert F_v in (7.12) and invoke the invariance property of MLEs to obtain the MLE $\hat{s}$ in terms of $\hat{q}_1$ and $\hat{q}_2$. This estimator is given in the following proposition along with its CRLB (Ribeiro and Giannakis 2006b).

Proposition 7.2.1 *Consider estimating s in (7.10), based on binary observations constructed from the regions defined in (7.11).*

(a) *The MLE of s is*

$$\hat{s} = \frac{F_v^{-1}(\hat{q}_2)\tau_1 - F_v^{-1}(\hat{q}_1)\tau_2}{F_v^{-1}(\hat{q}_2) - F_v^{-1}(\hat{q}_1)}, \tag{7.14}$$

with F_v^{-1} denoting the inverse function of F_v, and $\hat{q}_1$, $\hat{q}_2$ given by (7.13).

(b) *The variance of any unbiased estimator of s, $\text{var}(\hat{s})$, based on $\{b_1(n)\}_{n=0}^{N-1}$ is bounded by*

$$B(s) := \frac{2\sigma^2}{N} \left(\frac{\Delta_1 \Delta_2}{\Delta_2 - \Delta_1}\right)^2 \left[\frac{q_1(1 - q_1)}{p_v^2(\Delta_1)\Delta_1^2} + \frac{q_2(1 - q_2)}{p_v^2(\Delta_2)\Delta_2^2}\right] \tag{7.15}$$

where q_k is given by (7.12), and

$$\Delta_k := \frac{\tau_k - s}{\sigma}, \quad k = 1, 2, \tag{7.16}$$

is the σ-distance between s and the threshold τ_k.

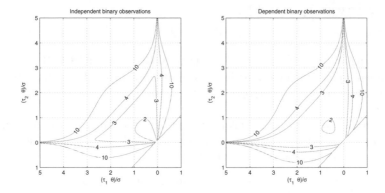

Figure 7.2 Per bit CRLB when the binary observations are independent and dependent, respectively. In both cases, the variance increase with respect to the sample mean estimator is small when the σ-distances are close to 1, being slightly better for the case of dependent binary observations (Gaussian noise).

Eq. (7.15) is reminiscent of (7.6), suggesting that the variances of the estimators they bound are related. This implies that even when the known noise pdf contains unknown parameters, the variance of $\hat{s}$ can come close to the variance of the clairvoyant estimator $\bar{x}$, provided that the thresholds τ_1, τ_2 are chosen close to s relative to the noise standard deviation (so that Δ_1, Δ_2, and $\Delta_2 - \Delta_1$ in (7.16) are ≈ 1). For the Gaussian pdf, Figure 7.2 shows the contour plot of $B(s)$ in (7.15) normalized by $\sigma^2/N := \mathrm{var}(\bar{x})$. Notice that in the low Q-SNR regime Δ_1, $\Delta_2 \approx 1$, and the relative variance increase $B(s)/\mathrm{var}(\bar{x})$ is less than 3. This is illustrated by the simulations shown in Figure 7.3 for two different sets of σ-distances, Δ_1, Δ_2, corroborating the values predicted by (7.15) and the fact that the performance loss with respect to the clairvoyant sample mean estimator, $\bar{x}$, is indeed small.

Dependent binary observations

In the previous subsection, we restricted the sensors to transmit only 1 bit per $x(n)$ datum, and divided the sensors in two classes each quantizing $x(n)$ using a different threshold. A related approach is to let each sensor use two thresholds:

$$B_1(n) := B_1 = (\tau_1, \infty), \qquad n = 0, 1, \ldots, N-1,$$

$$B_2(n) := B_2 = (\tau_2, \infty), \qquad n = 0, 1, \ldots, N-1 \qquad (7.17)$$

where $\tau_2 > \tau_1$. We define the per sensor vector of binary observations $\mathbf{b}(n) := [b_1(n), b_2(n)]^T$, and the vector Bernoulli parameter $\mathbf{q} := [q_1(n), q_2(n)]^T$, whose components are as in (7.12).

Note the subtle differences between (7.11) and (7.17). While each of the N sensors generates 1 binary observation according to (7.11), each sensor creates 2 binary observations as per (7.17). The total number of bits from all sensors in the former case is N, but in the latter $N \log_2 3$, since our constraint $\tau_2 > \tau_1$ implies that the realization $\mathbf{b} = (0, 1)$ is impossible. In addition, all bits in the former case are independent, whereas correlation is present in the latter since $b_1(n)$ and $b_2(n)$ come from the same $x(n)$. Even though one

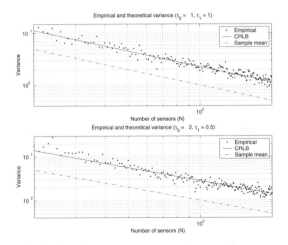

Figure 7.3 Noise of unknown power estimator. The simulation corroborates the close to clairvoyant variance prediction of (7.15) ($\sigma = 1$, $s = 0$, Gaussian noise).

would expect this correlation to complicate matters, a property of the binary observations defined as per (7.17), summarized in the next lemma, renders estimation of s based on them feasible.

Lemma 7.2.2 *The MLE of* $\mathbf{q} := (q_1(n), q_2(n))^T$ *based on the binary observations* $\{\mathbf{b}(n)\}_{n=0}^{N-1}$ *constructed according to* (7.17) *is given by*

$$\hat{\mathbf{q}} = \frac{1}{N} \sum_{n=0}^{N-1} \mathbf{b}(n). \tag{7.18}$$

Interestingly, (7.18) coincides with (7.13), proving that the corresponding estimators of s are identical; i.e., (7.14) yields also the MLE $\hat{s}$ even in the correlated case. However, as the following proposition asserts, correlation affects the estimator's variance and the corresponding CRLB (Ribeiro and Giannakis 2006b).

Proposition 7.2.3 *Consider estimating* s *in* (7.10), *when* σ *is unknown, based on binary observations constructed from the regions defined in* (7.17). *The variance of any unbiased estimator of* s, var($\hat{s}$), *based on* $\{b_1(n), b_2(n)\}_{n=0}^{N-1}$ *is bounded by*

$$B_D(s) := \frac{\sigma^2}{N} \left(\frac{\Delta_1 \Delta_2}{\Delta_2 - \Delta_1} \right)^2 \left[\frac{q_1(1-q_1)}{p_v^2(\Delta_1)\Delta_1^2} + \frac{q_2(1-q_2)}{p_v^2(\Delta_2)\Delta_2^2} - \frac{q_2(1-q_1)}{p_v(\Delta_1)p(\Delta_2)\Delta_1\Delta_2} \right], \tag{7.19}$$

where the subscript D *in* $B_D(s)$ *is used as a mnemonic for the dependent binary observations this estimator relies on [c.f.* (7.15)*].*

Unexpectedly, (7.19) is similar to (7.15). Actually, a fair comparison between the two requires compensating for the difference in the total number of bits used in each case. This

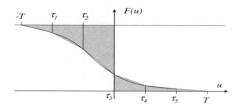

Figure 7.4 When the noise pdf is unknown numerically integrating the CCDF using the trapezoidal rule yields an approximation of the mean.

can be accomplished by introducing the per-bit CRLBs for the independent and correlated cases respectively,

$$C(s) = NB(s), \qquad C_D(s) = N \log_2(3) B_D(s), \tag{7.20}$$

which lower bound the corresponding variances achievable by the transmission of a single bit.

Evaluation of $C(s)/\sigma^2$ and $C_D(s)/\sigma^2$ follows from (7.15), (7.19) and (7.20) and is depicted in Figure 7.2 for Gaussian noise and σ-distances Δ_1, Δ_2 having amplitude as large as 5. Somewhat surprisingly, both approaches yield very similar bounds with the one relying on dependent binary observations being slightly better in the achievable variance; or correspondingly, in requiring a smaller number of sensors to achieve the same CRLB.

7.3   Unknown Noise pdf

In certain applications it may not be reasonable to assume knowledge about the noise pdf $p_w(w)$. These cases require *nonparametric* approaches as the one pursued in this section.

We assume that $p_w(w)$ has zero mean so that s in (7.1) is identifiable. Let $p_x(x)$ and $F_x(x)$ denote the pdf and CCDF of the observations $x(n)$. As s is the mean of $x(n)$, we can write

$$s := \int_{-\infty}^{+\infty} x p_x(x) \, dx = -\int_{-\infty}^{+\infty} x \frac{\partial F_x(x)}{\partial x} \, dx = \int_0^1 F_x^{-1}(v) \, dv, \tag{7.21}$$

where in establishing the second equality we used the fact that the pdf is the negative derivative of the CCDF, and in the last equality we introduced the change of variables $v = F_x(x)$. But note that the integral of the inverse CCDF can be written in terms of the integral of the CCDF as (see also Figure 7.4)

$$s = -\int_{-\infty}^0 [1 - F_x(u)] \, du + \int_0^{+\infty} F_x(u) \, du, \tag{7.22}$$

allowing one to express the mean s of $x(n)$ in terms of its CCDF. To avoid carrying out integrals with infinite range, let us assume that $x(n) \in (-T, T)$ which is always practically satisfied for T sufficiently large, so that we can rewrite (7.22) as

$$s = \int_{-T}^T F_x(u) \, du - T. \tag{7.23}$$

Numerical evaluation of the integral in (7.23) can be performed using a number of known techniques. Let us consider an ordered set of interior points $\{\tau_k\}_{k=1}^K$ along with end-points $\tau_0 = -T$ and $\tau_{K+1} = T$. Relying on the fact that $F_x(\tau_0) = F_x(-T) = 1$ and $F_x(\tau_{K+1}) = F_x(T) = 0$, application of the trapezoidal rule for numerical integration yields (see also Figure 7.4)

$$s = \frac{1}{2} \sum_{k=1}^K (\tau_{k+1} - \tau_{k-1}) F_x(\tau_k) - T + e_a, \tag{7.24}$$

with e_a denoting the approximation error. Certainly, other methods like Simpson's rule, or the broader class of Newton-Cotes formulas, can be used to further reduce e_a.

Whichever the choice, the key is that binary observations constructed from the region $B_k := (\tau_k, \infty)$ have Bernoulli parameters

$$q_k := \Pr\{x(n) > \tau_k\} = F_x(\tau_k). \tag{7.25}$$

Inserting the nonparametric estimators $\hat{F}_x(\tau_k) = \hat{q}_k$ in (7.24), our parameter estimator when the noise pdf is unknown takes the form:

$$\hat{s} = \frac{1}{2} \sum_{k=1}^K \hat{q}_k (\tau_{k+1} - \tau_{k-1}) - T. \tag{7.26}$$

Since $\hat{q}_k$'s are unbiased, (7.24) and (7.26) imply that $E(\hat{s}) = s + e_a$. Being biased, the proper performance indicator for $\hat{s}$ in (7.26) is the mean squared error (MSE), not the variance.

Maintaining the bandwidth constraint of 1 bit per sensor (i.e., $K = 1$), we divide the N sensors in K subgroups containing N/K sensors each, and define the regions

$$B_1(n) := B_k = (\tau_k, \infty), \quad n = (k-1)(N/K), \ldots, k(N/K) - 1; \tag{7.27}$$

Region $B_1(n)$ will be used by sensor n to construct and transmit the binary observation $b_1(n)$. Herein, the unbiased estimators of q_k are

$$\hat{q}_k = \frac{1}{(N/K)} \sum_{n=(k-1)(N/K)}^{k(N/K)-1} b_1(n), \quad k = 1, \ldots, K, \tag{7.28}$$

and are used in (7.26) to estimate s. It is easy to verify that $\text{var}(\hat{q}_k) = q_k(1 - q_k)/(N/K)$, and that $\hat{q}_{k_1}$ and $\hat{q}_{k_2}$ are independent for $k_1 \neq k_2$.

The resultant MSE, $E[(s - \hat{s})^2]$, can be bounded as follows (Ribeiro and Giannakis 2006b).

Proposition 7.3.1 *Consider $\hat{s}$ given by (7.26), with $\hat{q}_k$ as in (7.28). Assume that for T sufficiently large and known $p_x(x) = 0$, for $|x| \geq T$, the noise pdf has bounded derivative $\dot{p}_w(u) := \partial p_w(w)/\partial w$; and define $\tau_{\max} := \max_k\{\tau_{k+1} - \tau_k\}$ and $\dot{p}_{\max} := \max_{u \in (-T,T)} \{\dot{p}_w(u)\}$. The MSE is given by*

$$E[(s - \hat{s})^2] = |e_a|^2 + \text{var}(\hat{s}), \tag{7.29}$$

with the approximation error e_a and var($\hat{s}$), satisfying

$$|e_a| \leq \frac{T \dot{p}_{max}}{6} \tau_{max}^2,$$ (7.30)

$$\text{var}(\hat{s}) = \sum_{k=1}^{K} \frac{(\tau_{k+1} - \tau_{k-1})^2}{4} \frac{q_k(1 - q_k)}{N/K},$$ (7.31)

with $\{\tau_k\}_{k=1}^{K}$ a grid of thresholds in $(-T, T)$ and $\{q_k\}_{k=1}^{K}$ as in (7.25).

Note from (7.31) that the larger contributions to var($\hat{s}$) occur when $q_k \approx 1/2$, since this value maximizes the coefficients $q_k(1 - q_k)$; equivalently, this happens when the thresholds satisfy $\tau_k \approx s$ [cf. (7.25)]. Thus, as with the case where the noise pdf is known, when s belongs to an a-priori known interval $[s_1, s_2]$, this knowledge must be exploited in selecting thresholds around the likeliest values of s.

On the other hand, note that the var($\hat{s}$) term in (7.29) will dominate $|e_a|^2$ because $|e_a|^2 \propto \tau_{max}^4$ as per (7.30). To clarify this point, consider an equispaced grid of thresholds with $\tau_{k+1} - \tau_k = \tau = \tau_{max}$, $\forall k$, such that $\tau_{max} = 2T/(K + 1) < 2T/K$. Using the (loose) bound $q_k(1 - q_k) \leq 1/4$, the MSE is bounded by [cf. (7.29)–(7.31)]

$$\text{\quad\quad} \quad\quad E[(s - \hat{s})^2] < \frac{4T^6 \dot{p}_{max}^2}{9K^4} + \frac{T^2}{N}.$$ (7.32)

The bound in (7.32) is minimized by selecting $K = N$, which amounts to having *each sensor use a different region* to construct its binary observation. In this case, $|e_a|^2 \propto N^{-4}$ and its effect becomes practically negligible. Moreover, most pdfs have relatively small derivatives; e.g., for the Gaussian pdf we have $\dot{p}_{max} = (2\pi e\sigma^4)^{-1/2}$. The integration error can be further reduced by resorting to a more powerful numerical integration method, although its difference with respect to the trapezoidal rule will not have noticeable impact in practice.

Since $K = N$, the selection $\tau_{k+1} - \tau_k = \tau$, $\forall k$, yields

$$\hat{s} = \tau \sum_{n=0}^{N-1} b_1(n) - T = T \left[\frac{2}{N + 1} \sum_{n=0}^{N-1} b_1(n) - 1 \right],$$ (7.33)

that *does not require knowledge of the threshold* used to construct the binary observations at the FC of a WSN. This feature allows each sensor to randomly select its threshold without using values pre-assigned by the FC; see also (Luo 2005a) for related random quantization algorithms which also yielded *universal* (in the noise variance) parameter estimators based on severely quantized WSN data.

Remark 1 While $e_a^2 \propto T^6$ seems to dominate var($\hat{s}$) $\propto T^2$ in (7.32), this is not true for the operational low-to-medium Q-SNR range for distributed estimators based on binary observations. This is because the support $2T$ over which $F_x(x)$ in (7.23) is non-zero depends on σ and the dynamic range $|S_1 - S_2|$ of the parameter s. And as the Q-SNR decreases, $T \propto \sigma$. But since $\dot{p}_{max} \propto \sigma^{-2}$, $e_a^2 \propto \sigma^2/N^4$ which is negligible when compared to the term var($\hat{s}$) $\propto \sigma^2/N$.

Apart from providing useful bounds on the finite-sample performance, eqs. (7.30), (7.31), and (7.32) establish asymptotic optimality of the $\hat{s}$ estimators in (7.26) and (7.33) as summarized in the following:

Corollary 7.3.2 *Under the assumptions of Propositions 7.3.1 and the conditions: i)* $\tau_{max} \propto K^{-1}$; *and ii)* $T^2/N, T^6/K^4 \to 0$ *as* $T, K, N \to \infty$, *the estimators* $\hat{s}$ *in* (7.26) *and* (7.33) *are asymptotically (as* $K, N \to \infty$*) unbiased and consistent in the mean-square sense.*

The estimators in (7.26) and (7.33) are consistent even if the support of the data pdf is infinite, as long as we guarantee a proper rate of convergence relative to the number of sensors and thresholds.

Remark 2 To compare the estimators in (7.5) and (7.33), consider that $s \in [S_1, S_2] = [-\sigma, \sigma]$, and that the noise is Gaussian with variance σ^2, yielding a Q-SNR $\gamma = 4$. No estimator can have variance smaller than $\text{var}(\bar{x}) = \sigma^2/N$; however, for the (medium) $\gamma = 4$ Q-SNR value they can come close. For the known pdf estimator in (7.5), the variance is $\text{var}(\hat{s}) \approx 2\sigma^2/N$. The unknown pdf estimator in (7.33) requires an assumption about the essentially non-zero support of the Gaussian pdf. If we suppose that the noise pdf is non-zero over $[-2\sigma, 2\sigma]$, the corresponding variance becomes $\text{var}(\hat{s}) \approx 9\sigma^2/N$. The penalties due to the transmission of a single bit per sensor with respect to $\bar{x}$ are approximately 2 and 9. While the increasing penalty is expected as the uncertainty about the noise pdf increases, the relatively small loss is rather unexpected.

Figure 7.5 depicts theoretical bounds and simulated variances for the estimators (7.5) and (7.33) for an example Q-SNR $\gamma = 4$. The sample mean estimator variance, $\text{var}(\bar{x}) = \sigma^2/N$, is also depicted for comparison purposes. The simulations corroborate the implications of Remark 3, reinforcing the assertion that for low to medium Q-SNR problems quantization to a single bit per observation leads to minimal losses in variance performance. Note that for this particular example, the unknown pdf variance bound, (7.32), overestimates the variance by a factor of roughly 1.2 for the uniform case and roughly 2.6 for the Gaussian case.

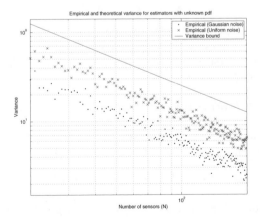

Figure 7.5 The variance of the estimators in (7.5) and (7.33) are close to the sample mean estimator variance ($\sigma^2 := \text{E}[w^2(n)] = 1$, $T = 3$, $s \in [-1, 1]$).

7.3.1   Lower Bound on the MSE

In Section 7.2 we derived the CRLB offering the fundamental *lower* bound on the achievable variance and the MLE that approaches this bound as N increases. In contrast, (7.32) is an *upper* bound on the MSE of the estimator in (7.33). The counterpart of the CRLB for estimation based on binary observations when the pdf is unknown is a lower bound in the MSE achievable by any estimator.

To obtain this bound we start from the CRLB when the noise pdf is known that we introduced in (7.6). We then maximize this CRLB with respect to the noise pdf and the local quantization rules to obtain a lower bound on the MSE performance of any estimator when the pdf is unknown. The result is summarized in the following proposition (Xiao et al. 2007).

Proposition 7.3.3 *Consider the signal model in (7.1); $x(n)$ observations belonging to the interval $(-T, T)$; i.e., $x(n) \in [-T, T]$; and let each sensor communicate one binary observation $b(n)$ as per (7.2). Then, for any estimator $\hat{s}$ of s relying on $\{b(n)\}_{n=0}^{N-1}$ there exists a noise pdf such that*

$$E[(s - \hat{s})^2] \geq \frac{T^2}{4N}. \tag{7.34}$$

Proposition 7.3.3 implies that no estimator based on quantized samples down to a single bit per sensor can attain an MSE smaller than $T^2/4N$. Comparing (7.32) with (7.34) we deduce that the estimator in (7.33) is optimal up to a constant factor of 4.

7.4   Estimation of Vector Parameters

Consider now the case of a physical phenomenon characterized by a set of p parameters that we lump in to the vector $\mathbf{s} := [s_1, \ldots, s_p]^T$. As before, we wish to find $\mathbf{s}$, by deploying a WSN composed of N sensors $\{S_n\}_{n=0}^{N-1}$, with each sensor observing $\mathbf{s}$ through a linear transformation

$$\mathbf{x}(n) = \mathbf{H}_n \mathbf{s} + \mathbf{w}(n), \tag{7.35}$$

where $\mathbf{x}(n) := [x_1(n), \ldots, x_M(n)]^T \in \mathbf{R}^M$ is the measurement vector at sensor S_n, $\mathbf{w}(n) \in \mathbf{R}^M$ is zero-mean additive noise with pdf $p_{\mathbf{w}}(\mathbf{w})$ and the matrices $\mathbf{H}_n \in \mathbf{R}^{M \times P}$.

As in (7.2), we define the binary observation $b_k(n)$ as the indicator function of $\mathbf{x}(n)$ belonging to the region $B_k(n) \subset \mathbf{R}^M$:

$$b_k(n) = \mathbf{1}\{\mathbf{x}(n) \in B_k(n)\}, \qquad k \in [1, K], \tag{7.36}$$

We then define the per sensor vector of binary observations $\mathbf{b}(n) := [b_1(n), \ldots, b_K(n)]^T$, and note that since its entries are binary, realizations $\mathbf{y}$ of $\mathbf{b}(n)$ belong to the set

$$\mathcal{B} := \{\boldsymbol{\beta} \in \mathbf{R}^K \mid [\boldsymbol{\beta}]_k \in \{0, 1\}, \ k \in [1, K]\}, \tag{7.37}$$

where $[\boldsymbol{\beta}]_k$ denotes the kth component of $\boldsymbol{\beta}$. With each $\boldsymbol{\beta} \in \mathcal{B}$ and each sensor we now associate the region

$$\mathbf{B}_{\boldsymbol{\beta}}(n) := \bigcap_{[\boldsymbol{\beta}]_k=1} B_k(n) \bigcap_{[\boldsymbol{\beta}]_k=0} \overline{B}_k(n), \tag{7.38}$$

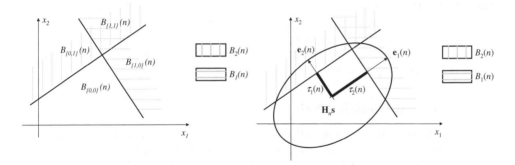

Figure 7.6 (Left): The vector of binary observations **b** takes on the value $\{y_1, y_2\}$ if and only if $x(n)$ belongs to the region $B_{\{y_1, y_2\}}(n)$; (Right): Selecting the regions $B_k(n)$ perpendicular to the covariance matrix eigenvectors results in independent binary observations when the noise is Gaussian.

where $\overline{B}_k(n)$ denotes the set-complement of $B_k(n)$ in $\mathbf{R}^M$. Note that the definition in (7.38) implies that $x(n) \in \mathbf{B}_\beta(n)$ if and only if $\mathbf{b}(n) = \beta$; see also Figure 7.6 (Left) for an illustration in $\mathbf{R}^2$ ($M = 2$). The corresponding probabilities are

$$q_\beta(n) := \Pr\{\mathbf{b}(n) = \beta\} = \int_{\mathbf{B}_\beta(n)} p_\mathbf{w}[\mathbf{u} - \mathbf{H}_n \mathbf{s}] \, d\mathbf{u}. \tag{7.39}$$

Using definitions (7.39) and (7.37), we can write the pertinent log-likelihood function as

$$L(\mathbf{s}) = \sum_{n=0}^{N-1} \sum_{\beta \in \mathcal{B}} \delta(\mathbf{b}(n) - \beta) \ln q_\beta(n), \tag{7.40}$$

and the MLE of **s** as

$$\hat{\mathbf{s}} = \arg \max_\mathbf{s} L(\mathbf{s}). \tag{7.41}$$

The nonlinear search needed to obtain $\hat{\mathbf{s}}$ could be challenging. Fortunately, as the following proposition asserts, under certain conditions that are usually met in practice, $L(\mathbf{s})$ is concave which implies that computationally efficient search algorithms can be invoked to find its global maximum (Ribeiro and Giannakis 2006b).

Proposition 7.4.1 *If the MLE problem in (7.41) satisfies the conditions:*

[c1] *The noise pdf $p_\mathbf{w}(\mathbf{w})$ is log-concave (Boyd and Vandenberghe 2004, p.104)*

[c2] *The regions $B_k(n)$ are chosen as half-spaces.*

then $L(\mathbf{s})$ in (7.40) is a concave function of **s**.

Note that [c1] is satisfied by common noise pdfs, including the multivariate Gaussian (Boyd and Vandenberghe 2004, p.104). On the other hand, [c2] places a constraint in the regions defining the binary observations, which is simply up to the designer's choice. The merits of having a concave log-likelihood function are summarized in the following remark.

Remark 3 The numerical search needed to obtain $\hat{\mathbf{s}}$ could be challenged either by the multimodal nature of $L(\mathbf{s})$ or by numerical ill-conditioning caused by e.g., saddle points. But when the log-concavity conditions in Proposition 7.4.1 are satisfied, computationally efficient search algorithms like e.g., Newton's method are guaranteed to converge to the global maximum (Boyd and Vandenberghe 2004, Chap. 2).

7.4.1   Colored Gaussian Noise

Analyzing the performance of the MLE in (7.41) is only possible asymptotically (as N or SNR go to infinity). Notwithstanding, when the noise is Gaussian, simplifications render variance analysis tractable and lead to interesting guidelines for constructing the estimator $\hat{\mathbf{s}}$.

Restrict $p_{\mathbf{w}}(\mathbf{w})$ to the class of multivariate Gaussian pdfs, and let $\mathbf{C}(n)$ denote the noise covariance matrix at sensor n. Assume that $\{\mathbf{C}(n)\}_{n=0}^{N-1}$ are known and let $\{(\mathbf{e}_m(n),$ $\sigma_m^2(n))\}_{m=1}^{M}$ be the set of eigenvectors and associated eigenvalues

$$\mathbf{C}(n) = \sum_{m=1}^{M} \sigma_m^2(n)\mathbf{e}_m(n)\mathbf{e}_m^T(n). \tag{7.42}$$

For each sensor, we define a set of $K = M$ regions $B_k(n)$ as half-spaces whose borders are hyper-planes perpendicular to the covariance matrix eigenvectors; i.e.,

$$B_k(n) = \{\mathbf{x} \in \mathbf{R}^M \mid \mathbf{e}_k^T(n)\mathbf{x} \geq \tau_k(n)\}, \quad k = 1, \ldots, K = M, \tag{7.43}$$

Figure (7.6) (Right) depicts the regions $B_k(n)$ in (7.43) for $M = 2$. Note that since each entry of $\mathbf{x}(n)$ offers a distinct scalar observation, the selection $K = M$ amounts to a bandwidth constraint of 1 *bit per sensor per dimension*.

The rationale behind this selection of regions is that the resultant binary observations $b_k(n)$ are independent, meaning that $\Pr\{b_{k_1}(n)b_{k_2}(n)\} = \Pr\{b_{k_1}(n)\}\Pr\{b_{k_2}(n)\}$ for $k_1 \neq k_2$. As a result, we have a total of MN independent binary observations to estimate $\mathbf{s}$.

Herein, the Bernoulli parameters $q_k(n)$ take on a particularly simple form in terms of the Gaussian tail function

$$q_k(n) = \int_{\mathbf{e}_k^T(n)\mathbf{u} \geq \tau_k(n)} p_{\mathbf{w}}(\mathbf{u} - \mathbf{H}_n\mathbf{s}) \, d\mathbf{u} = Q\left(\frac{\tau_k(n) - \mathbf{e}_k^T(n)\mathbf{H}_n\mathbf{s}}{\sigma_k(n)}\right) := Q[\Delta_k(n)], \tag{7.44}$$

where we introduced the σ-*distance* between $\mathbf{H}_n\mathbf{s}$ and the corresponding threshold $\Delta_k(n) := [\tau_k(n) - \mathbf{e}_k^T(n)\mathbf{H}_n\mathbf{s}]/\sigma_k(n)$.

Due to the independence among binary observations we have $p(\mathbf{b}(n)) = \prod_{k=1}^{K}[q_k(n)]^{b_k(n)}$ $[1 - q_k(n)]^{1-b_k(n)}$, leading to

$$L(\mathbf{s}) = \sum_{n=0}^{N-1} \sum_{k=1}^{K} b_k(n)\ln q_k(n) + [1 - b_k(n)]\ln[1 - q_k(n)], \tag{7.45}$$

whose NK *independent* summands replace the $N2^K$ *dependent* terms in (7.40).

Since the regions $B_k(n)$ are half-spaces, Proposition 7.4.1 applies to the maximization of (7.45) and guarantees that the numerical search for the $\hat{s}$ estimator in (7.45) is well-conditioned and will converge to the global maximum. More important, it will turn out that these regions render finite sample performance analysis of the MLE in (7.41), tractable. In particular, it is possible to derive a closed-form expression for the Fisher Information Matrix (FIM) (Kay 1993, p.44), as we outline next; see (Ribeiro and Giannakis 2006b) for detailed derivations.

Proposition 7.4.2 *The FIM,* **I***, for estimating* **s** *based on the binary observations obtained from the regions defined in (7.43), is given by*

$$
\mathbf{I} = \sum_{n=0}^{N-1} \mathbf{H}_n^T \left[\sum_{k=1}^{K} \frac{e^{-\Delta_k^2(n)} \mathbf{e}_k(n) \mathbf{e}_k^T(n)}{2\pi \sigma_k^2(n) Q(\Delta_k(n))[1 - Q(\Delta_k(n))]} \right] \mathbf{H}_n. \tag{7.46}
$$

Inspection of (7.46) shows that the variance of the MLE in (7.41) depends on the signal function containing the parameter of interest (via $\mathbf{H}_n$), the noise structure and power (via the eigenvalues and eigenvectors), and the selection of the regions $B_k(n)$ (via the σ-distances). Among these three factors only the last one is inherent to the bandwidth constraint, the other two being common to the estimator that is based on the original $\mathbf{x}(n)$ observations.

The last point is clarified if we consider the FIM $\mathbf{I}_x$ for estimating **s** given the unquantized vector $\mathbf{x}(n)$. This matrix can be shown to be (Ribeiro and Giannakis 2006b, Appendix. D),

$$
\mathbf{I}_x = \sum_{n=0}^{N-1} \mathbf{H}_n^T \left[\sum_{m=1}^{M} \frac{\mathbf{e}_m(n) \mathbf{e}_m^T(n)}{\sigma_m^2(n)} \right] \mathbf{H}_n^T. \tag{7.47}
$$

If we define the equivalent noise powers as

$$
\rho_k^2(n) := \frac{2\pi Q(\Delta_k(n))[1 - Q(\Delta_k(n))]}{e^{-\Delta_k^2(n)}} \sigma_k^2(n), \tag{7.48}
$$

we can rewrite (7.46) in the form

$$
\mathbf{I} = \sum_{n=0}^{N-1} \mathbf{H}_n^T \left[\sum_{k=1}^{K} \frac{\mathbf{e}_k(n) \mathbf{e}_k^T(n)}{\rho_k^2(n)} \right] \mathbf{H}_n^T, \tag{7.49}
$$

which except for the noise powers has form identical to (7.47). Thus, comparison of (7.49) with (7.47) reveals that from a performance perspective, *the use of binary observations is equivalent to an increase in the noise variance* from $\sigma_k^2(n)$ to $\rho_k^2(n)$, while the rest of the problem structure remains unchanged. Since we certainly want the equivalent noise increase to be as small as possible, minimizing (7.48) over $\Delta_k(n)$ calls for this distance to be set to zero, or equivalently, to select thresholds $\tau_k(n) = \mathbf{e}_k^T(n) \mathbf{H}_n \mathbf{s}$. In this case, the equivalent noise power is

$$
\rho_k^2(n) = \frac{\pi}{2} \sigma_k^2(n). \tag{7.50}
$$

Surprisingly, even in the vector case a judicious selection of the regions $B_k(n)$ can result in a very small penalty ($\pi/2$) in terms of the equivalent noise increase. Similar to Section 7.2, we can thus claim that while requiring the transmission of 1 bit per sensor per dimension,

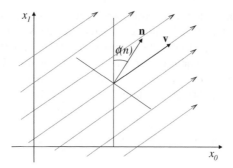

Figure 7.7 The vector flow **v** incises over a certain sensor capable of measuring the normal component of **v**.

the variance of the MLE in (7.41), based on $\{\mathbf{b}(n)\}_{n=0}^{N-1}$, yields a variance close to the clairvoyant estimator's variance – which is based on $\{\mathbf{x}(n)\}_{n=0}^{N-1}$ – for low-to-medium Q-SNR problems.

Example 7.4.3 *Suppose we wish to estimate a vector flow using incidence observations. With reference to Figure 7.7, consider the flow vector* $\mathbf{v} := (v_0, v_1)^T$, *and a sensor positioned at an angle* $\phi(n)$ *with respect to a known reference direction. We will rely on a set of so called incidence observations* $\{x(n)\}_{n=0}^{N-1}$ *measuring the component of the flow normal to the corresponding sensor*

$$x(n) := \langle \mathbf{v}, \mathbf{n} \rangle + w(n) = v_0 \sin[\phi(n)] + v_1 \cos[\phi(n)] + w(n), \qquad (7.51)$$

where $\langle, \rangle$ *denotes inner product,* $w(n)$ *is zero-mean AWGN, and* $n = 0, 1, \ldots, N-1$ *is the sensor index. The model (7.51) applies to the measurement of hydraulic fields, pressure variations induced by wind and radiation from a distant source (Mainwaring et al. 2002).*

Estimating **v** *fits the framework presented in this section requiring the transmission of a single binary observation per sensor,* $b_1(n) = \mathbf{1}\{x(n) \geq \tau_1(n)\}$. *The FIM in (7.49) is easily found to be*

$$\mathbf{I} = \sum_{n=0}^{N-1} \frac{1}{\rho_1^2(n)} \begin{pmatrix} \sin^2[\phi(n)] & \sin[\phi(n)]\cos[\phi(n)] \\ \sin[\phi(n)]\cos[\phi(n)] & \cos^2[\phi(n)] \end{pmatrix}. \qquad (7.52)$$

Furthermore, since $x(n)$ *in (7.51) is linear in* **v** *and the noise pdf is log-concave (Gaussian) the log-likelihood function is concave as asserted by Proposition 7.4.1.*

Suppose that we are able to place the thresholds optimally as implied by $\tau_1(n) = v_0 \sin[\phi(n)] + v_1 \cos[\phi(n)]$, *so that* $\rho_1^2(n) = (\pi/2)\sigma^2$. *If we also make the reasonable assumption that the angles are random and uniformly distributed,* $\phi(n) \sim U[-\pi, \pi]$, *then the average FIM turns out to be:*

$$\bar{\mathbf{I}} = \frac{2}{\pi\sigma^2} \begin{pmatrix} N/2 & 0 \\ 0 & N/2 \end{pmatrix}. \qquad (7.53)$$

But according to the law of large numbers $\mathbf{I} \approx \bar{\mathbf{I}}$, *and the estimation variance will be approximately*

$$\text{var}(v_0) = \text{var}(v_1) = \frac{\pi\sigma^2}{N}. \qquad (7.54)$$

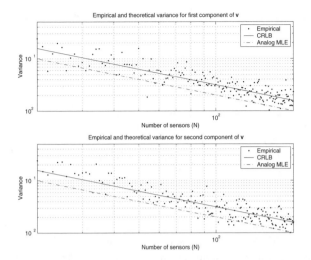

Figure 7.8 Average variance for the components of **v**. The empirical as well as the bound (7.54) are compared with the analog observations based MLE (**v** $= (1, 1)$, $\sigma = 1$).

Figure 7.8 depicts the bound (7.54), as well as the simulated variances $\text{var}(\hat{v}_0)$ *and* $\text{var}(\hat{v}_1)$ *in comparison with the clairvoyant MLE based on* $\{x(n)\}_{n=0}^{N-1}$, *corroborating our analytical expressions.*

7.5 Maximum a Posteriori Probability Estimation

The parameter of interest **s** was so far assumed deterministic. Consequently, the MLE was considered as the optimum estimator and the CRLB as the ultimate performance limit. An alternative formulation is to use available a priori knowledge to model **s** as a random vector parameter with a priory pdf $p_{\mathbf{s}}(\mathbf{s})$, estimate **s** using a maximum a posteriori (MAP) probability estimator, and regard the MSE as the performance indicator. We will show in this section that despite the different formulation we can obtain results similar to those described in Section 7.4.

Let us recall the observation model in (7.35), denote the mean of **s** as $\text{E}(\mathbf{s}) := \boldsymbol{\mu}_{\mathbf{s}}$ and suppose the noise vector is white and Gaussian i.e., $\text{E}[\mathbf{w}(n)\mathbf{w}^T(n)] = \text{diag}[\sigma_1^2(n), \ldots, \sigma_M^2(n)]$. In this case, we write $\mathbf{H}_n := [\mathbf{h}_{n1}, \ldots, \mathbf{h}_{nM}]^T$ and define the (independent) binary observations $\mathbf{b}(n) := [b_1(n), \ldots, b_M(n)]$ as

$$b_k(n) := \mathbf{1}\{x_k(n) > \mathbf{h}_{nk}^T \boldsymbol{\mu}_{\mathbf{s}}\} , \tag{7.55}$$

for $k \in [1, M]$. The resemblance with the problem of Section 7.4 is clear and not surprisingly the following proposition holds true (Shah et al. 2005).

Proposition 7.5.1 *Consider a vector parameter* **s**, *with log-concave prior distribution* $p_{\mathbf{s}}(\mathbf{s})$, *the model in (7.35) with* $p_{\mathbf{w}}(\mathbf{w})$ *white Gaussian with* $\text{E}[\mathbf{w}(n)\mathbf{w}^T(n)] = \text{diag}[\sigma_1^2(n), \ldots,$

$\sigma_M^2(n)$]; *and binary messages* $\{\mathbf{b}(n)\}_{n=0}^{N-1}$ *as in (7.55). Then, if we define the per sensor log-likelihood* $L_n(\mathbf{s})$ *as*

$$L_n(\mathbf{s}) = \sum_{k=1}^{M} \ln Q\left(\frac{b_k(n)\mathbf{h}_{nk}^T\left[\boldsymbol{\mu}_\mathbf{s} - \mathbf{s}\right]}{\sigma_k(n)}\right). \tag{7.56}$$

(a) The MAP estimator of $\mathbf{s}$ *based on* $\{\mathbf{b}(n)\}_{n=0}^{N-1}$ *is given by*

$$\hat{\mathbf{s}}_{\text{MAP}} = \arg\max\left[\sum_{n=0}^{N-1} L_n(\mathbf{s})\right] + \ln[p_\mathbf{s}(\mathbf{s})] := \arg\max L(\mathbf{s}). \tag{7.57}$$

(b) The log-likelihood $L(\mathbf{s})$ *is a concave function of* $\mathbf{s}$.

Proposition 7.5.1 establishes that at least for white Gaussian noise the comments in Remark 3 carry over to MAP based parameter estimation. In fact, Proposition 7.5.1 has been established under much more general assumptions, including the case of colored Gaussian noise (Shah et al. 2005).

7.5.1 Mean-Squared Error

For estimation of random parameters bounds on the MSE can be obtained by computing the pertinent Fisher Information Matrix (FIM) $\mathbf{J}$ that can be expressed as the sum of two parts (Van Trees 1968, p. 84):

$$\mathbf{J} = \mathbf{J}_D + \mathbf{J}_P, \tag{7.58}$$

where $\mathbf{J}_D$ represents information obtained from the data, and $\mathbf{J}_P$ captures *a priori* information. The MSE of the ith component of $\mathbf{s}$ is bounded by the ith diagonal element of $\mathbf{J}$; i.e.,

$$\text{MSE}(\hat{s}_i) \geq \left[\mathbf{J}^{-1}\right]_{ii}. \tag{7.59}$$

Also, note that for any FIM, $[\mathbf{J}^{-1}]_{ii} \geq 1/[\mathbf{J}]_{ii}$ (Kay 1993). This property yields a different bound on $\text{MSE}(\hat{s}_i)$

$$\text{MSE}(\hat{s}_i) \geq \frac{1}{[\mathbf{J}]_{ii}}, \tag{7.60}$$

which is easier to compute although not tight in general.

The following proposition provides a bound (exact value) on $[\mathbf{J}]_{ii}$ when binary (analog-amplitude) observations are used (Shah et al. 2005).

Proposition 7.5.2 *Consider the signal model in (7.35) with* $\mathbf{w}(n)$ *white Gaussian with covariance matrix* $\text{E}[\mathbf{w}(n)\mathbf{w}^T(n)] = \text{diag}[\sigma_1^2(n), \ldots, \sigma_M^2(n)]$ *and Gaussian prior distribution with covariance* $\text{E}[\mathbf{ss}^T] = \mathbf{C}_\mathbf{s}$. *Write (7.35) componentwise as* $x_k(n) = \mathbf{h}_{nk}^T\mathbf{s} + w_k(n)$. *Then, the* i^{th} *diagonal element of the FIM* $\mathbf{J}$ *in (7.58) satisfies:*

(a) when binary observations as in (7.55) are used

$$[\mathbf{J}]_{ii} \geq \frac{2}{\pi}\sum_{n=0}^{N-1}\sum_{k=1}^{M}\frac{h_{nki}^2}{\sigma_k(n)\sqrt{\sigma_k^2(n) + \mathbf{h}_{nk}^T\mathbf{C}_\mathbf{s}\mathbf{h}_{nk}}} + \left[\mathbf{C}_\mathbf{s}^{-1}\right]_{ii} \tag{7.61}$$

(b) when analog-amplitude observations are used

$$[\mathbf{J}_{CV}]_{ii} = \sum_{n=0}^{N-1} \sum_{k=1}^{M} \frac{h_{nki}^2}{\sigma_k^2(n)} + [\mathbf{C}_s^{-1}]_{ii} \,. \tag{7.62}$$

Comparing (7.61) with (7.62) the analogy with the result in Proposition 7.4.2 becomes clear. Indeed, we can define the equivalent noise powers as

$$\rho_k^2(n) = \frac{\pi}{2} \sigma_w^2 \sqrt{1 + \frac{\mathbf{h}_{nk}^T \mathbf{C}_s \mathbf{h}_{nk}}{\sigma_k^2(n)}} \tag{7.63}$$

so that we can express the bound in (7.61) as

$$[\mathbf{J}_{CV}]_{ii} = \sum_{n=0}^{N-1} \sum_{k=1}^{M} \frac{h_{nki}^2}{\rho_k^2(n)} + [\mathbf{C}_s^{-1}]_{ii} \,. \tag{7.64}$$

As in the case of deterministic parameters, the effect of quantization in MSE is equivalent to a noise power increase from $\sigma_k^2(n)$ to $\rho_k^2(n)$ [cf. (7.62) and (7.64)]. In the case of random signals, the average SNR of the observations $x_k(n)$ is well defined and given by $\gamma_{nk} := \mathbf{h}_{nk}^T \mathbf{C}_s \mathbf{h}_{nk} / \sigma_k^2(n)$. Using the latter and (7.64), we infer that the equivalent noise increase is

$$\mathcal{L}_k(n) := \frac{\rho_k^2(n)}{\sigma_k^2(n)} = \frac{\pi}{2} \sqrt{1 + \gamma_{nk}}. \tag{7.65}$$

Note that as $\gamma_{nk} \to 0$, the information loss $\mathcal{L}_k(n) \to \pi/2$ corroborating the results in Section 7.4 for deterministic parameter estimation. In any event, it is worth re-iterating the remarkable fact that for low to medium SNR γ, the equivalent noise increase $\mathcal{L}_K$ is small.

7.6 Dimensionality Reduction for Distributed Estimation

In this section, we consider linear distributed estimation of random signals when the sensors observe and transmit analog-amplitude data. Consider the WSN depicted in Figure 7.9, comprising N sensors linked with an FC. Each sensor, say the nth one, observes an $M_n \times 1$ vector $\mathbf{x}_n$ that is correlated with a $p \times 1$ random signal of interest $\mathbf{s}$. Through a $k_n \times M_n$ fat matrix $\mathbf{C}_n$, each sensor transmits a compressed $k_n \times 1$ vector $\mathbf{C}_n \mathbf{x}_n$, using e.g., multicarrier modulation with one entry riding per subcarrier. Low-power and bandwidth constraints at the sensors encourage transmissions with $k_n \ll M_n$, while linearity in compression and estimation are well motivated by low-complexity requirements. Furthermore, we assume that:

(a1) No information is exchanged among sensors, and each sensor-FC link comprises a $k_n \times k_n$ full rank fading multiplicative channel matrix $\mathbf{D}_n$ along with zero-mean additive FC noise $\mathbf{z}_n$, which is uncorrelated with $\mathbf{x}_n$, $\mathbf{D}_n$, and across channels; i.e., noise covariance matrices satisfy $\boldsymbol{\Sigma}_{z_{n_1} z_{n_2}} = \mathbf{0}$ for $n_1 \neq n_2$. Matrices $\{\mathbf{D}_n, \boldsymbol{\Sigma}_{z_n z_n}\}_{n=0}^{N-1}$ are available at the FC but not at the sensors.

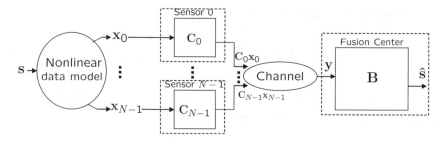

Figure 7.9 Distributed setup for estimating a random signal **s**.

(a2) Data $\mathbf{x}_n$ and the signal of interest $\mathbf{s}$ are zero-mean with full rank auto- and cross-covariance matrices $\boldsymbol{\Sigma}_{ss}$, $\boldsymbol{\Sigma}_{sx_n}$ and $\boldsymbol{\Sigma}_{x_{n_1}x_{n_2}}$ $\forall\, n_1, n_2 \in [0, N-1]$, all of which are available at the FC.

If sensors communicate with the FC using multicarrier modulation, full rank of the channel matrices $\{\mathbf{D}_n\}_{n=0}^{N-1}$ is ensured if sensors do not transmit over subcarriers with zero channel gain. Matrices $\{\mathbf{D}_n\}_{n=0}^{N-1}$ can be acquired via training, and likewise the signal and noise covariances in (a1) and (a2) can be estimated via sample averaging as usual. With multicarrier (and generally any orthogonal) sensor access, the noise uncorrelatedness across channels is also well justified.

Sensors transmit over orthogonal channels so that the FC separates and concatenates the received vectors $\{\mathbf{y}_n(\mathbf{C}_n) = \mathbf{D}_n\mathbf{C}_n\mathbf{x}_n + \mathbf{z}_n\}_{n=0}^{N-1}$, to obtain the $\sum_{n=0}^{N-1} k_n \times 1$ vector

$$\mathbf{y}(\mathbf{C}_0, \ldots, \mathbf{C}_{N-1}) = \mathrm{diag}(\mathbf{D}_0\mathbf{C}_0, \ldots, \mathbf{D}_{N-1}\mathbf{C}_{N-1})\mathbf{x} + \mathbf{z}, \qquad (7.66)$$

Left multiplying $\mathbf{y}$ by a $p \times (\sum_{n=0}^{N-1} k_n)$ matrix $\mathbf{B}$, we form the linear estimate $\hat{\mathbf{s}}$ of $\mathbf{s}$. For a prescribed power P_n per sensor, our problem is to obtain under (a1)-(a2) MSE optimal matrices $\{\mathbf{C}_n^o\}_{n=0}^{N-1}$ and $\mathbf{B}^o$; i.e., we seek (tr denotes matrix trace)

$$(\mathbf{B}^o, \{\mathbf{C}_n^o\}_{n=0}^{N-1}) = \arg\min_{\mathbf{B}, \{\mathbf{C}_n\}_{n=0}^{N-1}} E[\|\mathbf{s} - \mathbf{B}\mathbf{y}(\mathbf{C}_0, \ldots, \mathbf{C}_{N-1})\|^2],$$

$$\text{s. to} \quad \mathrm{tr}(\mathbf{C}_n \boldsymbol{\Sigma}_{x_n x_n} \mathbf{C}_n^T) \leq P_n, \quad n \in \{0, \ldots, N-1\}. \qquad (7.67)$$

The FC finds and communicates $\{\mathbf{C}_n^o\}_{n=0}^{N-1}$ to the sensors for them to form $\mathbf{C}_n^o\mathbf{x}_n$. This communication takes place during the startup phase or whenever the data (cross-) correlations change. Note that $\mathbf{C}_n^o\mathbf{x}_n$ involves a matrix-vector multiplication whose complexity is $\mathcal{O}(k_n M_n)$ with $k_n < M_n$, and can be afforded by the sensors.

7.6.1 Decoupled Distributed Estimation-Compression

We consider first the case where $\boldsymbol{\Sigma}_{x_n x_m} \equiv \mathbf{0}$, $\forall n \neq m$, which shows up e.g., when matrices $\{\mathbf{H}_n\}_{n=0}^{N-1}$ in the linear model $\mathbf{x}_n = \mathbf{H}_n\mathbf{s} + \mathbf{w}_n$ are mutually uncorrelated and also uncorrelated with $\mathbf{w}_n$. Then, the multi-sensor optimization task in (7.67) reduces to a set of N decoupled problems. Specifically, it is easy to show that the cost function in (7.67) can be written as (Schizas et al. 2007)

$$J(\mathbf{B}, \{\mathbf{C}_n\}_{n=0}^{N-1}) = \sum_{n=0}^{N-1} E[\|\mathbf{s} - \mathbf{B}_n(\mathbf{D}_n\mathbf{C}_n\mathbf{x}_n + \mathbf{z}_n)\|^2] - (N-1)\mathrm{tr}(\boldsymbol{\Sigma}_{ss}) \qquad (7.68)$$

where $\mathbf{B}_n$ is the $p \times k_n$ submatrix of $\mathbf{B} := [\mathbf{B}_0 \ldots \mathbf{B}_{N-1}]$. As the nth non-negative summand depends only on $\mathbf{B}_n$ and $\mathbf{C}_n$, the MSE optimal matrices are given by

$$(\mathbf{B}_n^o, \ \mathbf{C}_n^o) = \arg\min_{\mathbf{B}_n, \mathbf{C}_n} E[\|\mathbf{s} - \mathbf{B}_n(\mathbf{D}_n \mathbf{C}_n \mathbf{x}_n + \mathbf{z}_n)\|^2],$$

$$\text{s. to} \quad \text{tr}(\mathbf{C}_n \mathbf{\Sigma}_{x_n x_n} \mathbf{C}_n^T) \leq P_n, \quad n \in \{0, \ldots, N-1\}. \tag{7.69}$$

Since the cost function in (7.69) corresponds to a single-sensor setup ($N = 1$), we will drop the subscript n for notational brevity and write $\mathbf{B}_n = \mathbf{B}, \mathbf{C}_n = \mathbf{C}, \mathbf{x}_n = \mathbf{x}, \mathbf{z}_n = \mathbf{z}, P_n = P$ and $k_n = k$. The Lagrangian for minimizing (7.68) can be easily written as:

$$J(\mathbf{B}, \mathbf{C}, \mu) = J_o + \text{tr}(\mathbf{B}\mathbf{\Sigma}_{zz}\mathbf{B}^T) + \mu[\text{tr}(\mathbf{C}\mathbf{\Sigma}_{xx}\mathbf{C}^T) - P]$$

$$+ \text{tr}[(\mathbf{\Sigma}_{sx} - \mathbf{B}\mathbf{D}\mathbf{C}\mathbf{\Sigma}_{xx})\mathbf{\Sigma}_{xx}^{-1}(\mathbf{\Sigma}_{xs} - \mathbf{\Sigma}_{xx}\mathbf{C}^T\mathbf{D}^T\mathbf{B}^T)], \tag{7.70}$$

where $J_o := \text{tr}(\mathbf{\Sigma}_{ss} - \mathbf{\Sigma}_{sx}\mathbf{\Sigma}_{xx}^{-1}\mathbf{\Sigma}_{xs})$ is the minimum attainable MMSE for linear estimation of $\mathbf{s}$ based on $\mathbf{x}$.

In what follows, we derive a simplified form of (7.70) the minimization of which will provide closed-form solutions for the MSE optimal matrices $\mathbf{B}^o$ and $\mathbf{C}^o$. Aiming at this simplification, consider the SVD $\mathbf{\Sigma}_{sx} = \mathbf{U}_{sx}\mathbf{S}_{sx}\mathbf{V}_{sx}^T$, and the eigen-decompositions $\mathbf{\Sigma}_{zz} = \mathbf{Q}_z\mathbf{\Lambda}_z\mathbf{Q}_z^T$ and $\mathbf{D}^T\mathbf{\Sigma}_{zz}^{-1}\mathbf{D} = \mathbf{Q}_{zd}\mathbf{\Lambda}_{zd}\mathbf{Q}_{zd}^T$, where $\mathbf{\Lambda}_{zd} := \text{diag}(\lambda_{zd,1} \cdots \lambda_{zd,k})$ and $\lambda_{zd,1} \geq \cdots \geq \lambda_{zd,k} > 0$. Notice that $\lambda_{zd,i}$ captures the SNR of the ith entry in the received signal vector at the FC. Further, define $\mathbf{A} := \mathbf{Q}_x^T\mathbf{V}_{sx}\mathbf{S}_{sx}^T\mathbf{S}_{sx}\mathbf{V}_{sx}^T\mathbf{Q}_x$ with $\rho_a := \text{rank}(\mathbf{A}) = \text{rank}(\mathbf{\Sigma}_{sx})$, and $\mathbf{A}_x := \mathbf{\Lambda}_x^{-1/2}\mathbf{A}\mathbf{\Lambda}_x^{-1/2}$ with corresponding eigen-decomposition $\mathbf{A}_x = \mathbf{Q}_{ax}\mathbf{\Lambda}_{ax}\mathbf{Q}_{ax}$, where $\mathbf{\Lambda}_{ax} = \text{diag}(\lambda_{ax,1}, \cdots, \lambda_{ax,\rho_a}, 0, \cdots, 0)$ and $\lambda_{ax,1} \geq \ldots \geq \lambda_{ax,\rho_a} > 0$. Moreover, let $\mathbf{V}_a := \mathbf{\Lambda}_x^{-1/2}\mathbf{Q}_{ax}$ denote the invertible matrix which simultaneously diagonalizes the matrices $\mathbf{A}$ and $\mathbf{\Lambda}_x$. Since matrices $(\mathbf{Q}_{zd}, \mathbf{Q}_x, \mathbf{V}_a, \mathbf{U}_{sx}, \mathbf{\Lambda}_{zd}, \mathbf{Q}_{zd}, \mathbf{D}, \mathbf{\Sigma}_{zz})$ are all invertible, for every matrix $\mathbf{C}$ (or $\mathbf{B}$) we can clearly find a unique matrix $\mathbf{\Phi}_C$ (correspondingly $\mathbf{\Phi}_B$) that satisfies:

$$\mathbf{C} = \mathbf{Q}_{zd}\mathbf{\Phi}_C\mathbf{V}_a^T\mathbf{Q}_x^T, \quad \mathbf{B} = \mathbf{U}_{sx}\mathbf{\Phi}_B\mathbf{\Lambda}_{zd}^{-1}\mathbf{Q}_{zd}^T\mathbf{D}^T\mathbf{\Sigma}_{zz}^{-1}, \tag{7.71}$$

where $\mathbf{\Phi}_C := [\phi_{c,ij}]$ and $\mathbf{\Phi}_B$ have sizes $k \times M$ and $p \times k$, respectively. Using (7.71), the Lagrangian in (7.70) becomes

$$J(\mathbf{\Phi}_C, \mu) = J_o + \text{tr}(\mathbf{\Lambda}_{ax}) + \mu(\text{tr}(\mathbf{\Phi}_C\mathbf{\Phi}_C^T) - P) \tag{7.72}$$

$$- \text{tr}\left((\mathbf{\Lambda}_{zd}^{-1} + \mathbf{\Phi}_C\mathbf{\Phi}_C^T)^{-1}\mathbf{\Phi}_C\mathbf{\Lambda}_{ax}\mathbf{\Phi}_C^T\right).$$

Applying the well-known Karush-Kuhn-Tucker (KKT) conditions (e.g., (Boyd and Vandenberghe 2004, Ch. 5)) that must be satisfied at the minimum of (7.72), it can be shown that the matrix $\mathbf{\Phi}_C^o$ minimizing (7.72), is diagonal with diagonal entries (Schizas et al. 2007)

$$\phi_{c,ii}^o = \begin{cases} \pm\sqrt{\left(\dfrac{\lambda_{ax,i}}{\mu^o\lambda_{zd,i}}\right)^{1/2} - \dfrac{1}{\lambda_{zd,i}}}, & 1 \leq i \leq \kappa \\ 0, & \kappa+1 \leq i \leq k \end{cases} \tag{7.73}$$

where κ is the maximum integer in $[1, k]$ for which $\{\phi_{c,ii}^o\}_{i=1}^{\kappa}$ are strictly positive, or, $\text{rank}(\mathbf{\Phi}_C^o) = \kappa$; and μ^o is chosen to satisfy the power constraint $\sum_{i=1}^{\kappa}(\phi_{c,ii}^o)^2 = P$ as

$$\mu^o = \frac{(\sum_{i=1}^{\kappa}(\lambda_{ax,i}\lambda_{zd,i}^{-1})^{1/2})^2}{(P + \sum_{i=1}^{\kappa}\lambda_{zd,i}^{-1})^2}. \tag{7.74}$$

When $k > \rho_a$, the MMSE remains invariant (Schizas et al. 2007); thus, it suffices to consider $k \in [1, \rho_a]$. Summarizing, it has been established that:

Proposition 7.6.1 *Under (a1), (a2), and for $k \leq \rho_a$, the matrices minimizing $J(\mathbf{B}_{p \times k}, \mathbf{C}_{k \times M}) = E[\|\mathbf{s} - \mathbf{B}_{p \times k}(\mathbf{D}\mathbf{C}_{k \times M}\mathbf{x} + \mathbf{z})\|^2]$, subject to $tr(\mathbf{C}_{k \times M}\boldsymbol{\Sigma}_{xx} \mathbf{C}_{k \times M}^T) \leq P$, are:*

$$\mathbf{C}^o = \mathbf{Q}_{zd}\boldsymbol{\Phi}_C^o\mathbf{V}_a^T\mathbf{Q}_x^T, \tag{7.75}$$

$$\mathbf{B}^o = \boldsymbol{\Sigma}_{sx}\mathbf{Q}_x\mathbf{V}_a\boldsymbol{\Phi}_C^{o\,T}\left(\boldsymbol{\Phi}_C^o\boldsymbol{\Phi}_C^{o\,T} + \boldsymbol{\Lambda}_{zd}^{-1}\right)^{-1}\boldsymbol{\Lambda}_{zd}^{-1}\mathbf{Q}_{zd}^T\mathbf{D}^T\boldsymbol{\Sigma}_{zz}^{-1},$$

where $\boldsymbol{\Phi}_C^o$ is given by (7.73), and the corresponding Lagrange multiplier μ^o is specified by (7.74). The MMSE is

$$J_{\min}(k) = J_o + \sum_{i=1}^{\rho_a} \lambda_{ax,i} - \sum_{i=1}^{k} \frac{\lambda_{ax,i}(\phi_{c,ii}^o)^2}{\lambda_{zd,i}^{-1} + (\phi_{c,ii}^o)^2}. \tag{7.76}$$

According to Proposition 7.6.1, the optimal weight matrix $\boldsymbol{\Phi}_C^o$ in $\mathbf{C}^o$ distributes the given power across the entries of the pre-whitened vector $\mathbf{V}_a^T\mathbf{Q}_x\mathbf{x}$ at the sensor in a waterfilling-like manner so as to balance channel strength and additive noise variance at the FC with the degree of dimensionality reduction that can be afforded. It is worth mentioning that (7.73) dictates a minimum power per sensor. Specifically, in order to ensure that rank$(\boldsymbol{\Phi}_C^o) = \kappa$ the power must satisfy

$$P > \frac{\sum_{i=1}^{\kappa}(\lambda_{ax,i}\lambda_{zd,i}^{-1})^{1/2}}{\sqrt{\lambda_{ax,\kappa}\lambda_{zd,\kappa}}} - \sum_{i=1}^{\kappa}\lambda_{zd,i}^{-1}. \tag{7.77}$$

The optimal matrices in Proposition 7.6.1 can be viewed as implementing a two-step scheme, where: i) $\mathbf{s}$ is estimated based on $\mathbf{x}$ at the sensor using the LMMSE estimate $\hat{\mathbf{s}}_{LM} = \boldsymbol{\Sigma}_{sx}\boldsymbol{\Sigma}_{xx}^{-1}\mathbf{x}$; and
ii) compress and reconstruct $\hat{\mathbf{s}}_{LM}$ using the optimal matrices $\mathbf{C}^o$ and $\mathbf{B}^o$ implied by Proposition 7.6.1 after replacing $\mathbf{x}$ with $\hat{\mathbf{s}}_{LM}$. For this estimate-first compress-afterwards (EC) interpretation, (Schizas et al. 2007) have proved that:

Corollary 7.6.2 *For $k \in [1, \rho_a]$, the $k \times M$ matrix in (7.75) can be written as $\mathbf{C}^o = \hat{\mathbf{C}}^o\boldsymbol{\Sigma}_{sx}\boldsymbol{\Sigma}_{xx}^{-1}$, where $\hat{\mathbf{C}}^o$ is the $k \times p$ optimal matrix obtained by Proposition 7.6.1 when $\mathbf{x} = \hat{\mathbf{s}}_{LM}$. Thus, the EC scheme is MSE optimal in the sense of minimizing (7.68).*

Another interesting feature of the EC scheme implied by Proposition 7.6.1 is that the MMSE $J_{\min}(k)$ is non-increasing with respect to the reduced dimensionality k, given a limited power budget per sensor. Specifically, (Schizas et al. 2007) have shown that that:

Corollary 7.6.3 *If $\mathbf{C}_{k_1 \times M}^o$ and $\mathbf{C}_{k_2 \times M}^o$ are the optimal matrices determined by Proposition 7.6.1 with $k_1 < k_2$, under the same channel parameters $\lambda_{zd,i}$ for $i = 1, \ldots, k_1$, and common power P, the MMSE in (7.76) is non-increasing; i.e., $J_{\min}(k_1) \geq J_{\min}(k_2)$ for $k_1 < k_2$.*

Notice that Corollary 7.6.3 advocates the efficient power allocation that the EC scheme performs among the compressed components. To assess the difference in handling noise effects, it is useful to compare the EC scheme with the methods in (Zhu et al. 2005) which we abbreviate as C$'$E, and (Zhang et al. 2003) abbreviated as EC-d. Although C$'$E and EC-d have been derived under ideal link conditions, they can be modified here to account for $\mathbf{D}_n$. The comparisons will further include an option we term CE, which compresses first the data and reconstructs them at the FC using $\mathbf{C}^o$ and $\mathbf{B}^o$ found by (7.75) after setting $\mathbf{s} = \mathbf{x}$, and then estimates $\mathbf{s}$ based on the reconstructed data vector $\hat{\mathbf{x}}$. For benchmarking purposes, we also depict J_o, achieved when estimating $\mathbf{s}$ based on uncompressed data transmitted over ideal links. Figure 7.10 (Left) depicts the MMSE versus k for J_o, EC, CE, C$'$E and EC-d for a linear model $\mathbf{x} = \mathbf{H}\mathbf{s} + \mathbf{w}$, where $M = 50$ and $p = 10$. The matrices $\mathbf{H}$, $\mathbf{\Sigma}_{ss}$ and $\mathbf{\Sigma}_{ww}$, are selected randomly such that $\text{tr}(\mathbf{H}\mathbf{\Sigma}_{ss}\mathbf{H}^T)/\text{tr}(\mathbf{\Sigma}_{ww}) = 2$, while $\mathbf{s}$ and $\mathbf{w}$ are uncorrelated. We set $\mathbf{\Sigma}_{zz} = \sigma_z^2 \mathbf{I}_k$, and select P such that $10\log_{10}(P/\sigma_z^2) = 7$dB. As expected J_o benchmarks all curves, while the worst performance is exhibited by C$'$E. Albeit suboptimal, CE comes close to the optimal EC. Contrasting it with the increase EC-d exhibits in MMSE beyond a certain k, we can appreciate the importance of coping with noise effects. This increase is justifiable since each entry of the compressed data in EC-d is allocated a smaller portion of the given power as k grows. In EC, however, the quality of channel links and the available power determine the number of the compressed components, and allocate power optimally among them.

7.6.2 Coupled Distributed Estimation-Compression

In this section, we allow the sensor observations to be correlated. Because $\mathbf{\Sigma}_{xx}$ is no longer block diagonal, decoupling of the multi-sensor optimization problem cannot be effected in this case. The pertinent MSE cost is [cf. (7.67)]

$$J(\{\mathbf{B}_n, \mathbf{C}_n\}_{n=0}^{N-1}) = E[\|\mathbf{s} - \sum_{n=0}^{N-1} \mathbf{B}_n(\mathbf{D}_n\mathbf{C}_n\mathbf{x}_n + \mathbf{z}_n)\|^2]. \quad (7.78)$$

Minimizing (7.78) does not lead to a closed-form solution and incurs complexity that grows exponentially with N (Luo et al. 2005). For this reason, we resort to iterative alternatives which converge at least to a stationary point of the cost in (7.78). To this end, let us suppose temporarily that matrices $\{\mathbf{B}_l\}_{l=0,l\neq n}^{N-1}$ and $\{\mathbf{C}_l\}_{l=0,l\neq n}^{N-1}$ are fixed and satisfy the power constraints $\text{tr}(\mathbf{C}_l\mathbf{\Sigma}_{x_lx_l}\mathbf{C}_l^T) = P_l$, for $l = 0, \ldots, N-1$ and $l \neq n$. Upon defining the vector $\bar{\mathbf{s}}_n := \mathbf{s} - \sum_{l=0,l\neq n}^{N-1}(\mathbf{B}_l\mathbf{D}_l\mathbf{C}_l\mathbf{x}_l + \mathbf{B}_l\mathbf{z}_l)$ the cost in (7.78) becomes

$$J(\mathbf{B}_n, \mathbf{C}_n) = E[\|\bar{\mathbf{s}}_n - \mathbf{B}_n\mathbf{D}_n\mathbf{C}_n\mathbf{x}_n - \mathbf{B}_n\mathbf{z}_n\|^2], \quad (7.79)$$

which being a function of $\mathbf{C}_n$ and $\mathbf{B}_n$ only, falls under the realm of Proposition 7.6.1. This means that when $\{\mathbf{B}_l\}_{l=0,l\neq n}^{N-1}$ and $\{\mathbf{C}_l\}_{l=0,l\neq n}^{N-1}$ are given, the matrices $\mathbf{B}_n$ and $\mathbf{C}_n$ minimizing (7.79) under the power constraint $\text{tr}(\mathbf{C}_n\mathbf{\Sigma}_{x_nx_n}\mathbf{C}_n^T) \leq P_n$ can be directly obtained from (7.75), after setting $\mathbf{s} = \bar{\mathbf{s}}_n$, $\mathbf{x} = \mathbf{x}_n$, $\mathbf{z} = \mathbf{z}_n$ and $\rho_a = \text{rank}(\mathbf{\Sigma}_{\bar{s}_nx_n})$ in Proposition 7.6.1. The corresponding auto- and cross-covariance matrices needed must also be modified as $\mathbf{\Sigma}_{ss} = \mathbf{\Sigma}_{\bar{s}_n\bar{s}_n}$ and $\mathbf{\Sigma}_{sx_n} = \mathbf{\Sigma}_{\bar{s}_nx_n}$. The following result can thus be established for coupled sensor observations:

Proposition 7.6.4 *If (a1) and (a2) are satisfied, and further $k_n \leq rank(\mathbf{\Sigma}_{\bar{s}_nx_n})$, then for given matrices $\{\mathbf{B}_l\}_{l=0,l\neq n}^{N-1}$ and $\{\mathbf{C}_l\}_{l=0,l\neq n}^{N-1}$ satisfying $tr(\mathbf{C}_l\mathbf{\Sigma}_{x_lx_l}\mathbf{C}_l^T) = P_l$, the optimal $\mathbf{B}_n^o$ and $\mathbf{C}_n^o$ matrices minimizing $E[\|\mathbf{s} - \sum_{l=0}^{N-1}\mathbf{B}_l(\mathbf{D}_l\mathbf{C}_l\mathbf{x}_l + \mathbf{z}_l)\|^2]$ are provided by Proposition 7.6.1, after setting $\mathbf{x} = \mathbf{x}_n$, $\mathbf{s} = \bar{\mathbf{s}}_n$ and applying the corresponding covariance modifications.*

Algorithm 1 :

Initialize randomly the matrices $\{\mathbf{C}_n^{(0)}\}_{n=0}^{N-1}$ and $\{\mathbf{B}_n^{(0)}\}_{n=0}^{N-1}$, such that $\mathrm{tr}(\mathbf{C}_n^{(0)}\boldsymbol{\Sigma}_{x_n x_n}\mathbf{C}_n^{(0)^T}) = P_n$.

$i = 0$

repeat

$\quad i = i + 1$

$\quad$**for** $n = 0, N - 1$ **do**

$\qquad$Given the matrices $\mathbf{C}_0^{(i)}, \mathbf{B}_0^{(i)}, \ldots, \mathbf{C}_{n-1}^{(i)}, \mathbf{B}_{n-1}^{(i)}, \mathbf{C}_{n+1}^{(i-1)}, \mathbf{B}_{n+1}^{(i-1)}, \ldots, \mathbf{C}_{N-1}^{(i-1)}, \mathbf{B}_{N-1}^{(i-1)}$

$\qquad$determine $\mathbf{C}_n^{(i)}, \mathbf{B}_n^{(i)}$ via Proposition 7.6.1

$\quad$**end for**

until $|\mathrm{MSE}^{(i)} - \mathrm{MSE}^{(i-1)}| < \epsilon$ for given tolerance ϵ

Proposition 7.6.4 suggests Algorithm 1 for distributed estimation in the presence of fading and FC noise. Notice that Algorithm 1 belongs to the class of block coordinate descent iterative schemes. At every step n during the ith iteration, it yields the optimal pair of matrices $\mathbf{C}_n^o, \mathbf{B}_n^o$, treating the rest as given. Thus, the $\mathrm{MSE}^{(i)}$ cost per iteration is non-increasing and the algorithm always converges to a stationary point of (7.78). Beyond its applicability to possibly non-Gaussian and nonlinear data models, it is the only available algorithm for handling fading channels and generally colored FC noise effects in distributed estimation.

Next, we illustrate through a numerical example the MMSE performance of Algorithm 1 in a 3-sensor setup using the same linear model as in Section 7.6.1, while setting $M_0 = M_1 = 17$ and $M_2 = 16$. FC noise $\mathbf{z}_n$ is white with variance $\sigma_{z_n}^2$. The power P_n and variance $\sigma_{z_n}^2$ are chosen such that $10\log_{10}(P/\sigma_{z_n}^2) = 13\mathrm{dB}$, for $n = 0, 1, 2$, and $\epsilon = 10^{-3}$. Figure 7.10 (Right) depicts the MMSE as a function of the total number $k_{tot} = \sum_{n=0}^{2} k_n$ of compressed entries across sensors for: i) a centralized EC setup for which a single (virtual) sensor ($N = 1$) has available the data vectors of all three sensors; ii) the estimator returned by Algorithm 1; iii) the decoupled EC estimator which ignores sensor correlations; iv) the C'E estimator and v) an iterative estimator developed in (Schizas et al. 2007), denoted here as EC-d, which similar to C'E accounts for fading but ignores noise. Interestingly, the decentralized Algorithm 1 comes very close to the hypothetical single-sensor bound of the centralized EC estimator, while outperforming the decoupled EC one.

7.7 Distortion-Rate Analysis

In contrast to the previous section, here we consider digital-amplitude data transmission (bits) from the sensors to the FC. In such a setup, all the sensors must adhere to a rate constraint. In order to determine the minimum possible distortion (MSE) between the signal of interest and its estimate at the FC, under encoding rate constraints, we perform Distortion-Rate (D-R) analysis and determine bounds for the D-R function.

Figure 7.11 (Left) depicts a WSN comprising N sensors that communicate with an FC. Each sensor, say the nth, observes an $M_n \times 1$ vector $\mathbf{x}_n(t)$ which is correlated with a $p \times 1$

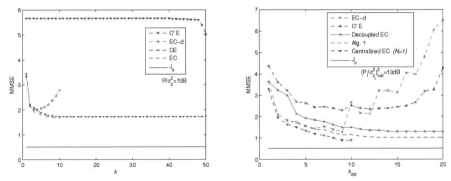

Figure 7.10 MMSE comparisons versus k for a centralized, $N = 1$ (Left), and a distributed 3-sensor setup (Right).

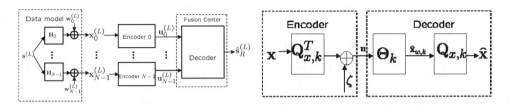

Figure 7.11 (Left): Distributed setup; (Right): Test channel for **x** Gaussian in a point-to-point link.

random signal (parameter vector) of interest $\mathbf{s}(t)$, where t denotes discrete time. Similar to (Oohama 1998; Pandya et al. 2004; Viswanathan and Berger 1997), we assume that:

(a3) No information is exchanged among sensors and the links with the FC are noise-free.

(a4) The random vector $\mathbf{s}(t)$ is generated by a stationary Gaussian vector memoryless source with $\mathbf{s}(t) \sim \mathcal{N}(\mathbf{0}, \boldsymbol{\Sigma}_{ss})$; the sensor data $\{\mathbf{x}_n(t)\}_{n=0}^{N-1}$ adhere to the linear-Gaussian model $\mathbf{x}_n(t) = \mathbf{H}_n \mathbf{s}(t) + \mathbf{w}_n(t)$, where $\mathbf{w}_n(t)$ denotes additive white Gaussian noise (AWGN); i.e., $\mathbf{w}_n(t) \sim \mathcal{N}(\mathbf{0}, \sigma^2 \mathbf{I})$; noise $\mathbf{w}_n(t)$ is uncorrelated across sensors, time and with **s**; and $\mathbf{H}_n$ as well as (cross-) covariance matrices $\boldsymbol{\Sigma}_{ss}$, $\boldsymbol{\Sigma}_{sx_n}$ and $\boldsymbol{\Sigma}_{x_n x_m}$ are known $\forall\, n, m \in \{0, \ldots, N-1\}$.

Notice that (a3) assumes that sufficiently strong channel codes are used; while whiteness of $\mathbf{w}_n(t)$ and the zero-mean assumptions in (a4) are made without loss of generality. The linear model in (a4) is commonly encountered in estimation and in a number of cases it even accurately approximates non-linear mappings; e.g., via a first-order Taylor expansion in target tracking applications. Although confining ourselves to Gaussian vectors $\mathbf{x}_n(t)$ is of interest on its own, it can be shown, similarly to (Berger 1971, p. 134), that the D-R functions obtained for Gaussian data bound from above their counterparts for non-Gaussian sensor data $\mathbf{x}_n(t)$.

Blocks $\mathbf{x}_n^{(L)} := \{\mathbf{x}_n(t)\}_{t=1}^{L}$, comprising L consecutive time instantiations of the vector $\mathbf{x}_n(t)$, are encoded per sensor to yield each encoder's output $\mathbf{u}_n^{(L)} = \mathbf{f}_n^{(L)}(\mathbf{x}_n^{(L)})$, $n = 0, \ldots, N-1$. These outputs are communicated through ideal orthogonal channels to the FC. There, $\mathbf{u}_n^{(L)}$'s are decoded to obtain an estimate of $\mathbf{s}^{(L)} := \{\mathbf{s}(t)\}_{t=1}^{L}$ denoted as $\hat{\mathbf{s}}_R^{(L)}(\mathbf{u}_0^{(L)}, \ldots, \mathbf{u}_{N-1}^{(L)}) = \mathbf{g}_R^{(L)}(\mathbf{x}_0^{(L)}, \ldots, \mathbf{x}_{N-1}^{(L)})$, since $\mathbf{u}_n^{(L)}$ is a function of $\mathbf{x}_n^{(L)}$. The rate constraint is imposed through a bound on the cardinality of the range of the sensor encoding functions, i.e., the cardinality of the range of $\mathbf{f}_n^{(L)}$ must be no greater than $2^{L R_n}$, where R_n is the available rate at the encoder of the nth sensor. The sum rate satisfies the constraint $\sum_{n=0}^{N-1} R_n \leq R$, where R is the total available rate shared by the N sensors. This setup is precisely the vector Gaussian CEO problem in its most general form without any restrictions in the number of observations and the number of parameters (Berger et al. 1996). Under this rate constraint, we want to determine the minimum possible MSE distortion $(1/L) \sum_{t=1}^{L} E[\|\mathbf{s}(t) - \hat{\mathbf{s}}_R(t)\|^2]$ for estimating $\mathbf{s}$ in the limit of infinite block-length L. When $N = 1$, a single-letter information theoretic characterization is known for the latter, but no simplification is known for the distributed multi-sensor scenario.

7.7.1 Distortion-Rate for Centralized Estimation

Let us first specify the D-R function for estimating $\mathbf{s}(t)$ in a *single-sensor* setup. The single-letter characterization of the D-R function in this setup allow us to drop the time index. Here, all $\{\mathbf{x}_n\}_{n=0}^{N-1} := \mathbf{x}$ are available to a single sensor, and $\mathbf{x} = \mathbf{H}\mathbf{s} + \mathbf{w}$. We let $\rho := \text{rank}(\mathbf{H})$ denote the rank of matrix $\mathbf{H}$. The D-R function in such a scenario provides a lower (non-achievable) bound on the MMSE that can be achieved in a multi-sensor distributed setup, where each $\mathbf{x}_n$ is observed by a different sensor. Existing works treat the case $M = p$ (Sakrison 1968; Wolf and Ziv 1970), but here we look for the D-R function regardless of M, p, in the linear-Gaussian model framework.

D-R analysis for reconstruction

The D-R function for encoding a vector $\mathbf{x}$, with pdf $p(\mathbf{x})$, using rate R at an individual sensor, and reconstructing it (in the MMSE sense) as $\hat{\mathbf{x}}$ at the FC, is given by (Cover and Thomas 1991, p. 342):

$$D_x(R) = \min_{p(\hat{\mathbf{x}}|\mathbf{x})} E_{p(\hat{\mathbf{x}},\mathbf{x})}[\|\mathbf{x} - \hat{\mathbf{x}}\|^2], \quad \text{s. to } I(\mathbf{x}; \hat{\mathbf{x}}) \leq R \tag{7.80}$$

where $\mathbf{x} \in \mathbb{R}^M$ and $\hat{\mathbf{x}} \in \mathbb{R}^M$, and the minimization is w.r.t. the conditional pdf $p(\hat{\mathbf{x}}|\mathbf{x})$. Let $\boldsymbol{\Sigma}_{xx} = \mathbf{Q}_x \boldsymbol{\Lambda}_x \mathbf{Q}_x^T$ denote the eigenvalue decomposition of $\boldsymbol{\Sigma}_{xx}$, where $\boldsymbol{\Lambda}_x = \text{diag}(\lambda_{x,1} \cdots \lambda_{x,M})$ and $\lambda_{x,1} \geq \cdots \geq \lambda_{x,M} > 0$.

For $\mathbf{x}$ Gaussian, $D_x(R)$ can be determined by applying reverse water-filling (rwf) to the pre-whitened vector $\mathbf{x}_w := \mathbf{Q}_x^T \mathbf{x}$ (Cover and Thomas 1991, p. 348). For a prescribed rate R, it turns out that $\exists\, k$ such that the first k entries $\{\mathbf{x}_w(i)\}_{i=1}^{k}$ of $\mathbf{x}_w$ are encoded and reconstructed independently from each other using rate $\{R_i = 0.5 \log_2 (\lambda_{x,i}/d(k, R))\}_{i=1}^{k}$, where $d(k, R) = \left(\prod_{i=1}^{k} \lambda_{x,i}\right)^{1/k} 2^{-2R/k}$ with $R = \sum_{i=1}^{k} R_i$; and the last $M - k$ entries of $\mathbf{x}_w$ are assigned no rate; i.e., $\{R_i = 0\}_{i=k+1}^{M}$. The corresponding MMSE for encoding $\mathbf{x}_w(i)$, the ith entry of $\mathbf{x}_w$, under a rate constraint R_i, is $D_i = E[\|\mathbf{x}_w(i) - \hat{\mathbf{x}}_w(i)\|^2] = d(k, R)$

when $i = 1, \ldots, k$; and $D_i = \lambda_{x,i}$ when $i = k+1, \ldots, M$. The resultant MMSE is

$$D_x(R) = E[\|\mathbf{x} - \hat{\mathbf{x}}\|^2] = E[\|\mathbf{x}_w - \hat{\mathbf{x}}_w\|^2] = kd(k, R) + \sum_{i=k+1}^{M} \lambda_{x,i}. \quad (7.81)$$

Especially for $d(k, R)$, it follows that $\max(\{\lambda_{x,i}\}_{i=k+1}^{M}) \leq d(k, R) < \min\{\lambda_{x,1}, \ldots, \lambda_{x,k}\}$. Intuitively, $d(k, R)$ is a threshold distortion determining which entries of $\mathbf{x}_w$ are assigned with nonzero rate. The first k entries of $\mathbf{x}_w$ with variance $\lambda_{x,i} > d(k, R)$ are encoded with non-zero rate, but the last $M - k$ ones are discarded in the encoding procedure (are set to zero).

Associated with the rwf principle is the so called test channel; see e.g., (Cover and Thomas 1991, p. 345). The encoder's MSE optimal output is $\mathbf{u} = \mathbf{Q}_{x,k}^T \mathbf{x} + \boldsymbol{\zeta}$, where $\mathbf{Q}_{x,k}$ is formed by the first k columns of $\mathbf{Q}_x$, and $\boldsymbol{\zeta}$ models the distortion noise that results due to the rate-constrained encoding of $\mathbf{x}$. The zero-mean AWGN $\boldsymbol{\zeta}$ is uncorrelated with $\mathbf{x}$ and its diagonal covariance matrix $\boldsymbol{\Sigma}_{\zeta\zeta}$ has entries $[\boldsymbol{\Sigma}_{\zeta\zeta}]_{ii} = \lambda_{x,i} D_i / (\lambda_{x,i} - D_i)$. The part of the test channel that takes as input $\mathbf{u}$ and outputs $\hat{\mathbf{x}}$, models the decoder. The reconstruction $\hat{\mathbf{x}}$ of $\mathbf{x}$ at the decoder output is

$$\hat{\mathbf{x}} = \mathbf{Q}_{x,k} \boldsymbol{\Theta}_k \mathbf{u} = \mathbf{Q}_{x,k} \boldsymbol{\Theta}_k \mathbf{Q}_{x,k}^T \mathbf{x} + \mathbf{Q}_{x,k} \boldsymbol{\Theta}_k \boldsymbol{\zeta}, \quad (7.82)$$

where $\boldsymbol{\Theta}_k$ is a diagonal matrix with non-zero entries $[\boldsymbol{\Theta}_k]_{ii} = (\lambda_{x,i} - D_i)/\lambda_{x,i}, i = 1, \ldots, k$.

D-R analysis for estimation

The D-R function for estimating a source $\mathbf{s}$ given observation $\mathbf{x}$ (where the source and observation are probabilistically drawn from the joint pdf $p(\mathbf{x}, \mathbf{s})$) with rate R at an individual sensor, and reconstructing it (in the MMSE sense) as $\hat{\mathbf{x}}$ at the FC is given by (Berger 1971, p. 79)

$$D_s(R) = \min_{p(\hat{\mathbf{s}}_R|\mathbf{x})} E_{p(\hat{\mathbf{s}}_R, \mathbf{s})}[\|\mathbf{s} - \hat{\mathbf{s}}_R\|^2], \quad \text{s. to } I(\mathbf{x}; \hat{\mathbf{s}}_R) \leq R \quad (7.83)$$

where $\mathbf{s} \in \mathbb{R}^p$ and $\hat{\mathbf{s}}_R \in \mathbb{R}^p$, and the minimization is w.r.t. the conditional pdf $p(\hat{\mathbf{s}}_R|\mathbf{x})$. Different from (7.80), the mutual information $I(\mathbf{x}; \hat{\mathbf{s}}_R)$ and the pdf $p(\hat{\mathbf{s}}_R|\mathbf{x})$ in (7.83) depend on $\mathbf{x}$ and not on $\mathbf{s}$, which is the signal of interest. The latter reflects the fact that each sensor observes a distorted version of the source signal, captured in the observation vector $\mathbf{x}$. Thus, (7.83) denotes the minimum possible estimation MSE that can be achieved for $\mathbf{s}$ using the information incorporated in $\mathbf{x}$, and with an available encoding rate R. In order to achieve the D-R function, one might be tempted to first compress $\mathbf{x}$ by applying rwf at the sensor, without taking into account the data model relating $\mathbf{s}$ with $\mathbf{x}$, and subsequently use the reconstructed $\hat{\mathbf{x}}$ to form the MMSE conditional expectation estimate $\hat{\mathbf{s}}_{ce} = E[\mathbf{s}|\hat{\mathbf{x}}]$ at the FC. An alternative option would be to first form the MMSE estimate $\hat{\mathbf{s}} = E[\mathbf{s}|\mathbf{x}]$, encode the latter using rwf at the sensor, and after decoding at the FC, obtain the reconstructed estimate $\hat{\mathbf{s}}_{ec}$. Referring as before the former option as *Compress-Estimate* (CE), and to the latter as *Estimate-Compress* (EC), we are interested in determining which one yields the smallest MSE under a rate constraint R. Another interesting question is whether any of the CE and EC schemes enjoys MMSE optimality (i.e., achieves (7.83)). With subscripts ce and ec corresponding to these two options, let us also define the errors $\tilde{\mathbf{s}}_{ce} := \mathbf{s} - \hat{\mathbf{s}}_{ce}$ and $\tilde{\mathbf{s}}_{ec} := \mathbf{s} - \hat{\mathbf{s}}_{ec}$.

For CE, we depict in Figure 7.12 (Top) the test channel for encoding $\mathbf{x}$ via rwf, followed by MMSE estimation of $\mathbf{s}$ based on $\hat{\mathbf{x}}$. Suppose that when applying rwf to $\mathbf{x}$ with prescribed

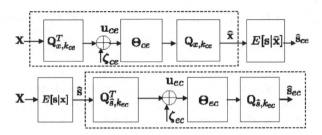

Figure 7.12 (Top): Test channel for the CE scheme.; (Bottom): Test channel for the EC scheme.

rate R, the first k_{ce} components of $\mathbf{x}_w$ are assigned with non-zero rate and the rest are discarded. The MMSE optimal encoder's output for encoding $\mathbf{x}$ is $\mathbf{u}_{ce} = \mathbf{Q}_{x,k_{ce}}^T \mathbf{x} + \zeta_{ce}$. The covariance matrix of ζ_{ce} has diagonal entries $[\Sigma_{\zeta_{ce}\zeta_{ce}}]_{ii} = \lambda_{x,i} D_i^{ce}/(\lambda_{x,i} - D_i^{ce})$ for $i = 1, \ldots, k_{ce}$, where $D_i^{ce} := E[(\mathbf{x}_w(i) - \hat{\mathbf{x}}_w(i))^2]$. Since $D_i^{ce} = \left(\prod_{i=1}^{k_{ce}} \lambda_{x,i}\right)^{1/k_{ce}} 2^{-2R/k_{ce}}$ when $i = 1, \ldots, k_{ce}$ and $D_i^{ce} = \lambda_{x,i}$, when $i = k_{ce} + 1, \ldots, M$, the reconstructed $\hat{\mathbf{x}}$ in CE is [cf. (7.82)]:

$$\hat{\mathbf{x}} = \mathbf{Q}_{x,k_{ce}} \Theta_{ce} \mathbf{Q}_{x,k_{ce}}^T \mathbf{x} + \mathbf{Q}_{x,k_{ce}} \Theta_{ce} \zeta_{ce}, \tag{7.84}$$

where $[\Theta_{ce}]_{ii} = (\lambda_{x,i} - D_i^{ce})/\lambda_{x,i}$, for $i = 1, \ldots, k_{ce}$. Letting $\check{\mathbf{x}} := \mathbf{Q}_x^T \hat{\mathbf{x}} = [\check{\mathbf{x}}_1^T \ \mathbf{0}_{1 \times (M-k_{ce})}]^T$, with $\check{\mathbf{x}}_1 := \Theta_{ce} \mathbf{Q}_{x,k_{ce}}^T \mathbf{x} + \Theta_{ce} \zeta_{ce}$, we have for the MMSE estimate $\hat{\mathbf{s}}_{ce} = E[\mathbf{s}|\hat{\mathbf{x}}]$

$$\hat{\mathbf{s}}_{ce} = E[\mathbf{s}|\mathbf{Q}_x^T \hat{\mathbf{x}}] = E[\mathbf{s}|\check{\mathbf{x}}_1] = \Sigma_{s\check{x}_1} \Sigma_{\check{x}_1\check{x}_1}^{-1} \check{\mathbf{x}}_1, \tag{7.85}$$

since $\mathbf{Q}_x^T$ is unitary and the last $M - k_{ce}$ entries of $\check{\mathbf{x}}$ are useless for estimating $\mathbf{s}$. It has been shown in (Schizas et al. 2005) that the covariance matrix $\Sigma_{\tilde{s}_{ce}\tilde{s}_{ce}} := E[(\mathbf{s} - \hat{\mathbf{s}}_{ce})(\mathbf{s} - \hat{\mathbf{s}}_{ce})^T] = \Sigma_{ss} - \Sigma_{s\check{x}_1} \Sigma_{\check{x}_1\check{x}_1}^{-1} \Sigma_{\check{x}_1 s}$ of $\hat{\mathbf{s}}_{ce}$ is

$$\Sigma_{\tilde{s}_{ce}\tilde{s}_{ce}} = \Sigma_{ss} - \Sigma_{sx} \Sigma_{xx}^{-1} \Sigma_{xs} + \Sigma_{sx} \mathbf{Q}_x \Delta_{ce} \mathbf{Q}_x^T \Sigma_{xs}, \tag{7.86}$$

where $\Delta_{ce} := \text{diag}\left(D_1^{ce} \lambda_{x,1}^{-2} \cdots D_N^{ce} \lambda_{x,M}^{-2}\right)$.

In Figure 7.12 (Bottom) we depict the test channel for the EC scheme. The MMSE estimate $\hat{\mathbf{s}} = E[\mathbf{s}|\mathbf{x}]$ is followed by the test channel that results when applying rwf to a pre-whitened version of $\hat{\mathbf{s}}$, with rate R. Let $\Sigma_{\hat{s}\hat{s}} = \mathbf{Q}_{\hat{s}} \Lambda_{\hat{s}} \mathbf{Q}_{\hat{s}}^T$ be the eigenvalue decomposition for the covariance matrix of $\hat{\mathbf{s}}$, where $\Lambda_{\hat{s}} = \text{diag}(\lambda_{\hat{s},1} \cdots \lambda_{\hat{s},p})$ and $\lambda_{\hat{s},1} \geq \cdots \geq \lambda_{\hat{s},p}$. Suppose now that the first k_{ec} entries of $\hat{\mathbf{s}}_w = \mathbf{Q}_{\hat{s}}^T \hat{\mathbf{s}}$ are assigned with non-zero rate and the rest are discarded. The MSE optimal encoder's output is given by $\mathbf{u}_{ec} = \mathbf{Q}_{\hat{s},k_{ec}}^T \hat{\mathbf{s}} + \zeta_{ec}$, and the estimate $\hat{\mathbf{s}}_{ec}$ is

$$\hat{\mathbf{s}}_{ec} = \mathbf{Q}_{\hat{s},k_{ec}} \Theta_{ec} \mathbf{Q}_{\hat{s},k_{ec}}^T \hat{\mathbf{s}} + \mathbf{Q}_{\hat{s},k_{ec}} \Theta_{ec} \zeta_{ec}, \tag{7.87}$$

where $\mathbf{Q}_{\hat{s},k_{ec}}$ is formed by the first k_{ec} columns of $\mathbf{Q}_{\hat{s}}$. For the $k_{ec} \times k_{ec}$ diagonal matrices $\mathbf{s}_{ec}$ and $\Sigma_{\zeta_{ec}\zeta_{ec}}$ we have $[\mathbf{s}_{ec}]_{ii} = (\lambda_{\hat{s},i} - D_i^{ec})/\lambda_{\hat{s},i}$ and $[\Sigma_{\zeta_{ec}\zeta_{ec}}]_{ii} = \lambda_{\hat{s},i} D_i^{ec}/(\lambda_{\hat{s},i} - D_i^{ec})$, where $D_i^{ec} := E[(\hat{\mathbf{s}}_w(i) - \hat{\mathbf{s}}_{ec,w}(i))^2]$, and $\hat{\mathbf{s}}_{ec,w} := \mathbf{Q}_{\hat{s}}^T \hat{\mathbf{s}}_{ec}$. Recall also that $D_i^{ec} = \left(\prod_{i=1}^{k_{ec}} \lambda_{\hat{s},i}\right)^{1/k_{ec}} 2^{\frac{-2R}{k_{ec}}}$ when $i = 1, \ldots, k_{ec}$ and $D_i^{ec} = \lambda_{\hat{s},i}$, for $i = k_{ec} + 1, \ldots, p$. Upon defining $\mathbf{\Delta}_{ec} := \text{diag} \left(D_1^{ec} \cdots D_p^{ec} \right)$, the covariance matrix of $\tilde{\mathbf{s}}_{ec}$ is given by (Schizas et al. 2005)

$$\mathbf{\Sigma}_{\tilde{s}_{ec}\tilde{s}_{ec}} = \mathbf{\Sigma}_{ss} - \mathbf{\Sigma}_{sx}\mathbf{\Sigma}_{xx}^{-1}\mathbf{\Sigma}_{xs} + \mathbf{Q}_{\hat{s}}\mathbf{\Delta}_{ec}\mathbf{Q}_{\hat{s}}^T. \tag{7.88}$$

The MMSE associated with CE and EC is given, respectively, by [cf. (7.86) and (7.88)]

$$D_{ce}(R) := \text{tr}(\mathbf{\Sigma}_{\tilde{s}_{ce}\tilde{s}_{ce}}) = J_o + \epsilon_{ce}(R),$$

$$D_{ec}(R) := \text{tr}(\mathbf{\Sigma}_{\tilde{s}_{ec}\tilde{s}_{ec}}) = J_o + \epsilon_{ec}(R), \tag{7.89}$$

where $\epsilon_{ce}(R) := \text{tr}(\mathbf{\Sigma}_{sx}\mathbf{Q}_x\mathbf{\Delta}_{ce}\mathbf{Q}_x^T\mathbf{\Sigma}_{xs})$, $\epsilon_{ec}(R) := \text{tr}(\mathbf{Q}_{\hat{s}}\mathbf{\Delta}_{ec}\mathbf{Q}_{\hat{s}}^T)$, and the quantity $J_o := \text{tr}(\mathbf{\Sigma}_{ss} - \mathbf{\Sigma}_{sx}\mathbf{\Sigma}_{xx}^{-1}\mathbf{\Sigma}_{xs})$ is the MMSE achieved when estimating $\mathbf{s}$ based on $\mathbf{x}$, without source encoding ($R \to \infty$). Since J_o is common to both EC and CE it is important to compare $\epsilon_{ce}(R)$ with $\epsilon_{ec}(R)$ in order to determine which estimation scheme achieves the smallest MSE. The following proposition provides such an asymptotic comparison:

Proposition 7.7.1 *If*

$$R > R_{th} := \frac{1}{2}\max\left\{\log_2\left(\left(\prod_{i=1}^{\rho}\lambda_{x,i}\right)/\sigma^{2\rho}\right), \log_2\left(\left(\prod_{i=1}^{\rho}\lambda_{\hat{s},i}\right)/(\lambda_{\hat{s},\rho})^\rho\right)\right\},$$

then it holds that $\epsilon_{ce}(R) = \gamma_1 2^{-2R/M}$ and $\epsilon_{ec}(R) = \gamma_2 2^{-2R/\rho}$, where γ_1 and γ_2 are constants.

An immediate consequence of Proposition 7.7.1 is that the MSE for EC converges as $R \to \infty$ to J_o with rate $O(2^{-2R/\rho})$. The MSE of CE converges likewise, but with rate $O(2^{-2R/M})$. For the typical case $M > \rho$, EC approaches the lower bound J_o faster than CE, implying correspondingly a more efficient usage of the available rate R. This is intuitively reasonable since CE compresses $\mathbf{x}$, which contains the noise $\mathbf{w}$. Since the last $M - \rho$ eigenvalues of $\mathbf{\Sigma}_{xx}$ equal the noise variance σ^2, part of the available rate is consumed to compress the noise. On the contrary, the MMSE estimator $\hat{\mathbf{s}}$ in EC suppresses significant part of the noise. For the special case of a scalar data model ($M = p = 1$) it has been shown (Schizas et al. 2005) that $D_{ec}(R) = D_{ce}(R)$, while for the vector and matrix models ($M > 1$ and/or $p > 1$) we have determined appropriate threshold rates R_{th} have been determined such that $D_{ce}(R) > D_{ec}(R)$ for $R > R_{th}$.

If the SNR is defined as $\text{SNR} = \text{tr}(\mathbf{H}\mathbf{\Sigma}_{ss}\mathbf{H}^T)/M\sigma^2$, it is possible to compare the MMSE when estimating $\mathbf{s}$ using the CE and EC schemes; see Figure 7.13 (Left). With $\mathbf{\Sigma}_{ss} = \sigma_s^2\mathbf{I}_p$, $p = 4$ and $M = 40$, we observe that beyond a threshold rate, the distortion of EC converges to J_o faster than that of CE, which corroborates Proposition 7.7.1.

The analysis so far raises the question whether EC is MSE optimal. We have seen that this is the case when estimating $\mathbf{s}$ with a given rate R without forcing any assumption about M and p. A related claim has been reported in (Sakrison 1968; Wolf and Ziv 1970) for $M = p$. The extension to $M \neq p$ established in (Schizas et al. 2005) can be summarized as follows:

Proposition 7.7.2 *The D-R function when estimating $\mathbf{s}$ based on $\mathbf{x}$ can be expressed as*

$$D_s(R) = \min_{\substack{p(\hat{s}_R|\mathbf{x}) \\ I(\mathbf{x};\hat{s}_R)\leq R}} E[\|\mathbf{s} - \hat{\mathbf{s}}_R\|^2] = E[\|\tilde{\mathbf{s}}\|^2] + \min_{\substack{p(\hat{s}_R|\hat{s}) \\ I(\hat{s};\hat{s}_R)\leq R}} E[\|\hat{\mathbf{s}} - \hat{\mathbf{s}}_R\|^2], \tag{7.90}$$

where $\hat{\mathbf{s}} = \mathbf{\Sigma}_{sx}\mathbf{\Sigma}_{xx}^{-1}\mathbf{x}$ is the MMSE estimator, and $\tilde{\mathbf{s}}$ is the corresponding MMSE.

Proposition 7.7.2 reveals that the optimal means of estimating $\mathbf{s}$ is to first form the optimal MMSE estimate $\hat{\mathbf{s}}$ and then apply optimal rate-distortion encoding to this estimate. The lower bound on this distortion when $R \to \infty$ is $J_o = E[\|\tilde{\mathbf{s}}\|^2]$, which is intuitively appealing. The D-R function in (7.90) is achievable, because the rightmost term in (7.90) corresponds to the D-R function for reconstructing the MMSE estimate $\hat{\mathbf{s}}$ that is known to be achievable using random coding; see e.g., (Berger 1971, p. 66).

7.7.2   Distortion-Rate for Distributed Estimation

Let us now consider the D-R function for estimating $\mathbf{s}$ in a multi-sensor setup, under a total available rate R which has to be shared among all sensors. Because the analytical specification of the D-R function in this case remains intractable, we will present an alternative algorithm that numerically determines an achievable upper bound for it. Combining this upper bound with the non-achievable lower bound corresponding to an equivalent single-sensor setup, and applying the MMSE optimal EC scheme, will provide a region wherein the D-R function lies. For simplicity in exposition, we confine ourselves to a two-sensor setup, but the results apply to any finite $N > 2$.

Consider the following single-letter characterization of the upper bound on the D-R function:

$$\overline{D}(R) = \min_{\{p(\mathbf{u}_n|\mathbf{x}_n)\}_{n=0}^{1}, \hat{\mathbf{s}}_R} E_{p(\mathbf{s}, \{\mathbf{u}_n\}_{n=0}^{1})}[\|\mathbf{s} - \hat{\mathbf{s}}_R\|^2], \quad \text{s. to } I(\mathbf{x}; \{\mathbf{u}_n\}_{n=0}^{1}) \le R, \qquad (7.91)$$

where the minimization is w.r.t. $\{p(\mathbf{u}_n|\mathbf{x}_n)\}_{n=0}^{1}$ and $\hat{\mathbf{s}}_R := \hat{\mathbf{s}}_R(\mathbf{u}_0, \mathbf{u}_1)$. Achievability of $\overline{D}(R)$ can be established by readily extending to the vector case the scalar results in (Chen et al. 2004, Theorem 3). Details of this extension can be found in (Schizas et al. 2005). To carry out the minimization in (7.91), we will develop an alternating scheme whereby $\mathbf{u}_1$ is treated as side information that is available at the decoder when optimizing (7.91) w.r.t. $p(\mathbf{u}_0|\mathbf{x}_0)$ and $\hat{\mathbf{s}}_R(\mathbf{u}_0, \mathbf{u}_1)$. The side information $\mathbf{u}_1$ is considered as the output of an optimal rate-distortion encoder applied to $\mathbf{x}_1$ to estimate $\mathbf{s}$, without taking into account $\mathbf{x}_0$. Since $\mathbf{x}_1$ is Gaussian, the side information will have the form (cf. subsection 7.7.1) $\mathbf{u}_1 = \mathbf{Q}_1\mathbf{x}_1 + \boldsymbol{\zeta}_1$, where $\mathbf{Q}_1 \in \mathbb{R}^{k_1 \times M_1}$ and $k_1 \le M_1$, due to the rate constrained encoding of $\mathbf{x}_1$. Recall that $\boldsymbol{\zeta}_1$ is uncorrelated with $\mathbf{x}_1$ and Gaussian; i.e., $\boldsymbol{\zeta}_1 \sim \mathcal{N}(\mathbf{0}, \boldsymbol{\Sigma}_{\zeta_1\zeta_1})$.

Based on $\boldsymbol{\psi} := [\mathbf{x}_0^T \ \mathbf{u}_1^T]^T$, the optimal estimator for $\mathbf{s}$ is the MMSE one: $\hat{\mathbf{s}} = E[\mathbf{s}|\boldsymbol{\psi}] = \boldsymbol{\Sigma}_{s\psi}\boldsymbol{\Sigma}_{\psi\psi}^{-1}\boldsymbol{\psi} = \mathbf{L}_0\mathbf{x}_0 + \mathbf{L}_1\mathbf{u}_1$, where $\mathbf{L}_0, \mathbf{L}_1$ are $p \times M_0$ and $p \times k_1$ matrices so that $\boldsymbol{\Sigma}_{s\psi}\boldsymbol{\Sigma}_{\psi\psi}^{-1} = [\mathbf{L}_0 \ \mathbf{L}_1]$. If $\tilde{\mathbf{s}}$ is the corresponding MSE, then $\mathbf{s} = \hat{\mathbf{s}} + \tilde{\mathbf{s}}$, where $\tilde{\mathbf{s}}$ is uncorrelated with $\boldsymbol{\psi}$ due to the orthogonality principle. Noticing also that $\hat{\mathbf{s}}_R(\mathbf{u}_0, \mathbf{u}_1)$ is uncorrelated with $\tilde{\mathbf{s}}$ because it is a function of $\mathbf{x}_0$ and $\mathbf{u}_1$, we have $E[\|\mathbf{s} - \hat{\mathbf{s}}_R(\mathbf{u}_0, \mathbf{u}_1)\|^2] = E[\|\hat{\mathbf{s}} - \hat{\mathbf{s}}_R(\mathbf{u}_0, \mathbf{u}_1)\|^2] + E[\|\tilde{\mathbf{s}}\|^2]$, or,

$$E[\|\mathbf{s} - \hat{\mathbf{s}}_R(\mathbf{u}_0, \mathbf{u}_1)\|^2] = E[\|\mathbf{L}_0\mathbf{x}_0 - (\hat{\mathbf{s}}_R(\mathbf{u}_0, \mathbf{u}_1) - \mathbf{L}_1\mathbf{u}_1)\|^2] + E[\|\tilde{\mathbf{s}}\|^2]. \qquad (7.92)$$

Clearly, it holds that $I(\mathbf{x}; \mathbf{u}_0, \mathbf{u}_1) = R_1 + I(\mathbf{x}_0; \mathbf{u}_0) - I(\mathbf{u}_1; \mathbf{u}_0)$, where $R_1 := I(\mathbf{x}; \mathbf{u}_1)$ is the rate consumed to form the side information $\mathbf{u}_1$ and the rate constraint in (7.91) becomes $I(\mathbf{x}; \mathbf{u}_0, \mathbf{u}_1) \le R \Leftrightarrow I(\mathbf{x}_0; \mathbf{u}_0) - I(\mathbf{u}_1; \mathbf{u}_0) \le R - R_1 := R_0$. The new signal of interest in (7.92) is $\mathbf{L}_0\mathbf{x}_0$; thus, $\mathbf{u}_0$ has to be a function of $\mathbf{L}_0\mathbf{x}_0$. Using also the fact that $\mathbf{x}_0 \to \mathbf{L}_0\mathbf{x}_0 \to \mathbf{u}_0$ constitutes a Markov chain, it is possible to obtain from (7.91) the D-R upper bound

(Schizas et al. 2005):

$$\overline{\overline{D}}(R_0) = E[\|\tilde{\mathbf{s}}\|^2] + \min_{\substack{p(\mathbf{u}_0|\mathbf{L}_0\mathbf{x}_0),\hat{\mathbf{s}}_R \\ I(\mathbf{L}_0\mathbf{x}_0;\mathbf{u}_0)-I(\mathbf{u}_0;\mathbf{u}_1)\leq R_0}} E[\|\mathbf{L}_0\mathbf{x}_0 - \tilde{\mathbf{s}}_{R,01}(\mathbf{u}_0, \mathbf{u}_1)\|^2], \qquad (7.93)$$

where $\tilde{\mathbf{s}}_{R,01}(\mathbf{u}_0, \mathbf{u}_1) := \hat{\mathbf{s}}_R(\mathbf{u}_0, \mathbf{u}_1) - \mathbf{L}_1\mathbf{u}_1$. Through (7.93) we can determine an achievable D-R region, having available rate R_0 at the encoder and side information $\mathbf{u}_1$ at the decoder. Since $\mathbf{x}_0$ and $\mathbf{u}_1$ are jointly Gaussian, the Wyner-Ziv result applies (Wyner and Ziv 1976), which allows one to consider that $\mathbf{u}_1$ is available both at the decoder and the encoder. This, in turn, permits re-writing (7.93) as (Schizas et al. 2005)

$$\overline{\overline{D}}(R_0) = \min_{\substack{p(\hat{\mathbf{s}}_{R,01}|\tilde{\mathbf{s}}_0) \\ I(\tilde{\mathbf{s}}_0;\hat{\mathbf{s}}_{R,01})\leq R_0}} E[\|\tilde{\mathbf{s}}_0 - \hat{\mathbf{s}}_{R,01}(\mathbf{u}_0, \mathbf{u}_1)\|^2] + E[\|\tilde{\mathbf{s}}\|^2], \qquad (7.94)$$

where $\hat{\mathbf{s}}_{R,01}(\mathbf{u}_0, \mathbf{u}_1) = \hat{\mathbf{s}}_R(\mathbf{u}_0, \mathbf{u}_1) - \mathbf{L}_1\mathbf{u}_1 - E[\mathbf{L}_0\mathbf{x}_0|\mathbf{u}_1]$ and $\tilde{\mathbf{s}}_0 = \mathbf{L}_0\mathbf{x}_0 - E[\mathbf{L}_0\mathbf{x}_0|\mathbf{u}_1]$.

Notice that (7.94) is the D-R function for reconstructing the MSE $\tilde{\mathbf{s}}_0$ with rate R_0. Since $\tilde{\mathbf{s}}_0$ is Gaussian, we can readily apply rwf to the pre-whitened $\mathbf{Q}_{\tilde{s}_0}^T\tilde{\mathbf{s}}_0$ to determine $\overline{\overline{D}}(R_0)$ and the corresponding test channel that achieves $\overline{\overline{D}}(R_0)$. Through the latter, and considering the next eigenvalue decomposition $\mathbf{\Sigma}_{\tilde{s}_0\tilde{s}_0} = \mathbf{Q}_{\tilde{s}_0} \text{diag}(\lambda_{\tilde{s}_0,1} \cdots \lambda_{\tilde{s}_0,p})\mathbf{Q}_{\tilde{s}_0}^T$, it follows that the first encoder's output that minimizes (7.91) has the form:

$$\mathbf{u}_0 = \mathbf{Q}_{\tilde{s}_0,k_0}^T\mathbf{L}_0\mathbf{x}_0 + \boldsymbol{\zeta}_0 = \mathbf{Q}_0\mathbf{x}_0 + \boldsymbol{\zeta}_0, \qquad (7.95)$$

where $\mathbf{Q}_{\tilde{s}_0,k_0}$ denotes the first k_0 columns of $\mathbf{Q}_{\tilde{s}_0}$, k_0 is the number of $\mathbf{Q}_{\tilde{s}_0}^T\tilde{\mathbf{s}}_0$ entries that are assigned with non-zero rate, and $\mathbf{Q}_0 := \mathbf{Q}_{\tilde{s}_0,k_0}^T\mathbf{L}_0$. The $k_0 \times 1$ AWGN $\boldsymbol{\zeta}_0 \sim \mathcal{N}(\mathbf{0}, \mathbf{\Sigma}_{\zeta_0\zeta_0})$ is uncorrelated with $\mathbf{x}_0$. Additionally, we have $[\mathbf{\Sigma}_{\zeta_0\zeta_0}]_{ii} = \lambda_{\tilde{s}_0,i}D_i^0/(\lambda_{\tilde{s}_0,i} - D_i^0)$, where $D_i^0 = \left(\prod_{i=1}^{k_0}\lambda_{\tilde{s}_0,i}\right)^{1/k_0}2^{-2R_0/k_0}$, for $i = 1, \ldots, k_0$, and $D_i^0 = \lambda_{\tilde{s}_0,i}$ when $i = k_0 + 1, \ldots, p$. This way, we are able to determine also $p(\mathbf{u}_0|\mathbf{x}_0)$. The reconstruction function has the form:

$$\hat{\mathbf{s}}_R(\mathbf{u}_0, \mathbf{u}_1) = \mathbf{Q}_{\tilde{s}_0,k_0}\boldsymbol{\Theta}_0\mathbf{u}_0 + \mathbf{L}_0\mathbf{\Sigma}_{x_0u_1}\mathbf{\Sigma}_{u_1u_1}^{-1}\mathbf{u}_1 + \mathbf{L}_1\mathbf{u}_1$$

$$- \mathbf{Q}_{\tilde{s}_0,k_0}\boldsymbol{\Theta}_0\mathbf{Q}_{\tilde{s}_0,k_0}^T\mathbf{L}_0\mathbf{\Sigma}_{x_0u_1}\mathbf{\Sigma}_{u_1u_1}^{-1}\mathbf{u}_1, \qquad (7.96)$$

where $[\boldsymbol{\Theta}_0]_{ii} = \lambda_{\tilde{s}_0,i}D_i^0/(\lambda_{\tilde{s}_0,i} - D_i^0)$, and the MMSE is $\overline{\overline{D}}(R_0) = \sum_{j=1}^p D_j^0 + E[\|\tilde{\mathbf{s}}\|^2]$.

The approach in this subsection can be applied in an alternating fashion from sensor to sensor in order to determine appropriate $p(\mathbf{u}_n|\mathbf{x}_n)$, for $n = 0, 1$, and $\hat{\mathbf{s}}_R(\mathbf{u}_0, \mathbf{u}_1)$ that at best globally minimize (7.93). The conditional pdfs can be determined by finding the appropriate covariances $\mathbf{\Sigma}_{\zeta_n\zeta_n}$. Furthermore, by specifying the optimal $\mathbf{Q}_0$ and $\mathbf{Q}_1$, characterization of the encoders' structure is obtained. In Figure 7.13 (Right), we plot the non-achievable lower bound which corresponds to one sensor having available the entire $\mathbf{x}$ and using the optimal EC scheme. Moreover, we plot an achievable D-R upper bound determined by letting the n-th sensor form its local estimate $\hat{\mathbf{s}}_n = E[\mathbf{s}|\mathbf{x}_n]$, and then apply optimal rate-distortion encoding to $\hat{\mathbf{s}}_n$. If $\hat{\mathbf{s}}_{R,0}$ and $\hat{\mathbf{s}}_{R,1}$ are the reconstructed versions of $\hat{\mathbf{s}}_0$ and $\hat{\mathbf{s}}_1$, respectively, then the decoder at the FC forms the final estimate $\hat{\mathbf{s}}_R = E[\mathbf{s}|\hat{\mathbf{s}}_{R,0}, \hat{\mathbf{s}}_{R,1}]$. We also plot the achievable D-R region determined numerically by Algorithm 2. For each rate, the smallest distortion is recorded after 500 executions of the algorithm simulated

---

Algorithm 2 :

Initialize $\mathbf{Q}_0^{(0)}$, $\mathbf{Q}_1^{(0)}$, $\boldsymbol{\Sigma}_{\zeta_0\zeta_0}^{(0)}$, $\boldsymbol{\Sigma}_{\zeta_1\zeta_1}^{(0)}$ by applying optimal D-R encoding to each sensor's test channel independently. For a total rate R, generate J random increments $\{r(m)\}_{m=0}^{J}$, such that $0 \leq r(m) \leq R$ and $\sum_{m=0}^{M} r(m) = R$. Set $R_0(0) = R_1(0) = 0$.

for $j = 1, J$ **do**

    Set $R(j) = \sum_{l=0}^{j} r(l)$

    for $n = 0, 1$ **do**

        $\bar{n} = |n - 1|$ %The complementary index

        $R_0(j) = I(\mathbf{x}; \mathbf{u}_{\bar{n}}^{(j)})$

        We use $\mathbf{Q}_{\bar{n}}^{(j-1)}$, $\boldsymbol{\Sigma}_{\zeta_{\bar{n}}\zeta_{\bar{n}}}^{(j-1)}$, $R(j)$, $R_0(j)$ to determine $\mathbf{Q}_n^{(j)}$, $\boldsymbol{\Sigma}_{\zeta_n\zeta_n}^{(j)}$ and $\overline{\overline{D}}(R_n(j))$

    end for

    Update matrices $\mathbf{Q}_l^{(j)}$, $\boldsymbol{\Sigma}_{\zeta_l\zeta_l}^{(j)}$ that result the smallest distortion $\overline{\overline{D}}(R_l(j))$, with $l \in [0, 1]$

    Set $R_l(j) = R(j) - I(\mathbf{x}; \mathbf{u}_{\bar{l}}^{(j)})$ and $R_{\bar{l}}(j) = I(\mathbf{x}; \mathbf{u}_{\bar{l}}^{(j)})$.

end for

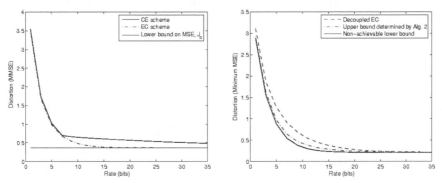

Figure 7.13 (Left): D-R region for EC and CE at SNR $= 2$; (Right): Distortion-rate bounds for estimating $\mathbf{s}$; here $N = 2$.

with $\boldsymbol{\Sigma}_{ss} = \mathbf{I}_p$, $p = 4$, and $M_0 = M_1 = 20$, at SNR $= 2$. We observe that the algorithm provides a tight upper bound of the achievable D-R region, which combined with the non-achievable lower bound (solid line) effectively reduces the 'uncertainty region' where the D-R function lies.

7.7.3 D-R Upper Bound via Convex Optimization

In this subsection we outline an alternative approach which relies on convex optimization techniques to obtain numerically an upper bound of the D-R region (Xiao et al. 2005). The idea is to calculate the Berger-Tung achievable D-R region (Berger 1977) for the vector Gaussian CEO problem, and subsequently determine the minimum sum rate $R_\Sigma = \sum_{n=0}^{N-1} R_n$ such that the estimation MSE satisfies $\mathrm{tr}(E[(\mathbf{s} - \hat{\mathbf{s}}_R)(\mathbf{s} - \hat{\mathbf{s}}_R)^T]) < D$, where $\hat{\mathbf{s}}_R = E[\mathbf{s}|\{\mathbf{u}_n\}_{n=0}^{N-1}]$ and D is the desired upper bound on the distortion. The Berger-Tung achievable region is calculated after having the encoders' output in form $\mathbf{u}_n = \mathbf{x}_n + \boldsymbol{\zeta}_n$, where $\boldsymbol{\zeta}_n \sim \mathcal{N}(\mathbf{0}, \boldsymbol{\Sigma}_{\zeta_n\zeta_n})$ are independent of $\mathbf{x}_n$, for $n = 0, 1, \ldots, N - 1$. Furthermore, the sum rate can be expressed as a function of $\mathbf{H}_n$ and $\boldsymbol{\Sigma}_{\zeta_n\zeta_n}$ (Xiao et al. 2005)

$$R_\Sigma = 0.5 \log \left(\det \left(\mathbf{I}_p + \sum_{n=0}^{N-1} \mathbf{H}_n^T (\mathbf{I}_{M_n} + \boldsymbol{\Sigma}_{\zeta_n\zeta_n})^{-1} \mathbf{H}_n \right) \prod_{n=0}^{N-1} \det(\mathbf{I}_{M_n} + \boldsymbol{\Sigma}_{\zeta_n\zeta_n}^{-1}) \right).$$

The D-R upper bound is obtained as the optimal solution of the following minimization problem ($\succeq$ denotes positive semidefiniteness)

$$\min_{\{\boldsymbol{\Sigma}_{\zeta_n\zeta_n}\}_{n=0}^{N-1}} R_\Sigma, \text{ s. to } \boldsymbol{\Sigma}_{\zeta_n\zeta_n} \succeq \mathbf{0}, \text{ tr}(\boldsymbol{\Sigma}_{\tilde{s}_R\tilde{s}_R}) \leq D, \tag{7.97}$$

where $\boldsymbol{\Sigma}_{\tilde{s}_R\tilde{s}_R} := E[(\mathbf{s} - \hat{\mathbf{s}}_R)(\mathbf{s} - \hat{\mathbf{s}}_R)^T] = (\mathbf{I}_p + \sum_{n=0}^{N-1} \mathbf{H}_n^T(\mathbf{I}_{M_n} + \boldsymbol{\Sigma}_{\zeta_n\zeta_n})^{-1}\mathbf{H}_n)^{-1}$.

Although, the minimization problem in (7.97) is not convex, (Xiao et al. 2005) has shown that (7.97) is equivalent to the following convex formulation:

$$\min_{\boldsymbol{\Sigma}_{\tilde{s}_R\tilde{s}_R}, \{\boldsymbol{\Sigma}_{\zeta_n\zeta_n}\}_{n=0}^{N-1}} -0.5 \log \det(\boldsymbol{\Sigma}_{\tilde{s}_R\tilde{s}_R}) + 0.5 \sum_{n=0}^{N-1} \log \det(\mathbf{I}_{M_n} + \boldsymbol{\Sigma}_{\zeta_n\zeta_n}^{-1}), \tag{7.98}$$

subject to $\text{tr}(\boldsymbol{\Sigma}_{\tilde{s}_R\tilde{s}_R}) \leq D, \boldsymbol{\Sigma}_{\zeta_n\zeta_n} \succeq \mathbf{0}, \boldsymbol{\Sigma}_{\tilde{s}_R\tilde{s}_R} \succeq \mathbf{0},$

$$(\mathbf{I}_p + \sum_{n=0}^{N-1} \mathbf{H}_n^T(\mathbf{I}_{M_n} + \boldsymbol{\Sigma}_{\zeta_n\zeta_n})^{-1}\mathbf{H}_n)^{-1} \preceq \boldsymbol{\Sigma}_{\tilde{s}_R\tilde{s}_R},$$

which is solved numerically using the interior point method (Boyd and Vandenberghe 2004).

7.8 Conclusion

We considered the problem of distributed estimation using wireless sensor networks and demonstrated that under limited resources the seemingly unrelated problems of dimensionality reduction, compression, quantization and estimation are actually connected due to the distributed nature of the WSN.

We started with parameter estimation under severe bandwidth constraints that were adhered to by *quantizing* each sensor's observation to one or a few bits. By jointly accounting for the unique quantization-estimation tradeoffs present, these bit(s) per sensor were first used to derive distributed maximum likelihood estimators (MLEs) for scalar mean-location estimation in the presence of generally non-Gaussian noise when the noise pdf is completely known; subsequently, when the pdf is known except for a number of unknown parameters; and finally, when the noise pdf is unknown. In all three cases, the resulting estimators turned out to exhibit comparable variances that can come surprisingly close to the variance of the clairvoyant estimator which relies on unquantized observations. This happens when the SNR capturing both quantization and noise effects assumes low-to-moderate values. Analogous claims were established for practical generalizations that were pursued in the multivariate and colored noise cases for distributed estimation of vector deterministic and random parameters. Therein, MLE and MAP estimators were formed via numerical search but the log-likelihoods were proved to be concave thus ensuring fast convergence to the unique global maximum.

We also pursued a related but distinct approach where the bandwidth constraint is adhered to by reduced-dimensionality observations. We dealt with non-ideal channel links that are characterized by multiplicative fading and additive noise. When data across sensors are uncorrelated, we established global MSE optimal schemes in closed-form and proved that they implement estimation followed by compression per sensor. For correlated sensor observations, we presented a block coordinate descent algorithm which guarantees convergence at least to a stationary point of the associated mean-square error cost. The optimal estimators allocate properly the prescribed power following a waterfilling-like principle. Fundamental MSE limits were finally studied through the D-R function to estimate a random vector in a single-sensor setup, while optimality of the estimate-first compress-afterwards approach was established. An alternating algorithm was also outlined to determine numerically an D-R upper bound in the distributed multi-sensor setup. Using this upper bound in conjunction with the non-achievable lower bound, determined through the single-sensor D-R function, yielded a tight region, where the D-R function for distributed estimation lies.

7.9   Further Reading

The problem of estimation based on quantized observations was studied in early works by (Gubner 1993), and (Lam and Reibman 1993). The context of distributed estimation using WSNs was first revisited by (Papadopoulos et al. 2001). The material in Sections 7.1–7.4 is based on results that appeared in (Ribeiro and Giannakis 2006a) and (Ribeiro and Giannakis 2006b), while the material in Section 7.5 has been reported in (Shah et al. 2005). When the noise pdf is unknown, the problem of estimation based on severely quantized data has been also studied by (Luo 2005a), (Luo 2005b), and (Luo and Xiao 2005) where the notion of universal estimators was introduced. A recent extension of the material covered in these sections to state estimation of dynamical stochastic processes is the Sign of Innovation Kalman Filter (SOI-KF) introduced by (Ribeiro et al. 2006).

Distributed estimation via dimensionality reduction has been also considered in (Zhu et al. 2005), (Gastpar et al. 2006) and (Zhang et al. 2003) for ideal channel links and/or Gaussian data models. Detailed derivations of what is presented in Section 7.6 can be found in (Schizas et al. 2007). The estimation schemes in Section 7.6 are intended for WSNs following a star topology. When it comes to rate constrained distributed estimation D-R bounds for the Gaussian CEO setup, results are due to (Oohama 1998) and (Chen et al. 2004) when $M = p$. The results in Section 7.7, are from (Schizas et al. 2005) and (Xiao et al. 2005), and correspond to arbitrary M and p. Maximum likelihood multiterminal estimation under rate constraints is treated in (Han et al. 1995); see also (Zhang et al. 1988).

A different approach to reducing communication costs in distributed estimation is to allow communication between one-hop neighbors only and let the sensors converge to a common estimate. In (Xiao and Boyd 2004) estimation is considered tantamount to convergence to the steady state distribution of a Markov chain. In (Schizas et al. 2006) estimation is shown to be equivalent to distributed optimization of a convex argument. Yet another approach in (Barbarossa and Scuttari 2006) is to model the WSN as a network of coupled oscillators. A different estimation approach using hidden Markov fields is reported in (Dogandžić 2006).

Bibliography

S. Barbarossa and G. Scutari, "Decentralized maximum likelihood estimation for sensor networks composed of nonlinearly coupled dynamical systems," *IEEE Trans. on Signal Proc.,* vol. 55, pp. 3456–3470, July 2007.

T. Berger 1971, *Rate Distortion Theory: A Mathematical Basis for Data Compression.* Prentice Hall, Englewood Cliffs, NJ.

T. Berger 1977, Multiterminal source coding, in Lectures presented at CISM Summer School on the Info. Theory Approach to Comm., July.

T. Berger, Z. Zhang, and Viswanathan H. 1996, The CEO problem, *IEEE Transactions on Information Theory,* vol. 42, pp. 406–411, May.

S. Boyd and Vandenberghe L. 2004, *Convex Optimization.* Cambridge University Press, Cambridge.

J. Chen, Zhang X., Berger T. and Wicker S. B. 2004, An upper bound on the sum-rate distortion function and its corresponding rate allocation schemes for the CEO problem, *IEEE Journal on Selected Areas in Communications,* pp. 406–411, August.

T. Cover and Thomas J. 1991, *Elements of Information Theory.* John Wiley and Sons, Chichester, 2nd edition.

A. Dogandžić and Zhang B. 2006, Distributed estimation and detection for sensor networks using hidden Markov random field models . *IEEE Trans. on Signal Processing,* vol. 54, pp. 3200–3215, August.

M. Gastpar, Draggoti P. L., and Vetterli M. 2006, The distributed Karhunen-Loève transform, *IEEE Transactions on Information Theory,* vol. 52, pp. 5177–5196, December.

J. Gubner 1993, Distributed estimation and quantization, *IEEE Transactions on Information Theory,* vol. 39, pp. 1456–1459.

T. S. Han, and Amari S. 1995, Statistical inference under multiterminal data compression, *IEEE Transactions on Information Theory,* vol. 41, pp. 2300–2324, November.

S. M. Kay 1993, *Fundamentals of Statistical Signal Processing : Estimation Theory.* Prentice Hall, Englewood Cliffs, NJ.

W. Lam and Reibman A. 1993, Quantizer design for decentralized systems with communication constraints, *IEEE Transactions on Communications,* vol. 41, pp. 1602–1605, Aug.

Z.-Q. Luo 2005, An isotropic universal decentralized estimation scheme for a bandwidth constrained ad hoc sensor network, *IEEE Journal on Selected Areas in Communications,* vol. 23, pp. 735–744, April.

Z.-Q. Luo 2005, Universal decentralized estimation in a bandwidth constrained sensor network, *IEEE Transactions on Information Theory,* vol. 51, pp. 2210–2219, June.

Z.-Q. Luo, Giannakis G. B., and Zhang S. 2005, Optimal linear decentralized estimation in a bandwidth constrained sensor network, in *Proc. of the Intl. Symp. on Info. Theory,* pp. 1441–1445, Adelaide, Australia, Sept. 4–9 2005.

Z.-Q. Luo and J.-J. Xiao 2005, Decentralized estimation in an inhomogeneous sensing environment, *IEEE Transactions on Information Theory,* vol. 51, pp. 3564–3575, October.

A. Mainwaring, Culler D., Polastre J., Szewczyk R., and Anderson J. 2002, Wireless sensor networks for habitat monitoring, in *Proc. of the 1st ACM International Workshop on Wireless Sensor Networks and Applications,* vol. 3, pp. 88–97, Atlanta, Georgia.

Y. Oohama 1998, The rate-distortion function for the quadratic gaussian CEO problem, *IEEE Transactions on Information Theory,* pp. 1057–1070, May.

A. Pandya, Kansal A., Pottie G., and Srivastava M. 2004, Fidelity and resource sensitive data gathering, in *Proc. of the 42nd Allerton Conference,* Allerton, IL, September.

H. Papadopoulos, Wornell G., and Oppenheim A. 2001, Sequential signal encoding from noisy mea-surements using quantizers with dynamic bias control, *IEEE Transactions on Information Theory*, vol. 47, pp. 978–1002.

A. Ribeiro and G. B. Giannakis 2006a, Bandwidth-constrained distributed estimation for wire-less sensor networks, Part I: Gaussian case, *IEEE Transactions on Signal Processing*, vol. 54, pp. 1131–1143, March 2006a.

A. Ribeiro and Giannakis G. B. 2006b, Bandwidth-constrained distributed estimation for wire-less sensor networks, Part II: unknown pdf, *IEEE Transactions on Signal Processing*, vol. 54, pp. 2784–2796, July.

A. Ribeiro, Giannakis G. B., and Roumeliotis S. I. 2006, SOI-KF: distributed Kalman filtering with low-cost communications using the sign of innovations, *IEEE Transactions on Signal Processing*, vol. 54, pp. 4782–4795, December.

D. J. Sakrison 1968, Source encoding in the presence of random disturbance, *IEEE Transactions on Information Theory*, pp. 165–167, January.

I. D. Schizas, Giannakis G. B., and Jindal N. 2005, Distortion-rate analysis for distributed estimation with wireless sensor networks, in *Proc. of 43rd Allerton Conf.*, Univ. of Illinois at U-C, Monticello, IL, Sept. 28–30.

I. D. Schizas, G. B. Giannakis, and Z.-Q. Luo 2007, "Distributed estimation using reduced dimen-sionality sensor observations," *IEEE Transactions on Signal Processing,* vol. 55, pp. 4284–4299, August 2007.

I. D. Schizas, Ribeiro A., and Giannakis G. B. 2006, Distributed estimation with ad hoc wireless sensor networks, *Proc. of XIV European Sign. Proc. Conf.,* Florence, Italy, Sept. 4–8.

A. F. Shah, Ribeiro A., and Giannakis G. B. 2005, Bandwidth-constrained MAP estimation for wireless sensor networks, *Conference Record of the Thirty-Ninth Asilomar Conference on Signals, Systems and Computers*, Pacific Grove, CA, October 28–November 1, 2005, pp.: 215–219.

H. L. Van Trees 1968, *Detection, Estimation, and Modulation Theory.* John Wiley and Sons, Chich-ester, first ed.,

H. Viswanathan and Berger T. 1997, The quadratic gaussian CEO problem, *IEEE Transactions on Information Theory*, pp. 1549–1559, September.

J. Wolf and Ziv J. 1970, Transmission of noisy information to a noisy receiver with minimum distortion, *IEEE Transactions on Information Theory*, pp. 406–411, July.

A. Wyner and Ziv J. 1976, The rate-distortion function for source coding with side information at the decoder, *IEEE Trans. on Info. Theory*, pp. 1–10, January.

J.-J. Xiao, Luo Z.-Q. and Giannakis G. B. 2007, Performance bounds for the rate-constrained univer-sal decentralized estimators, *IEEE Signal Processing Letters*, vol. 14, no. 1, pp. 47–50, January.

J.-J. Xiao and Luo Z.-Q. 2005, Optimal rate and power allocation in Gaussian vector CEO problem. In *IEEE Int. Workshop Comp. Advances in Multi-Sensor Adaptive Processing*, Puerto Vallarta, Mexico, 13–15 December.

L. Xiao and Boyd S. 2004, Fast linear iterations for distributed averaging. *Systems and Control Letters*, vol. 53, pp. 65–78.

K. Zhang, Li X. R., Zhang P., and Li H.2003, Optimal linear estimation fusion – Part VI: Sensor data compression. In *Proc. of the Intl. Conf. on Info. Fusion*, pp. 221–228, Queensland, Australia.

Z. Zhang, and Berger T. 1988, Estimation via compressed information. *IEEE Transactions on Infor-mation Theory*, vol. 34, pp. 198–211, March.

Y. Zhu, Song E., Zhou J., and You Z.2005, Optimal dimensionality reduction of sensor data in multisensor estimation fusion, *IEEE Transactions on Signal Processing*, vol. 53, pp. 1631–1639, May.

8

Distributed Learning in Wireless Sensor Networks

Joel B. Predd, Sanjeev R. Kulkarni, and H. Vincent Poor

8.1 Introduction

Wireless sensor networks have attracted considerable attention in recent years (Akyildiz et al. 2002). Research in this area has focused on two separate aspects of such networks: networking issues, such as capacity, delay, and routing strategies; and applications issues. This chapter is concerned with the second of these aspects of wireless sensor networks, and in particular with the problem of distributed inference. Wireless sensor networks are *a fortiori* designed for the purpose of making inferences about the environments that they are sensing, and they are typically characterized by limited communication capabilities due to tight energy and bandwidth limitations. Thus, distributed inference[1] is a major issue in the study of such networks.

Distributed inference has a rich history within the information theory and signal processing communities, especially in the framework of parametric models. Recall that in parametric settings, the statistics of the phenomena under observation are assumed to be known to the system designer. Under such assumptions, research has typically focused on determining how the capacity of the sensor-to-fusion center channel fundamentally limits the quality of estimates (e.g., rate-distortion tradeoffs: Berger et al. 1996; Gastpar et al.

---

[1]The terms distributed inference and decentralized inference are used somewhat interchangeably in the literature, and often qualified as decentralized (distributed) detection or estimation depending on the context. In the present article, we also use the terms interchangeably, and apply an author's own convention when discussing her or his work.

Wireless Sensor Networks: Signal Processing and Communications Perspectives   A. Swami, Q. Zhao, Y.-W. Hong and L. Tong

2006; Han and Amari 1998; Viswanathan and Berger 1997), on determining delay-sensitive optimal (under various criteria) sensor decision rules and fusion strategies under unreliable bandwidth constrained channels (e.g., Chen et al. 2006; Varshney 1996; Vishwanathan and Varshney 1997), on characterizing the performance of large networks relative to their centralized communication-unconstrained counterparts (e.g., Chamberland and Veeravalli 2004; Negi et al. 2007), or on developing message-passing algorithms through which globally optimal estimates are computed with only local inter-sensor communications (e.g., Delouille et al. 2004). As this diverse yet non-exhaustive list of issues suggests, the literature on decentralized inference is massive and growing. See, for example, Blatt and Hero (2004); Blum et al. (1997); D'Costa and Sayeed (2003); D'Costa et al. (2004); Gastpar (2007); Kotecha et al. (2005); Li et al. (2002); Nowak (2003); Ribeiro and Giannakis (2006a); Ribeiro et al. (2007); Servetto (2002); Tsitsiklis (1993); Veeravalli and Chamberland (2007) and references thereto and therein for entry points.

From a theoretical perspective, parametric models enable a rigorous examination of many fundamental questions for inference under communication constraints. However, practically speaking, such strong assumptions should be motivated by data or prior application-specific domain knowledge. If, instead, data is sparse and prior knowledge is limited, robust *nonparametric methods* for decentralized inference are generally preferred.

The anticipated applications for wireless sensor networks range broadly from homeland security and surveillance to habitat and environmental monitoring. Indeed, advances in microelectronics and wireless communications have made wireless sensor networks the predicted panacea for attacking a host of large-scale decision and information-processing tasks. As the demand for these devices increases, one cannot expect that the necessary data or domain knowledge will always be available to support a parametric approach. Consequently, applications of wireless sensor networks provide an especially strong motivation for the study of nonparametric methods for decentralized inference.

Recognizing this demand, a variety of researchers have taken steps to relax the need to make strong statistical assumptions about phenomena under observation, moving toward a nonparametric approach to distributed inference. For example, requiring only weak assumptions about the underlying distribution, Nasipuri and Tantaratana (1997) and Vishwanathan and Ansari (1989) consider schemes based on the Wilcoxon signed-rank test statistic, Al-Ibrahim and Varshney (1989) and Han et al. (1990) study the sign detector, and Barkat and Varshney (1989) and Hussaini et al. (1995) address constant-false-alarm-rate detection, all in a distributed environment. Schemes for universal decentralized detection and estimation are surveyed in Xiao et al. (2006), and are studied in detail in Luo (2005a,b); Ribeiro and Giannakis (2006b); Xiao and Luo (2005a,b). From a practical perspective, these approaches are attractive not only because they permit the design of robust networks with provable performance guarantees, but also because in principle, they support the design of 'isotropic' sensors, i.e., sensors that may be deployed for multiple applications without reprogramming.

In this chapter, our focus is on an alternative nonparametric approach, the learning-theoretic approach (e.g., Vapnik 1991). Frequently associated with pattern recognition (e.g., Devroye et al. 1996; Duda et al. 2001), nonparametric regression (e.g., Gyorfi et al. 2002), and neural networks (e.g., Anthony and Bartlett 1999), learning-theoretic methods are aimed precisely at decision problems in which prior knowledge is limited and information about the underlying distributions is available via a limited number of observations, i.e. via a training data set. Researchers in computer science, statistics, electrical engineering, and other communities have been united in the field of machine learning, in which computationally tractable and statistically sound methods for nonparametric inference have been developed. Powerful tools such as boosting (e.g., Freund and Schapire 1997) and kernel methods (e.g., Schölkopf and Smola 2002) have been successfully employed in real-world applications ranging from handwritten digit recognition to functional genomics, and are well understood statistically and computationally. A general research question arises: can the power of these tools be tapped for inference in wireless sensor networks?

As we discuss below, the classical limits of and algorithms for nonparametric learning are not always applicable in wireless sensor networks, in part because the classical models from which they are derived have abstracted away the communication involved in data-acquisition. This observation provides inspiration for distributed learning in wireless sensor networks, and leads to a variety of fundamental questions. How is distributed learning in sensor networks different from centralized learning? In particular, what fundamental limits on learning are imposed by constraints on energy and bandwidth? In light of such limits, can existing learning algorithms be adapted? These questions are representative of a larger thrust within the sensor network community which invites engineers to consider signal processing and communications jointly.

Though the impetus for nonparametric distributed learning has been recognized in a variety of fields, the literature immediately relevant to sensor networks is small and is not united by a single model or paradigm. Indeed, distributed learning is a relatively young area, as compared to (parametric) decentralized detection and estimation, wireless sensor networks, and machine learning. Thus, an exhaustive literature review would necessarily focus on numerous disparate papers rather than aggregate results organized by model. In the interest of space, this chapter divides the literature on distributed learning according to two general research themes: *distributed learning in wireless sensor networks with a fusion center*, where the focus is on how learning is effected when communication constraints limit access to training data; and *distributed learning in wireless sensor networks with in-network processing*, where the focus is on how inter-sensor communications and local processing may be exploited to enable communication-efficient collaborative learning. We discuss these themes within the context of several papers from the field. Though the result is a survey unquestionably biased toward the authors' own interests, our hope is to provide the interested reader with an appreciation of a set of fundamental issues within distributed signal processing and an entrée to a growing body of literature.

The remainder of this chapter is organized as follows. In Section 8.2 we review the classical supervised learning model, and discuss kernel methods, a popular and well-studied class of learning algorithms. In Section 8.3, we discuss distributed learning, contrasting it with its classical counterpart and highlighting its relevance to and the challenges posed by wireless sensor networks. In Section 8.4, we examine research aimed at distributed learning in wireless sensor networks with a fusion center. In Section 8.5, we study distributed learning in wireless sensor networks with in-network processing; in particular, we examine several message-passing algorithms for collaboratively training least-squares regression estimators. Finally, we end with conclusions in Section 8.6. Given that nonparametric methods often permit parametric (e.g., Bayesian) interpretations, various connections exist between the parametric analyses previously cited and those cast formally within the learning framework. As appropriate, we highlight such connections in the discussion to follow.

8.2 Classical Learning

In this section, we summarize the supervised learning model which is often studied in learning theory, nonparametric statistics and statistical pattern recognition. For a thorough introduction to classical learning models and algorithms, we refer the reader to the review paper (Kulkarni et al. 1998) and references therein, and to standard books, e.g., (Anthony and Bartlett 1999; Devroye et al. 1996; Duda et al. 2001; Gyorfi et al. 2002; Hastie et al. 2001; Mitchell 1997).

8.2.1 The Supervised Learning Model

Let X and Y be $\mathcal{X}$-valued and $\mathcal{Y}$-valued random variables, respectively. $\mathcal{X}$ is known as the feature, input, or observation space; $\mathcal{Y}$ is known as the label, output, target, or parameter space. Attention in this chapter is restricted to detection and estimation, i.e., we consider two cases corresponding to binary classification ($\mathcal{Y} = \{0, 1\}$) and regression ($\mathcal{Y} = \Re$). To ease exposition, we assume that $\mathcal{X} \subseteq \Re^d$.

Given a loss function $l : \mathcal{Y} \times \mathcal{Y} \to \Re$, we seek a *decision rule* $g : \mathcal{X} \to \mathcal{Y}$ that minimizes expected loss,

$$\mathbf{E}\{l(g(X), Y)\}. \qquad (8.1)$$

In the binary classification setting, the criterion of interest is the probability of misclassification, which corresponds to the zero-one loss function $l(y, y') = 1_{\{y \neq y'\}}(y, y')$. In the context of estimation, the squared error $l(y, y') = |y - y'|^2$ is the metric of choice. In parametric statistics, one assumes prior knowledge of a joint probability distribution $\mathbf{P}_{XY}$ that describes the stochastic relationship between inputs and outputs. Under this assumption, the structure of the loss minimizing decision rule is well understood. The regression function

$$g(x) = \mathbf{E}\{Y \,|\, X = x\}$$

achieves the minimal expected squared error, and the maximum *a posteriori* (MAP) decision rule

$$g(x) = \begin{cases} 1 & \text{if } \mathbf{P}\{Y = 1 \,|\, X = x\} > \mathbf{P}\{Y = 0 \,|\, X = x\} \\ 0, & \text{otherwise} \end{cases}$$

is Bayes optimal for binary classification (Devroye et al. 1996) under the zero-one loss. In the sequel, we use

$$L^\star = \min_g \mathbf{E}\{l(g(X), Y)\}$$

to denote the loss achieved by the loss-minimizing decision rule.

In the supervised learning model, prior knowledge of the joint distribution $\mathbf{P}_{XY}$ is not available and thus, computing the MAP decision rule or the regression function is not possible. Instead, one is provided a collection of training data

$$S_n = \{(x_i, y_i)\}_{i=1}^n \subset \mathcal{X} \times \mathcal{Y},$$

i.e., a set of exemplar input-output pairs, to use in designing a decision rule. In order to characterize the statistical limits of learning or to analyze specific learning algorithms, S_n is often assumed to be generated from some stochastic process. Here, we make the standard assumption that $S_n = \{(X_i, Y_i)\}_{i=1}^n$ is independently and identically distributed (i.i.d.) with $(X_i, Y_i) \sim \mathbf{P}_{XY}$.

In this chapter, a *learning rule* (or learning algorithm) is taken to be a sequence $\{g_n\}_{n=1}^\infty$ of data-dependent decision rules $g_n : \mathcal{X} \times (\mathcal{X} \times \mathcal{Y})^n \to \mathcal{Y}$, thought to be designed without making additional, unverifiable assumptions on $\mathbf{P}_{XY}$. The process of constructing the decision rule $g_n(\cdot, S_n)$ is called *training*, and any procedure for doing so is called a *training algorithm*. When a learning rule $\{g_n\}_{n=1}^\infty$ is implicit, we use

$$L_n = L_n(S_n) = \mathbf{E}\{l(g_n(X, S_n), Y) \mid S_n\}.$$

to denote its expected loss conditioned on the (random) training data set S_n.

8.2.2 Kernel Methods and the Principle of Empirical Risk Minimization

Kernel methods constitute a popular class of learning rules that have been developed in the context of the supervised learning model. The kernel approach to learning can be summarized as follows. First, design a kernel, i.e., a positive semi-definite function $K : \mathcal{X} \times \mathcal{X} \to \mathfrak{R}$, as a similarity measure for inputs. Though kernel design is an active area of research, it is generally an art, typically guided by application-specific domain knowledge; Table 8.1 lists several commonly used kernels for $\mathcal{X} = \mathfrak{R}^d$. Then, construct a real-valued decision rule g_n as follows:

$$g_n(x) = g_n(x, S_n) = \begin{cases} \frac{\sum_{i=i}^n K(x, x_i) y_i}{\sum_{i=i}^n K(x, x_i)} & \text{if } \sum_{i=i}^n K(x, x_i) > 0 \\ 0, & \text{otherwise} \end{cases} \tag{8.2}$$

In words, $g_n(x)$ associates with each input $x \in \mathcal{X}$ a weighted average of the training data outputs, with the weights determined by how 'similar' the corresponding inputs are to x. With the naive kernel, $g_n(x)$ is analogous to the Parzen-window rule for density estimation.

The wisdom behind the kernel approach is demonstrated by Stone's Theorem (Stone 1977), which establishes that kernel learning rules can be made universally consistent. Intuitively, a learning rule is universally consistent if in the limit of large amounts of data, its loss is expected to be as small as if one had known $\mathbf{P}_{XY}$ in advance. This notion is formalized with the following definition:

Definition 8.2.1 *A learning rule* $\{g_n\}_{n=1}^\infty$ *is* universally consistent *if and only if* $\mathbf{E}\{L_n\} \to L^\star$ *for all distributions* $\mathbf{P}_{XY}$ *with* $\mathbf{E}\{Y^2\} < \infty$.

Table 8.1 Common kernels.

$K(x, x')$	Name
$1_{\{\|x - x'\|_2 \leq r_n\}}$	Naive kernel
$x^T x'$	Linear kernel
$(1 + x^T x')^d$	Polynomial kernel
$\exp^{-\|x - x'\|_2^2}$	Gaussian kernel
$(1 - \|x - x'\|_2^2)_+$	Epanechnikov kernel
$1/(1 + \|x - x'\|_2^{d+1})$	Cauchy kernel

Stated in full generality, Stone's Theorem establishes that a large class of 'weighted average' learning rules can be made universally consistent, including kernel rules with a Gaussian kernel, nearest neighbor rules, and histogram estimators. Described in detail in Devroye et al. (1996) and in Gyorfi et al. (2002), the following theorem prescribes sufficient conditions for the naive kernel rule to be universally consistent in a least-squares setting. Interestingly, under identical assumptions, the binary decision rule induced by thresholding (8.2) at one-half is universally consistent for binary classification under the zero-one loss (see Devroye et al. 1996).

Theorem 8.2.2 (Stone) *Suppose that $\{g_n\}_{n=1}^{\infty}$ is as in (8.2) with the naive kernel (Table 8.1). If $r_n \to 0$ and $nr_n^d \to \infty$, then $\{g_n\}_{n=1}^{\infty}$ is universally consistent under the squared-error criterion.*

Though this seminal result is promising, there is a catch. It is well known that without additional assumptions on $\mathbf{P}_{XY}$, the convergence rate of $\mathbf{E}\{L_n\}$ may be arbitrarily slow. Moreover, even with appropriate assumptions, the rate of convergence is typically exponentially slow in d, the dimensionality of the input space. These caveats have inspired the development of practical learning rules that recognize the finite-data reality and the so-called curse of dimensionality.

Many popular learning rules are inspired by the principle of empirical risk minimization (Vapnik 1991), which requires the learning rule to minimize a data-dependent approximation of the expected loss (8.1). For example, consider the learning rule defined as follows:

$$g_n^{\lambda} = \arg \min_{f \in \mathcal{F}} \left[\frac{1}{n} \sum_{i=1}^{n} l(f(x_i), y_i) + \lambda \|f\|_{\mathcal{F}}^2 \right]. \tag{8.3}$$

The first term in the objective function (8.3) is the empirical loss[2] of a decision rule $f : \mathcal{X} \to \mathfrak{R}$, and serves as a measurement of how well f 'fits the data'; the second term acts as a complexity control and regularizes the optimization. $\lambda \in \mathfrak{R}$ is a parameter that governs the trade-off between these two terms. The optimization variable (i.e., function) is f, which is constrained to be in a Hilbert space $\mathcal{F}$ with norm $\| \cdot \|_{\mathcal{F}}$.

Intuitively, when n is large and λ is small, the objective function will closely approximate the expected loss of f over $\mathcal{F}$. The hope is that the decision rule g_n^{λ} will then approximately minimize the expected loss. The principle of empirical risk minimization is well understood, but unfortunately a more thorough introduction is beyond the scope of this chapter. We refer the reader to standard references for additional information; see, for example, Vapnik (1991).

Reproducing kernel methods generalize the simple kernel rule (8.2), while employing the principle of empirical risk minimization. In particular, reproducing kernel methods follow (8.3), taking $\mathcal{F} = \mathcal{H}_K$ to be the *reproducing kernel Hilbert space* (RKHS) induced by a kernel $K(\cdot, \cdot)$. More precisely, for any positive definite function $K(\cdot, \cdot)$, we can

---

[2]In practice, for various statistical and computational reasons, the empirical loss is often measured using a convex loss function which may bound or otherwise approximate the loss criterion of interest. See Schölkopf and Smola (2002) for discussion and examples.

construct a unique collection of functions $\mathcal{H}_K$ such that

$$K_t = K(\cdot, t) \in \mathcal{H}_K \ \forall t \in \mathcal{X}$$

$$\sum_{i=1}^{n} \alpha_i K_{t_i} \in \mathcal{H}_K \ \forall \{\alpha_i\}_{i=1}^{n} \subset \mathfrak{R}, \ n < \infty. \tag{8.4}$$

If we equip $\mathcal{H}_K$ with an inner-product defined by $< K_s, K_t > = K(s, t)$, extend $\mathcal{H}_K$ using linearity to all functions of the form (8.4), and include the point-wise limits, then $\mathcal{H}_K$ is called an RKHS. Note that

$$< K_x, f > = f(x),$$

for all $x \in \mathcal{X}$ and all $f \in \mathcal{H}_K$; this identity is the reproducing property from which the name is derived. Henceforth, we use $\| \cdot \|_{\mathcal{H}_K}$ to denote the norm associated with $\mathcal{H}_K$.

Returning to the discussion on learning with kernel methods, note that the inner-product structure of $\mathcal{H}_K$ implies the following 'Representer Theorem', proved in a least-squares context in Kimeldorf and Wahba (1971); a generalization appears in Schölkopf and Smola (2002).

Theorem 8.2.3 (Representer Theorem) *The minimizer $g_n^\lambda \in \mathcal{H}_K$ of (8.3) admits a representation of the form*

$$g_n^\lambda(\cdot) = \sum_{i=1}^{n} c_{n,i}^\lambda K(\cdot, x_i), \tag{8.5}$$

for some $\mathbf{c}_n^\lambda \in \mathfrak{R}^n$.

Theorem 8.2.3 is significant because it highlights that while the optimization in (8.3) is defined over a potentially infinite dimensional Hilbert space, the minimizer must lie in a finite dimensional subspace. It also highlights a sense in which reproducing kernel methods generalize their more naive counterpart, since (8.2) can be expressed as (8.5) for a particular choice of $\mathbf{c}_n^\lambda$. To emphasize the significance of the Representer Theorem, note that in least-squares estimation it implies that $\mathbf{c}_n^\lambda$ is the solution to a system of n linear equations. In particular, it satisfies

$$\mathbf{c}_n^\lambda = (K + \lambda I)^{-1} \mathbf{y}, \tag{8.6}$$

where $K = (k_{ij})$ is the kernel matrix ($k_{ij} = K(x_i, x_j)$).

The statistical behavior of kernel methods is well understood under various assumptions on the stochastic process that generates the examples in S_n (see, e.g., Gyorfi et al. 2002; Schölkopf and Smola 2002). Moreover, this highly successful technique has been verified empirically in applications ranging from bioinformatics to hand-written digit recognition.

8.2.3 Other Learning Algorithms

As the reader may well be aware, the scope of supervised learning rules extends far beyond kernel methods, and includes, for example, neural networks (e.g., Anthony and Bartlett 1999), nearest-neighbor rules (e.g., Gyorfi et al. 2002), decision-trees (e.g., Quinlan 1992),

Bayesian and Markov networks (e.g., Jordan 1999; Pearl 1988), and boosting (e.g., Freund and Schapire 1997). Many of these algorithms are well understood computationally and statistically; together they comprise an indispensable toolbox for learning. At this point, we leave classical supervised learning in general and kernel methods in particular, referring the interested reader to previously cited references for additional information.

8.3   Distributed Learning in Wireless Sensor Networks

To illustrate how learning is relevant to wireless sensor networks and to motivate the problem of distributed learning, let us consider the following toy example.

Suppose that the feature space $\mathcal{X}$ models a set of observables measured by sensors in a wireless network. For example, the components of an element $x \in \mathcal{X} = \Re^3$ may model coordinates in a (planar) environment and time. $\mathcal{Y} = \Re$ may represent the space of temperature measurements. A decision maker may wish to know the temperature at some point in space-time; to reflect that these coordinates and the corresponding temperature are unknown, let us model them with the random variable (X, Y). A joint distribution $\mathbf{P}_{XY}$ may model the spatio-temporal correlation structure of a temperature field. If the field's structure is well understood, i.e., if $\mathbf{P}_{XY}$ can be assumed known a priori, then an estimate may be designed within the parametric framework (e.g., Poor 1994). However, if such prior information is unavailable, an alternative approach is necessary.

Suppose that sensors are randomly deployed about the environment, and collectively acquire a set $S_n \subset \mathcal{X} \times \mathcal{Y}$ of temperature measurements at various points in space-time.[3] The set S_n is akin to the training data described in Section 8.2, and thus supervised learning algorithms seem naturally applicable to this field-estimation problem. However, the supervised learning model has abstracted away the process of data acquisition, and generally does not incorporate communication constraints that may limit a learning algorithm's access to data. Indeed, the theory and methods discussed in Section 8.2 depend critically on the assumption that the training data is entirely available to a single processor. However, in wireless sensor networks, the energy and bandwidth required to collect the sensors' raw measurements may be prohibitively large. Thus, training centralized learning rules may limit the sensors' battery life, may waste bandwidth, and may ultimately preclude one from realizing the potential of wireless sensor networks.

Sensors in WSNs are typically equipped with on-board processing capabilities, and this fact has important implications for distributed inference. In particular, with the ability to locally process information, the sensors are more than mere data-collectors, as the name 'sensor' may suggest. Rather, sensors are full-fledged information processors, and therefore the network is better viewed as a distributed information processing system. However, to suggest that a sensor network is a collection of processors is an oversimplification. Indeed, inter-sensor communications are severely limited by tight constraints on energy and bandwidth, and this fundamentally distinguishes distributed learning from being an application of parallel computing.

One might employ multi-terminal data compression, in hopes of exploiting correlation amongst sensors' measurements to send the training set to a central processing site;

---

[3]A host of localization algorithms have been developed to enable sensors to measure their location; see, for example, Gezici et al. (2005); Nguyen et al. (2005a); Patwari et al. (2005).

ostensibly, such an approach would be a significant improvement over sending 'raw' data. However, the goal of such an approach would be to reconstruct the complete training set S_n at a central location, and subsequently train a classical supervised learning algorithm. In short, the goal is the same as before – to learn a general decision rule – and thus, one is inclined to skip the intermediate step of sending data, and proceed directly to learning. Indeed, the sensors' ability to compute allows us to push intelligence to the outer extremes of the network, and may ultimately provide an increase in communication efficiency. Therein lies the motivation for *distributed learning* in wireless sensor networks, and the inspiration for many fundamental questions.

In particular, how does communication fundamentally influence learning? Can we design learning algorithms to respect constraints on energy and bandwidth? In the next section, we introduce a model for distributed learning that will be the focus of analysis in the ensuing chapters.

Before proceeding, note that the simplicity of the preceding example should not temper the promise for wireless sensor networks, nor should it mask the fundamental importance of distributed learning. In particular, note that $\mathcal{X}$ may model more than position or time, and may represent a space of multi-modal sensor measurements that commonly occur in wireless sensor network applications. $\mathcal{Y}$ may model any number of quantities of interest, for example, the strength of a signal emitted from a target, a force measured by a strain gauge, or an intensity level assessed by an acoustic sensor; $\mathcal{Y}$ may even be multidimensional. In general, each sensor or the fusion center aims to design a decision rule $g : \mathcal{X} \to \mathcal{Y}$ using the data observed by the sensor network.

8.3.1 A General Model for Distributed Learning

As a starting point to studying the aforementioned questions, consider a model for distributed learning in wireless sensor networks. Suppose that in a network of m sensors, sensor j has acquired a unique set of measurements, i.e., training data, $S_n^j \subseteq S_n = \cup_{j=1}^m S_n^j$. In the example above, S_n^j may represent a stationary sensor's measurements of temperature over the course of a day, or a mobile sensor's readings at various points in space-time. Suppose further that the sensors form a wireless network, whose topology is specified by a graph. For example, consider the models depicted pictorially in Figures 8.1 and 8.2.

Each node in the graph represents a sensor and its locally observed data; an edge in the graph posits the existence of a wireless link between sensors. A fusion center can be modeled as an additional node in the graph, perhaps with larger capacity links between itself and the sensors, to reflect its larger energy supply and computing power. A priori, this model makes no assumptions on the topology of the network (e.g., the graph is not necessarily connected), though the promise of distributed learning may in fact depend on such properties.

Much of the existing research on distributed learning can be categorized according to its focus on one of two classes of networks. Depicted in Figure 8.1, the parallel network supposes a collection of sensors that communicate directly with a fusion center over a multiple-access channel; or viewed differently, that a single agent has a bandwidth-limited channel over which it may access all the data. In this setting, the question is: how is learning fundamentally affected when bandwidth constraints limit our access to the data? This architecture is relevant to wireless sensor networks whose primary purpose is data collection.

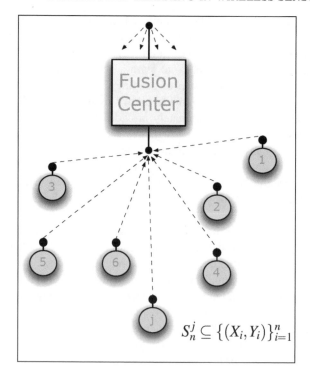

$$S_n^j \subseteq \{(X_i, Y_i)\}_{i=1}^n$$

Figure 8.1 A parallel network with a fusion center.

A second class of networks retain an ad-hoc structure like the network in Figure 8.2. A typical assumption is that the topology of these networks is dynamic and perhaps unknown prior to deployment; a fusion center may exist, but the sensors are largely autonomous and may make decisions independently of the single, coordinating authority (e.g., to track a moving object). Research on these types of structures typically focuses on how localized processing and inter-sensor communication can be exploited to allow communication-efficient learning by enabling collaboration.

Much of the work in distributed learning differs in the way that the capacity of the links is modeled. Given that learning is already a complex problem, simple application-layer abstractions are typically preferred over detailed physical layer models. The links are often assumed to support the exchange of 'simple' real-valued messages, where simplicity is assessed relative to the application (e.g., sensors share summary statistics rather than entire data sets). Lacking a formal communication model, quantifying the efficiency of various methods from an energy and bandwidth perspective is not always straightforward.

Transcending many analyses and implementations of wireless networks is the importance of local communication. Loosely speaking, local communications are those that occur between neighboring sensors in a communication network. In wireless networks, topology of the network is in correspondence with the topology of the environment, which is to say that a sensor's network neighborhood is roughly equal to its physical neighborhood. The energy required for two sensors to communicate decreases as the distance between

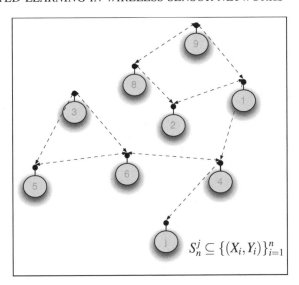

$$S_n^j \subseteq \{(X_i, Y_i)\}_{i=1}^n$$

Figure 8.2 An ad-hoc network with in-network processing. © 2006 IEEE. Reprinted with permission, from Distributed Learning in Wireless Sensor Networks by Joel B. Predd, Sanjeev R. Kulkani and H. Vincent Poor, *IEEE Signal Processing Magazine*, Vol 23(4), 56–69 pp. July 2006.

them decreases, often according to an inverse square law; by the same law, multiple-access interference decreases as the distance between pairs of communicating nodes increases. Thus, by minimizing energy expenditure and by enabling spectral reuse, local communications are often an efficient mode of information transport in WSNs.[4]

The foregoing observation is the starting point for many studies on distributed learning (and indeed, on distributed inference more generally). Rather than formalize a detailed physical layer communication model, which may or may not be relevant to any specific WSN application, studies of distributed learning often posit a model for local communication and then study how sensor-to-sensor (sensor-to-fusion center) interactions can improve learning by enabling collaboration. Ultimately, assumptions about the efficiency of local communications must be justified, perhaps by formalizing a physical-layer communication model or perhaps through scaling law analyses. However, application-layer abstractions of local communication are nonetheless a reasonable starting point to investigate the fundamental limits of distributed learning. We will revisit these ideas in the discussion to follow, after instantiating the preceding models with more specific assumptions about sensor-to-sensor (sensor-to-fusion center) communication.

In the next few sections, we review recent work aimed at understanding these issues in the context of wireless sensor networks. Though statistical and machine learning are rife with results relevant to distributed learning in general, to our knowledge surprisingly little research has addressed learning in wireless sensor networks in particular. Thus, before

---

[4]These tradeoffs are studied more formally, for example, in the literature on scaling laws in WSNs; see, e.g., Gupta and Kumar (2000); Kulkarni and Viswanath (2004) and references therein and thereto.

proceeding, let us highlight several areas of machine learning research that are relevant to distributed learning, if not to wireless sensor networks, and that may have a bearing on future studies in distributed learning in wireless sensor networks.

8.3.2   Related Work

Within the context of wireless sensor networks, Nguyen et al. (2005b) develop a nonparametric kernel-based methodology for decentralized detection. As in centralized learning, a training set is assumed available offline to a single processor. The data is used to train a learning rule that solves an optimization problem similar to (8.3), with the additional constraint that the resulting decision rule lies within a restricted class which is deployable across a sensor network; the powerful notion of a marginal kernel is exploited in the process. This setting is fundamentally different from the present context in that the data is centralized. Thus, one might distinguish the former topic of *centralized learning for decentralized inference* from the present topic of distributed learning for decentralized inference.

Ensemble methods have attracted considerable attention within machine learning. Examples of these techniques include bagging, boosting, mixtures of experts, and others (see, e.g., Breiman 1996; Freund and Schapire 1997; Freund et al. 1997; Jacobs et al. 1991; Kittler et al. 1998). Intuitively, these methods allocate portions of the training database to different learning algorithms, which are independently trained. The predictions of the individual learning rules are subsequently aggregated, sometimes by employing a different training algorithm and sometimes using feedback from the training phase of the individual learning algorithms. One might cast such methods within a framework of distributed learning in a parallel network, but ensemble methods are generally designed within the classical model for supervised learning, and fundamentally assume that the training set S_n is available to a single coordinating processor. In general, the focus of ensemble learning is on the statistical and algorithmic advantages of learning with an ensemble and not on the nature of learning under communication constraints. Nevertheless, many fundamental insights into learning have arisen from ensemble methods; future research in distributed learning stands to benefit.

Related to ensemble methods, and inspired by the availability of increasingly large data sets, an active area of machine learning research focuses on 'scaling up' existing learning rules to handle massive training databases; see, for example, Bordes et al. (2005); Chawla et al. (2004); Graf et al. (2005); Provost and Hennessy (1996) and references thereto and therein. One approach is to decompose the training set into smaller 'chunks', and subsequently parallelize the learning process by assigning distinct processors/agents to each of the chunks. In this setting, sometimes termed parallel learning, the communication constraints arise as parameters to be tweaked, rather than from resources to be conserved; this difference in perspective often limits the applicability of the underlying communication model to applications like sensor networks. However, in principle, algorithms for parallelizing learning may be useful for distributed learning with a fusion center and vice versa.

Population learning is an early model for distributed learning (Kearns and Seung 1995; Nakamura et al. 1998; Yamanishi 1997). Once again, a parallel network is considered, but in contrast to ensemble methods and parallel learning in which the predictions of individual learning rules are combined, it is assumed that the 'sensors' transmit a complete description of their locally trained decision rules to the fusion center. The fusion center's task is to observe the response of the network to infer a more accurate rule. The original model (Kearns and Seung 1995) was parametric (i.e., 'distribution specific' learning), and was constructed in the spirit of the PAC 'probably approximately correct' framework (Valiant 1984). Generalizations considered in Nakamura et al. (1998) relaxed such assumptions, but the results ultimately depend on strong assumptions about a class of hypotheses that generate the data. The utility of these results to wireless sensor networks may be limited by these strong assumptions, or by the demands of communicating a complete description of the rule. Nevertheless, population learning may provide insights for distributed learning with a fusion center.

The online learning framework also appears relevant to distributed learning in wireless sensor networks with a fusion center (see, e.g., Cesa-Bianchi et al. 1997; Freund et al. 1997; Littlestone and Warmuth 1994). In that setting, a panel of experts (i.e., a network of sensors) provide predictions (one can imagine that, as in ensemble and parallel learning, predictions arose from independently trained estimators, but such assumptions are unnecessary). In contrast to ensemble learning, online learning occurs in repeated trials. At each trial, a centralized agent makes its own prediction by combining expert predictions through a weighted average; after learning the 'truth' (i.e., Y), the agent suffers a loss (e.g., squared error), and attempts to 'track' the best expert by updating the weights of its weighted average by taking into account the past performance of each expert. Under minimal assumptions on the evolution of these trials, bounds are derived that compare the trial-averaged performance of the central agent with that of the best (weighted combination of) expert(s). Communication constraints enter online learning implicitly, since the information that the individual experts use is not needed by the centralized agent. This framework may be relevant to aggregation problems that arise in wireless sensor networks, however, to our knowledge such applications have not been made.

At a higher level, the field of data mining has explored distributed learning in the context of distributed databases. Here, communication constraints arise when various agents have access to distinct training databases but are unable to share their data due to security, privacy, or legal concerns (the fraud detection application is illustrative of a relevant scenario in which corporations have access to large databases of customer transactions and wish to collaborate to identify fraudulent interactions). Though the communication constraints arise for very different reasons, the problem bears resemblance to the ad-hoc structure of distributed learning in sensor networks. In the data mining context, a distributed boosting algorithm is studied in Lazarevic and Obradovic (2001); a similar algorithm is analyzed in the context in secure multi-party computation in Gambs et al. (2005).

8.4 Distributed Learning in WSNs with a Fusion Center

In this section, we discuss distributed learning in wireless sensor networks with a fusion center, which focuses on the parallel network depicted in Figure 8.1. Recall that in this setting, each sensor in the network acquires a set of data. In the running example, the data may constitute the sensors' temperature measurements at discrete points in space-time. The fusion center would like to use the locally observed data to construct a global estimate of the continuously varying temperature field.

8.4.1 A Clustered Approach

The naive approach in this setting would require the sensors to send all of their data to the fusion center. As has been discussed, this approach would be costly in terms of energy and bandwidth. A more principled methodology might designate a small subset of nodes to send data. If the number of nodes is small, and the data (or the nodes) are wisely chosen, then such a strategy may be effective in optimizing learning performance while keeping communication costs to a minimum.

For example, one may partition the sensors into subgroups, and assign each a 'cluster head'.[5] Cluster heads may retrieve the data from sensors within its group; since the sensors within a cluster are nearby, this exchange may be inexpensive since it involves only local communications. Then, the cluster head may filter this data and send the fusion center a summary, which might include a locally learned rule or data that is particularly informative (e.g., 'support vectors'). Clustered approaches have been considered frequently within parametric frameworks for detection and estimation (e.g., D'Costa et al. 2004).

Nguyen et al. (2005a) considered a clustered approach to address sensor network localization. There, the feature space $\mathcal{X} = \Re^2$ models points in a planar terrain, and the output space $\mathcal{Y} = \{0, 1\}$ models whether or not a point belonged to a (specifically designed) convex region within the terrain. Training data is acquired from a subset of sensors (base stations) whose positions were estimated using various physical measurements. The fusion center uses reproducing kernel methods for learning, with a kernel designed using signal-strength measurements. The output is a rule for determining whether any sensor (i.e., non-base stations) lay in the convex region using only a vector of signal-strength measurements. We refer the reader to the paper for additional details, and reports on several real-world experiments. However, we highlight this as an example of the clustered approach to distributed learning with a fusion center, a methodology which is broadly applicable.

8.4.2 Statistical Limits of Distributed Learning

Stone's Theorem is a seminal result in statistical pattern recognition which established the existence of universally consistent learning rules (see Theorem 8.2.2). Many efforts have extended this result to address the consistency of Stone-type learning rules under various sampling processes; for example, Devroye et al. (1996), Gyorfi et al. (2002) and references therein, Cover (1968); Greblicki and Pawlak (1987); Krzyżak (1986); Kulkarni and Posner (1995); Kulkarni et al. (2002); Morvai et al. (1999); Nobel (1999); Nobel and Adams (2001); Nobel et al. (1998); Roussas (1967); Stone (1977); Yakowitz (1989, 1993). These results extend Theorem 8.2.2 by considering various dependency structures within the training data (e.g., Markovian data). However, all of these works are in the centralized setting and assume that the training database is available to a single processor.

Predd et al. (2006a) attempted to characterize the limits of distributed learning with a fusion center, by overlaying several simple communication models onto the classical model for supervised learning. In particular, extensions of Stone's Theorem were considered in light of the following question: with sensors that have each acquired a small set of training data and that have some limited ability to communicate with the fusion center, can enough information be exchanged to enable universally consistent learning?

---

[5]Distributed clustering algorithms have been developed with such applications in mind; see Bandyopadhyay and Coyle (2003), for example.

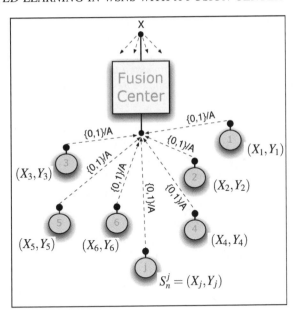

Figure 8.3 The model studied in Predd et al. (2004, 2006a). © 2006 IEEE. Reprinted with permission, from Distributed Learning in Wireless Sensor Networks by Joel B. Predd, Sanjeev R. Kulkani and H. Vincent Poor, *IEEE Signal Processing Magazine*, Vol 23(4), 56–69 pp. July 2006.

To address this question, Predd et al. (2006a) supposes that each sensor acquires just one training example, i.e., $S_n^j = \{(X_j, Y_j)\}$. Communication was modeled as follows: when the fusion center observes a new observation $X \sim \mathbf{P}_X$, it broadcasts the observation to the network in a request for information. At this time, bandwidth constraints limit each sensor to responding with at most one bit. That is, each sensor chooses whether or not to respond to the fusion center's request for information; if it chooses to respond, a sensor sends either a 1 or a 0 based on its local decision algorithm. Upon observing the response of the network, the fusion center combines the information to create an estimate of Y.

A refined depiction of the architecture of this model is depicted in Figure 8.3. To emphasize its structure, note that the fusion center has a broadcast channel back to the sensor (for requesting information on X), and each sensor has a point-to-point wireless uplink channel over which they can send one bit. Since each sensor may *abstain* from voting altogether, the sensors' uplink channels have a slightly larger capacity than is suggested by this mere one bit that we have allowed them to physically transmit to the fusion center. Indeed, sensor-to-fusion center communication occurs even when a sensor abstains from voting.

Taken together, the sensors' local decision algorithms and the fusion center's combining rule form a distributed data-dependent decision rule. Can this decision rule be designed to simultaneously satisfy the communication constraints of the model and enable universally consistent learning? The following theorem settles the question in this model for distributed learning with abstention.

Theorem 8.4.1 (*Predd et al. (2006a)*) *[Classification and Estimation with Abstention] Suppose that the sensors' data $S_n = \cup_{j=1}^{m} S_n^j$ are i.i.d. with $(X_i, Y_i) \sim \mathbf{P}_{XY} \forall i \in \{1, \ldots, m\}$, and that each sensor has knowledge of m, the size of the sensor network. Then, in binary classification under the zero-one loss, and in estimation under the squared-error criterion, there exist sensor decision algorithms and a fusion rule that enable universally consistent distributed learning with abstention.*

In this model, each sensor decision rule can be viewed as a selection of one of *three* states: abstain, vote and send 0, and vote and send 1. With this observation, Theorem 8.4.1 can be interpreted as follows: $\log_2(3)$ bits per sensor per decision is sufficient to enable universally consistent learning in this model for distributed learning with abstention. In this view, it is natural to ask whether these $\log_2(3)$ bits are necessary. That is, can consistency be achieved by communicating at lower bit rates?

To answer this question, a revised model was considered, precisely the same as above, except that in response to the fusion center's request for information, each sensor must respond with 1 or 0; abstention is not an option and thus, each sensor responds with exactly one bit per decision. Can the sensors communicate enough information to the fusion center to enable universally consistent distributed learning *without abstention*? The following theorems settle this question.

Theorem 8.4.2 (*Predd et al. (2006a)*) *[Classification without Abstention] Suppose that the sensors' data $S_n = \cup_{j=1}^{m} S_n^j$ are i.i.d. with $(X_i, Y_i) \sim \mathbf{P}_{XY} \forall i \in \{1, \ldots, m\}$, and that each sensor has knowledge of m, the size of the sensor network. Then, in binary classification under the zero-one loss, there exist sensor decision rules and a fusion rule that enable universally consistent distributed learning without abstention.*

Theorem 8.4.3 (*Predd et al. (2006a)*) *[Estimation without Abstention] Suppose that the sensors' data $S_n = \cup_{j=1}^{m} S_n^j$ are i.i.d. with $(X_i, Y_i) \sim \mathbf{P}_{XY} \forall i \in \{1, \ldots, m\}$, that each sensor has knowledge of m, and that the fusion rule satisfies a set of regularity conditions.[6] Then, for any sensor decision rule that obeys the constraints of distributed learning without abstention, there does not exist a regular fusion rule that is consistent for every distribution $\mathbf{P}_{XY}$ with $\mathbf{E}\{Y^2\} < \infty$ under the squared-error criterion.*

Theorems 8.4.1 and 8.4.2 are proved by construction; sensor decision algorithms and fusion rules are specified that simultaneously satisfy the communication constraints of the respective models and are provably universally consistent. Theorem 8.4.3 is proved via a counter-example, and thereby establishes the impossibility of universal consistency in distributed regression without abstention for a restricted, but reasonable class of wireless sensor networks.

Theorems 8.4.1, 8.4.2, and 8.4.3 establish fundamental limits for distributed learning in wireless sensor networks, by addressing the issue of whether or not the guarantees provided by Stone's Theorem in centralized environments hold in distributed settings. However, the applicability of these results may be limited by the appropriateness of the model. For example, in practice, the training data observed by a sensor network may not be i.i.d.; in the field estimation problem, data may be corrupted by correlated noise (see studies in Son

---

[6]Predd et al. (2006a) assumes that the fusion rule is invariant to the order of bits received from the sensor network and Lipschitz continuous in the average Hamming distance.

et al. 2005b; Sung et al. 2005, 2006). Moreover, the process by which sensors acquire data may differ from the process observed by the fusion center; for example, sensors may be deployed uniformly about a city, despite the fusion center's interest in a particular district. In the context of binary classification, Predd et al. (2004) established the achievability of universally consistent distributed learning with abstention under a class of sampling processes which model such an asymmetry. In general, extending the above results to realistic sampling processes is of practical importance.

In these models, the assumption that each sensor acquires only one training example appears restrictive. However, the results hold for training sets of any finite (and fixed) size. Thus, these results have examined an asymptote not often considered in statistical learning, corresponding to the limit of the number of learning agents. One can argue that if the number of examples per sensor grows large, then universally consistent learning is possible within most reasonable communication models. Thus, communication-constrained sensor networks with finite training sets is an interesting case.

Finally, note that these models generalize, in a sense, models recently considered in universal decentralized detection and estimation (e.g., Luo 2005a,b; Ribeiro and Giannakis 2006b; Xiao and Luo 2005a,b). The communication and network models in that setting are nearly identical to those considered here. However, there the fusion center is interested in making a binary decision or in estimating a real-valued parameter, whereas in the present setting, the fusion center estimates a function.

8.5 Distributed Learning in Ad-hoc WSNs with In-network Processing

In this section, we turn our attention to distributed learning in wireless sensor networks with in-network processing, considering networks with the ad-hoc structure depicted in Figure 8.2. One should note that in doing so, we do not exclude the possibility of there being a fusion center. Our shift represents merely a change in focus. We consider how in-network processing and local inter-sensor communication may improve learning by enabling collaboration.

Many classical learning rules are infeasible in wireless sensor networks, because constraints on energy and bandwidth constraints preclude one from accessing the entire training set. One approach to extending classical learning rules to distributed learning, and in particular to wireless sensor networks, focuses on developing communication-efficient *training algorithms*. While recognizing the strong theoretical foundation on which existing learning rules are designed, this approach interprets communication constraints as imposing computational limits on training, and assumes there is methodology for assessing the efficiency of training algorithms from an energy and bandwidth perspective.

In Section 8.3.1, we discussed how the importance of local communication transcends many analyses and implementations of wireless networks. This observation has motivated the development and analysis of many so-called *local message-passing algorithms* for distributed inference in wireless sensor networks, and for distributed computation more generally (Giridhar and Kumar 2007). Roughly speaking, message-passing algorithms are those that use only local communications to achieve the same end (or approximately the same end) as 'global' (i.e., centralized) algorithms that require sending 'raw' data to a central processing facility. Message-passing algorithms are thought to be efficient by virtue of their exploitation of local communications. In practice, such intuitions must be formally justified. In theory, application-layer abstractions of local communication constitute a reasonable framework for studying distributed inference in general, and for developing communication-efficient training algorithms for distributed learning in particular.

Message-passing algorithms are a hot topic in many fields, wireless communications and machine learning notwithstanding. This surge in popularity is inspired in part by the powerful graphical model framework that has enabled many exciting applications and inspired new theoretical tools (see, e.g., Aji and McEliece 2000; Jordan 1999; Kschischang et al. 2001; Loeliger 2004; Paskin and Lawrence 2003; Pearl 1988; Plarre and Kumar 2004). These tools are often applicable to signal processing in wireless sensor networks, since often the correlation structure of the phenomenon under observation (e.g., a temperature field) can be represented using a graphical model (e.g., Markov networks) and since inter-sensor communications are envisioned to occur over similar graphical structures. Indeed, graphical models form a broad topic in their own right, and applications to sensor networks are deserving of a separate article (e.g., Cetin et al. 2007; Ihler et al. 2005). Here, our focus is specifically on how message-passing algorithms, broadly construed, may be applied to develop training algorithms for distributed learning in wireless sensor networks. The learning formalism aside, various connections may exist between the work we now discuss and the previously cited studies.

8.5.1   Message-passing Algorithms for Least-Squares Regression

To simplify the exposition, let us restrict ourselves to a least-squares estimation problem, and consider the reproducing kernel estimator discussed in Section 8.2. Also to simplify our discussion, assume that each sensor measures a single training example, i.e., $S_n^j = (x_j, y_j)$. Finally, assume that each sensor has been pre-programmed with the same kernel K.

Recall, reproducing kernel methods take as input a training set $\cup_{j=1}^m S_n^j = S_n = \{(x_j, y_j)\}_{i=1}^m$ and in the least-squares regression setting output a function $g_n^\lambda : \mathcal{X} \to \mathcal{Y}$ which solves the optimization problem

$$\min_{f \in \mathcal{H}_K} \left[\sum_{j=1}^m (f(x_j) - y_j)^2 + \lambda \|f\|_{\mathcal{H}_K}^2 \right]. \tag{8.7}$$

As discussed in Section 8.3, solving (8.7) is infeasible in wireless sensor networks, since the data in S_n is distributed about the network of sensors.

For learning rules motivated by the principle of empirical risk minimization, a training algorithm often must solve an optimization problem, e.g., (8.7). As a result, distributed and parallel optimization, fields with rich histories in their own right (see, e.g., Bertsekas and Tsitsiklis 1997; Censor and Zenios 1997), have an immediate bearing on distributed learning. Indeed, many tools from distributed and parallel optimization have been applied to develop tools for distributed inference; see, for example, Delouille et al. (2004); Moallemi and Roy (2004); Predd et al. (2005, 2006b); Rabbat and Nowak (2004, 2006); Son et al. (2005a). We now discuss three approaches to developing distributed training algorithms that differ by the structure that they exploit and by the messages that sensors exchange.

Training Distributively by Exploiting Sparsity

One class of distributed training algorithms is constructed to exploit an assumed relationship between the topology of the wireless network and the correlation structure of sensors' measurements. In the toy example of Section 8.3, for example, the temperature field may be slowly varying in space-time and thus it may be reasonable to assume that physically nearby sensors have similar temperature measurements. Since the sensors 'exist' in the space-time feature space $\mathcal{X}$, the network topology is intimately related to the topology of feature space, and hence the correlation structure of the temperature field. Many algorithms for distributed estimation using graphical models rely on formalizations of this powerful intuition, e.g., Delouille et al. (2004).

To see how such a relationship may be exploited in developing a distributed training algorithm, recall that in least-squares estimation, the Representer Theorem shows that the minimizer g_n^λ of (8.7) is implied by the solution to a system of linear equations,

$$(K + \lambda I)\mathbf{c}_\lambda = \mathbf{y} \qquad (8.8)$$

where $K = (K_{ij})$ is the kernel matrix with $K_{ij} = K(x_i, x_j)$. If each sensor acquires a single training datum so that there is a one-to-one correspondence between training examples and sensors, then K is a matrix of sensor-to-sensor similarity measurements. For many kernels, K is sparse. Various algorithms are available for efficiently solving sparse systems of linear equations, some of which admit message-passing implementations (e.g., Golub and Loan 1989; Paskin and Lawrence 2003). When the sparsity structure of K 'corresponds' in a convenient way with the topology of the network, often the messages are passed between neighboring nodes in the network, and the result is a training algorithm that implements a classical learning rule in a distributed way.

Guestrin et al. (2004) developed a distributed algorithm based on a distributed Gaussian elimination algorithm executed on a cleverly engineered junction tree. A detailed description of this algorithm requires familiarity with the junction tree formalism and knowledge of a distributed Gaussian elimination algorithm, which unfortunately are beyond the scope of the present chapter. Notably, the algorithm has provable finite-time convergence guarantees, and arrives at the globally optimal solution to (8.7). Because the system in Guestrin et al. (2004) is developed within a very general framework for distributed inference in sensor networks (Paskin et al. 2005), this approach is applicable in many cases when the intuition we have described fails (e.g., when sparsity is prevalent, but may not 'correspond' in an intuitive way the network topology). Nevertheless, the approach appears maximally efficient from an energy and bandwidth perspective when the intuition bears credibility. We refer the reader to Guestrin et al. (2004) for additional detail and a description of several interesting experiments.

Training distributively using incremental subgradient methods

Assumptions that couple the network and the correlation structure of the sensors' observations are powerful, but may be of limited use, since it is easy to envision examples where those assumptions break down. For example, sensors deployed about a city may observe correlated measurements of traffic flow, despite being unable to communicate due to a signal-obstructing skyscraper. In general, there is no fundamental, application-independent reason to assume a correspondence between the topology of the feature space $\mathcal{X}$ and the topology of the network.

A second approach to developing distributed training algorithms exploits the additive structure of the regularized empirical loss functional. To illustrate, suppose that agent (sensor) j has access to a single training datum $(x_j, y_j) \in S_n$, and for reasons that will soon become clear, let us rewrite (8.7) as

$$\min_{f \in \mathcal{H}_K} \left[\sum_{j=1}^{m} (f(x_j) - y_j)^2 + \sum_{j=1}^{m} \lambda_j \|f\|^2_{\mathcal{H}_K} \right]. \tag{8.9}$$

When $\sum_{j=1}^{m} \lambda_j = \lambda$, the (unique) minimizer of (8.9) is clearly equivalent to the minimizer of (8.7).

Gradient and subgradient methods (e.g., gradient descent) are popular iterative algorithms for solving optimization problems. In a centralized setting, the gradient descent algorithm for solving (8.7) defines a sequences of estimates

$$\hat{f}^{(k+1)} = \hat{f}^{(k)} - \alpha_k \frac{\partial F}{\partial f}(\hat{f}^{(k)})$$

where $F(f) = \sum_{j=1}^{m} (f(x_j) - y_j)^2 + \lambda \|f\|^2_{\mathcal{H}_K}$ is the objective function, and $\frac{\partial F}{\partial f}$ denotes its functional derivative. Note that $\frac{\partial F}{\partial f}(f^{(k)})$ factors due to its additive structure. *Incremental subgradient methods* exploit this additivity to define an alternative set of update equations:

$$j = k \mod m \tag{8.10}$$

$$\hat{f}^{(k+1)} = \hat{f}^{(k)} - \alpha_k \frac{\partial G_j}{\partial f}(\hat{f}^{(k)}), \tag{8.11}$$

where $G_j = (f(x_j) - y_j)^2 + \lambda_j \|f\|^2_{\mathcal{H}_K}$. In short, the update equations iterate over the m terms in F. Incremental subgradient algorithms have been studied in detail in Nedic and Bertsekas (1999, 2000). Under reasonable regularity (e.g., bounded $\|\frac{\partial G_j}{\partial f}\|$), one can show that if $\alpha_k \to 0$, then $\|\hat{f}^{(k+1)} - g_n^\lambda\|_{\mathcal{H}_K} \to 0$; with a constant step size (i.e., $\alpha_k = \alpha$), one can bound the number of iterations required to make $\|\hat{f}^{(k)} - g_n^\lambda\|_{\mathcal{H}_K} \le \epsilon$.

These ideas were exploited in Rabbat and Nowak (2004, 2006) to develop a message-passing algorithm that may be applied as a distributed training algorithm. After noting that the update equation at iteration k depends only on the data observed by sensor $k \mod m$, a two-step process is proposed. First, a path is established that visits every sensor. Then, the incremental subgradient updates are executed by iteratively visiting each sensor along the path. For example, sensor one may initialize $\hat{f}^{(0)} = 0 \in \mathcal{H}_K$ and then compute $\hat{f}^1$ according to the update equations (which depend on sensor one's only training datum). Once finished, sensor one passes $\hat{f}^1$ on to the second sensor in the path, which performs a similar update before passing its estimate onto the third sensor. The process continues over multiple passes through the network, at each stage, data is not exchanged – only the current estimates. By the comments above, only a finite number of iterations are required for *each* sensor to arrive at an estimate f with $\|f - g_n^\lambda\|_{\mathcal{H}_K} \le \epsilon$. The algorithm is depicted pictorially in Figure 8.4.

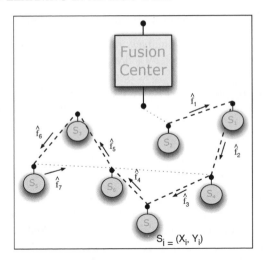

Figure 8.4 An incremental subgradient approach to training distributively (Rabbat and Nowak 2004, 2006). Reproduced by permission of © 2006 IEEE.

Notably, the present setting is slightly different than the one originally conceived in Rabbat and Nowak (2004, 2006). First, more general non-quadratic objective functions were considered. Secondly, there the optimization variable was a real-valued (i.e., real-valued parameter estimation); here we estimate a function. From a theoretical perspective, the differences are primarily technical. However, practically speaking the second difference is important. In particular, one can show that the functional derivative is given by

$$\frac{\partial G_j}{\partial f} = 2(f(x_j) - y_j)K(\cdot, x_j) + 2\lambda_j f(\cdot).$$

In consequence, communicating $\hat{f}^{(k)}$ ultimately requires the sensors to communicate the data, since exchanging (x_j, y_j) is necessary to share $\frac{\partial G_j}{\partial f}$ (assuming that the sensors are preprogrammed with the kernel). This is precisely what we were trying to avoid in the first place. Thus, in the general case, the incremental subgradient approach may have limited use for reproducing kernel methods. However, often $\mathcal{H}_K$ admits a lower dimensional parameterization; for example, this is the case for the linear kernel when $\mathcal{H}_K$ is the space of linear functions on $\mathcal{X} = \Re^d$. In that case, messages may be communicated more efficiently to the tune of considerable energy savings. The energy-accuracy trade-off is discussed in the full paper (Rabbat and Nowak 2006).

Note that unlike the sparsity-driven approach, the incremental subgradient-based algorithm is independent of modeling assumptions which link the kernel to the topology of the network. Indeed, the distributed training algorithms depends only on there being a path through the network; the kernel and the network are distinct objects. Finally, note that Son et al. (2005a) addressed a generalization of the incremental subgradient message-passing methodology by considering a clustered network topology.

Training distributively using alternating projection algorithms

A final approach to solving (8.7) distributively relies on sensors to locally (and iteratively) share data, not entire functions, and thereby addresses the practical weakness that sometimes limits the incremental subgradient approach. To construct the algorithm, assume that sensor j can query its neighbors' data (x_i, y_i) for all $i \in N_j$ (where $N_j \subseteq \{1, \ldots, m\}$ denotes the neighbors of sensor j), and may use this local data to compute a global estimate for the field by solving

$$\min_{f \in \mathcal{H}_K} \left[\sum_{i \in N_j} (f(x_i) - y_i)^2 + \lambda_j \|f\|_{\mathcal{H}_K}^2 \right]. \tag{8.12}$$

Presumably, each sensor can compute such an estimate; thus, in principle, one could iterate through the network allowing each sensor to compute a global estimate using only local data. The key idea behind an algorithm presented in Predd et al. (2005, 2006b) is to couple this iterative process using a set of message variables. Specifically, sensor j maintains an auxiliary message variable $z_j \in \Re$, which is interpreted as an estimate of the field at X_j. Each sensor initializes its message variable according to its initial field measurement, i.e., $z_j = y_j$ to start.

Subsequently, the sensors perform a local computation in sequential order. At its turn, sensor j *queries* its neighbors' message variables and computes $f_j \in \mathcal{H}_K$ as the solution to (8.12) using $\{(x_i, z_i)\}_{i \in N_j}$ as training data. Then, sensor j *updates* its neighbors' message variables, setting $z_i = f_j(x_i)$ for all $i \in N_j$. Since sensor j's neighbors may pass along their newly updated message variables to other sensors, the algorithm allows local information to propagate globally.

Two additional modifications are needed to fully specify the algorithm. First, multiple passes (in fact, T iterations) through the network are made; for convenience, denote sensor j's global estimate at iteration t by $f_{j,t} \in \mathcal{H}_K$. Secondly, each sensor controls the 'intertia' of the algorithm, by modifying the complexity term in (8.12). Specifically, at iteration t, $f_{j,t} \in \mathcal{H}_K$ is found to minimize

$$\min_{f \in \mathcal{H}_K} \left[\sum_{i \in N_j}^{n} (f(x_i) - z_i)^2 + \lambda_j \|f - f_{j,t-1}\|_{\mathcal{H}_K}^2 \right]. \tag{8.13}$$

The resulting algorithm is summarized more concisely in Table 8.2, and depicted pictorially in Figure 8.5.

Here, the algorithm has been derived through an intuitive argument. However, Predd et al. (2005, 2006b) introduce this approach as an application of successive orthogonal projection (SOP) algorithms (Censor and Zenios 1997) applied to a geometric topology-dependent relaxation of the centralized kernel estimator (8.7). Using standard analysis of SOP algorithms, Predd et al. (2005) prove that the algorithm converges in the limit of the number of passes through the network (i.e., as $T \to \infty$) and characterizes the point of convergence as an *approximation* to the globally optimal solution to the centralized problem (8.7). For additional detail on this general approach, we refer the interested reader to the full paper.

A few comments are in order. First, note that as was the case for the incremental subgradient approach, this algorithm is independent of assumptions that couple the kernel

Table 8.2 Training distributively with alternating projections (Predd et al. 2005, 2006b)

Initialization:	Neighboring sensors share training data *inputs*: sensor s stores $\{x_j\}_{j \in N_s}$ Sensor s initializes $z_s = y_s$, $f_{s,0} = 0 \in \mathcal{H}_K$

Train:	for $t = 1, \ldots, T$
	    for $s = 1, \ldots, n$
	        Sensor s:
	            Queries z_j $\forall j \in N_s$
	            $f_{s,t} := \arg\min_{f \in \mathcal{H}_K} \left[\sum_{j \in N_s} (f(x_j) - z_j)^2 \right.$
	                           $\left. + \lambda_s \| f - f_{s,t-1} \|_{\mathcal{H}_K}^2 \right]$
	            Updates $z_j \leftarrow f_{s,t}(x_j)$ $\forall j \in N_s$
	        end
	    end

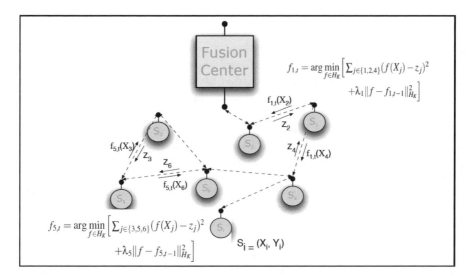

Figure 8.5 Training distributively with alternating projections (Predd et al. 2005, 2006b). © 2006 IEEE. Reprinted with permission, from Distributed Learning in Wireless Sensor Networks by Joel B. Predd, Sanjeev R. Kulkani and H. Vincent Poor, *IEEE Signal Processing Magazine*, Vol 23(4), 56–69 pp. July 2006.

matrix K with the network topology. Thus, prior domain knowledge about $\mathbf{P}_{XY}$ can be encoded in the kernel; the training algorithm approximates the centralized estimator as well as the communication constraints allow. In a simple example discussed in Predd et al. (2006b), the statistical behavior of individual sensor's estimates is shown to depend on the relationship between the topology of the sensor network and *the representational capacity of the underly reproducing kernel Hilbert space*. This fact is in contrast to other analyses which depend on the relationship between the correlation structure encoded by the kernel and the network topology.

Second, in contrast to the previous approach, sensors share data, i.e., real-valued evaluations of functions, and not the functions themselves. This significantly broadens the scope of problems where the approach is applicable. Notably, just as in the incremental approach, each sensor derives a global estimate, despite having access to only local data; this is useful when the sensors are autonomous (e.g., mobile), and may make predictions on their own independent of a fusion center. Next, sensor i can compute $f_{i,t} \in \mathcal{H}_K$ in a manner similar to (8.8); the calculation requires solving an $|N_i|$-dimensional system of linear equations. As stated in Table 8.2, the algorithm assumes that the sensors perform their local computations in sequence. As discussed in the full paper, the computations can be parallelized, insofar as none of the message variables is updated by multiple sensors simultaneously. Finally, experiments in Predd et al. (2005) suggest that the algorithm may converge quickly in practice; this is promising since for energy efficiency, the number of iterations (i.e., T) must be bounded. Additional experiments suggest that this approach to passing data considerably enhances the accuracy of individual sensors' estimates.

8.5.2 Other Work

Many other learning algorithms implicitly solve (or approximately solve) an optimization problem similar to (8.7), perhaps with a different loss function and perhaps over a different class of functions. Thus, though the discussion has focused exclusively on least-squares kernel regression, the key ideas are more broadly applicable, increasing their relevance to distributed learning in sensor networks.

In the context of boundary estimation in wireless sensor networks, Nowak and Mitra (2003) derived a hierarchical processing strategy by which sensors collaboratively prune a regression tree. The algorithm exploits additivity in the objective function of the complexity penalized estimator (i.e., an optimization similar in structure to (8.3)), and enables an interesting energy-accuracy analysis. Nowak (2003) derives a distributed EM algorithm for density estimation in sensor networks. Though formally parametric, EM is popular for clustering problems and thus the approach may be broadly applicable. Finally, He et al. (2005) uses a learning-theoretic approach to study change detection in sensor networks.

Before proceeding, we note the preceding discussion has contrasted various distributed training algorithms by the structure that the algorithms exploit and by the type of messages that sensors exchanged. A key component of future work will be to compare these methods in the context of real-world applications, and in terms of sensor network-relevant metrics such as energy-efficiency, latency, routing requirements, etc.

8.6 Conclusion

This chapter has surveyed the problem of distributed learning in wireless sensor networks. Motivated by the anticipated breadth of applications of wireless sensor networks, we first discussed how parametric methods for distributed signal processing may be inappropriate in those applications where data is sparse and prior knowledge is limited. Then, inspired by the success of machine learning in classical, centralized signal processing applications, we sought to understand whether and how the power of existing learning models and algorithms could be leveraged for nonparametric distributed signal processing in wireless sensors networks. After identifying the challenges that bandwidth and energy constraints impose on learning and posing a general model for distributed learning, we considered two general themes of existing and future research: distributed learning in networks with a fusion center, and distributed learning in ad-hoc networks with in-network processing. Subsequently, we discussed recent research within these themes. In doing so, we hope that this chapter has usefully described a set of fundamental issues for nonparametric distributed signal processing and provided an entry point to a larger body of literature.

Acknowledgements

An earlier version of this article appeared in Predd et al. (2006c), and was completed while J. B. Predd was a Ph.D. candidate at Princeton University. This research was supported in part by the Army Research Office under Grant DAAD19-00-1-0466, in part by Draper Laboratory under IR&D 6002 Grant DL-H-546263, in part by the National Science Foundation under Grants CCR-02055214 and CCR-0312413, and in part by the U. S. Army Pantheon Project.

Bibliography

Aji SM and McEliece RJ 2000 The generalized distributive law. *IEEE Transactions on Information Theory* **46**(2), 325–343.

Akyildiz IF, Su W, Sankarasubramaniam Y and Cayirci E 2002 A survey on sensor networks. *IEEE Communications Magazine* **40**(8), 102–114.

Al-Ibrahim MM and Varshney PK 1989 Nonparametric sequential detection based on multisensor data. In *Proceedings of the 23rd Annual Conference on Information Science and Systems*, pp. 157–162, The Johns Hopkins University, Baltimore, MD.

Anthony M and Bartlett P 1999 *Neural Network Learning: Theoretical Foundations*. Cambridge University, Cambridge, UK.

Bandyopadhyay S and Coyle E 2003 An energy efficient hierarchical clustering algorithm for wireless sensor networks. In *Proceedings of the 22nd Annual Joint Conference of the IEEE Computer and Communications Societies (Infocom)*, vol. 3, pp. 1713–1723, San Francisco, CA.

Barkat M and Varshney PK 1989 Decentralized CFAR signal detection. *IEEE Transactions on Aerospace and Electronic Systems* **25**(2), 141–149.

Berger T, Zhang Z and Vishwanathan H 1996 The CEO problem. *IEEE Transactions on Information Theory* **42**(3), 887–902.

Bertsekas DP and Tsitsiklis JN 1997 *Parallel and Distributed Computation: Numerical Methods*. Athena Scientific, Belmont, MA.

Blatt D and Hero A 2004 Distributed maximum likelihood estimation for sensor networks. In *Proceedings of the International Conference on Acoustics, Speech, and Signal Processing*, pp. 929–932, Montreal, Quebec, Canada.

Blum R, Kassam SA and Poor HV 1997 Distributed detection with multiple sensors: Part II – Advanced topics. *Proceedings of the IEEE* **85**(1), 64–79.

Bordes A, Ertekin S, Weston J and Bottou L 2005 Fast kernel classifiers with online and active learning. *Journal of Machine Learning Research* **6**, 1579–1619.

Breiman L 1996 Bagging predictors. *Machine Learning* **26**(2), 123–140.

Censor Y and Zenios SA 1997 *Parallel Optimization: Theory, Algorithms, and Applications*. Oxford University Press, New York.

Cesa-Bianchi N, Freund Y, Haussler D, Helmbold DP, Schapire RE and Warmuth MK 1997 How to use expert advice. *Journal of the ACM* **44**(3), 427–485.

Cetin M, Chen L, Fisher JW, Ihler AT, Kreidl OP, Moses RL, Wainwright MJ, Williams JL and Willsky AS 2007 Graphical models and fusion in sensor networks In *Wireless Sensor Networks: Signal Processing and Communications Perspectives* (ed. Swami A, Zhao Q, Hong YW and Tong L) Wiley, New York.

Chamberland JF and Veeravalli VV 2004 Asymptotic results for decentralized detection in power constrained wireless sensor networks. *IEEE Journal on Selected Areas in Communications* **22**(6), 1007–1015.

Chawla NV, Hall LO, Bowyer KW and Kegelmeyer WP 2004 Learning ensembles from bites: A scalable and accurate approach. *Journal of Machine Learning Research* **5**, 421–451.

Chen B, Tong L and Varshney PK 2006 Channel aware distributed detection in wireless sensor networks. *IEEE Signal Processing Magazine, Special Issue on Distributed Signal Processing in Sensor Networks*.

Cover TM 1968 Rates of convergence for nearest neighbor procedures. In *Proceedings of the Hawaii International Conference on Systems Sciences*, pp. 413–415, Honolulu, HI.

D'Costa A and Sayeed AM 2003 Collaborative signal processing for distributed classification in sensor networks. In *Proceedings of Second International Workshop on Information Processing in Sensor Networks*, pp. 193–208, Palo Alto, CA.

D'Costa A, Ramachandran V and Sayeed AM 2004 Distributed classification of Gaussian space-time sources in wireless sensor networks. *IEEE Journal of Selected Areas of Communication, Special Issue on Fundamental Performance Limits of Wireless Sensor Networks* **22**(6), 1026–1036.

Delouille V, Neelamani R and Baraniuk R 2004 Robust distributed estimation in sensor networks using the embedded polygons algorithm. In *Proceedings of the Third International Symposium on Information Processing in Sensor Networks*, pp. 405–413, Berkeley, CA.

Devroye L, Györfi L and Lugosi G 1996 *A Probabilistic Theory of Pattern Recognition*. Springer, New York.

Duda R, Hart P and Stork D 2001 Pattern Classification. 2nd edn. Wiley-Interscience, New York.

Freund Y and Schapire RE 1997 A decision-theoretic generalization of on-line learning and an application to boosting. *Computer and System Sciences* **55**(1), 119–139.

Freund Y, Schapire RE, Singer Y and Warmuth MK 1997 Using and combining predictors that specialize. In *Proceedings of the Twenty-Ninth Annual ACM Symposium on the Theory of Computing*, pp. 334–343, El Paso, Texas.

Gambs S, Kégl B and Aïmeur E 2005 Privacy-preserving boosting. *preprint*.

Gastpar M 2007 Information-theoretic bounds on sensor network performance. In *Wireless Sensor Networks: Signal Processing and Communications Perspectives* (ed. Swami A, Zhao Q, Hong YW and Tong L) Wiley, New york.

Gastpar M, Vetterli M and Dragotti PL 2006 Sensing reality and communicating bits: A dangerous liaison. *IEEE Signal Processing Magazine, Special Issue on Distributed Signal Processing in Sensor Networks*.

Gezici S, Tian Z, Giannakis GB, Kobayashi H, Molisch AF, Poor HV and Sahinoglu Z 2005 Localization via ultra-wideband radios. *IEEE Signal Processing Magazine* **22**(4), 70–84.

Giridhar A and Kumar PR 2007 In-network computation in wireless sensor networks. In *Wireless Sensor Networks: Signal Processing and Communications Perspectives* (ed. Swami A, Zhao Q, Hong YW and Tong L) Wiley, New York.

Golub G and Loan CV 1989 *Matrix Computations*. Johns Hopkins University Press, Baltimore, MD.

Graf HP, Cosatto E, Bottou L, Dourdanovic I and Vapnik V 2005 Parallel support vector machines: The Cascade SVM. In *Advances in Neural Information Processing Systems* (ed. Saul L, Weiss Y and Bottou L), vol. 17, pp. 521–528. MIT Press, Cambridge, MA.

Greblicki W and Pawlak M 1987 Necessary and sufficient conditions for Bayes risk consistency of recursive kernel classification rule. *IEEE Transactions on Information Theory* **33**(3), 408–412.

Guestrin C, Bodi P, Thibau R, Paskin M and Madde S 2004 Distributed regression: An efficient framework for modeling sensor network data. In *Proceedings of the Third International Symposium on Information Processing in Sensor Networks*, pp. 1–10, Berkeley, CA.

Gupta P and Kumar PR 2000 Capacity of wireless networks. *IEEE Transactions on Information Theory* **46**(2), 388–401.

Gyorfi L, Kohler M, Krzyzak A and Walk H 2002 *A Distribution-Free Theory of Nonparametric Regression*. Springer, New York.

Han J, Varshney PK and Vannicola VC 1990 Some results on distributed nonparametric detection. In *Proceedings of the 29th Conference on Decision and Control*, pp. 2698–2703, Honolulu, HI.

Han TS and Amari S 1998 Statistical inference under multiterminal data compression. *IEEE Transactions on Information Theory* **44**(6), 2300–2324.

Hastie T, Tibshirani R and Friedman J 2001 *The Elements of Statistical Learning: Data Mining, Inference, and Prediction*. Springer, New York.

He T, Ben-David S and Tong L 2005 Nonparametric change detection and estimation in large scale sensor networks. to appear in *IEEE Transactions on Signal Processing*.

Hussaini EK, Al-Bassiouni AAM and El-Far YA 1995 Decentralized CFAR signal detection. *Signal Processing* **44**(3), 299–307.

Ihler A, Fisher J, Moses R and Willsky A 2005 Nonparametric belief propagation for sensor network self-calibration. *IEEE Journal of Selected Areas in Communication*. **23**(4), 809–19.

Jacobs R, Jordan MI, Nowlan S and Hinton GE 1991 Adaptive mixtures of local experts. *Neural Computation* **3**(1), 125–130.

(ed.) Jordan M 1999 *Learning in Graphical Models*. MIT Press, Cambridge, MA.

Kearns M and Seung HS 1995 Learning from a population of hypotheses. *Machine Learning* **18**(2-3), 255–276.

Kimeldorf G and Wahba G 1971 Some results on Tchebycheffian spline functions. *Journal of Mathematical Analysis and Applications* **33**(1), 82–95.

Kittler J, Hatef M, Duin PW and Matas J 1998 On combining classifiers. *IEEE Transactions Pattern Analysis and Machine Intelligence* **20**(3), 226–239.

Kotecha JH, Ramachandran V and Sayeed A 2005 Distributed multi-target classification in wireless sensor networks. *IEEE Journal on Selected Areas of Communications, Special Issue on Self-Organizing Distributed Collaborative Sensor Networks* **23**(4), 703–713.

Krzyżak A 1986 The rates of convergence of kernel regression estimates and classification rules. *IEEE Transactions on Informations Theory* **32**(5), 668–679.

Kschischang FR, Frey BJ and Loeliger HA 2001 Factor graphs and the sum-product algorithm. *IEEE Transactions on Information Theory* **47**(2), 498–519.

Kulkarni SR and Posner SE 1995 Rates of convergence of nearest neighbor estimation under arbitrary sampling. *IEEE Transactions Information Theory* **41**(4), 1028–1039.

Kulkarni SR and Viswanath P 2004 A deterministic approach to throughput scaling in wireless networks. *IEEE Transactions on Information Theory* **50**(6), 1041–1049.

Kulkarni SR, Lugosi G and Venkatesh SS 1998 Learning pattern classification: a survey. *IEEE Transactions on Information Theory* **44**(6), 2178–2206.

Kulkarni SR, Posner SE and Sandilya S 2002 Data-dependent k_n-nn and kernel estimators consistent for arbitrary processes. *IEEE Transactions on Information Theory* **48**(10), 2785–2788.

Lazarevic A and Obradovic Z 2001 The distributed boosting algorithm. In *Proceedings of the Seventh ACM SIGKDD International Conference on Knowledge Discovery and Data Mining*, pp. 311–316. ACM Press, San Francisco, CA.

Li D, Wong K, Hu YH and Sayeed A 2002 Detection, classification, and tracking of targets. *IEEE Signal Processing Magazine* **19**(2), 17–29.

Littlestone N and Warmuth M 1994 The weighted majority algorithm. *Information and Computation* **108**(2), 212–261.

Loeliger HA 2004 An introduction to factor graphs. *IEEE Signal Processing Magazine* **21**(1), 28–41.

Luo ZQ 2005a An isotropic universal decentralized estimation scheme for a bandwidth constrained ad hoc sensor network. *IEEE Journal on Selected Areas of Communications, Special Issue on Self-Organizing Distributed Collaborative Sensor Networks* **23**(4), 735–744.

Luo ZQ 2005b Universal decentralized estimation in a bandwidth constraint sensor network. *IEEE Transactions on Information Theory* **51**(6), 2210–2219.

Mitchell T 1997 *Machine Learning*. McGraw-Hill, New York.

Moallemi CC and Roy BV 2004 Distributed optimization in adaptive networks. In *Advances in Neural Information Processing Systems 16* (ed. Thrun S, Saul L and Schölkopf B) MIT Press, Cambridge, MA.

Morvai G, Kulkarni SR and Nobel AB 1999 Regression estimation from an individual stable sequence. *Statistics* **33**, 99–118.

Nakamura A, Takeuchi J and Abe M 1998 Efficient distribution-free learning of simple concepts. *Annals of Mathematics and Artificial Intelligence* **23**(1-2), 53–82.

Nasipuri A and Tantaratana S 1997 Nonparametric distributed detection using Wilcoxin statistics. *Signal Processing* **57**(2), 139–146.

Nedic A and Bertsekas D 1999 Incremental subgradient methods for nondifferentiable optimization. Technical Report LIDS-P-2460, Massachusetts Institute of Technology, Cambridge, MA.

Nedic A and Bertsekas D 2000 Convergence rate of incremental subgradient algorithms. In *Stochastic Optimization: Algorithms and Applications* (ed. Uryasev S and Pardalos PM) Kluwer, Dordrecht, The Netherlands, pp. 263–304.

Negi R, Rachlin Y and Khosla P 2007 The sensing capacity of sensor networks. In *Wireless Sensor Networks: Signal Processing and Communications Perspectives* (ed. Swami A, Zhao Q, Hong YW and Tong L) Wiley, New York.

Nguyen X, Jordan MI and Sinopoli B 2005a A kernel-based learning approach to ad hoc sensor network localization. *ACM Transactions on Sensor Networks* **1**(1), 134–152.

Nguyen X, Wainwright MJ and Jordan MI 2005b Nonparametric decentralized detection using kernel methods. *IEEE Transactions on Signal Processing* **53**(11), 4053–4066.

Nobel AB 1999 Limits to classification and regression estimation from ergodic processes. *Annals of Statistics* **27**(1), 262–273.

Nobel AB and Adams TM 2001 Estimating a function from ergodic samples with additive noise. *IEEE Transactions on Information Theory* **47**(7), 2895–2902.

Nobel AB, Morvai G and Kulkarni S 1998 Density estimation from an individual sequence. *IEEE Transactions on Information Theory* **44**(2), 537–541.

Nowak RD 2003 Distributed EM algorithms for density estimation and clustering in sensor networks. *IEEE Transactions on Signal Processing* **51**(8), 2245–2253.

Nowak R and Mitra U 2003 Boundary estimation in sensor networks: Theory and methods. In *Proceedings of Second International Workshop on Information Processing in Sensor Networks*, pp. 80–95, Palo Alto, CA.

Paskin MA and Lawrence GD 2003 Junction tree algorithms for solving sparse linear systems. Technical Report UCB/CSD-03-1271, EECS Department, University of California, Berkeley.

Paskin MA, Guestrin CE and McFadden J 2005 A robust architecture for inference in sensor networks. In *Proceedings of the Fourth International Symposium on Information Processing in Sensor Networks*, pp. 55–62, UCLA, Los Angeles, CA.

Patwari N, Ash JN, Kyperountas S, Hero AO, Moses RL and Correal NS 2005 Locating the nodes: Cooperative localization in wireless sensor networks. *IEEE Signal Processing Magazine* **22**(4), 54–69.

Pearl J 1988 *Probabilistic Reasoning in Intelligent Systems: Networks of Plausible Inference*. Morgan Kaufmann, San Francisco.

Plarre K and Kumar PR 2004 Extended message passing algorithm for inference in loopy Gaussian graphical models. *Ad Hoc Networks* **2**(2), 153–169.

Poor HV 1994 *An Introduction to Signal Detection and Estimation*. Springer Verlag, New York.

Predd JB, Kulkarni SR and Poor HV 2004 Consistency in a model for distributed learning with specialists *Proceedings of the 2004 IEEE International Symposium on Information Theory*, Chicago, IL.

Predd JB, Kulkarni SR and Poor HV 2005 Regression in sensor networks: Training distributively with alternating projections. In *Proceedings of the SPIE Conference on Advanced Signal Processing Algorithms, Architectures, and Implementations XV* (invited), San Diego, CA.

Predd JB, Kulkarni SR and Poor HV 2006a Consistency in models for distributed learning under communication constraints. *IEEE Transactions on Information Theory* **52**(1), 52–63.

Predd JB, Kulkarni SR and Poor HV 2006b Distributed kernel regression: An algorithm for training collaboratively. In *Proceedings of the 2006 IEEE Information Theory Workshop*, Punta del Este, Uruguay.

Predd JB, Kulkarni SR and Poor HV 2006c Distributed learning in wireless sensor networks. *IEEE Signal Processing Magazine*. **23**(4), 56–69.

Provost F and Hennessy DN 1996 Scaling up: Distributed machine learning with cooperation. In *Proceedings of the Thirteenth National Conference on Artificial Intelligence*, pp. 74–79, Portland, OR.

Quinlan JR 1992 *C4.5: Programs for Machine Learning*. Morgan Kauffman, San Mateo, CA.

Rabbat M and Nowak R 2004 Distributed optimization in sensor networks. In *Proceedings of the Third International Symposium on Information Processing in Sensor Networks*, pp. 20–27, Berkeley, CA.

Rabbat MG and Nowak RD 2006 Quantized incremental algorithms for distributed optimization. *IEEE Journal of Special Areas of Communication* **23**(4), 798–808.

Ribeiro A and Giannakis GB 2006a Bandwidth-constrained distributed estimation for wireless sensor networks, part I: Gaussian PDF. *IEEE Transactions on Signal Processing* **54**(3), 1131–1143.

Ribeiro A and Giannakis GB 2006b Bandwidth-constrained distributed estimation for wireless sensor networks, part II: Unknown PDF. *IEEE Transactions on Signal Processing*. **54**(7), 2784–2796.

Ribeiro A, Schizas ID, Xiao JJ, Giannakis GB and Luo ZQ 2007 Distributed estimation under bandwidth and energy constraints. In *Wireless Sensor Networks: Signal Processing and Communications Perspectives* (ed. Swami A, Zhao Q, Hong YW and Tong L) Wiley, New York.

Roussas G 1967 Nonparametric estimation in Markov processes. *Annals of the Institute for Statistical Mathematics* **21**, 73–87.

Schölkopf B and Smola A 2002 *Learning with Kernels*, 1st edn. MIT Press, Cambridge, MA.

Servetto SD 2002 On the feasibility of large scale wireless sensor networks. In *Proceedings of the 40th Annual Allerton Conference on Communication, Control, and Computing*, Monticello, IL

Son SH, Chiang M, Kulkarni SR and Schwartz SC 2005a The value of clustering in distributed estimation for sensor networks. In *Proceedings of the IEEE International Conference on Wireless Networks, Communications, and Mobile Computing*, vol. 2, pp. 969–974, Maui, HI.

Son SH, Kulkarni SR, Schwartz SC and Roan M 2005b Communication-estimation tradeoffs in wireless sensor networks. In *Proceedings of the IEEE International Conference on Acoustics, Speech, and Signal Processing*, vol. 5, pp. 1065–1068, Philadelphia, PA.

Stone CJ 1977 Consistent nonparametric regression. *Annals of Statistics* **5**(4), 595–645.

Sung Y, Tong L and Poor HV 2005 A large deviations approach to sensor scheduling for detection of correlated random fields. In *Proceedings of the IEEE International Conference on Acoustics, Speech, and Signal Processing*, Philadelphia, PA.

Sung Y, Tong L and Poor HV 2006 Neyman-Pearson detection of Gauss-Markov signals in noise: Closed-form error exponent and properties. *IEEE Transactions on Information Theory* **52**(4), 1354–1365.

Tsitsiklis JN 1993 Decentralized detection. In *Advances in Statistical Signal Processing* (ed. Poor HV and Thomas JB) JAI Press, Greenwich, CT pp. 297–344.

Valiant LG 1984 A theory of the learnable. *Communications of the ACM* **27**(11), 1134–1142.

Vapnik VN 1991 *The Nature of Statistical Learning Theory*. Springer-Verlag, New York.

Varshney PK 1996 *Distributed Detection and Data Fusion*. Springer, New York.

Veeravalli VV and Chamberland JF 2007 Detection in sensor networks. In *Wireless Sensor Networks: Signal Processing and Communications Perspectives* (ed. Swami A, Zhao Q, Hong YW and Tong L) Wiley, New York.

Vishwanathan R and Ansari A 1989 Distributed detection of a signal in generalized Gaussian noise. *IEEE Transactions on Acoustic, Speech, and Signal Processing* **37**(5), 775–778.

Vishwanathan R and Varshney PK 1997 Distributed detection with multiple sensors: Part I – Fundamentals. *Proceedings of the IEEE* **85**(1), 54–63.

Viswanathan H and Berger T 1997 The quadratic Gaussian CEO problem. *IEEE Transactions on Information Theory* **43**(5), 1549–1559.

Xiao JJ and Luo ZQ 2005a Universal decentralized detection in a bandwidth constraint sensor network. *IEEE Transactions on Signal Processing* **53**(8), 2617–2624.

Xiao JJ and Luo ZQ 2005b Universal decentralized estimation in an inhomogeneous sensing environment. *IEEE Transactions on Information Theory* **51**(10), 3564–3575.

Xiao JJ, Ribeiro A, Luo ZQ and Giannakis GB 2006 Distributed compression-estimation using wireless sensor networks. *IEEE Signal Processing Magazine, Special Issue on Distributed Signal Processing in Wireless Sensor Networks*.

Yakowitz S 1989 Nonparametric density and regression estimation from Markov sequences without mixing assumptions. *Journal of Multivariate Analysis* **30**(1), 124–136.

Yakowitz S 1993 Nearest neighbor regression estimation for null-recurrent Markov time series. *Stochastic Processes and their Applications* **48**(2), 311–318.

Yamanishi K 1997 Distributed cooperative Bayesian learning. In *Proceedings of the Conference on Learning Theory*, pp. 250–262, Nashville, TN.

9

Graphical Models and Fusion in Sensor Networks

Müjdat Çetin, Lei Chen, John W. Fisher III, Alexander T. Ihler, O. Patrick Kreidl, Randolph L. Moses, Martin J. Wainwright, Jason L. Williams, and Alan S. Willsky

9.1 Introduction

Graphical models provide a rich framework for capturing the statistical relationships among large numbers of variables, some of which may be measured and others of which are to be estimated or inferred from the available data. In addition, the natural algorithms for performing such inference tasks involve parallel message-passing procedures – essentially the passing of estimated statistical likelihoods – for the fusion of information across the entire graphical model. Because of both their graphical structure and the distributed nature of the inference procedures they admit, graphical models provide a natural framework to investigate fusion algorithms for sensor networks, an observation that has motivated a number of researchers, including the authors. Moreover, since the building or learning of such models, the analysis of such inference algorithms, and the development of enhanced algorithms are a rich, active, and growing field, there is a rich foundation on which to build methodologies for sensor network fusion algorithms.

The objective of this chapter is to confirm this observation and also to make clear that there are additional issues that arise in the context of sensor networks that require new questions to be asked that have not typically been part of the line of inquiry for graphical

Wireless Sensor Networks: Signal Processing and Communications Perspectives   A. Swami, Q. Zhao, Y.-W. Hong and L. Tong
© 2007 John Wiley & Sons, Ltd

models. The result is a new and very rich research area that has already added some important new results for sensor networks and, interestingly, also for graphical models.

In the next section of this chapter we present a concise introduction to some of the basic ideas underlying graphical models and their inference algorithms, as well as an important extension that generalized particle filtering to graphical models. Following this, we introduce two sensor network applications that we use as vehicles to illustrate the methods. These applications are self-localization in sensor networks and distributed data association and object tracking. The discussion of these applications allows us to discuss and illustrate a concept of primary importance. Specifically, the mapping of a sensor network fusion problem to a graphical model is far from unique and different choices of graphical representations can have drastically different implications for the organization and effectiveness of the fusion algorithms that result.

The core of this chapter is the description of research that deals with several of the new and key issues that arise in adapting and extending graphical model inference algorithms to sensor networks. In particular, we describe two approaches – message censoring and efficient communication of particle-based messages – aimed at addressing the fact that communication resources in sensor networks are generally severely limited. Such methods introduce errors into the transmitted messages (in order to conserve communication resources), and we summarize new results analyzing the impact of such errors on overall fusion performance, providing a complete audit trail from bits to fusion accuracy. We then examine a second very important issue, namely the fact that there is considerable flexibility in how inference computations are distributed among sensor nodes, leading (especially in tracking applications) to problems of the handoff of inference responsibilities. We do this in the context of power-constrained resource allocation, in which we must account not only for the power consumed in taking and communicating messages but also the power required for sensor handoff. All of the methods described so far involve the approximation of standard message-passing algorithms to accommodate power constraints. The last line of research we describe is that of completely redesigning message passing by taking into account from the start that there are bit-limitations on transmissions. This work, which makes contact with the field of decentralized team theory, also makes clear that there is a communication cost if sensors must organize themselves to achieve joint objectives. Our presentation of these methods is necessarily concise, and we provide ample references to complete treatments of each of these lines of investigation. We close the chapter with a brief discussion of some research directions that we believe are critical for the way forward.

9.2 Graphical Models

We begin with a brief discussion of graphical models, focusing on aspects related to inference and linking these to distributed inference in sensor networks in later sections. In this chapter we focus exclusively on the class of graphical models defined on undirected graphs (i.e., in which there is no parent-child relationship between the nodes connected by any edges). Such models are also commonly referred to as Markov random fields. There is also a very important class of graphical models defined on directed graphs – sometimes referred to as Bayesian networks – and we refer the reader to the literature (e.g., Pearl (1988)) for development for these models. Also, our discussion here is, perforce, brief and focused on the characteristics and results we require in our development. We refer the reader to the references, especially to the survey article and books Jordan (1998, 2004, in preparation); Lauritzen (1996); Whittaker (1990), for complete developments of this very important area.

9.2.1 Definitions and Properties

A *graphical model* is specified by an undirected graph, $G = (V, E)$, consisting of a vertex or node set V, and a set of edges $E \subset V \times V$. Associated with each node $v \in V$ is a random variable X_v; the full collection $X = \{X_v, v \in V\}$ of random variables must collectively satisfy a set of Markov properties with respect to G. Specifically, for any subset U of V, let $X_U = \{X_v, v \in U\}$. We say that the random vector X is Markov with respect to G if for any partition of V into disjoint sets A, B, C, in which B *separates* A and C (i.e., all paths in G from A to C include vertices in B), the random vectors X_A and X_C are conditionally independent given X_B. For the 'graph' associated with time series – i.e., consecutive points in time with each point connected to its immediate predecessor and successor – this corresponds to the usual notion of temporal Markovianity (i.e., that the past and future are conditionally independent given the present). For general graphs, however, the Markov property requires a far richer set of conditional independencies and associated challenges in both specifying such distributions and in performing inference using them. By way of example, consider the graph of Figure 9.7(a) (used for subsequent analysis) in which the variables x and y are conditionally independent given variables w and z. However, the w and z are *not* conditionally independent given x and y due to the edge between w and z.

The celebrated Hammersley-Clifford Theorem (Brémaud 1991) provides a sufficient condition (also necessary for strictly positive probability distributions) for the form that the joint distribution must take in order to be Markov with respect to G. Specifically, let $\mathcal{C}$ denote the set of all cliques in G, where a subset of nodes C is a clique if it is fully connected (i.e., an edge exists between each pair of nodes in C). The random vector X is Markov with respect to G if (and only if for strictly positive probability distributions) its distribution admits a factorization as a product of functions of variables restricted to cliques of the form

$$p(x) = \frac{\prod_{C \in \mathcal{C}} \psi_C(x_C)}{Z}, \qquad \text{where} \qquad Z \triangleq \sum_x \prod_{C \in \mathcal{C}} \psi_C(x_C) \qquad (9.1)$$

is the *partition function*, and the $\psi_C(x_C)$ are so-called *compatibility functions*. The logarithms of these compatibility functions are commonly referred to as *potentials* or *potential functions*.

For simplicity we will assume for the remainder of this chapter that each of the nonzero potentials (or equivalently each compatibility function in equation (9.1) that is not constant) is a function either of the variable at a single node of the graph (*node potentials*) or of the variables at a pair of nodes corresponding to an edge in E (*edge potentials*). In this case, (9.1) takes the form:

$$p(x) = \frac{\left(\prod_{s \in V} \psi_s(x_s)\right)\left(\prod_{(s,t) \in E} \psi_{s,t}(x_s, x_t)\right)}{Z} \qquad (9.2)$$

Note that any graphical model can be put into this form by appropriate node aggregation (Brémaud 1991). While all of the ideas that are presented here can be extended to the more general case, pairwise potentials are sufficient for the specific applications considered in this chapter. Moreover, the communication interpretation of the so-called message-passing algorithms used herein are more easily explained in this context.

As long as G is a relatively sparse graph, the factorizations (9.1) or (9.2) represent parsimonious means to describe the joint distribution of a large number of random variables – in the same way that specifying an initial (or final) distribution and a set of one-step transition distributions is a compact way in which to specify a Markov chain. Morever, for many inference and estimation problems (including those described in this chapter), such a specification is readily available. The challenge, however, is that unless the graph has very special properties – such as in the case of Markov chains – the compatibility functions do not readily describe the quantities of most interest, such as the marginal distribution of the variables at individual (or small sets of) nodes or the overall peak of the distribution jointly optimized over all nodes. Indeed, for discrete-valued random variables the computation of such quantities for general graphs is NP-Hard.

9.2.2   Sum-Product Algorithms

For graphs without cycles (Markov chains and, more generally, graphical models on trees), computation of the marginal distributions is relatively straightforward. In this case, the node and pair-wise potentials of the joint distribution in (9.2) for any cycle-free graph can be expressed in terms of the marginal probabilities at individual nodes and joint probabilities of pairs of nodes connected by edges (Cowell et al. 1999; Wainwright et al. 2003):

$$p(x) = \prod_{s \in V} p_s(x_s) \prod_{(s,t) \in E} \frac{p_{st}(x_s, x_t)}{p_s(x_s) p_t(x_t)}, \tag{9.3}$$

That is, $\psi_s(x_s) = p_s(x_s)$ (or $\psi_s(x_s) = p_s(x_s) p(y_s|x_s)$ when there is a measurement y_s associated with x_s) and $\psi(x_s, x_t) = \frac{p(x_s, x_t)}{p(x_s) p(x_t)}$. Marginal probabilities can be efficiently calculated in a *distributed* fashion by so-called sum-product algorithms. Specifically, as shown in (Pearl 1988), the marginal probabilities at any node s in the graph can be expressed in terms of the local potential ψ_s at node s, along with a set of so-called *messages* from each of its neighbors in the set $\mathcal{N}(s) = \{t \in V \mid (s, t) \in E\}$. The message from node t to node s is a function $M_{ts}(x_s)$ that (up to normalization) represents the likelihood function of x_s based on the subtree rooted at t and extending away from s. In particular, the marginal distribution p_s takes the form

$$p_s(x_s) \propto \psi_s(x_s) \prod_{t \in \mathcal{N}(s)} M_{ts}(x_s). \tag{9.4}$$

Furthermore, in the absence of cycles, these messages are related to each other via a sum-product formula:

$$M_{ts}(x_s) \propto \sum_{x_t} \psi_{st}(x_s, x_t) \psi_t(x_t) \prod_{u \in \mathcal{N}(t) \backslash s} M_{ut}(x_t). \tag{9.5}$$

The product operation embedded in the message computation from node t to s combines the information in the subtree rooted at node t, combining the likelihood information from all neighbors of node t other than s with the local potential at node t. This yields a likelihood function for the random variable X_t at node t. This is then converted to a likelihood for the random variable X_s at node s by multiplying by the compatibility function between these two nodes and then 'summing' or integrating out the variable at node t in a fashion analogous to the Chapman-Kolmogorov equation in a Markov chain.

Together equations (9.4) and (9.5) relating messages throughout the loop-free graph represent a set of fixed-point equations that can be solved in a variety of ways corresponding to different message-passing algorithms. For example, one can solve these equations explicitly, much as in Gaussian elimination, by starting at leaf nodes, working inward toward a 'root' node, and then propagating back toward the leaves – this is a generalization of two-pass smoothing algorithms for Markov chains. An alternative is to solve these equations iteratively: we begin with guesses (often taken simply to be constant) of all of the messages and iteratively update messages by substitution into the fixed-point equations. Each step of this procedure involves passing the current guess of messages among neighboring nodes. While there is great flexibility in how one schedules these messages, the happy fact remains that after a sufficient number of iterations (enough so that information propagates from every node to every other), the correct messages are obtained from which the desired probabilities can then be computed.

9.2.3 Max-Product Algorithms

Interestingly, for loop-free graphs, a variant of this approach also yields the solution to the problem of computing the overall *maximum a posteriori* (MAP) configuration for the entire graphical model. For such graphical models, there is an alternative factorization of $p(x)$ in terms of so-called *max-marginals*. As their name would suggest, these quantities are defined by eliminating variables through maximization (as opposed to summation); in particular, we define

$$q_s(x_s) := \max_{x_u, u \in V \setminus s} p(x_1, \ldots, x_n) \tag{9.6a}$$

$$q_{st}(x_s, x_t) := \max_{x_u, u \in V \setminus \{s,t\}} p(x_1, \ldots, x_n). \tag{9.6b}$$

It is a remarkable fact that for a tree-structured graph, the distribution (9.1) can also be factorized in terms of these max-marginals, namely:

$$p(x) \propto \prod_{s \in V} q_s(x_s) \prod_{(s,t) \in E} \frac{q_{st}(x_s, x_t)}{q_s(x_s) q_t(x_t)}. \tag{9.7}$$

Furthermore, there are equations analogous to those for the sum-product algorithm that show how these quantities can be computed in terms of node potentials and messages, where the fixed point equations involve maximization rather than summation (yielding what are known as *max-product* algorithms). The solution of these fixed point equations can be computed via leaf-root-leaf message passing (corresponding to dynamic programming/Viterbi algorithms Forney (1973)) or by iterative message passing with more general message scheduling.

9.2.4 Loopy Belief Propagation

While the representation of marginal distributions or max-marginals at individual nodes in terms of messages holds only if the graph is cycle-free, equations (9.4) and (9.5) are well-defined for any graph, and one can consider applying the sum-product algorithm to arbitrary graphs which contain cycles (often referred to as *loopy belief propagation*), this corresponds to fusing information based on assumptions that are not precisely valid. For example, the product operation in equations (9.4) as well as (9.5) corresponds to the fusion of information from the different neighbors of a node, t, assuming that the information contained in the messages from these different neighbors are conditionally independent given the value of x_t, something that is valid for trees but is decidedly *not* true if there are cycles.

Despite this evident suboptimality, this loopy form of the sum-product algorithm has been extremely successful in certain applications, most notably in decoding of low-density parity check codes (Kschischang 2003; Richardson and Urbanke 2001) which can be described by graphs with long cycles. In contrast, many sensor network applications (as well as others) involve graphs with relatively short cycles. This has led to a considerable and still growing body of literature on the analysis of these algorithms on arbitrary graphs, as well as the development of new ones that yield superior performance. For arbitrary loopy graphs, the reparameterization perspective on these algorithms (Wainwright et al. 2003), in conjunction with a new class of efficiently computable bounds (Wainwright et al. 2005b) on the partition function Z, provides computable bounds on the error incurred via application of sum-product to loopy graphs. Similar analysis is also applicable to the max-product updates (Freeman and Weiss 2001; Wainwright et al. 2004). The fact that the max-product algorithm may yield incorrect (i.e., non-MAP) configurations motivates the development of a new class of *tree-reweighted max-product algorithms* (TRMP) (Wainwright et al. 2005a) for which – in sharp contrast to the ordinary max-product updates – there are a set of testable conditions to determine if the solution is indeed the MAP configuration. Tight performance guarantees can be provided for TRMP for specific classes of graphs (Feldman et al. 2005; Kolmogorov and Wainwright 2005). More broadly, we refer the reader to various research and survey papers, e.g. (Loeliger 2004; Wainwright and Jordan 2005; Yedidia et al. 2005), as well as citations at the end of this chapter, which provide only a sampling of this rapidly growing literature.

9.2.5 Nonparametric Belief Propagation

We close with a brief discussion of an algorithm which will play an important role in later sections. Whether applied to loop-free or loopy graphs, message-passing algorithms corresponding to the iterative solution of (9.5) require the transmission of a full likelihood function, parameterized by the variable at the receiving node. When those variables are discrete (take on finitely many values), the message can be represented as a vector of numbers; when the variables are jointly Gaussian, they can be described by means and covariances. However, for non-Gaussian continuous variables, in principle we need to transmit an entire continuous function. One common approach is to discretize the underlying continuous variables. Unfortunately, this can often lead to high (and unwarranted) computational complexity (and in sensor networks, a high communication overhead as well). In nonlinear, non-Gaussian systems defined on Markov chains, particle filtering is often used to avoid these high computational and representation costs. Here, we describe a recently developed generalization of particle filtering to more complex graphical models, called *Nonparametric Belief Propagation*, or NBP (Sudderth et al. 2003).

Particle filtering also involves an equation similar to (9.5), with one very important simplification: there is no 'product', since there is only one 'other' neighbor of node t. The algorithm takes a set of particles, representing samples from the single message corresponding to the product term in (9.5), weights these particles by the local node's compatibility function (and perhaps resamples from this weighted collection of particles), and then performs the 'sum' operation by simulating the transition dynamics from node t to s, i.e., for each sample at node t we sample from the transition distribution to generate a sample at node s.

All of these steps, with one significant exception, apply equally well to message passing in general graphical models. The difficulty, however, arises from the fact that we have the product of *several* particle-based messages (one from each neighbor) at each node. As developed in (Sudderth et al. 2003), in order to make sense of this product, one can interpret each particle-based message as a nonparametric estimate of the likelihood function/probability density corresponding to the exact message, which can be smoothed to create a well-defined product. For example, if we use Gaussian kernels for these nonparametric densities, the set of particles corresponds to a Gaussian sum. As a result, the problem of generating samples from the product of messages, each represented by a set of particles, reduces to that of drawing samples from a density defined by the product of a collection of Gaussian sums.

This operation presents some computational difficulty: unless we are careful, we may have a geometrically growing number of terms in the aforementioned Gaussian sums, too many to work with efficiently. A key to drawing samples efficiently is to draw them without ever explicitly constructing the true product. Specifically, one can generate a sample in two steps: first, choose one of the Gaussian mixture components in the product from which to sample (i.e., the set of labels corresponding to one component in each of the incoming messages), and then drawing a sample from the Gaussian corresponding to that set of labels. Since this latter step is straightforward, the challenge is in the former (sampling the labels). As one solution, importance and Gibbs sampling methods can be applied to draw a label without enumerating all possible combinations.

Additionally, the sampling process can be made dramatically more efficient by using a multi-resolution representation based on the so-called k-dimensional (or KD-) trees (a structure that will also play a key role in efficient coding and communication of messages in Section 9.4.2). Given a set of k-dimensional particles (i.e., x_s is a k-dimensional real-valued random variable) representing one of the messages in (9.5), we create a multi-resolution hierarchical clustering of the particles, beginning with all particles clustered together at the root of the tree. Proceeding downward through the tree, at each node we subdivide that node's cluster into two sub-clusters (associated with the children of the current node) by splitting the data along one of the k dimensions (and cycle through each of these dimensions as we proceed down the tree) until at the finest level of the tree we have individual particles. The sampling operations described previously can then be accomplished very efficiently using this coarse-to-fine representation; we refer the reader to (Ihler et al. 2003) for details.

9.3   From Sensor Network Fusion to Graphical Models

In this section we describe how one maps a fusion problem involving a network of sensors into an inference problem on a graphical model. Such a mapping might at first seem trivial, as a natural graph already exists, defined by the sensor nodes and the inter-sensor communication structure. However, it is the informational structure of the inference problem – involving the relationships between sensed information and the variables about which we wish to perform estimation – that forms the basis of the mapping, and is just as critical as the communication structure of the problem. We illustrate this mapping by describing two sensor networks applications, focusing primarily on their informational structure.

9.3.1   Self-Localization in Sensor Networks

A crucial first step for many sensor network applications is that of sensor localization – determining the locations of sensor nodes in a network. Figure 9.1 illustrates the localization problem. Each node corresponds to a sensor, and the random vector at that node describes sensor location, and also perhaps other calibration variables such as the sensor's orientation (e.g., if the node provides directional-sensing capability), and the time offset of its internal clock (e.g., if inter-sensor time-of-flight measurements are used by the network).

To simplify the discussion we will consider only location variables, although the framework we describe extends immediately to a more general setting. We consider a case in which the available information for estimating sensor locations consists of: (i) uncertain prior information about the location of a subset of the sensors (e.g., if any of the sensors are provided with GPS); (ii) how likely sensors are to 'hear' one another and attempt to measure their inter-sensor distance (typically only possible for nearby pairs of sensors); and (iii) any distance measurements so obtained. Specifically, let us denote by $\rho_s(x_s)$ the prior location probability distribution for sensor s, if any, let $Pr(x_s, x_t)$ be the probability

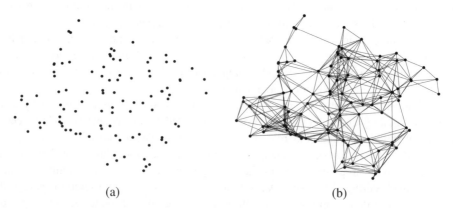

(a)   (b)

Figure 9.1 Sensor localization: (a) the physical location of a collection of sensors may be represented as (b) a graphical model in which each node corresponds to a sensor's location variables and each edge corresponds to observed information, such as the inter-sensor distance measurements between a pair of sensors. Reproduced by permission of © 2006 IEEE.

of observing a distance measurement between two sensors s, t located at x_s and x_t, and let $\rho_L(l_{st}|x_s, x_t)$ be the probability distribution the measured distance l_{st} given that the true sensor positions are x_s and x_t. Notice that all three sources of information involve only the variables at a single sensor or at pairs of sensors; thus we may use a pairwise graphical model to describe the joint distribution of sensor locations, with each node in the graph associated with one sensor and its location or calibration variables of interest. One may thus immediately write that the joint probability distribution has the form of (9.2), with

$$\psi_s(x_s) = \rho_s(x_s) \tag{9.8a}$$

$$\psi_{st}(x_s, x_t) = \begin{cases} Pr(x_s, x_t)\rho_L(l_{st} \mid x_s, x_t) & \text{; if } l_{st} \text{ is observed} \\ 1 - Pr(x_s, x_t) & \text{; otherwise} \end{cases} \tag{9.8b}$$

If we momentarily ignore the information content of aspect (ii) (as is common in many localization problems, though we shall discuss its inclusion subsequently) we immediately obtain the graphical model shown in Figure 9.1(b), in which an edge connects each pair of sensors which are able to measure their distance. In this case, single-node potentials are included for those sensors with prior location information, and edge potentials represent the likelihood function for the locations of the two sensors based on the observed distance between them. Note that the above graphical model structure depends on the information structure of the measurements, rather than on the communication structure of the network.

The sensor localization problem is then precisely one of computing the best estimates of all sensor locations given the prior and measured information. As an optimization problem, this has been well studied by others, (Moses et al. 2003; Patwari et al. 2003; Thrun 2006), often under the assumption that the distributions (ρ_s, ρ_L) involved are Gaussian so that its solution entails the nonlinear optimization of a quadratic cost function to obtain a point estimate. However, by formulating this problem as one of inference for a graphical model (in which localization corresponds to computing the marginal distributions of variables at each node), we directly obtain message-passing algorithms (such as sum-product) that distribute the computations across the network. The resulting solution is an estimate of the marginal distribution of the location variables, rather than a point estimate or confidence interval, and thereby provides a detailed statistical description of the localization solution.

Moreover, the graphical model formulation allows us to easily include features which bring additional realism, accuracy, and aptness to the problem (Ihler et al. 2005b; Ihler 2005). In particular, anomalous range measurements can occur with nontrivial probability; for example, anomalies can occur when the direct path between sensors is obscured, or when malicious signals are sent to thwart a localization process. In the graphical model framework, anomalous estimates are easily handled through a simple modification of the edge potential likelihood functions to capture such measurement errors. Also, location distributions for sensors can be multi-modal even when measurements are Gaussian – for example, receiving perfect range measurements from two neighboring sensors whose locations are known perfectly yields two possible locations. The NBP formulation finds an estimate of the node location probability distributions – as opposed to a point estimate such as the distribution maximum – from which multi-modal, non-Gaussian, or other characteristics are readily seen. Representing such multi-modality at the individual node level is one of the strengths of the NBP algorithm, and is far easier than dealing with this at a centralized level.

The graphical model formulation of the localization problem also provides a natural mechanism for including uncertain information about *which* pairs of sensors are able to measure distance. This information can improve the aptness and reduce estimation error in the location estimates. For example, the topmost sensor node in Figure 9.1 forms a location estimate using distance measurements from two very closely-located sensors. Thus, the probability distribution for the location estimate will have large values in an entire circular region centered at the midpoint of these two other sensors. However, if we also account for the fact that the node on the other side of these two sensors *cannot* hear (or can hear only with small probability) the topmost sensor, this circular ambiguity is considerably reduced. Incorporating the absence of a measurement as an indicator of sensors being more distant from each other can be difficult to accommodate in the traditional optimization framework (for example, it is certainly not a simple quadratic cost). In this graphical model formulation, however, this information can be included simply by adding edges to the graph. Specifically, we include 2-step (or more generally, n-step) edges, where a 2-step edge is one between sensors that are heard by a common sensor but cannot hear each other (Ihler et al. 2005b). Including these additional edges increases the complexity of inference slightly, although experimentally it appears that a few edges are often sufficient to resolve most of the ambiguities which may arise.

The fact that the random variables in the localization graphical model are continuous-valued – leading to our use of the particle-based NBP algorithm – raises a number of questions unique to sensor networks related to how best to use scarce communication resources. For example, how many particles do we need to send, and how can we encode them efficiently? Moreover, how do we adapt these choices as estimates evolve dynamically (e.g., as ambiguities due to multi-modality in distributions resolve themselves)? These are among the issues considered in Section 9.4.

9.3.2 Multi-Object Data Association in Sensor Networks

A second application for sensor networks is that of multi-sensor, multi-object tracking. This is a challenging problem even for centralized algorithms in large part because of the embedded problem of data association, i.e., of determining which measurements from different sensors correspond to the same object. For sensor networks there are additional challenges, due to the need for distributed implementation, but typical networks also have structure; e.g. sensors have limited sensing range overlapping the range of a limited number of other sensors. This suggests new approaches for solving data association problems that are computationally feasible and fully distributed.

Figure 9.2 depicts a notional example of the problem. Here 25 sensors cover a region of interest with overlapping areas of regard. Several targets are located within the region, each sensed by one or more sensors. In this case, the mapping of the inference problem to a graphical model is not unique. Different mappings reflect computational tradeoffs leading to different solutions than in typical centralized multi-target tracking approaches. In particular, a widely used centralized approach (Kurien 1990) organizes data association hypotheses based on so-called track hypotheses, leading to data structures – and corresponding graphical models – in which, roughly speaking, the nodes correspond to targets.

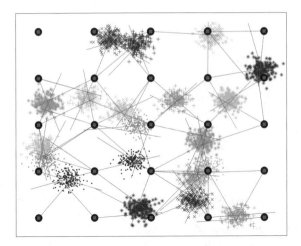

Figure 9.2 A snapshot of a typical data association scenario in a sensor network. 25 sensors (circle nodes) and the bearing-only measurements (line segments) are shown. Prior distributions of target locations are represented nonparametrically with samples from the individual marginal target distributions (appearing as clusters). Reproduced by permission of © 2006 IEEE.

In a sensor network, however, it is advantageous to organize the representation around sensors rather than targets. For centralized processing, such measurement-oriented approaches have been discarded for the same basic reason that purely sensor-based representations do not work here. In particular, consider a simple situation in which we know how many targets are present and we know which sensors see which targets. If we wish to use a model in which the nodes are in 1-1 correspondence with the sensors, the variable to be estimated at each node is simply the association vector that describes which measurement from that sensor goes with which target, which measurements are false alarms, and which targets in its area of regard it fails to detect. The problem with such a graphical model is that if multiple targets are seen by the same set of sensors, the likelihoods of these sets of associations (across sensors and targets) are coupled, implying that in the representation in (9.1), we must include cliques of size larger than two. In principle, these cliques can be quite large, and it is precisely for this reason that the association problem is NP-Hard.

By taking advantage of the sparse structure of sensor networks; i.e., the fact that each sensor has only a limited field of view and thus has only a modest number of other sensors with which it interacts and small number of targets within its measurement range – one can readily construct a hybrid representation comprised of two types of nodes. *Sensor nodes* capture the assignment of groups of measurements to *multi-target nodes* in addition to assignments that do not have any multi-sensor/target contention. Multi-target nodes corresponding to sets of targets seen by the same set of 3 or more sensors. In such a model, the variable at each sensor node captures the assignment of groups of measurements to each of these multi-target nodes as well as any assignments that do not have such a

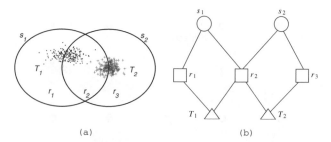

(a)   (b)

Figure 9.3 (a) A piece of a partially organized sensor network, where the sensor-target coverage relationship is ambiguous. Two sensors with their surveillance regions (s_1 and s_2). Prior distributions of *two* targets are represented nonparametrically as samples from and individual marginal target distributions (T_2 and T_2). The surveillance area is divided into three non-overlapping subregions (r_1–r_3), each of which is covered by a distinct subset of sensors. (b) The graphical model for the scenario in (a); circles, squares, and triangles correspond to sensors, subregions, and targets, respectively. Reproduced by permission of © 2006 IEEE.

level of multi-sensor/target contention. Moreover, the resulting graphical model yields a representation as in (9.2) with only pairwise potentials (Chen et al. 2005b). Furthermore, while we have described the idea for the case in which we already know which targets are seen by which sets of sensors, it is also possible to formulate graphical models that deal with the problem of also determining which targets are seen by which subsets of sensors. We do so by introducing virtual nodes representing regions of space corresponding to overlaps in areas of regard of multiple sensors, as shown in Figure 9.3. In addition, although we have discussed the data association problem at a single time point here for simplicity, the tracking problem is of course dynamic, and our framework can be generalized to incorporate data from multiple time slices by using a multiple hypothesis tracking-like approach (Chen et al. 2005a).

For the data association problem, both the computation of marginal probabilities and of the overall MAP estimate are of interest. The MAP estimate is of importance because it captures consistency of association across multiple sensors and targets (for example, capturing the fact that one measurement cannot correspond to multiple targets). As a result, algorithms such as sum-product and max-product are both of interest, as is the issue of communications-sensitive message passing, a topic to which we turn in the next section.

9.4 Message Censoring, Approximation, and Impact on Fusion

The sensor network applications in the previous section are two of many that can be naturally cast as problems of inference in graphical models – at least in part, as there are other issues, including power conservation and careful use of scarce communication resources, that must be considered. Using the two previous applications as examples, we describe approaches to dealing with such power and communication issues.

9.4.1 Message Censoring

As described in the preceding section, multi-object data association in sensor networks can be formulated as a problem either of computing the marginal probabilities or the overall MAP estimate for a graphical model whose variables are discrete and represent various assignments of measurements to objects or spatial regions. In our work we have applied a variety of different algorithms to solve this problem, including the sum-product algorithm (Kschischang et al. 2001) for the computation of approximate marginals, the max-product algorithm (Wainwright et al. 2004) for the computation of approximate MAP estimates and the TRMP algorithm (Wainwright et al. 2004) for the computation of the true MAP estimate.

One issue with all of these algorithms is that messages are, in principle, continually transmitted among all of the nodes in the graphical model. In typical graphical model applications some type of global stopping rule is applied to decide when the inference iterations should be terminated. Also, it is often the case that convergence behavior can depend strongly on the *message schedule* – i.e., the order in which messages are created, transmitted, and processed. For sensor network applications, convergence criteria or message schedules that require centralized coordination are out of the question. Rather, what is needed are local rules by which individual nodes can decide, at each iteration, whether it has sufficient new information to warrant the transmission of a new message, with the understanding that the receiving node will simply use the preceding message if it does not get a new one, and will use a default value corresponding to a noninformative message if it has not received any prior message. This formulation also allows for transmission erasures.

A simple local rule for message censoring is the following: first, we interpret each message as a probability distribution on the state of the node to which the message is to be sent; easily accomplished by normalizing the message. We then compute the Kullback-Leibler Divergence (KLD) between each message and its successor,

$$D\left(M_{ts}^k \| M_{ts}^{k-1}\right) = \sum_{x_s} M_{ts}^k(x_s) \log \frac{M_{ts}^k(x_s)}{M_{ts}^{k-1}(x_s)}, \tag{9.9}$$

as a measure of novel information and send M_{ts}^k only if $D\left(M_{ts}^k \| M_{ts}^{k-1}\right)$ exceeds a threshold ϵ. This procedure is completely local to each node and provides for network adaptivity, as these rules lead to data-dependent message scheduling; indeed, it is quite common for a node to become silent for one or more iterations and then to restart sending messages as sufficiently new information reaches it from elsewhere in the network.

When applying the above method to algorithms such as sum-product, we observe that major savings in communication (hence power) can be achieved with modest performance loss as compared to standard message passing algorithms. An example is shown in Figure 9.4, where the data are obtained by simulating tracking of 50 targets in a 25 sensor network and censored versions of max-product are readily compared to max-product and TRMP. In the figure, both communication cost and data association error are shown for censored versions of the max-product algorithm as the censoring threshold is varied. Also shown are the communication and association error for the standard max-product and TRMP algorithms. It can be seen that for certain thresholds the data association performance loss is very small, while the amount of communication is dramatically reduced.

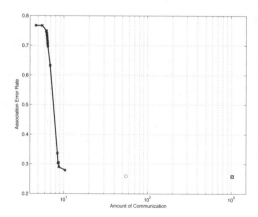

Figure 9.4 Performance-communication trade-off with varying thresholds for message censoring. Inference performance is evaluated by the association error rate, which is defined as the ratio of the number of measurements that are assigned to wrong targets to the total number of measurements. The amount of communication is defined as the number of messages sent by each node on average. Max-product (circle) and TRMP (square) are plotted for comparison. Reproduced by permission of © 2006 IEEE.

This shows that censored message passing can provide significant communication savings together with near-optimal performance. In addition, we have found examples in which this algorithm yields *better* performance than one without message censoring. We conjecture this is related to the so-called 'rumor propagation' behavior of algorithms such as sum-product in which the repeated propagation of messages around loops leads to incorrect corroboration of hypotheses; by censoring messages, this corroboration is attenuated. The communication-fusion performance tradeoff, a topic to which we will also return in Section 9.5, is examined more thoroughly in the next section.

9.4.2   Trading Off Accuracy for Bits in Particle-Based Messaging

The NBP algorithm for inference in graphical models involves exchanging particle-based representations of messages between nodes of the graph. When these nodes correspond to separate sensors in a network, this raises a number of questions, including (a) how does one efficiently represent and transmit these messages; and (b) how many particles are required – i.e., how can one decide what level of accuracy is required to represent the true, continuous message, and how can one send particles that provide that level of accuracy?

The heart of the first question is the following. We would like to transmit a probability distribution $q(x)$ from one sensor to another, where $q(x)$ is represented by a collection of particles $\{x_i\}$ which can be viewed as a set of i.i.d. samples from $q(x)$. Although this may seem to be a standard problem in communications and information theory, there is a key distinction – we have no interest in preserving the *order* of the samples, nor even truly in the accuracy of the transmission of those individual particles; our interest is only in obtaining an accurate reconstruction of $q(x)$ at the receiver. The fact that the set representation is

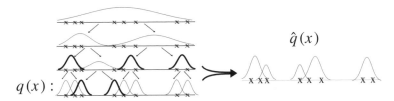

Figure 9.5 Approximating a message (or density estimate) $q(x)$ using a hierarchical, KD-tree structure. The same hierarchical structure can also be used to encode the approximation, providing a trade-off between bits and approximation error. Reproduced by permission of © 2006 IEEE.

order-invariant immediately suggests communications protocols which could be used to reduce the number of bits required.

For example, consider a one-dimensional distribution $q(x)$, so that the x_i are scalar values. In this case, one canonical ordering of the samples is from smallest to largest. Both sender and receiver can apply the knowledge that each subsequent particle will be larger than the previous particle to save a considerable number of bits. Indeed, in this case the optimal communications rate is defined not by the entropy of $q(x)$ but by the entropy of its order statistics, saving approximately a constant number of bits *per sample* (Ihler et al. 2004).

Another possible technique for encoding (either scalar or vector) particles to obtain these savings is to employ the same KD-tree structure used in NBP to provide a multi-resolution communications protocol (see Figure 9.5). Recall that a KD-tree provides a hierarchical clustering of the particles $\{x_i\}$ into a binary tree structure, starting from a root node (representing all the particles) and refined at each level by splitting the cluster into two subsets, corresponding to the particles to the left and right of their median value along one of the dimensions. We may also use this data structure to create a multi-resolution representation of the distribution $q(x)$ by creating simple density approximations to the clusters at each node of the KD-tree.

These representations can then be used to trade off between message accuracy and the representation cost of the message in bits. In particular, by selecting any cut through this tree, we obtain a mixture approximation $\hat{q}(x)$ to the finest-scale distribution $q(x)$. Moreover, the tree structure allows us to efficiently estimate the KL-divergence between $q(x)$ and any of these approximations. We can also use the tree structure to obtain an efficient method of transmitting the distribution at any particular resolution, by transmitting the mean and variances of its parent and then applying the knowledge (also available at the receiver due to the protocol used) that the left-child distribution will have a mean smaller than the parent mean, and the right-child will have a mean that is larger. A simple predictive encoder can be used to capture this information (for example, the left or right half of the Gaussian distribution associated with the parent node). This allows the transmitter to easily compute the cost, in bits, of sending any particular approximation.

The hierarchical structure of approximations within the KD-tree allows us to adaptively trade off quality versus communications cost. For example, a maximum allowable message distortion can be used to determine a specific cut through the tree (and corresponding

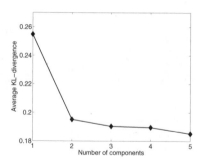

Figure 9.6 Reducing communications via message approximation in the sensor localization problem. Approximating each message using a single Gaussian can cause errors since multimodalities in the messages may be obscured, but using even a few components is typically sufficient to accurately represent the information.

message approximation), which in turn leads to a protocol for efficiently communicating that approximation. Conversely, a maximum allowable communications cost can be used to determine the most accurate approximation whose representation cost is less than the bound.

Figure 9.6 illustrates the performance of this method in the context of the sensor localization problem. The messages and beliefs computed during localization can be multimodal, particularly at the initial stages of the algorithm when little information has been exchanged. However, as the iterations of sum-product proceed, more sensor locations are resolved and fewer messages consist of multi-modal distributions. As a result, an adaptive algorithm for message communication often initially sends several mixture components, but over time may require only coarse, single-mode distributions which require fewer bits to communicate. Combining this behavior with the message censoring approach described previously provides a sensor localization algorithm which is resource-aware, and trades off the cost of communications with the quality and quantity of messages sent.

9.5 The Effects of Message Approximation

The previous section described a number of ways in which limited communications resources can be conserved by allowing certain approximations or errors to occur in the sum-product algorithm, whether from censoring (using the old version of a message in place of a new one) or from explicitly approximating particle-based messages. However, the metric we care most about is not the individual message quality, but rather the accuracy of the final, fused estimates having used those approximate messages. The relationship between message approximations and the ultimate errors in the estimated beliefs is examined in detail in (Ihler et al. 2005a).

The analysis described in (Ihler et al. 2005a) considers two possible metrics for quantifying the difference between an exact and an approximate message. One of these metrics is the Kullback-Leibler divergence (as used in Section 9.4.1); the other is a measure of the 'dynamic range' in this difference, equivalent to a norm on the log-messages. In particular, the dynamic range is

$$d\left(M_{ts}, \hat{M}_{ts}\right) = \sup_{x,x'} \left(\frac{M_{ts}(x)\, \hat{M}_{ts}(x')}{\hat{M}_{ts}(x)\, M_{ts}(x')}\right)^{1/2} \tag{9.10}$$

which can be shown to be equivalent (Ihler et al. 2005a), in the log-domain, to

$$\log d\left(M_{ts}, \hat{M}_{ts}\right) = \inf_{\alpha} \sup_{x} |\log \alpha + \log M_{ts}(x) - \log \hat{M}_{ts}(x)| \tag{9.11}$$

The measure $d(\cdot)$ has several very important properties. First, as with KLD, it is insensitive to the (irrelevant) scaling of entire messages. Indeed, $\log d(\cdot)$ turns out to be a sup-norm on the quotient space defined by this rescaling, and can be related to the usual sup-norm between two log-messages. However, the most important property for our purposes is that the effect of errors measured by the dynamic range can be bounded through each of the two steps of the sum-product algorithm (9.5). Specifically, one can show that $\log d(\cdot)$ behaves subadditively with respect to the 'product' operation, so that the log of dynamic range of the product of several approximate messages is at most the sum of the logs of the dynamic ranges of the messages in the product. Furthermore, the dynamic range allows one to establish a minimum rate for the mixing produced by the 'sum' operation, i.e., the convolution of the product of incoming messages with the potential function relating one node to another. This mixing behavior causes the sum operation to act as a contraction, attenuating the total error in the outgoing message. Specifically, if one measures the strength of a potential ψ_{ts} as

$$S(\psi_{ts}) = \sup_{a,b,c,d} \sup_{\psi_s,\psi_t} \frac{\psi_{ts}(a,b)}{\psi_t(a)\psi_s(b)} \frac{\psi_t(c)\psi_s(d)}{\psi_{ts}(c,d)}$$

and denotes the product of incoming messages and local potential as M, one may show that

$$d\left(\sum \psi_{ts} M, \sum \psi_{ts} \hat{M}\right) \leq \frac{S(\psi_{ts}) d\left(M, \hat{M}\right) + 1}{S(\psi_{ts}) + d\left(M, \hat{M}\right)}. \tag{9.12}$$

Combining (9.12) with the subadditivity relationship between M and its component incoming messages provides bounds on how errors propagate at each iteration of the sum-product algorithm. The behavior over multiple iterations can then be analyzed as well, and is most easily visualized via a computation tree. Figure 9.7 shows the computation tree for a simple graphical model with cycles; the messages that a particular node receives after several iterations involve a large number of paths, many of which may include loops. In this way, it is possible to compute a bound on the log dynamic range after any number of iterations of sum-product. Interestingly, this analysis not only yields bounds on behavior in the presence of message approximations, but also provides important results for sum-product in general, such as the best conditions known to date for algorithm convergence, and bounds on the distance between any two fixed points of sum-product.

Since the analysis just described yields bounds, there are cases in which the results it provides can be conservative. As an alternative, we have also developed an approximation (not a bound) built on the same principle as that used in roundoff noise analysis of digital

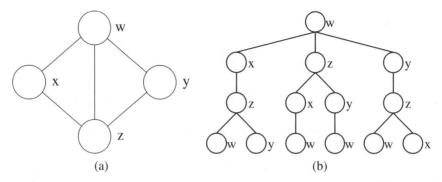

(a)   (b)

Figure 9.7 (a) A small graphical model and (b) the associated computation tree with root w. The messages received at node w after 3 iterations of loopy sum-product in the original graph is equivalent to the messages received at the root node of the tree after 3 (or more) iterations. Reproduced by permission of © 2006 IEEE.

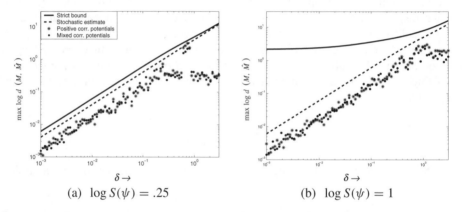

(a) $\log S(\psi) = .25$ (b) $\log S(\psi) = 1$

Figure 9.8 Maximum errors incurred as a function of the quantization error δ in a quantized implementation of sum-product. The scatterplot indicates the maximum error measured in the graph for each of 200 Monte Carlo runs, as compared to our upper bound (solid) and approximation (dashed). Reproduced by permission of © 2006 IEEE.

filters, namely that at each point at which messages are approximated, the log dynamic range error can be modeled as a white noise perturbation. This assumption can then be used to provide easily computed approximations to the variance in message errors.

Figure 9.8 depicts the results of these analyses for two problems in which errors are introduced to sum-product via message quantization. The first problem involves relatively weak edge potentials (ones for which convergence of sum-product is guaranteed); the second involves stronger edge potentials (typical of cases in which one would expect our strict bounds to be conservative). These curves depict resulting error measures (bound, stochastic approximation, and results of simulations) as functions of the quantization error introduced at each iteration of sum-product. Note that even in the strong-potentials case the approximation generally provides an accurate (and still slightly conservative) estimate of resulting error. Most importantly, when these results are combined with methods for balancing communication costs and message approximation error, we can create a complete 'audit trail' from bits used to message errors incurred and finally to resulting errors in the final estimated beliefs.

9.6 Optimizing the Use of Constrained Resources in Network Fusion

The developments in the preceding sections have provided a picture of how the message-passing structure of inference algorithms on graphical models can provide the basis for effective distributed network fusion algorithms. Of course this is not the entire story: in most if not all sensor network applications, we must also deal with limitations on available resources (for sensing, communication, and computation). Section 9.5 provided a first look at the implications of these limitations in terms of the errors that arise from the fact that messages cannot be sent perfectly and, in some cases, may not be sent at all. While that analysis is important in understanding communication-performance tradeoffs, it does not address the problem of making optimum use of the available resources so that the network, as a whole, can achieve its overall (perhaps distributed) objectives.

Consideration of constrained resources highlights the fact that there are generally two sides to communication in fusion networks. The first, which is the only one present in standard message-passing algorithms, is *information push*: the transmitting node sends bits it decides are important. The second is *information pull*: the receiving node lets the transmitting node know what bits would be of most value. Furthermore, 'information pull' methods, in which a processor with access to a sufficient statistic for past observations decides which resources to activate, may provide some advantage over 'information push' methods, in which local sensors with access to only their own observations decide when to transmit their information.

In this section we take a look at two research directions that are aimed precisely at viewing a sensor network as a resource-constrained *team* and developing strategies for optimal coordinated use of the team's resources. The first of the two directions focuses on distributed target tracking. As we saw in Section 9.3.2, there are systematic methods that allow us to transform data association and tracking problems into message-passing algorithms for graphical models. Here, we generalize these ideas to include two additional constraints. The first is simply that there is a power cost to sensing and computation. The second issue is that, as can be seen in Figure 9.3, the graphical models that arise in data association and tracking problems have some nodes that correspond to sensors but others that correspond to other 'hidden' variables (regions, target states, etc.). Of course those computations need to be performed at one of the sensing nodes, and this raises the question of who takes the lead – i.e., to whom are the available measurements sent for fusion with the current estimates of target states? However, there is also a power cost in communicating which raises two of its own issues, both related to the fact that communications power required increases with distance. The first is the obvious one, namely there is a distance-related cost in transmitting raw measurements from sensor to leader. The second is more involved but at least as important. Specifically, as a target moves, there are changes in which sensors are the best to task, and when this happens, the cost of communication also requires that we consider changing the choice of leader, i.e. that we consider the problem of handoff of responsibilities for maintenance of target state estimates. Of course there is a cost for this, namely the cost of transmitting the current target information state – e.g., in particle form using the methods described in Section 9.4.2 – from the old leader to the new. The development of a dynamic resource allocation method for balancing all of these choices and objectives – track accuracy, power conservation, choices of nodes to task for sensing and decisions on when and to whom to hand off leader responsibilities is the subject of Section 9.6.1.

The methods of decentralized team theory provide a natural setting to investigate a push-pull notion of *self-organization* as an optimal equilibrium strategy for communications-constrained network fusion. This second line of inquiry, described in Section 9.6.2, takes as the starting point the fact that there are distributed objectives and communication constraints in a sensor network, and seeks to develop message-passing strategies to optimize those objectives subject to those constraints. In so doing, the network addresses and achieves *self-organization*, in which each node learns how to interpret the bits received (or perhaps *not* received) from other nodes and how to generate bits that are most informative to other nodes. Moreover, the iterative computation we describe to find such equilibria is itself a distributed message-passing algorithm to be executed 'offline' (i.e., before actual measurements are processed): at each stage in the iteration, one node adjusts its local decision rule (for subsequent 'online' processing) based on incoming messages from its neighbors and, in turn, sends adjusted outgoing messages to its neighbors. Some of the offline messages received by each node define, in the context of its local objectives (e.g., target detection, cost of communication), a statistical model for the information it may receive online (e.g., 'what do my neighbor's bits mean to me?'), while the other offline messages define, in the context of all other nodes' objectives, the value of information it may transmit online (e.g., 'what do my bits mean to my neighbors, their neighbors, and so on?'). The result of these offline iterations can be viewed as a *fusion protocol* to maintain online decision-making performance in the face of anticipated network resource constraints, emphasizing a critical point for the practitioner: if such self-organization is expected to be done in situ, as is often discussed for ad-hoc sensor networks, it must of necessity expend only a small fraction of the available power at each node.

9.6.1   Resource Management for Object Tracking in Sensor Networks

Here, we consider an object tracking problem in which we seek to tradeoff estimation performance with energy consumed by sensing and communication. We approach the trade off between these two quantities by maximizing estimation performance subject to a constraint on energy cost, or the dual of this, i.e., minimize energy cost subject to a constraint on estimation performance. We assign to each operation (sensing, communication, etc.) an energy cost, and then we seek to develop a mechanism which will allow us to choose only those actions for which the resulting estimation gain received outweighs the energy cost incurred. We assume that sensing and communication are orders of magnitude more costly than computation, and hence that the cost of computation can be neglected.

In order to operate in this regime, we must be able to calculate both the benefit of a particular action in terms of estimation performance, and the corresponding energy cost, without consuming significant communication energy. Indeed, needing to communicate in order to determine how best to use our communication resources would represent quite a conundrum. This requirement effectively dictates that we should maintain the PDF for the object under track at a single node in the network at each time step, as proposed by previous authors (e.g., (Jones et al. 2002; Liu et al. 2003)). By ensuring that all of the information which comprises this PDF is located at a single node, we can calculate future expectations without expending energy. In the context of single object tracking, we refer to this node as the leader node. Of course, the leader node may change at each time step; this is yet another action over which we plan, taking into account the corresponding communication cost.

Entropy is a measure of uncertainty which has been applied to a wide variety of contexts including sensor management (e.g., (Chu et al. 2002; Zhao et al. 2002)). We measure the reward of obtaining a particular measurement, y_k, as the reduction in entropy that it yields in the quantity being estimated, x_k (e.g. the position and velocity of the object under track). This can be expressed equivalently as the mutual information between the observation and the quantity being estimated, conditioned on all previous measurements, $y_{0:k-1}$:

$$I(x_k; y_k | y_{0:k-1}) = H(x_k | y_{0:k-1}) - H(x_k | y_{0:k})$$

or, equivalently, the Kullback-Leibler divergence between the prior and posterior distributions of the quantity being estimated, $D(p(x_k | y_{0:k}) || p(x_k | y_{0:k-1}))$.

We assume that the energy cost of each operation is known by the leader node. The specification of these abstract energy costs allows great flexibility. For example, one could choose to implement a protocol in which the tasked sensor sends the measurement, after which the leader node confirms reception. If confirmation is not received, then the tasked sensor could re-send the measurement. The communication costs could easily incorporate the expected cost of operating this protocol. Similarly, one could design a protocol in which the information is sent once. One could then discount the expected information reward by the probability that the measurement will never be received. One could also envision a system which adaptively selects online between these two protocols, trading off the benefits and costs of each.

When one makes resource management decisions for sensor network object tracking, one is obliged to consider not only the instantaneous benefit of decisions at the current time, but also the long-term impact of the decisions. For example, suppose that one has the choice of either immediately obtaining information about an object from a distant sensor for which communication comes at a high price, or to wait a few time steps and then obtain essentially the same information from a nearer sensor for which communication is comparatively cheap. Clearly, the knowledge of future opportunities should impact the current decisions. Furthermore, each time one makes a decision and receives the resulting measurements, one then has the opportunity to alter subsequent decisions.

The optimal solution of this category of problem is a dynamic program (e.g., (Bertsekas 2000)). Unfortunately, since our observations are noise-corrupted measurements of portions of the quantities we are estimating (e.g., position and velocity of the object being tracked), the decision state of the dynamic program will incorporate the conditional PDF of these quantities, $p(x_k | y_{0:k-1})$, a structure similar to a partially observed Markov decision process. The complexity of this state space necessitates the use of suboptimal solution methods; the optimal method requires infinite computation and storage, and direct approximations thereof have exponential complexity.

A common sub-optimal control method used for situations involving partially observed processes is Open Loop Feedback Control (OLFC). In this scheme, at each time the controller constructs a plan of which action to choose at each step in the planning horizon. In doing so, the controller does not anticipate availability of any new information while the plan is being executed. After executing one or more steps of that plan, the controller then constructs a new plan which incorporates the new information received in the interim. We utilize OLFC in this way, at each step planning over the following N steps in time (referred to as a rolling horizon).

Greedy heuristics have been shown to perform well in many object-tracking applications. These methods select actions which maximize the instantaneous reward (e.g. the largest reduction in entropy), paying no regard to future opportunities. Indeed, if estimation performance as measured by mutual information is the only criterion, then one can prove that, in open loop, greedy methods produce a result with reward no less than one half that of the optimal solution (see Krause and Guestrin (2005); Williams et al. (2006a)). In the context of sensor networks this situation changes. If we operate the same greedy algorithm paying attention only to short-term information gain, the energy consumption may be arbitrarily high. Conversely, if we operate the greedy algorithm paying attention only to the immediate communication cost, the estimation performance may be arbitrarily poor. How one should modify this heuristic in order to *optimally* trade off these competing objectives is an open question.

We formulate the tradeoff between estimation performance and energy consumption through a constrained dynamic program, similarly to (Castañón 1997), where we either maximize estimation performance subject to a constraint on energy usage, or minimize energy usage subject to a constraint on estimation performance. Denoting the control policy for the next N steps as $\pi = \{\mu_k, \ldots, \mu_{k+N-1}\}$, and the decision state (i.e., the combination of conditional PDF of object state and the previous leader node) as $\mathbb{X}_k$ the underlying dynamic program becomes:

$$\min_{\pi} \mathrm{E} \left[\sum_{i=k}^{k+N-1} g(\mathbb{X}_i, \mu_i(\mathbb{X}_i)) \right]$$

$$\text{s.t. } \mathrm{E} \left[\sum_{i=k}^{k+N-1} G(\mathbb{X}_i, \mu_i(\mathbb{X}_i)) \right] \leq M \tag{9.13}$$

In the first formulation, which optimizes estimation performance subject to a constraint on energy cost, we set the per-stage cost to be

$$g(\mathbb{X}_k, u_k) = -I(x_k; y_k^{u_k} | \mathbb{X}_k) \tag{9.14}$$

so that the sum of the rewards is the total reduction in entropy over the planning horizon. We set the per-stage constraint contribution $G(\mathbb{X}_k, u_k)$ to be the energy cost of implementing the decisions u_k. In the dual formulation, which minimizes energy cost subject to a constraint on estimation performance, these two quantities are reversed.

By approaching the constrained optimization using a Lagrangian relaxation (a common approximation for integer programming problems), we find a principled method for trading off competing objectives using algorithms based on an extension of the basic greedy heuristic. Our approach exploits the additive structure of both information rewards (e.g., due to the additive decomposition of mutual information) and energy costs in order to find a per-stage cost of the form:

$$\overline{g}(\mathbb{X}_k, u_k, \lambda) = g(\mathbb{X}_k, u_k) + \lambda G(\mathbb{X}_k, u_k) \tag{9.15}$$

This can be motivated by applying a Lagrangian relaxation, defining the dual function:

$$J_k^D(\mathbb{X}_k, \lambda) = \min_\pi \mathrm{E}\left[\sum_{i=k}^{k+N-1} g(\mathbb{X}_i, \mu_i(\mathbb{X}_i)) + \lambda\left(\sum_{i=k}^{k+N-1} G(\mathbb{X}_i, \mu_i(\mathbb{X}_i)) - M \right) \right] \tag{9.16}$$

and solving the dual optimization problem involving this function:

$$J_k^L(\mathbb{X}_k) = \max_{\lambda \geq 0} J_k^D(\mathbb{X}_k \lambda) \tag{9.17}$$

We note that the dual function $J_k^D(\mathbb{X}_k, \lambda)$ takes the form of an unconstrained dynamic program with a modified per-stage cost, defined in (9.15). This motivates the use of an algorithm that is greedy with respect to this augmented cost, in which we choose to utilize a sensor only if the expected information of the sensor measurement outweighs the cost of obtaining the measurement. For example, suppose we want to select a subset of sensors to activate at a particular time step during operation. We proceed with greedy selection using this augmented cost, activating the sensor with the smallest augmented cost. When either all sensors are active, or all augmented costs are positive (indicating that, conditioned on the observations already chosen, the additional benefit of any remaining observation does not outweigh the cost of obtaining it), we terminate with the current subset.

In the context of sensor networks, in which communication is a primary source of energy expenditure, the cost of obtaining an observation is dependent upon the current choice of leader node, to which the sensor taking the measurement must transmit the observation. Conditioned on a particular choice of leader node trajectory (i.e., a choice of which node to select as leader at each time step in the planning horizon), the subset selection method above provides a means of selecting which sensors to activate at each time. However, in online operation, we simultaneously choose which sensors to activate, *and* which node to activate as leader. Our approach to this problem is a hierarchical decomposition, in which we consider different choices of leader node at each time, and then conditioned on these choices, we use the greedy subset selection. Obviously we cannot plan over all possible trajectories of leader node, so we utilize a scheme which adaptively prunes the search tree, at each level comparing the sequences ending with a particular node as leader and keeping the sequence with highest reward. A detailed description of our scheme can be found in Williams et al. (2005, 2006b).

The final question which must be addressed is how to choose the value of the Lagrange multiplier. This could be performed using a number of methods such as a line search or a subgradient search. However, in practice, we simply update the value once at each iteration of the algorithm depending on whether the constraint had slack or was exceeded. If the constraint was exceeded, then the Lagrange multiplier is increased (either additively or multiplicatively by a constant amount); if the constraint had slack, then the value is decreased. This provides an online approximation of a subgradient search.

In essence, the algorithm described performs long-term planning over trajectories of leader node and Lagrange multiplier value, and short-term planning over which subset

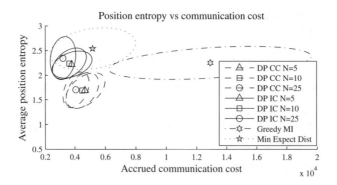

Figure 9.9 Position entropy versus communication cost for a source tracking problem using a network of 20 acoustic sensors, using several different tracking algorithms. Each ellipse shows the mean (center) and covariance of the communication cost and position entropy, obtained by averaging over 100 Monte-Carlo runs. The algorithms considered are: dynamic programming method with communication constraint (DP CC) and information constraint (DP IC) with different planning horizon lengths (N); and two greedy methods in which the leader node at each iteration is selected as the node with either the largest mutual information (greedy MI) or the smallest expected square distance to the object (min expect dist).

of sensors to activate conditioned on the long-term choice of leader node and Lagrange multiplier. In this way, we gain the efficiency of the greedy heuristic while still capturing the trade-off between the competing quantities of estimation performance and energy cost.

As an example, Figure 9.9 illustrates the performance of our method in which we track an object moving through a network of 20 randomly-placed acoustic sensors. Figure 9.9 demonstrates that the communication-constrained formulation provides a way of controlling sensor selection and leader node which reduces the communication cost and improves estimation performance substantially over the myopic single-sensor methods, which at each time activate and select as leader node the sensor with the measurement producing the largest expected reduction in entropy. The information-constrained formulation allows for an additional saving in communication cost while meeting an estimation criterion wherever possible.

The previous discussion has been in the context of tracking a single object. If multiple objects are moving independently and are never observed by the same sensor, then the algorithm can be executed in parallel for each object. If objects are close together and are observed by the same sensor, then the joint observation process induces conditional dependency between the objects. In this situation, one could choose to store the joint pdf of the two objects at a single node, utilizing a single instance of the algorithm previously described for joint estimation and planning for the two objects. Thereafter, at each step one could consider the control option to discard the joint representation in favor of two marginal pdfs, evaluating the information loss which would result against the communication saving it would yield. Alternatively, one could utilize a distributed representation such as that described in Section 9.3 with some prediction of the communication costs that will be necessary.

9.6.2 Distributed Inference with Severe Communication Constraints

The previous analyses demonstrate fundamental tradeoffs between achievable fusion performance and available communication resources in distributed sensor networks. The results in Section 9.4, in particular, establish that distributed message-passing algorithms for graphical models are robust to substantial message errors, but also suggest performance may be unsatisfactory as communication resources become severely constrained. For example, Figure 9.4 illustrates a catastrophic failure in data association performance when the censoring thresholds result in a very low amount of communication. Also observe that the upper bound and approximation curves in Figure 9.8 tend to a common positive slope as per-link quantization error becomes arbitrarily large, implying very low link capacities (i.e., reliable communication of only a few bits per desired estimate) which impact performance. That there are limits to the demonstrated robustness of conventional message-passing algorithms is not surprising, considering they are originally derived assuming ideal communications. In this subsection, we explicitly model the presence of low-rate, unreliable communication links and examine the extent to which message-passing algorithms other than those discussed in Section 9.2 can mitigate the potential loss in fusion performance.

Problem formulation

We begin with a graphical model (as described in Section 9.2) and focus on the fusion objective of estimating a discrete state process $X = \{X_s, s \in V\}$, based on a noisy measurement process $Y = \{Y_s, s \in V\}$. Our goal is to design a distributed fusion approach which minimizes the expected number of errors across all inference nodes, X (i.e., bit-error-rate in the case of binary state variables). Specifically, we denote by Γ the set of all functions γ that map the support of Y into the support of X, we seek the estimator $\hat{X} = \overline{\gamma}^*(Y)$ such that

$$J(\overline{\gamma}^*) = \min_{\gamma \in \Gamma} \underbrace{E[c(\gamma(Y), X)]}_{\equiv J(\gamma)}. \tag{9.18}$$

Here, so that J indeed measures the expected number of nodes in error, the numeric 'cost' $c(\hat{x}, x)$ associated with each possible realization of the joint process $(\hat{X}, X)$ is taken to be

$$c(\hat{x}, x) = \sum_{s \in V} c(\hat{x}_s, x_s) \quad \text{where, for each node } s, \quad c(\hat{x}_s, x_s) = \begin{cases} 0 & , \quad \hat{x}_s = x_s \\ 1 & , \quad \hat{x}_s \neq x_s \end{cases}. \tag{9.19}$$

Classical decision theory (Van Trees 1968) states that the estimator satisfying (9.18) for the costs in (9.19) is equivalent to finding the mode of every state variable's marginal distribution conditioned on *all* measurements i.e., per realization $Y = y$, compute $\hat{x} = \{\hat{x}_s, s \in V\}$ via

$$\hat{x}_s = \arg\max_{x_s} p(x_s|y), \quad \forall s \in V. \tag{9.20}$$

Indeed, as discussed in previous sections, the goal of sum-product algorithms is to obtain exactly these marginal distributions $p(x_s|y)$ at every node $s \in V$. It follows that the realization of each optimal estimate $\hat{x} = \overline{\gamma}^*(y)$ requires total communication overhead of *at least* $2|E|$ real-valued messages. In contrast, we embark upon realizing each estimate $\hat{x} = \gamma(y)$ under the restriction that total communication overhead is *at most* $|E|$ finite-alphabet messages, or *symbols* (e.g., just one bit per edge). Such severe constraints clearly render the optimal estimator in (9.20) infeasible, even for tree-structured graphical models, shifting the purpose of our problem formulation to finding a feasible (yet effective) estimator.

We view restrictions on communication as placing explicit constraints on the function space Γ in (9.18). Specifically, let the *network topology* be defined by a directed acyclic graph $N = (V, D)$. Assume each edge $(t, s) \in D$ represents a point-to-point communication link from node t to node s with known finite capacity (in bits per estimate), denoting by

$$\pi(s) = \{t \in V, (t, s) \in D\} \quad \text{and} \quad \chi(s) = \{t \in V, (s, t) \in D\}$$

the parents and children, respectively, of each node s. Sources of unreliable communication (e.g., multicast interference, dropped packets) at node s are modeled by a discrete-memoryless channel $p(z_s | x_s, m_{\pi(s) \to s})$: here, process Z_s denotes the information received by node s given the symbols $m_{\pi(s) \to s} = \{m_{t \to s}, t \in \pi(s)\}$ were transmitted to node s (see Figure 9.10). Moreover, fusion objectives with costly or selective (i.e.,censored) communication can be captured by augmenting the cost function $c(\hat{x}, x)$ to also depend on all transmitted symbols $m = \{m_{s \to t}, (s, t) \in D\}$. For example, suppose the alphabet of each symbol $m_{s \to t}$ includes a 'no-send' option and a numeric cost $c(m_s)$ equals the number of such symbols $m_s = \{m_{s \to t}, t \in \chi(s)\}$ selected by node s that do *not* exercise this option: then, choosing

$$c(m, \hat{x}, x) = \sum_{s \in V} c(\hat{x}_s, x_s) + \lambda c(m_s), \qquad \text{for a specified constant } \lambda \geq 0, \qquad (9.21)$$

results in J measuring a (λ-weighted) sum of the *node-error-rate* and *link-use-rate*.

Altogether, denoting by Γ_s the set of functions γ_s at node s by which any particular input (y_s, z_s) maps to an output $(m_s, \hat{x}_s)$, we seek the collection $\gamma^* = \{\gamma_s^*, s \in V\}$ such that

$$J(\gamma^*) \quad = \quad \min_{\gamma \in \Gamma} J(\gamma) \quad \text{subject to } \gamma \in \Gamma(N) = \Gamma_1 \times \cdots \times \Gamma_{|V|}. \qquad (9.22)$$

By definition, any collection of local decision processes $(M_s, \hat{X}_s) = \gamma_s(Y_s, Z_s)$ satisfying (9.22) will be both feasible and achieve the minimum loss in fusion performance. This formulation consolidates a number of problem variants studied in decentralized detection

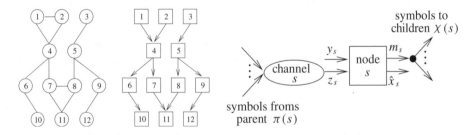

Figure 9.10 The two types of graphs in our formulation: (a) an undirected graph G underlying a probabilistic graphical model, and (b) a directed acyclic graph N, defining the parents $\pi(s)$ and children $\chi(s)$ of each node s in a network topology, along with (c) the implied fusion rule at each node, invoked in succession from parent-less nodes to child-less nodes in N.

(Tsitsiklis 1993; Varshney 1997; Viswanathan and Varshney 1997), including the allowance of an arbitrary network topology (Pete et al. 1996) and selective or unreliable communication (Chen et al. 2004; Papastavrou and Athans 1986; Pothiawala 1989; Rago et al. 1996).

Summary of algorithmic solutions

An important implication of our problem formulation is the distinction between the sensor network's *online* algorithm, or the implementation of any feasible fusion strategy $\gamma \in \Gamma(N)$, and *offline* algorithm, or the solution to the constrained optimization problem in (9.22). In particular, when online estimation is severely resource-constrained, the offline optimization by which performance loss can be mitigated introduces an additional tax on network resources. Given we anticipate intermittent reorganization by the network to stay connected (due to e.g., node dropouts, link failures), we must also anticipate the incentive for intermittent re-optimization. Hence, unless the offline algorithm itself admits an efficient distributed implementation, little hope exists for preserving satisfactory fusion performance without also rapidly diminishing the resources available for actual online measurement processing.

As highlighted earlier, the constrained optimization problem in (9.22) falls within the class of (discrete) decentralized decision problems, for which optimal solution is known to be NP-hard (Tsitsiklis and Athans 1985). Also known is an approximation to the problem, called person-by-person optimality in team theory (Marschak and Radner 1972), for which analytical progress can be made. It applies to our formulation provided the channel noise at every node is independent of the channel noise at all other nodes as well as the observation noise at all nodes (Kreidl and Willsky 2006a). The resulting online strategy is essentially equivalent to a collection of elementary Bayesian detectors, taking the form

$$\gamma_s^*(Y_s, Z_s) = \arg \min_{(m_s, \hat{x}_s)} \sum_{x_s} \theta_s^*(m_s, \hat{x}_s, x_s; Z_s) p(Y_s | x_s) \quad \text{with probability one,} \quad (9.23)$$

where parameters $\theta^* = \{\theta_s^*, s \in V\}$ (reducing to a set of likelihood-ratio thresholds in the case of binary state variables) are globally coupled through a nonlinear fixed-point equation,

$$\theta_s = f_s(\theta_{V \setminus s}) \quad \text{for } s \in V, \quad (9.24)$$

to be solved offline (Kreidl and Willsky 2006a,b; Tang et al. 1991; Tsitsiklis 1993). Moreover, given any initial parameters θ^0, iterating (9.24) in a node-by-node fashion (i.e., Gauss-Seidel iterations) guarantees a parameter sequence $\{\theta^k\}$ for which the associated cost sequence $\{J(\gamma^k)\}$ is convergent. In general, however, a distributed implementation of this offline algorithm (i) assumes all nodes are initialized with common global knowledge i.e., probabilities $p(x)$, costs $c(m, \hat{x}, x)$ and channels $\{p(z_s | x_s, m_{\pi(s) \to s}), s \in V\}$; and (ii) requires computation/communication overhead that scales exponentially with the number of nodes.

Provided the directed graph N is tree-structured (as in Figure 9.10(b)), it can be shown (Kreidl and Willsky 2006a,b; Tang et al. 1993) that (9.24) specializes to the form

$$\begin{aligned} P_{s \to \chi(s)} &= f_s^1(\theta_s, P_{\pi(s) \to s}) \\ \theta_s &= f_s^2(P_{\pi(s) \to s}, C_{\chi(s) \to s}) \qquad \text{for } s \in V. \quad (9.25) \\ C_{s \to \pi(s)} &= f_s^3(\theta_s, P_{\pi(s) \to s}, C_{\chi(s) \to s}) \end{aligned}$$

Here, for each link (s, t) in N, the *forward messages* $P_{s \to t}(m_{s \to t}|x_s)$ represent a likelihood function, giving global context to the symbol $m_{s \to t}$ from the receiving node's perspective, while the *backward messages* $C_{t \to s}(m_{s \to t}, x_s)$ represent an expected cost-to-go function, giving global context to the symbol $m_{s \to t}$ from the transmitting node's perspective. Observe that parameters θ are still globally coupled, but this coupling is now expressed only implicitly through the forward and backward message recursions. Indeed, convergence of the sequence $\{J(\gamma^k)\}$ is guaranteed by iterating the equations of (9.25) in any order corresponding to repeated forward-backward sweeps through the network (Kreidl and Willsky 2006a,b). Moreover, a distributed implementation of this offline message-passing algorithm (i) assumes each node s is initialized with local knowledge i.e., probabilities $p(x_{\pi(s)}, x_s)$, costs $c(m_s, \hat{x}_s, x_s)$ and channel $p(z_s|x_s, m_{\pi(s) \to s})$; and (ii) requires computation/communication overhead that scales linearly with the number of nodes. Note that this local message-passing algorithm can always be applied to arbitrary directed acyclic networks, but convergence is then in question.

An illustrative example

Consider the following instance of the problem described above, involving only three main parameters for ease of illustration. The joint distribution $p(x, y)$ is defined on the graph G shown in Figure 9.10(a): each node's state variable x_s and measurement variable y_s is binary-valued and (scalar) real-valued, respectively, where potential functions are taken to be

$$\psi_{st}(x_s, x_t) = \begin{cases} \eta & , \quad x_s = x_t \\ 1 - \eta & , \quad x_s \neq x_t \end{cases} \quad \text{and} \quad \psi_s(x_s, y_s) = \exp\left(-\frac{(y_s - (x_s - 0.5))^2}{2\sigma^2}\right).$$

The parameter $\eta \in (0, 1)$ captures the correlation between neighboring states (i.e., negative or positive for $\eta < 0.5$ or $\eta > 0.5$, respectively) while the parameter $\sigma \in (0, \infty)$ captures the measurement accuracy at every node. The network topology is the graph N shown in Figure 9.10(b) with each link taken to be an independent and identically distributed binary erasure channel together with a (reliable) 'no-send' option: each channel model is $p(z_s|x_s, m_{\pi(s) \to s}) = \prod_{t \in \pi(s)} p(z_{t \to s}|m_{t \to s})$ where

$$p(z_{t \to s}|m_{t \to s}) = \begin{cases} 1 - \epsilon & , \quad z_{t \to s} = m_{t \to s} \text{ when } m_{t \to s} \in \{0, 1\} \\ \epsilon & , \quad z_{t \to s} = \emptyset \text{ when } m_{t \to s} \in \{0, 1\} \\ 1 & , \quad z_{t \to s} = \emptyset \text{ when } m_{t \to s} = \emptyset \\ 0 & , \quad \text{otherwise} \end{cases}.$$

The parameter $\epsilon \in [0, 1]$ captures the reliability of the unit-rate communication network (i.e., perfect for $\epsilon = 0$), and the correct symbol is always received in the absence of an erasure.

We consider the cost function $c(m, \hat{x}, x)$ defined in (9.21), varying parameter $\lambda \geq 0$ in small increments (of size 3×10^{-4}) and repeatedly applying the offline message-passing algorithm until the sequence $\{J(\gamma^k)\}$ converges (using a tolerance of 10^{-3}). The initial parameters θ_s^0 at each node s are chosen to partition the local measurement space into $6|\chi(s)|$ intervals, assigning

(i) decision $\hat{x}_s = m_{s \to t} = 0$ to the interval $(-\infty, (1 - 3|\chi(s)|)\sigma^{-1})$

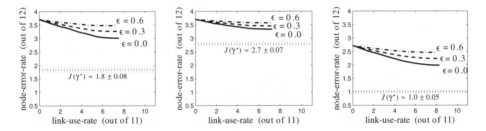

Figure 9.11 Optimized tradeoff curves, assuming the graphs shown in Figure 9.10, as a function of link unreliability (i.e., parameter ϵ) for (a) nominal state correlation and measurement noise, (b) low state correlation but nominal measurement noise, and (c) nominal state correlation but low measurement noise. Each curve is generated by incrementing parameter λ in (9.21) from zero, meaning online communication is cost-free, to a value in which the offline message-passing algorithm converges to the *myopic* strategy (i.e., each node acting as if in isolation), meaning any online communication is too costly relative to the potential improvement in node-error-rate performance. The link-use-rate is never at its maximum and the node-error-rate never exceeds that of the myopic strategy: that is, the optimized team strategy consistently regulates its exploitation of the 'no-send' option (i.e., 'no news is news,' even with an imperfect communication medium) and gracefully avoids catastrophic performance failure. Also shown in each plot (by the horizontal dotted lines) is a Monte-Carlo estimate (plus/minus one standard deviation) of the *unconstrained* optimal node-error-rate: observe that the offline message-passing algorithm can recover up to (a) 37%, (b) 40% and (c) 43% of the fusion performance lost by the myopic strategy.

(ii) decision $\hat{x}_s = m_{s \to t} = 1$ to the interval $(\sigma^{-1}(3|\chi(s)| - 1), \infty)$, and

(iii) all remaining decisions to equally-spaced sub-intervals of $[\sigma^{-1}(1 - 3|\chi(s)|), (3|\chi(s)| - 1)\sigma^{-1}]$.

Results are shown in Figure 9.11 and indicate that the optimized team strategy consistently hedges against all sources of uncertainty as we weigh between fusion inaccuracy (i.e., node-error-rate) and online communication overhead (i.e., link-use-rate). Table 9.1 indicates that offline convergence occurs in roughly five iterations, each such iteration assuming the reliable transmission of $12|D| = 132$ real numbers. Thus, while satisfactory fusion performance is achievable given online communication can be at most $|D| = 11$ bits (per estimate), it depends upon offline communication on the order of 660 real-valued messages (per reorganization).

9.7 Conclusion

In this chapter we have described a collection of research results and directions that build on the evocative connection between distributed fusion in sensor networks and message-passing algorithms in graphical models. Through two example applications (self-localization and data association/target tracking), we have shown how network fusion problems can be mapped to graphical models, allowing one, in principle to apply graphical model inference

Table 9.1 Overhead of message-passing algorithm to generate the curves of Figure 9.11.

Link Unreliability	Lowest Myopic Weight			Offline Iteration Count		
ϵ	(a)	(b)	(c)	(a)	(b)	(c)
0.0	0.1368	0.0774	0.1056	4.5 ± 0.74	4.4 ± 0.71	4.7 ± 0.65
0.3	0.0957	0.0543	0.0762	4.3 ± 0.74	4.2 ± 0.71	4.1 ± 0.29
0.6	0.0549	0.0312	0.0462	3.9 ± 0.89	3.9 ± 0.78	4.0 ± 0.08

'Lowest Myopic Weight' refers to the value of parameter λ at which convergence to the myopic strategy (always in exactly two iterations) first occurs; and 'Offline Iteration Count' refers to the average (plus/minus one standard deviation) number of iterations to convergence, taken over only the samples of λ below the lowest myopic weight. Note that, all other things being equal, the Lowest Myopic Weight is inversely related to Link Unreliability: with respect to achievable fusion accuracy, the offline algorithm captures the diminishing marginal value of online communication as link reliability degrades. Two other trends are also worth noting, but whether they apply beyond these particular examples are open questions: (i) relative to the nominal conditions (i.e., column (a)), lower state correlation or lower measurement noise (i.e., column (b) or column (c), respectively) also diminish the value of online communication; and (ii) the on-average rate of offline convergence increases as (online) link reliability degrades.

methods to sensor network fusion problems. Not unexpectedly, of course, there are issues of critical importance in sensor networks – in particular related to constraints on available power (for communication and/or sensing)-that require analysis beyond that found in the graphical model literature, and we have presented two lines of inquiry aimed at addressing these issues.

The first involved the idea of simply approximating or 'censoring' (i.e., not sending) messages in graphical model inference procedures. In the context of data association we described how censoring can lead to desirable adaptivity in network messaging ('only say something when you have something new to say'), and experiments show a sharp threshold in performance; i.e., there appear to exist critical communication rates in data association. Above this rate, performance varies only gradually with reduction in communication, but at this critical value, performance degrades precipitously. In the context of self-localization, in which messages represent 'particles' we described a framework for trading off bits in communicating these particle-based messages against overall fusion performance. Interestingly, our methodology for mapping message approximations to fusion performance has independent value in the context of graphical models, as it provides error bounds and convergence results for belief propagation algorithms.

The second line of inquiry involved the development of methods for network fusion that optimize overall fusion performance taking into account, from the start, that there are costs or constraints on power consumption or communication. In the context of target tracking we described the fact that communication costs are incurred for two distinct reasons: Communicating sensed measurements to a leader or fusion node and the communication required to hand off leader responsibility as an object moves through a sensor field. Combining these with the power cost of sensing, we have described a framework for dynamically optimizing the tradeoff between power consumption and tracking accuracy. The second part of our work on optimizing resource utilization formulated the problem of designing message-passing strategies to achieve overall network fusion objectives subject to constraints on node-to-node communication. As we discussed, the resulting message-passing strategy for each not only takes into account power concerns but also the objectives of those receiving the messages that are transmitted, capturing not only information-push from transmitter to receiver but also information-pull from receiver to transmitter. Importantly, the algorithms to achieve this coordinated team behavior have message-passing structure themselves and point to the fact that the process of network organization itself consumes some of the resources that the resulting strategies aim to conserve.

While the results in this chapter provide important components for a methodology for distributed fusion in sensor networks, they are far closer to the start than the end of the story, and there are many important lines of research that remain to be considered, some of which are being pursued by our team. The first of these involves the idea of developing different ways of organizing message passing that may be more appropriate for sensor networks. For example, one can imagine operational structures in which 'seed' nodes initiate messaging, propagating information radially outward, fusing information around these radially expanding regions as they meet, and then propagating information back inward toward the seed nodes. Such an algorithmic structure allows great flexibility (e.g., one can imagine allowing any sensor to act as a seed if it measures something of interest) and also leads to new algorithms with great promise for inference on graphical models more generally. We refer the reader to (Johnson and Willsky 2006, in review) for a first treatment of this approach. Also, the computation tree interpretation of the sum-product algorithm allows one to clearly see the computations that sum-product fails to make that a truly optimal algorithm would – computations that in essence take into account the dependencies between messages that sum-product neglects (Johnson et al. 2005). This suggests another line of research that focuses on one of the significant differences between standard graphical model inference problems and sensor networks. In particular, when viewed as a sensor network fusion algorithm sum-product has the property that it makes very little use of local node memory and computational power (all that is remembered from step to step are the most recent messages, and all that are computed are essentially the sum-product computations). Can we develop algorithms that use more memory and perform more local computation and that as a result reduce the number of messages that need to be sent? Several ideas along these lines are currently under investigation. Also, a standard feature in wireless networks is the inclusion of header bits that provide information on the path a message has taken from initiator to receiver. Can we take advantage of such header bits to capture dependencies between messages so that they can then be used to fuse messages in a manner that is not as naïve as assuming conditional independence? Of course using such header bits for what we term informational pedigree means that there are fewer bits available for the actual message, so that the message quantization error will be larger. How does the error incurred by such an increased quantization error compare to the additional fusion accuracy provided by providing these pedigree bits? Current research building on what we have presented here, is addressing this question.

Numerous other directions for further work also suggest themselves, such as allowing nodes to actively *request* information from other nodes. Indeed, if this is part of the self-organizing protocol of the sensor network, then each node not only will be able to extract information from such a request but also from the *absence* of such a request (i.e., 'no news is news'). Also, while some of the work in Section 9.6 can accommodate imperfect communication channels, there is a real need to expand the development to allow for more serious disruptions-e.g., the failure of a node, which requires that message-passing strategies be robust or adaptable to such failures. This touches on an even richer class of problems that arise when we allow the possibility that the structure of the network; i.e., the messaging graph – can be different from the statistical graph underlying the inference problem being solved by the network. As questions such as these make clear, there is much more to be done in an area that cuts across the areas of signal processing, estimation, communication, computation, and optimization in new and fascinating ways.

Bibliography

Bertsekas DP 2000 *Dynamic Programming and Optimal Control*, second edn. Athena Scientific, Belmont, MA.

Brémaud P 1991 *Markov Chains, Gibbs Fields, Monte Carlo Simulation, and Queues*. Springer, New York.

Castañon DA 1997 Approximate dynamic programming for sensor management. In *Proc 36th Conference on Decision and Control*, pp. 1202–1207. IEEE.

Chen B, Jiang R, Kasetkasem T and Varshney PK 2004 Channel aware decision fusion in wireless sensor networks. *IEEE Trans. Signal Processing* **52**(12), 3454–3458.

Chen L, Çetin M and Willsky AS 2005a Distributed data association for multi – target tracking in sensor networks. *Information Fusion 2005*, Philadelphia, PA.

Chen L, Wainwright M, Cetin M and Willsky A 2005b Data association based on optimization in graphical models with applications to sensor networks. *Mathematical and Computer Modeling (Special Issue on optimization and control for military applications)*. **43**(9-10), 1114–1135.

Chu M, Haussecker H and Zhao F 2002 Scalable information-driven sensor querying and routing for ad hoc heterogeneous sensor networks. *International Journal of High Performance Computing Applications* **16**(3), 293–313.

Cowell RG, Dawid AP, Lauritzen SL and Spiegelhalter DJ 1999 *Probabilistic Networks and Expert Systems*. Statistics for Engineering and Information Science. Springer-Verlag, New York.

Feldman J, Wainwright MJ and Karger DR 2005 Using linear programming to decode binary linear codes. *IEEE Transactions on Information Theory* **51**, 954–972.

Forney, Jr. GD 1973 The Viterbi algorithm. *Proc. IEEE* **61**, 268–277.

Freeman WT and Weiss Y 2001 On the optimality of solutions of the max-product belief propagation algorithm in arbitrary graphs. *IEEE Trans. Info. Theory* **47**, 736–744.

Heskes T, Albers K and Kappen B 2003 Approximate inference and constrained optimization *Uncertainty in Artificial Intelligence*, vol. 13, pp. 313–320.

Ihler A, Fisher III J and Willsky A 2005a Loopy belief propagation: Convergence and effects of message errors. *Journal of Machine Learning Research* **6**, 905–936.

Ihler A, Fisher III J, Moses R and Willsky A 2005b Nonparametric belief propagation for self-localization of sensor networks. *IEEE J. on Select Areas in Communication* **23**(4), 809–819.

Ihler AT 2005 *Inference in Sensor Networks: Graphical Models and Particle Methods* PhD thesis Massachusetts Institute of Technology, Cambridge, MA.

Ihler AT, Fisher III JW and Willsky AS 2004 Communication-constrained inference. Technical Report 2601, MIT, Laboratory for Information and Decision Systems, Cambridge, MA.

Ihler AT, Sudderth EB, Freeman WT and Willsky AS 2003 Efficient multiscale sampling from products of Gaussian mixtures *Neural Info. Proc. Systems 17*.

Johnson J and Willsky A 2006, in review A recursive model-reduction method for approximate inference in Gaussian Markov random fields. *IEEE Trans. Image Processing*.

Johnson J, Malioutov D and Willsky A 2005 Walk-sum interpretation and analysis of Gaussian belief propagation. *Adv. Neural Inf. Proc. Systems*.

Jones M, Mehrotra S and Park JH 2002 Tasking distributed sensor networks. *International Journal of High Performance Computing Applications* **16**(3), 243–257.

Jordan M 1998 *Learning in Graphical Models*. MIT Press, Cambridge, MA.

Jordan M 2004 Graphical models. *Statistical Science* **19**(1), 140–155.

Jordan M in preparation *An Introduction to Probabilistic Graphical Models*.

Kolmogorov V and Wainwright MJ 2005 On optimality properties of tree-reweighted message-passing. *Uncertainty in Artificial Intelligence*.

Krause A and Guestrin C 2005 Near-optimal nonmyopic value of information in graphical models. *UAI 2005*.

Kreidl OP and Willsky AS 2006a Efficient message-passing algorithms for optimizing decentralized detection networks. In *Proc. of the 45th IEEE Conf. on Decision and Control*. In review.

Kreidl OP and Willsky AS 2006b Inference with minimal communication: a decision-theoretic variational approach. *Adv. Neural Inf. Proc. Systems* vol. 19, MIT Press, Cambridge, MA.

Kschischang F 2003 Codes defined on graphs. *IEEE Signal Processing Magazine* **41**, 118–125.

Kschischang F, Frey B and Loeliger HA 2001 Factor graphs and the sum-product algorithm. *IEEE Trans. Info. Theory* **47**(2), 498–519.

Kurien T 1990 Issues in the design of practical multitarget tracking algorithms. In *Multitarget-Multisensor Tracking: Advanced Applications* (ed. Bar-Shalom Y), vol. 1, Artech House, Norwood, MA, pp. 43–83.

Lauritzen SL 1996 *Graphical Models*. Oxford University Press, Oxford.

Liu J, Reich J and Zhao F 2003 Collaborative in-network processing for target tracking. *EURASIP Journal on Applied Signal Processing* (4), 378–391.

Loeliger HA 2004 An introduction to factor graphs. *IEEE Signal Processing Magazine* **21**, 28–41.

Marschak J and Radner R 1972 *The Economic Theory of Teams*. Yale University Press, New Haven, CT.

Mooij JM and Kappen HJ 2005a On the properties of the Bethe approximation and loopy belief propagation on binary networks. *Journal of Statistical Mechanics: Theory and Experiment* **P11012**, 407–432.

Mooij JM and Kappen HJ 2005b Sufficient conditions for convergence of loopy belief propagation. Technical Report arxiv:cs.IT:0504030, University of Nijmegen. Submitted to IEEE Trans. Info. Theory.

Moses R, Krishnamurthy D and Patterson R 2003 Self-localization for wireless networks. *Eurasip Journal on Applied Signal Processing* pp. 348–358.

Papastavrou JD and Athans M 1986 A distributed hypothesis-testing team decision problem with communications cost. In *Proc. of the 25th IEEE Conf. on Decision and Control*, pp. 219–225.

Patwari N, Hero III AO, Perkins M, N. S. Correal N and O'Dea RJ 2003 Relative location estimation in wireless sensor networks. *IEEE Trans. Signal Processing* **51**(8), 2137–2148.

Pearl J 1988 *Probabilistic Reasoning in Intelligent Systems*. Morgan Kaufmann, San Mateo, CA.

Pete A, Pattipati KR and Kleinman DL 1996 Optimization of decision networks in structured task environments. *IEEE Trans. Systems, Man and Cybernetics* **26**(6), 739–748.

Pothiawala J 1989 Analysis of a two-sensor tandem distributed detection network Master's thesis, MIT, Department of Electrical Engineering and Computer Science, Cambridge, MA.

Rago C, Willett P and Bar-Shalom Y 1996 Censoring sensors: A low-communication-rate scheme for distributed detection. *IEEE Trans. Aero. and Elec. Systems* **32**(2), 554–568.

Richardson T and Urbanke R 2001 The capacity of low-density parity check codes under message-passing decoding. *IEEE Trans. Info. Theory* **47**, 599–618.

Rusmevichientong P and Roy BV 2000 An analysis of turbo decoding with Gaussian densities. *NIPS 12*, pp. 575–581. MIT Press, Cambridge, MA.

Sudderth EB, Ihler AT, Freeman WT and Willsky AS 2003 Nonparametric belief propagation. *Computer Vision and Pattern Recognition*.

Tang ZB, Pattipati KR and Kleinman DL 1991 An algorithm for determining the decision thresholds in a distributed detection problem. *IEEE Trans. Systems, Man and Cybernetics* **21**(1), 231–237.

Tang Z, Pattipati K and Kleinman D 1993 Optimization of detection networks: Part II – Tree structures. *IEEE Trans. Systems, Man and Cybernetics* **23**(1), 211–221.

Tatikonda S and Jordan MI 2002 Loopy belief propagation and Gibbs measures. In *Proc. Uncertainty in Artificial Intelligence*, vol. 18, pp. 493–500.

Thrun S 2006 Affine structure from sound. In *Advances in Neural Information Processing Systems 18* (ed. Weiss Y, Schölkopf B and Platt J) MIT Press, Cambridge, MA pp. 1355–1362.

Tsitsiklis JN 1993 Decentralized detection In *Advances in Statistical Signal Processing* (ed. Poor HV and Thomas JB) vol. 2, JAI Press, Greenwich, CT, pp. 297–344.

Tsitsiklis JN and Athans M 1985 On the complexity of decentralized decision making and detection problems. *IEEE Trans. Automatic Control* **30**(5), 440–446.

Van Trees HL 1968 *Detection, Estimation, and Modulation Theory* vol. 1. John Wiley & Sons, New York.

Varshney PK 1997 *Distributed Detection and Data Fusion*. Springer-Verlag, New York.

Viswanathan R and Varshney PK 1997 Distributed detection with multiple sensors: Part I – Fundamentals. *Proceedings of the IEEE* **85**(1), 54–63.

Wainwright MJ 2005 Stable message-passing and convex surrogates: Joint parameter estimation and prediction in coupled Gaussian mixture models. *IEEE Workshop on Statistical Signal Processing*.

Wainwright MJ and Jordan MI 2005 A variational principle for graphical models In *New Directions in Statistical Signal Processing*. MIT Press, Cambridge, MA.

Wainwright MJ, Jaakkola T and Willsky AS 2004 Tree consistency and bounds on the performance of the max-product algorithm and its generalizations. *Statistics and Computing* **14**, 143–166.

Wainwright MJ, Jaakkola TS and Willsky AS 2003 Tree – based reparameterization analysis of sum-product and its generalizations. *IEEE Trans. Info. Theory* **49**(5), 1120–1146.

Wainwright MJ, Jaakkola TS and Willsky AS 2005a Exact MAP estimates via agreement on (hyper)trees: Linear programming and message-passing. *IEEE Trans. Info. Theory*.

Wainwright MJ, Jaakkola TS and Willsky AS 2005b A new class of upper bounds on the log partition function. *IEEE Trans. Info. Theory* **51**(7), 2313–2335.

Weiss Y 2000 Correctness of local probability propagation in graphical models with loops. *Neural Computation* **12**, 1–41.

Weiss Y and Freeman WT 2000 Correctness of belief propagation in Gaussian graphical models of arbitrary topology. *NIPS 12*, pp. 673–679. MIT Press.

Welling M 2004 On the choice of regions for generalized belief propagation. *Uncertainty in Artificial Intelligence*.

Whittaker J 1990 *Graphical Models in Applied Multivariate Statistics*. Wiley, New York.

Wiegerinck W 2005 Approximations with reweighted generalized belief propagation. Workshop on Artificial Intelligence and Statistics.

Williams J, Fisher III J and Willsky A 2005 An approximate dynamic programming approach to a communication constrained sensor management problem. *Proceedings of 8th International Conference on Information Fusion*.

Williams JL, Fisher III JW and Willsky AS 2006a Approximate dynamic programming for communication-constrained sensor network management. Technical Report LIDS-TR-2637, Massachusetts Institute of Technology.

Williams JL, Fisher III JW and Willsky AS 2006b Performance guarantees in dynamic active learning. Technical Report LIDS-TR-2706, Massachusetts Institute of Technology.

Yedidia J, Freeman WT and Weiss Y 2005 Constructing free energy approximations and generalized belief propagation algorithms. *IEEE Trans. Info. Theory* **51**(7), 2282–2312.

Zhao F, Shin J and Reich J 2002 Information-driven dynamic sensor collaboration. *IEEE Signal Processing Magazine* **19**(2), 61–72.

Part III

Communications, Networking and Cross-Layered Designs

10

Randomized Cooperative Transmission in Large-Scale Sensor Networks

Birsen Sirkeci-Mergen and Anna Scaglione

10.1 Introduction

In wireless ad-hoc networks, communication is prone to strong signal attenuation due to channel fading and interference that is caused by multiple uncoordinated users. Wireless sensor networks (WSNs), a special case of wireless ad-hoc networks, have additional design constraints such as power efficiency and limited processing capability at the sensors. Hence, any protocol designed for WSNs should work with limited resources (power efficiency), should be robust to number of sensors (scalability), and should deal with issues of wireless medium.

In addition to data acquisition, a WSN has three essential tasks: (i) data compression (source coding problem); (ii) data processing and computation (signal processing problem); and (iii) data communication (channel coding problem). The problem of optimally performing all these tasks over multiple sensors is quite challenging. In the literature, the solutions that have emerged so far can be classified based on the options to perform any combination of the tasks jointly or separately. In either scenario, there exist trade-offs between different tasks. For example, consider the sensing and communication tasks where the sensors report their data directly to a fusion center. In this case, there is a trade-off between the network sensing range, which grows in the order of the network radius, and

---

the network communication range, which decreases proportionally to $R^{-\alpha}$, with α being the path loss exponent. In order to overcome this obvious limitation, multi-hop ad-hoc networking has been considered for transporting the sensor data. Unfortunately, multi-hop wireless networks do not scale (Gupta and Kumar 2000) and for this reason some have come to question the practical value of densely populated networks of low cost sensor devices (e.g. *Smartdust* vision (Warneke et al. 2001)). Other researchers have pointed out that the theoretical results in (Gupta and Kumar 2000) do not strictly apply, since sensors record large amounts of redundant and idle data. Furthermore, the information in most cases does not need to be shuttled in a point-to-point fashion. On the contrary, motivated by the observation that in a wireless network transmissions are overheard by several unintended recipients, a recent direction that has gained momentum is *cooperative transmission*. Cooperative transmission is a cross-layer approach bridging the network and physical layers with the objective of forwarding information that is available at multiple terminals more reliably.

In this chapter, we study the problem of broadcasting sensed information to a faraway fusion center or throughout the entire network leveraging on all the available wireless transmission resources. More specifically, the chapter is composed of two main parts. In the first part we introduce a class of cooperative transmission protocols that are decentralized and we derive their performance. The key characteristic of a decentralized cooperative protocol, compared to a centralized solution, is the randomization of the code construction. The ensuing performance analysis helps identify which design features are effective in providing diversity and coding gains comparable to those of a centralized assignment and which ones are not. The second part of the chapter analyzes the network dynamics when multiple groups (levels) of cooperative relays pass a message in succession, much like in a generalized version of the multi-hop model in which information is relayed by cooperative links. Using a continuum network model we obtain closed-form expressions that are accurate when the network density is high. We show that there exists a phase transition in the network behavior and if the power density is above a fixed threshold, a succession of cooperative transmissions can reach reliably to all network nodes.

10.2   Transmit Cooperation in Sensor Networks

Cooperative communication (Laneman and Wornell 2003; Stefanov and Erkip 2004) differs from traditional network solutions because it encodes data across different sources in the network. This allows the receivers to obtain diversity gains leading to performance improvements. In other words, cooperation allows multiple low-power sensors to act as a distributed powerful antenna arrays while forwarding a common set of data. In the following, we describe physical layer model for cooperative radios and then discuss transmit cooperation protocols in both single-hop and multi-hop scenarios.

10.2.1   Physical Layer Model for Cooperative Radios

In this section, we describe the complex discrete time baseband equivalent model for the transmit and receive signals of a set of cooperative radios. We denote as $X_i[k]$ the sequence transmitted by node i, $i = 1, 2, \ldots, N$, and assume that the nodes meet an average power constraint $\sum_{k=0}^{P-1} |X_i[k]|^2 \le P\mathcal{P}_i$, where P is the duration of the signal and $\mathcal{P}_i$ is the transmission power per symbol. At the receiver, the received signal is $Y_i[k]$, $i = 1, 2, \ldots, N$. Hardware limitations introduce the so-called *half-duplex* constraint, namely, the impossibility of concurrent radio transmission and reception. Incorporating this constraint, we model the discrete-time received signal at radio i and time sample k as:

$$Y_i[k] = \begin{cases} \sum_{j=1, j\ne i}^{N} H_{i,j}[k] \star X_j[k] + W_i[k] & \text{if radio } i \text{ receives at time } k \\ 0 & \text{if radio } i \text{ transmits at time } k \end{cases} \quad (10.1)$$

where $H_{i,j}[k]$ captures the combined effects of symbol asynchronism, frequency-selectiveness, quasi-static multi-path fading, shadowing, and path-loss between radios i and j; $W_i[k]$ is a sequence of mutually independent, circularly-symmetric, complex Gaussian random variables with common variance N_0 that models the thermal noise and other interference received at radio i. Note that $H_{i,j}[k]$ is assumed to be fixed during the block length. We assume that the i'th receiver can estimate the channels $H_{i,j}[k]$ but not the channels towards other destinations $H_{m,j}[k]$, for $m \ne i$, and $j = 1, 2, \ldots, N$. We also assume that, when transmitting, the i'th node does not know $H_{m,i}[k]$, $m = 1, 2, \ldots, N$, $m \ne i$, i.e. the channel state is not available at the transmitters. The $H_{i,j}[k]$ are assumed to be independent complex-valued random impulse responses for different j, which is reasonable for scenarios in which the radios are separated by a number of carrier wavelengths (in all cases, each transmitter has an independent random phase due to its local oscillator).

The key assumption is that cooperating nodes have a common data, represented by a vector of length M denoted by $\mathbf{s} = (S[0], \ldots, S[M-1])^T$. One can assume that the common message $\mathbf{s}$ was embedded in the received vector $\mathbf{y}_i = (Y_i[0], \ldots, Y_i[P-1])^T$ where node i is a (cooperative relay). The message possibly includes some errors or noise. Alternatively, the message can be generated directly at the nodes through the sensor measurements; instances of this last situation arises in data-driven sensor access protocols (Hong and Scaglione; Mergen and Tong 2006), when the data observed by each of the nodes in a group of sensors belongs to a set (or type) which is labeled with the same code $\mathbf{s}$.

General relaying is done by mapping the message $\mathbf{s}$ onto a matrix code where each column is the new relay signal. Specifically, each one of the N cooperating relay nodes transmits a column $\mathbf{x}_r = (X_r[0], \ldots X_r[P-1])^T$ of a $P \times N$ matrix code $\mathbf{X}(\mathbf{s})$ (Figure 10.1). Denoting by $\log_2(|S|)$ the number of bits per symbol, $(M/P)\log_2(|S|)$ is the spectral efficiency of the code. The number of columns N is the number of cooperating nodes. Different cooperative schemes correspond to different instantiations of the mapping $\mathbf{s} \rightarrow \mathbf{X}(\mathbf{s})$. Hence cooperative transmission is equivalent to a multi-input single output system (MISO) with a per antenna power constraint.

Most of the chapter will consider frequency flat fading channels, i.e. $H_{i,j}[k] = H_{i,j}\delta[k]$, where $\delta[k]$ is the dirac-delta function. This requires the assumption that the group of cooperative nodes transmit synchronously; this assumption, in general, requires not only small multi-path delay spread, but also sub-symbol time accuracy in the nodes' synchronization, which can be relaxed in several ways as argued in (Scaglione et al. 2006; Sharp et al. 2007). We further assume that the channels do not change during the course of the transmission of several blocks of data and that frequency drifts among transmissions from different nodes are negligible and the slow phase fluctuations caused by them can be accounted for in the slowly-time varying fading process characterizing the channel fluctuations.

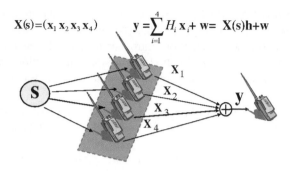

Figure 10.1 Codes for cooperative transmission.

Under the assumption of mutual synchronization among the cooperative nodes and flat fading channel, the received data vector $\mathbf{y}_i = (Y_i[0], \ldots, Y_i[P-1])^T$ is:

$$\mathbf{y}_i = \sum_{j=1}^{N} H_{i,j}\mathbf{x}_j + \mathbf{w}_i = \mathbf{X}(\mathbf{s})\mathbf{h}_i + \mathbf{w}_i, \tag{10.2}$$

where $\mathbf{h}_i = (H_{i,1}, \ldots, H_{i,N})^T$ is the vector of the relays' fading coefficients.

The parameter that easily captures the benefits of cooperation is the diversity gain, and is derived by the fact that the transmission occurs through multiple channels that are independently faded. We adopt following definition of diversity.

Definition: *The diversity order d^* of a scheme with average error probability $P_e(\text{SNR})$ is defined as $d^* = \lim_{\text{SNR} \to +\infty} \frac{\log P_e(\text{SNR})}{\log \text{SNR}}$. We say that a code $\mathbf{X}(\mathbf{s})$ achieves any diversity order d and coding gain $\mathcal{G}$ if $d \leq d^*$ and $P_e(\text{SNR}) \leq \mathcal{G}\text{SNR}^{-d^*}$, where SNR denotes the signal-to-noise ratio.*

10.2.2 Cooperative Schemes with Centralized Code Assignment

The simplest forms of cooperative relays are the so-called *amplify and forward* (AF) and *decode and forward* (DF). In the AF strategy, for each transmit symbol S the nodes retransmit a scaled version of the samples received over orthogonal channels. This can be expressed in our general model by the following coding rule when $\mathbf{s} = S$:

$$\mathbf{X}(S) = diag(\beta_1 Z_1, \ldots, \beta_N Z_N)$$
$$Z_r = \mathbf{h}_r^H \mathbf{y}_r, \ \beta_r \leq \sqrt{\frac{\mathcal{P}_r}{E\left\{\mathbf{h}_r^H \mathbf{y}_r \mathbf{y}_r^H \mathbf{h}_r\right\}}}, \tag{10.3}$$

where $\mathbf{y}_r$ in $Z_r = \mathbf{h}_r^H \mathbf{y}_r$ is the received vector containing the symbol S of the message. Note that the constraints on the scaling coefficients β_r guarantees that the node transmit power is $\mathcal{P}_r$. For the DF strategy the nodes decode each symbol of the message and transmit the decoded symbol. If the transmission are on orthogonal time intervals, the code matrix that corresponds to the DF is:

$$\mathbf{X}(S) = diag(\sqrt{\mathcal{P}_1}\hat{S}_1, \ldots, \sqrt{\mathcal{P}_T}\hat{S}_N), \tag{10.4}$$

where $\hat{S}_i$ is the estimate of common data S at the i'th node. In both cases (AF and DF), it is assumed that each relay transmits in an orthogonal channel, so $P = N$ and $M = 1$, resulting in a spectral efficiency equal to $(1/N) \log_2(|S|)$ that decreases with the number of nodes.

In order to achieve greater spectral efficiency while still obtaining the maximum diversity, similar to multiple-antenna systems, space-time codes have been proposed for distributed cooperative radios. For example, two cooperating nodes can utilize the Alamouti code given by the code matrix

$$\mathbf{X}(s) = \begin{bmatrix} \hat{S}_1[0] & \hat{S}_2[1] \\ \hat{S}_1^*[1] & -\hat{S}_2^*[0] \end{bmatrix}, \tag{10.5}$$

where $\hat{S}_i[n]$ is the estimate of symbol $S[n]$ at the i'th node and $\mathbf{s} = [S[0] \ S[1]]^T$ is the common data set. The space-time codes in cooperative networks are called distributed space-time codes since the antenna array is distributed in space. However, most of the so-called distributed space-time codes proposed for cooperative networks are not really the result of a distributed protocol and require some form of central control that assigns the space-time code matrix columns to the active relay nodes. Distributed space-time codes tend to attain diversity gains that are similar to those of DF but have both N and P growing in the same order. Note that in order to harvest the maximum diversity, the design of $\mathbf{X}_{P \times N}(\mathbf{s})$ requires $P \geq N$. This means that in order to support many nodes (large N), a greater latency or a wider bandwidth is required.

There are several alternative designs for transmission of cooperating nodes (Gamal and Aktas 2003; Hua et al. 2003; Jing and Hassibi 2004; Laneman and Wornell 2003; Yiu et al. 2006). The majority of them require some form of code assignment. Such designs can all be cast as a code matrix $\mathbf{X}(\mathbf{s})$, and the assignment of the columns of this matrix to relays makes the cooperative protocol centralized.

10.3 Randomized Distributed Cooperative Schemes

As explained in Section 10.2.2, in the presence of a central control unit each of N cooperative nodes is assigned a column $\mathbf{x}_l$ of a predetermined code matrix $\mathbf{X}_{P \times N}(\mathbf{s})$. When the nodes are unaware of how many nodes are going to cooperate and when there is no central code assignment, a randomized coding rule can replace the deterministic one; such rules can target a fixed maximum diversity order L, which is independent of the actual number of nodes cooperating.

10.3.1 Randomized Code Construction and System Model

At each node, the $\mathbf{s}$ is mapped onto a matrix $\mathbf{X}(\mathbf{s})$ equivalent to one of the options discussed previously in Section 10.2.2:

$$\mathbf{s} \rightarrow \mathbf{X}(\mathbf{s}).$$

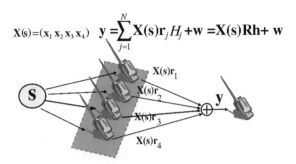

Figure 10.2 Randomized cooperative transmission.

Here, the number of columns L of $\mathbf{X}(\mathbf{s})$ denotes the number of antennas for which the underlying code is designed. For example, $\mathbf{X}(\mathbf{s})$ can be the Alamouti code, in which case $L = 2$. In randomized cooperation each node transmits a code formed as the random linear combination of the columns of $\mathbf{X}(s)$. Let $\mathbf{r}_j$ be the $L \times 1$ random vector that contains the linear combination coefficients for the j'th node, then $\tilde{\mathbf{x}}_j = \mathbf{X}(\mathbf{s})\mathbf{r}_j$ is the code transmitted by the j'th node. For example, a random antenna selection rule corresponds to a specific choice for the distribution of $\mathbf{r}_j$, which is discrete and has only one non-zero entry, picked at random. The following discussion clarifies the limitations of this choice compared to, for example, continuous distributions for the $\mathbf{r}_j$s.

The received vector is the mixture of each of these randomized codes convolved with their respective channel impulse response. Define $\tilde{\mathbf{X}} = [\tilde{\mathbf{x}}_1 \ \tilde{\mathbf{x}}_2 \dots \tilde{\mathbf{x}}_N]$ as the $P \times N$ random code matrix whose rows represent the time and columns represent the space. Under the assumption of flat fading (see Figure 10.2):

$$\mathbf{y}_i = \sum_{j=1}^{N} H_{i,j}\mathbf{X}(\mathbf{s})\mathbf{r}_j + \mathbf{w}_i = \mathbf{X}(\mathbf{s})\mathbf{R}\mathbf{h}_i + \mathbf{w}_i = \tilde{\mathbf{X}}\mathbf{h}_i + \mathbf{w}_i, \tag{10.6}$$

where $\mathbf{h}_i = (H_{i,1}, \dots, H_{i,N})^T$ and $\mathbf{R} = [\mathbf{r}_1 \ \mathbf{r}_2 \dots \mathbf{r}_N]$. Alternatively, one can view the randomization as a factor on the effective channel, that is

$$\mathbf{y}_i = \sum_{l=1}^{L} \left(\sum_{j=1}^{N} H_{ij}r_{jl} \right) \mathbf{x}_l + \mathbf{w}_i = \sum_{l=1}^{L} \tilde{H}_{il}\mathbf{x}_l + \mathbf{w}_i = \mathbf{X}(\mathbf{s})\tilde{\mathbf{h}}_i + \mathbf{w}_i \tag{10.7}$$

where $\tilde{\mathbf{h}}_i = \mathbf{R}\mathbf{h}_i = (\tilde{H}_{i1}, \dots, \tilde{H}_{iL})^T$, $\mathbf{x}_l$ is the l'th column of $\mathbf{X}(\mathbf{s})$ and r_{jl} is the (j, l)'th element of $\mathbf{R}$.

These two definitions express two critical interpretations of the proposed scheme. If $\mathbf{X}(\mathbf{s})\mathbf{R}$ is considered as a whole, then the scheme can be viewed as a randomized code $\tilde{\mathbf{X}}$ transmitted over channel $\mathbf{h}$. On the other hand, if $\mathbf{R}\mathbf{h}_i$ is considered as a whole, then the scheme can be viewed as a deterministic code $\mathbf{X}(\mathbf{s})$ transmitted over a randomized channel $\tilde{\mathbf{h}}_i$. The second interpretation is especially important for decoding purposes at the receiver. In order to perform coherent decoding, instead of estimating the channel vector $\mathbf{h}_i$ (which would require knowledge of the randomization matrix $\mathbf{R}$), the receiver can estimate the effective channel coefficients $\tilde{\mathbf{h}}_i$. To this end, the training data at the transmitters should use the same randomization procedure. Estimating the effective channel provides two main advantages: (i) decoders that have been already designed for multiple-antenna space-time codes can be directly used for randomized cooperative coding; and (ii) the complexity of the receiver is fixed and thus one can, for example, choose $L \leq N$ to have an effective channel vector $\tilde{\mathbf{h}}_i$ shorter than the actual channel vector $\mathbf{h}_i$ would be.

Omitting the index of the receiver, the signal model is therefore:

$$\mathbf{y} = \mathbf{X}(\mathbf{s})\mathbf{R}\mathbf{h} + \mathbf{w}, \tag{10.8}$$

where $\mathbf{w} \sim \mathcal{N}_c(\mathbf{0}, N_0\mathbf{I})$. For our analysis we will also assume that $\mathbf{h} \sim \mathcal{N}_c(\mathbf{0}, \Sigma_\mathbf{h})$. In the analysis of randomized codes, we will assume that the code is perfectly decoded at the nodes that are cooperating.

Note from (10.8) that randomized cooperative coding can be expressed as the double mapping:

$$\mathbf{s} \rightarrow \mathbf{X}(\mathbf{s}) \rightarrow \mathbf{X}(\mathbf{s})\mathbf{R}. \tag{10.9}$$

In the following, the $L \times N$ matrix $\mathbf{R}$ will be referred to as the *randomization matrix*. Since each node's processing is intended to be local, $\mathbf{r}_i$s should be independent for each $i = 1 \ldots N$, and we will also assume that they are identically distributed. This property allows the randomized cooperative coding to be implemented in a decentralized fashion. In other words, each node chooses a random set of linear combination coefficients from a given distribution, which does not depend on the node index.

There are several choices for the randomization matrix, we are going to discuss mainly the following examples:

1. Complex Gaussian Randomization: The elements of $L \times N$ randomization matrix $\mathbf{R}$ are drawn from a zero-mean, independent and complex Gaussian.

2. Uniform Phase Randomization: In this case the k'th column of $\mathbf{R}$ is $\mathbf{r}_k = a_k[e^{j\theta_i[0]}, \ldots$, $e^{j\theta_i[L]}]^t$ where each $\theta_i[N] \sim U(0, 2\pi)$ and $a_k \sim U(1 - \epsilon, 1 + \epsilon)$ for some small $\epsilon > 0$, where $U(a, b)$ denotes the uniform distribution in the interval (a, b) and all $\theta_i[N]$, a_k are independent of each other. With this design each node can easily enforce its transmit power constraint.

3. Complex Spherical Randomization: In this case the k'th column of $\mathbf{R}$, $\mathbf{r}_k$, is uniformly selected on the surface of a complex hyper-sphere of radius ρ, i.e., $||\mathbf{r}_k|| = \rho$. The $\mathbf{r}_i$s can be generated by creating zero-mean independent complex Gaussian vectors with covariance $\mathbf{I}$, and then normalized to have the norm $\rho = ||\mathbf{r}_i|| = 1$ (Marsaglia 1972; Muller 1959).

4. Random Selection Scheme: The random selection is the simplest among the randomized cooperation protocols since each node randomly selects one of the columns of a given code matrix $\mathbf{X}(\mathbf{s})$ at random. The performance of the scheme is limited by the possibility that two nodes selects the same antenna.

Example 10.3.1 (Cooperative Delay Diversity): *Under this strategy, each node randomly selects a delay and transmits accordingly (Scaglione and Y.-W.Hong 2003; Wei et al. 2004).*

In this way, one can achieve diversity by having the cooperative nodes behave intentionally as active multi-path scatterers and require no prior channel assignment. Choosing a specific delay amounts to selecting a column of the toeplitz matrix $\mathbf{X}(\mathbf{s})$ whose (i, j)'th element is:

$$\mathbf{X}_{ij}(\mathbf{s}) = X[i - j], \tag{10.10}$$

where the sequence $X[k]$ could simply equal the message $S[k]$ or could be encoded to guarantee the extraction of diversity from it. More specifically, (10.10) can be combined with spread spectrum techniques, or Orthogonal Frequency Division Multiplexing (OFDM) (see e.g. (Barbarossa and Scutari 2004)).

Cooperative delay diversity scheme (Wei et al. 2004) is a special case of the random selection scheme. Hence, the performance is limited by the event that all nodes choose the same delay in which case the scheme would yield no frequency diversity.

Example 10.3.2 (Randomized Space-Time Coding): *Another possible choice for $\mathbf{X}(\mathbf{s})$ is a space time code designed for a MISO system with L antennas. For instance, $\mathbf{X}(\mathbf{s})$ can be an orthogonal space-time code (Sirkeci-Mergen and Scaglione 2007). The advantage of this scheme is that the code can be fixed a priori and will not have to change dynamically depending on the number of the cooperating nodes.*

10.4   Performance of Randomized Cooperative Codes

The design of $\mathbf{X}(\mathbf{s})$ has been studied not only in the context of cooperative transmission (as described in Section 10.2.2), but also in the context of multiple-input multiple-output (MIMO) systems. There is a vast literature on the design of deterministic codes $\{\mathbf{X}(\mathbf{s})\}$, and the design of $\{\mathbf{X}(\mathbf{s})\}$ problem has been thoroughly investigated by many authors, starting from the seminal work in (Tarokh et al. 1998) and we refer the reader to (Tse and Viswanath 2005) for a comprehensive discussion on codes that harvest *diversity gains*. This chapter discusses, instead, the design of the randomization matrix $\mathbf{R}$, summarizing the results in (Sirkeci-Mergen and Scaglione 2005b, 2006a,b, 2007).

Without loss of generality, we assume that $P \geq L$ for the $P \times L$ deterministic code matrix $\mathbf{X}$. Define $\mathbf{X}_i \triangleq \mathbf{X}(\mathbf{s}_i)$. In the following, we will assume that the underlying deterministic code $\mathbf{X}$ satisfies the rank criterion (Tarokh et al. 1998):

C1) *The Rank Criterion for $\mathbf{X}$:* For any pair of code matrices $\{\mathbf{X}_k, \mathbf{X}_i\}$, the matrix $(\mathbf{X}_k - \mathbf{X}_i)$ is full-rank, i.e., of rank L.

This class of codes contains all orthogonal formulations (e.g. orthogonal DF and AF) and most of the space-time codes (Alamouti 1998; Tarokh et al. 1998).

10.4.1   Characterization of the Diversity Order

The performance degradation in fading channels results from the *deep fade event* (Tse and Viswanath 2005, Ch. 3). Diversity is the tool to combat this phenomenon. In the case of randomized codes, a deep fade event occurs when the effective channel coefficients $\tilde{\mathbf{h}}$ fades. The following lemma asserts the equivalence of analyzing the deep fade event in lieu of the average error probability P_e, to calculate the diversity of the scheme.

Lemma 10.4.1 *Let* $\{||\mathbf{Rh}||^2 \leq \mathrm{SNR}^{-1}\}$ *be the deep fade event, and*

$$P_{deep}(\mathrm{SNR}) \triangleq \Pr\{||\mathbf{Rh}||^2 \leq \mathrm{SNR}^{-1}\} \tag{10.11}$$

its probability. If the assumption C1) is satisfied, then the diversity order of P_e is the same as that of the deep fade event, i.e.,

$$d^* = \lim_{\mathrm{SNR}\to\infty} \frac{-\log P_{deep}(\mathrm{SNR})}{\mathrm{SNR}} \tag{10.12}$$

Proof. See (Sirkeci-Mergen and Scaglione 2007, Lemma 1).

An interesting corollary from the lemma is that the diversity order d^* is completely independent of the underlying code $\{\mathbf{X}_i\}$ as long as the underlying code is full rank. The main utility of Lemma 10.4.1 is that the diversity order of P_{deep} is much easier to analyze than that of P_e.

Let $\mathbf{U}^H \mathbf{\Lambda} \mathbf{U}$ denote the eigenvalue decomposition of $\Sigma_h^{1/2} \mathbf{R}^H \mathbf{R} \Sigma_h^{1/2}$, where $\mathbf{U}$ is a random Hermitian matrix and $\mathbf{\Lambda} = \mathrm{diag}(\lambda_1, \ldots, \lambda_\eta)$ is the diagonal matrix composed of the ordered eigenvalues (squared singular values of $\mathbf{R}\Sigma_h^{1/2}$). The following theorem provides a very general and clean characterization of the diversity order in terms of the distribution of the singular values of $\mathbf{R}\Sigma_h^{1/2}$. Let notation 0^- denote a negative real number that is close to zero and $\Gamma(\alpha_1, \ldots, \alpha_\eta)$ represent the following function:

$$\Gamma(\alpha_1, \ldots, \alpha_\eta) = \lim_{\mathrm{SNR}\to\infty} \frac{-\log \Pr(\lambda_1 \leq \mathrm{SNR}^{-\alpha_1}, \ldots, \lambda_\eta \leq \mathrm{SNR}^{-\alpha_\eta})}{\log \mathrm{SNR}}. \tag{10.13}$$

The parameters $\alpha_1, \ldots, \alpha_\eta$ are called the deep fade exponents of the singular values.

Theorem 10.4.2 *Under the assumption C1) the diversity order in (10.12) of the randomized code is*

$$d^* = \inf_{(\alpha_1, \ldots, \alpha_\eta)} \left(\Gamma(\alpha_1, \ldots, \alpha_\eta) + \sum_{i=1}^{\eta} (1 - \alpha_i) \right), \tag{10.14}$$

where the infimum is over $\alpha_i \in [0^-, 1]$, $i = 1, \ldots, \eta$.

Proof. See (Sirkeci-Mergen and Scaglione 2007, Th.1).

The interpretation of Theorem 10.4.2 is easier when $\Sigma_h = \mathbf{I}$. In this case, $\sqrt{\lambda_i}$'s are the singular values of the randomization matrix $\mathbf{R}$. In simpler terms, the theorem states that the deep fade event happens because of the simultaneous fades of the randomization matrix (i.e. $\mathbf{R}$ may be ill-conditioned) and the channel coefficients with exponents α_i's and $1 - \alpha_i$'s, respectively. This can be clearly seen in the structure of Eq. 10.14.

Theorem 10.4.2 completely characterizes the diversity order of a randomized code for a given $\mathbf{R}$; however, it is not obvious how to use Theorem 10.4.2 constructively and therefore how to optimize the choice of the distribution of $\mathbf{R}$. In fact, it is unclear how one can choose the singular vector and singular value distributions such that the singular value distribution has the local properties that are required to maximize d^* in (10.14) and, at the same time, the columns of $\mathbf{R}$ are statistically independent.

Good and bad design choices for the probability density of $\mathbf{R}$ can be determined from the conditions under which the λ_i's are, respectively, less or more likely to be small.

Since the eigenvalues values $\lambda_\eta \leq \cdots \leq \lambda_1$ are ordered, λ_η is the first to *fade*. Intuitively speaking, the λ_η fades if and only if the columns of $\mathbf{R}$ are completely confined to an $\eta - 1$ dimensional space or λ_n is almost zero relative to the rest of the eigenvalues.

Luckily, good designs are not hard to find. Towards the end of this section, we analyze a number of specific designs for $\mathbf{R}$ and conclude that the best designs have random column vectors in $\mathbf{R}$ which have the least probability of being linearly dependent. In fact, the design that performs best among the ones we examine has $\mathbf{R}$ with i.i.d. columns uniformly distributed in the complex unit sphere.

Choosing $\alpha_i = 0^-$, $\forall i$ in Theorem 10.4.2 one can see that the diversity order is always bounded by the minimum of the number of relays and the underlying code dimension, i.e.:

$$d^* \leq \eta = \min(L, N). \tag{10.15}$$

Example 10.4.3 (Fractional Diversity): *An interesting observation that can be made from Theorem 10.4.2 is that the diversity orders can be fractional depending on $\Gamma(\cdot)$. A concrete example of this is as follows. Assume that the k'th column of the $L \times N$ randomization matrix is $\mathbf{r}_k = [e^{j\theta_i[0]}, \ldots, e^{j\theta_i[L]}]^t$ where each $\theta_i[N] \sim U(0, 2\pi)$, and $U(a, b)$ denotes the uniform distribution in the interval (a, b). Let $L = N = 2$. Then, the eigenvalues of $\mathbf{R}\mathbf{R}^H$ are $\lambda_1 = 2 + \sqrt{2 + 2\cos(\theta)}$ and $\lambda_2 = 2 - \sqrt{2 + 2\cos(\theta)}$, where θ is a uniform random variable in the interval $[0, 2\pi)$. Note that $\lambda_1 \in [1, 4]$ with probability 1. Using Theorem 10.4.2 and the fact that $\lambda_1 \geq 1$, we can easily see that the optimal $\alpha_1 = 0^-$. Hence,*

$$d^* = \min_{\alpha_2} \Gamma(0^-, \alpha_2) + 2 - \alpha_2. \tag{10.16}$$

One can derive the distribution of λ_2 as

$$F_{\lambda_2}(\lambda) = \Pr\{\lambda_2 \leq \lambda\} = \frac{2}{\pi} \cos^{-1}\left(1 - \frac{\lambda}{2}\right), 0 \leq \lambda \leq 2.$$

Then, the behavior of the $F_{\lambda_2}(\lambda)$ around zero is given as $F_{\lambda_2}(\lambda) \approx \frac{2}{\pi}\sqrt{\lambda}$, as $\lambda \to 0$. The infimum in (10.16) is obtained when $\alpha_2 = 1$, which gives us a fractional value $d^* = 1.5$.

Example 10.4.4 (Random Antenna Selection) : *As described before, under this strategy each node randomly selects to serve as one of the antennas of a multi-antenna system. Using Theorem 10.4.2, we conclude that the diversity order of random selection scheme is $d^* = 1$ for $N < \infty$. This result might be discouraging; however, this simple method almost meets the ideal performance for SNR below a threshold SNR_t, which increases with node density. This can be easily seen by upper bounding the average error probability with a polynomial in $1/\mathrm{SNR}$. For example, let the relays chose randomly and uniformly one of the L codes in the $L \times L$ underlying code. Then,*

$$P_e \leq \sum_{m=1}^{\eta} \frac{C_m}{\mathrm{SNR}^m}, \tag{10.17}$$

where $\eta \triangleq \min\{N, L\}$ and C_m depends on Σ_h, underlying code and the number of nodes N. The expression (10.17) says that when the number of nodes is finite but sufficiently large, the probability of error curve changes its slope, but above a certain SNR threshold, the expected $O(1/\mathrm{SNR})$ behavior is obtained. The breaking points of the curve change and move towards higher SNRs as the number of nodes increases.

The diversity order for a large class of randomization matrices can be identified by simply comparing the number of nodes N and the size of the underlying code matrix L using the following theorem:

Theorem 10.4.5 *Let* $\mathbf{R}$ *be an* $L \times N$ *random complex matrix and* $p(\mathbf{R})$ *its probability density function. Assume that* $p(\mathbf{R})$ *is bounded and it satisfies the total power constraint:*

$$Tr(\mathbf{RR}^H) \leq \mathcal{P}_T < \infty \text{ with probability } 1 \qquad (10.18)$$

where $\mathcal{P}_T < \infty$ *is the total relay power available[1]. For* $N \neq L$, *if C1) holds and* $p(\mathbf{R})$ *is such that* $\mathbf{R}$ *is full rank with probability one, then the diversity order is:*

$$d^* = \begin{cases} N & \text{if } N \leq L - 1 \\ L & \text{if } N \geq L + 1 \end{cases} \qquad (10.19)$$

For $N = L$, *the diversity order is such that* $N - 1 \leq d^* \leq N$.

Proof. See (Sirkeci-Mergen and Scaglione 2007).

The above result shows that the randomized codes achieve the maximum diversity order N achievable by any scheme if $N < L$. It also indicates the diversity order saturates at L if the number of relay nodes is greater than or equal to $L + 1$. This problem can be solved by using codes with large enough dimensions. However, N may be random and may take large values in practical networks. In such cases, using smaller L may be preferred for decoding simplicity. For fixed L, randomized codes still give the highest order L for $N \geq L + 1$. Table 10.1 summarizes the diversity order of the schemes described previously for certain values of N and L.

10.4.2 Simulations and Numerical Evaluations

Given the relative complexity of the diversity expressions, it is hard to optimize over the randomization matrices that lead to the best symbol error rate performance. Resorting to numerical simulations sheds some light on what designs work best.

Observing the curves in Figure 10.3, one can note how the error probability is quite sensitive to the choice of the randomization matrix statistics when N is close to L. Different

Table 10.1 Diversity order for different schemes.

Distribution of $\mathbf{R}$	Condition on N and L	Diversity Order
Complex Gaussian	$N = L$	N
Complex Gaussian	$N \neq L$	$\min(N, L)$
Uniform Phase	$N \neq L$	$\min(N, L)$
Uniform Phase	$N = L = 2$	1.5
Complex Spherical Distribution	$N = L = 2$	2
Random selection	any N and L	1

---

[1]Note that there is no expectation in the power condition. We want it to be satisfied almost surely. Condition (10.18) implies that the pdf of $\mathbf{R}$ has bounded support.

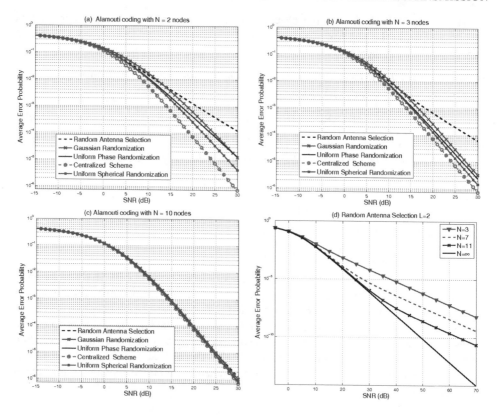

Figure 10.3 Average Probability of Error versus SNR (dB) (a) $L = 2$, $N = 2$ (upper-left); (b) $L = 2$, $N = 3$ (upper-right); (c) $L = 2$, $N = 10$ (lower-left); (d) $L = 2$, $N = 3, 7, 11, \infty$ (lower-right). The channel coefficient are assume i.i.d., i.e., $h_k \sim \mathcal{N}_c(0, 1)$. The transmission power of each node is $\mathcal{P}_t = 1/N$ for the centralized Alamouti, antenna selection, and spherical randomization schemes. For the Gaussian and uniform phase randomization schemes, $\mathcal{P}_t = 1/(NL)$.

randomization choices proposed previously are tested numerically using $\mathbf{X(s)}$ as the Alamouti scheme, i.e.

$$\mathbf{X(s)} = \begin{bmatrix} s_1 & s_2 \\ s_2^* & -s_1^* \end{bmatrix},$$

where $\mathbf{s} = [s_1 \ s_2]$ is the transmitted symbol vector and $s_i = \pm 1$ (BPSK symbols). In addition, we compare the performance of randomized strategies with the centralized Alamouti. In the centralized Alamouti, half of the nodes choose to serve as the first antenna, and the other half choose to serve as the second antenna (if N is odd, at one of the nodes the power is equally distributed between two antennas).

For $N = 2$, we can observe that the Gaussian and spherical randomization schemes have diversity order $d^* = 2$; the uniform phase randomization has diversity order $d^* < 2$ and the diversity order of the random antenna selection is 1. This is consistent with the analytical results. For $N = 3$, the Gaussian, uniform phase, and spherical randomization schemes achieve diversity order 2 similar to the centralized scheme. Observe that when $N > L$, the randomized scheme except antenna selection achieves the full diversity order L. This is consistent with the analytical proofs provided previously. Note also that there is a performance loss in coding gain of the decentralized schemes compared to the centralized one. Nevertheless, one can observe that as N increases, the performance of the distributed schemes approaches the centralized scheme in both the diversity and coding gains.

In Figure 10.3d, we plot the average error probability for random antenna selection scheme for different number of nodes (N). When N is odd and $L = 2$, the analytical expression simplifies to (Sirkeci-Mergen and Scaglione 2005b),

$$P_e = \frac{1}{2^N} \sum_{k=0}^{N} \binom{N}{k} \frac{g(k) - g(N-k)}{2k - N},$$
(10.20)

where $g(x) = \frac{x}{2}\left(1 - \sqrt{\frac{x \, \text{SNR}}{x \, \text{SNR}+1}}\right)$. Figure 10.3d illustrates a multi-slope behavior for the random code selection scheme. This is consistent with the discussion in Example 10.4.4. The P_e curves have a breaking point, which becomes more pronounced as N increases; beyond a certain SNR, they all have the same slope which corresponds to diversity order 1. For SNR values less than a threshold, the diversity order 2 is achieved. This can be clearly seen for $N = 11$ which has a breakpoint around SNR $= 35$ dB.

10.5 Analysis of Cooperative Large-scale Networks Utilizing Randomized Cooperative Codes

Cooperative protocols that consider *single- or two-hop* communication are illustrated in Figures 10.1 and 10.2, and discussed in the first part of this chapter. As mentioned before, randomized cooperative codes, which are a special class of cooperative codes, allow a group of transmitters to reach a destination reliably without any central controller. This feature makes randomized codes more attractive especially in large networks, where self-organization is critical. On the other hand, in large-scale networks, it is impossible for nodes that are further apart to communicate with each other in a few number of hops. The importance of *multi-hop* communication in large-scale setups has been acknowledged in (Boyer et al. 2004; Gastpar and Vetterli 2005; Hong and Scaglione 2003; Kramer et al. 2005; Maric and Yates 2004; Sirkeci-Mergen et al. 2005, 2006). In the following, we describe a generalized *multi-hop* communication strategy in which each hop is a group of transmitters instead of a single node and the message is passed from one group to another (see Figure 10.4). It is important to note that these groups are not predetermined; they are recruited during the communication flow based on their received signal quality. Thanks to randomized cooperative codes, the nodes in a given group do not require any coordination either. This leads to great simplification for the overall management of the cooperative links.

In the second part of the chapter, we analyze the transmission dynamics of randomized cooperative codes in large-scale extended networks. For brevity, we discuss only the broadcast scenario-the case where a single source aims to reach entire network (Sirkeci-Mergen

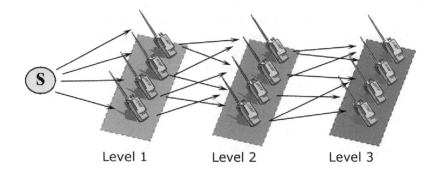

Level 1   Level 2   Level 3

Figure 10.4 Multistage cooperative transmission.

et al. 2006). In the following, the node index is replaced by the coordinates of the node. For simplicity, we consider a network where N relay nodes are distributed uniformly in the disc $\mathbb{S} = \{(x, y) : x^2 + y^2 \leq R^2\}$ and the source is located at the center $(0, 0)$. The modified version of Eqn. 10.8 is:

$$y(x, y) = \frac{1}{\sqrt{L}} \mathbf{X}(\mathbf{s}) \mathbf{Rh}_{\mathcal{L}}(x, y) + \mathbf{n}(x, y)^2 \qquad (10.21)$$

where $\mathcal{L} \subset \{1, \ldots, N\}$ denotes a subset of relay nodes that transmits and $\mathbf{h}_{\mathcal{L}}(x, y)$ denotes the channel impulse response vector from level set $\mathcal{L}$ to a hypothetical node at (x, y).

We consider codes $\mathbf{X}(\mathbf{s})$ that are orthogonal, i.e., $\mathbf{X}(\mathbf{s})^H \mathbf{X}(\mathbf{s}) = c\mathbf{I}$ for some scalar c. In this case the analysis is simplified since the symbol-by-symbol decoding is optimal (Larsson and Stoica 2003). The effective SNR is defined as

$$\gamma_{\mathcal{L}} = \frac{\mathcal{P}_r \|\mathbf{Rh}_{\mathcal{L}}(x, y)\|^2}{L}, \qquad (10.22)$$

where $\mathcal{P}_r$ is the relay per-relay power.

In the setup considered, a source node initiates the broadcast by transmitting a packet. The decision criterion of when to relay packets is a subtle issue. The simple criterion we choose to consider is based on the notion of matched-filter upper bound, which is directly linked to effective received SNR. Specifically, we consider a reception successful if $\gamma_{\mathcal{L}}$ exceeds a certain threshold τ. Considering this criterion, let $\mathcal{S} = \{(x_i, y_i) : i = 1, \ldots, N\}$ be the set of relay nodes that are randomly and uniformly distributed in $\mathbb{S}$. We will call the nodes who can hear the source with sufficient signal-to-noise ratio has *level-1 nodes*. The set of random locations corresponding to the nodes in level-1 is

$$\mathcal{S}_1 = \{(x, y) \in \mathcal{S} : \gamma_0 \geq \tau\},$$

where γ_0 is the effective SNR due to transmission of the source node. The level-1 nodes decode and retransmit the source after successful reception. This transmission excites a

---

[2]Note the normalization of power with 1/L.

second group of nodes and the retransmissions continue until every node who hears the others with sufficient SNR, retransmits once. The random locations of the level-k nodes form the set:

$$S_k = \{(x, y) \in S \setminus \bigcup_{i=1}^{k-1} S_i : \gamma_{\mathcal{L}_{k-1}} \geq \tau\}, \tag{10.23}$$

where $\mathcal{L}_{k-1} \subset \{1, \ldots, N\}$ is the index set of level $(k-1)$ nodes. See Figure 10.4 for an example.

An important property of the described protocol is its decentralized nature. The nodes use a simple *SNR threshold* criterion to decide if they are going to retransmit or not, i.e., every node monitors its received SNR and decodes and retransmits using randomized codes if and only if its SNR exceeds a certain pre-determined threshold. In this way, the network can operate in a distributed fashion, since the nodes only use the locally available received SNR information to make transmission decisions. We need to note that the nodes do not retransmit the same packet more than once in order to avoid cycles in the network. This can be handled locally at the node via checking a table of previously transmitted packets. We assume that this control is done with no error.

In the following, we present a framework to analyze the evolution of the stochastic process $S_k(x, y)$. We fix the total relay power $\mathcal{P}_r N$ and consider the asymptote as $N \to \infty$, $\mathcal{P}_r \to 0$. Here, the relay power density,

$$\overline{\mathcal{P}}_r = \frac{\mathcal{P}_r N}{\text{Area}(S)} = \mathcal{P}_r \rho,$$

is also fixed, where ρ denotes the node density. This asymptote will also be called *continuum model*. The analysis of the sets $S_k(x, y)$ in the continuum lets us understand the effect of network parameters such as the source/relay transmission powers, the decoding threshold and intended diversity order L on the performance.

We are interested in knowing how likely that a node in a certain location belongs to the k'th level. For this reason, we define the following family of probability functions:

Definition 10.5.1 $P_k(x, y)$ *denotes the probability that a node at location (x, y) joins level-k.*

As we noted in Section 10.3, there is an interesting tradeoff in choosing the code order L relative to the expected number of cooperative nodes N: if L is likely to be smaller than N the diversity saturates to L, but the scheme requires less bandwidth or latency and may be overall preferable to a scheme with L that grows in the order of N. Hence, for $N \gg 1$, we refer to the case where L is finite as the *narrowband* network, and to the case where $L = N$ as the *wideband* network. In mathematical terms, the two different regimes we consider are

     Case I – (Narrowband): L is fixed and $N \to \infty$

     Case II – (Wideband): $L = N$ and $N \to \infty$, $L \to \infty$

Theorem 10.5.2 *Let $\ell(x - x', y - y')$ denote the path-loss attenuation function between two nodes located at (x, y) and (x', y'). Let $\sigma_L = \mathbb{E}\{|h_i|^2\}$, where h_i is the i'th element of $\mathbf{h}_L(\cdot)$. Let r_{ij} denote the (i, j)'th element of the randomization matrix $\mathbf{R}$. Assume that r_{ij}*

are i.i.d., $\mathbb{E}\{r_{ij}\} = 0$, *and* $\mathbb{E}\{|r_{ij}|^2\} = 1$. *Under these assumptions, consider the continuum model. The* $P_k(x, y)$, $k = 1, 2, 3 \ldots$ *is given by a recursive set of equations*

$$P_{k+1}(x, y) = \Pr\{\gamma_k(x, y) \geq \tau\} \prod_{i=0}^{k-1} [1 - \Pr\{\gamma_i(x, y) \geq \tau\}], \qquad (10.24)$$

where $P_1(x, y) = \Pr\{\|\gamma_0(x, y)\|^2 \geq \tau\}$, $\gamma_0(x, y) = \mathcal{P}_s \|\mathbf{h}_0(x, y)\|^2$, $\mathcal{P}_s$ *is source transmission power, and* $\mathbf{h}_0(x, y)$ *is the channel coefficient from source to a node located at* (x, y). *The function* $\gamma_k(x, y)$ *in two different asymptotic regimes is given as*

$$\gamma_k(x, y) \sim \chi^2(2L, \sigma_k^2(x, y)/L) \quad narrowband$$

$$\gamma_k(x, y) = \sigma_k^2(x, y) \qquad\qquad wideband$$

where the notation $X \sim \chi^2(V, \sigma^2)$ *is equivalent to saying* X/σ^2 *is a chi-square random variable with V degrees of freedom and*

$$\sigma_k^2(x, y) = \iint_{\mathbb{S}} \overline{\mathcal{P}}_r P_k(x', y') \ell(x - x', y - y') \, dx' \, dy'. \qquad (10.25)$$

Proof. See Appendix for the proof.

The functions P_k, σ_k^2 define a non-linear dynamical system which evolves with k. Although the analytical solution of this system is hard to find, it can be evaluated numerically. Due to symmetric structure of the path-loss model and the network topology, the $P_k(x, y)$ and $\sigma_k^2(x, y)$ are only functions of $r = \sqrt{x^2 + y^2}$. For convenience, we will use the notations $P_k(x, y)$ and $P_k(r)$ interchangeably.

10.5.1 Numerical Evaluations and Further Discussions

For our numerical evaluations in this section, we will use the following path-loss model

$$\ell(r) \triangleq \begin{cases} 1/r^2 & r_0 \leq r \\ 1/r_0^2 & 0 \leq r \leq r_0, \end{cases} \qquad (10.26)$$

with a small $r_0 > 0$ to avoid the singularity in the integral (10.25). The squared-distance attenuation model $\ell(r) = 1/r^2$ comes from the free-space attenuation of electromagnetic waves, and it *does not* hold when r is very small (Rappaport 2001).

We evaluate (10.24) and (10.25) numerically for large R. In Figures 10.5a and 10.5b, we plot the level curves, $P_k(r), r \in \mathbb{R}$ for the narrowband network. We observe that there exists a critical threshold τ_c. For $\tau > \tau_c$, the transmissions eventually die out (Figure 10.5b), that is i.e.,

$$\sup_{r \in \mathbb{R}^2} P_k(r) \to 0 \quad \text{as} \quad k \to \infty.$$

Otherwise, the transmissions propagate to the whole network, while the level curves, $P_k(r)$, $r \in \mathbb{R}$, become wider as k increases (Figure 10.5a). We observe similar behavior for the wideband network (Figures 10.6a and 10.6b). Interestingly, the level curve $P_k(r)$ has a

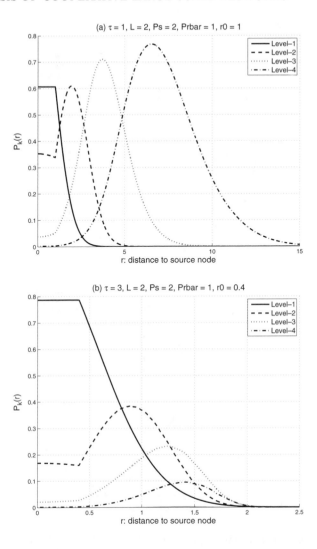

Figure 10.5 (a) Transmissions continue. $\tau = 1, \mathcal{P}_s = 2, \overline{\mathcal{P}}_r = 1, L = 2, r_0 = 1$ (b) Transmissions die out. $\tau = 3, \mathcal{P}_s = 2, \overline{\mathcal{P}}_r = 1, L = 2, r_0 = 0.4$

discrete structure in the wideband scenario, that is $P_k(r), k > k^*$, for some k^*, takes values either 0 or 1 for a given r. This can be understood by analyzing the analytical formulation: in the continuum asymptote for wideband networks $P_1(r)$ is a continuous function; however, $\Pr\{\gamma_k(x, y) \geq \tau\}$ only takes values 0 or 1. The effect of $P_1(x, y)$ taking values in the interval [0, 1] tends to cause negligible effect in Eqn. (10.24) as k increases.

In general, we do not have an explicit characterization of τ_c. The numerical analysis shows that there exists a phase transition in the network behavior, i.e. if the decoding threshold is below a fixed value (or equivalently the power density is above a fixed value),

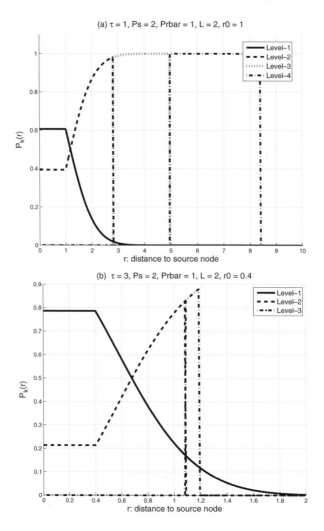

Figure 10.6 (a) The transmissions continue. $\tau = 1, \mathcal{P}_s = 2, \overline{\mathcal{P}}_r = 1, r_0 = 1$. (b) The transmissions die out. $\tau = 3, \mathcal{P}_s = 2, \overline{\mathcal{P}}_r = 1, r_0 = 0.4$.

a succession of cooperative transmissions can reach reliably all network nodes. Otherwise, part of the nodes do not get the source message.

To compare the message propagation behavior of both wideband and narrowband scenarios, in Figure 10.7 we show the corresponding $P_k(r)$ for both network asymptotes. For both wideband and narrowband models, $P_1(\cdot)$ is the same; however, the level curves $P_k(\cdot)$ differ significantly for large k. In particular, the level curves in the narrowband case move faster. This is a rather counter-intuitive result, because the orthogonal system with large L uses much more bandwidth, and the use of orthogonal channels is more reliable in the sense that the receiver eliminates/reduces the effects of fading via maximal ratio combining. Our

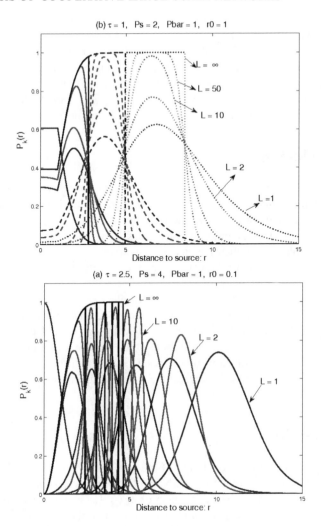

Figure 10.7 Wideband vs. narrowband transmission: large L corresponds to wideband: (a) low threshold regime; (b) high threshold regime.

result implies that the system with low diversity and therefore *with fading* provide faster message delivery on the average.

Clearly, a high dimensional code reduces the probability that the combined signal experiences a deep fade; however, this comes at the cost of reducing the probability that the signal experiences a favorable fade. In a dense network, favorable fading realizations are very valuable, because when the node density is high, although there is a small probability of having a good fading realization, there is always a fraction of nodes that experiences them. Once these lucky nodes receive and retransmit, the nodes neighboring them see a boost of signal power because of the properties of $\ell(r)$. In conclusion, the forefront of each level tends to be placed further when there is less diversity and the levels and the mode of the distribution of $P_k(r)$ for small values of L tends to be ahead compared to the forefront of the corresponding $P_k(r)$ for large L.

In the narrowband system, favorable fading realizations occur, when the phases of two or more simultaneously transmitting nodes happen to add coherently, or when one of the transmitting nodes experiences a very good effective channel to the receiver. Considering that finite L scenarios do not require infinite bandwidth, we conclude that the narrowband schemes are more advantageous also in terms of end-to-end delay.

10.6   Conclusion

In this chapter, we proposed a decentralized coding for distributed networks which is based on independent randomization done at each node. We analyzed its performance and proposed different designs that achieve the diversity order $(\min(N, L))$ when the number of nodes N is different than the number of antennas L in the underlying code. For $N = L$, we presented examples where the diversity order is fractional. In addition, we showed that the randomized schemes achieve the performance of a centralized code in terms of coding gain as the number of nodes increases.

Furthermore, we analyzed the behavior of a wireless network with cooperative broadcasting where each group utilizes randomized codes. The analysis is based on the idea of continuum approximation, which models networks with high node density. The interesting conclusion drawn from the analysis is that there exists a phase transition in the propagation of the message, which is a function of the node powers and the reception threshold. In addition, we presented a trade-off between the intended diversity order and the speed of propagation.

10.7   Appendix

In the following, we analyze the evolution of the functions $P_k(x, y)$ with k. First, we consider the asymptotic statistics of effective SNR in both narrowband and wideband regimes. Let $\sigma_{\mathcal{L}} = \mathbb{E}\{|h_i|^2\}$, where h_i is the i'th element of $\mathbf{h}_{\mathcal{L}}(\cdot)$. Let r_{ij} denote the (i, j)'th element of the randomization matrix $\mathbf{R}$. For simplicity, we assume that r_{ij}'s are i.i.d., $\mathbb{E}\{r_{ij}\} = 0$, and $\mathbb{E}\{|r_{ij}|^2\} = 1$. In the following, we consider the effect of path-loss attenuation and small-scale fading $(\alpha_{ij} \overset{\text{i.i.d.}}{\sim} \mathcal{N}_c(0, 1))$ on the flat fading channel coefficients. Under these assumptions, the the asymptotic statistics of the effective SNR $\gamma_{\mathcal{L}}$ (due to transmission of nodes in the set $\mathcal{L}$) are given as:

Case I – (Narrowband): Fix L and take $N \to \infty$:

$$\gamma_{\mathcal{L}} \overset{\text{d}}{\to} \chi^2(2L, \sigma_{\mathcal{L}}^2/L) \tag{10.27}$$

Case II – (Wideband): ($L = N$, and $N \to \infty$, $L \to \infty$):

$$\gamma_{\mathcal{L}} \overset{\text{p}}{\to} \sigma_{\mathcal{L}}^2. \tag{10.28}$$

Here $\overset{\text{d}}{\to}$ refers to convergence in distribution and $\overset{\text{p}}{\to}$ refers to convergence in probability.

The probability that a node at location (x, y) receives the source transmission successfully and joins level-1 is:

$$P_1(x, y) = \Pr\{||\gamma_0(x, y)||^2 \geq \tau\},$$

where $\gamma_0(x, y) = \mathcal{P}_s ||\mathbf{h}_0(x, y)||^2$, P_s is source transmission power, and $\mathbf{h}_0(x, y)$ is the channel coefficient from source to a node located at (x, y). Let $\gamma_k(x, y)$ denote the asymptotic SNR function due to transmission of k'th level at a node located at (x, y), $k = 1 \ldots$.

From the law of large numbers it follows that, for each set $\mathbb{U} \subset \mathbb{S}$, the number of level-1 nodes in $\mathbb{U}$ scales as $\iint_{\mathbb{U}} \rho P_1(x', y') \, dx' \, dy'$, i.e.,

$$\frac{|\mathbb{U} \cap \mathcal{S}_1|}{\iint_{\mathbb{U}} \rho P_1(x', y') \, dx' \, dy'} \to 1 \quad \text{as } \rho \to \infty \tag{10.29}$$

almost surely. When $\mathcal{P}_r \rho$ is fixed to $\overline{\mathcal{P}}_r$, the total transmit power of level-1 nodes $\mathcal{P}_r |\mathbb{U} \cap \mathcal{S}_1|$ converges to

$$\mathcal{P}_T = \iint_{\mathbb{S}} \overline{\mathcal{P}}_r P_1(x', y') \, dx' \, dy' \tag{10.30}$$

almost surely. Furthermore, the locations of level-1 nodes are distributed according to the density $\tilde{\rho}(x', y') \triangleq \frac{P_1(x', y')}{\iint_{\mathbb{S}} P_1(x', y') \, dx' \, dy'}$. Hence

$$\sigma_1^2(x, y) = \iint_{\mathbb{S}} \overline{\mathcal{P}}_r P_1(x', y') \ell(x - x', y - y') \, dx' \, dy'. \tag{10.31}$$

If this $\sigma_1^2(x, y)$ is substituted into Equations 10.27 and 10.28, we see that a node at (x, y) receives the level-1 transmission successfully with probability $\Pr\{\gamma_1(x, y) \geq \tau\}$. The probability that a node at (x, y) joins level-2 is

$$P_2(x, y) = \Pr\{\text{receives from level-1, does not receive from the source}\}$$

$$= \Pr\{\gamma_1(x, y) \geq \tau\} \, [1 - \Pr\{\gamma_0(x, y) \geq \tau\}]. \tag{10.32}$$

In both Case I and II the $\Pr\{\gamma_0(x, y) \geq \tau\}$ and $\Pr\{\gamma_1(x, y) \geq \tau\}$ are non-linear functions of $\sigma_0^2(x, y)$ and $\sigma_1^2(x, y)$ respectively. In particular, in Case II, $\Pr\{\gamma_1(x, y) \geq \tau\}$ can only be 1 or 0 depending on whether $\sigma_1^2(x, y) \lessgtr \tau$.

Now, we can generalize what is done so far for all values of k. Let $\sigma_k^2(x, y)$ be the sum of signal powers from level-k at location (x, y). For $k = 1, 2, 3, \ldots$, the equations are

$$P_{k+1}(x, y) = \Pr\{\gamma_k(x, y) \geq \tau\} \prod_{i=0}^{k-1} [1 - \Pr\{\gamma_i(x, y) \geq \tau\}], \tag{10.33}$$

$$\sigma_k^2(x, y) = \iint_{\mathbb{S}} \overline{\mathcal{P}}_r P_k(x', y') \ell(x - x', y - y') \, dx' \, dy'. \tag{10.34}$$

This completes the proof.

Bibliography

Alamouti SM 1998 A simple transmit diversity technique for wireless communications. *IEEE J. Select. Areas Commun.* **16**(8), 1451–1458.

Barbarossa S and Scutari G 2004 Distributed space-time coding for multihop networks. In *Proc. of IEEE International Conference on Communications*, vol. 2, pp. 916–920.

Boyer J, Falconer D.D, Yanikomeroglu H 2001 A theoretical characterization of the multihop wireless communications channel with diversity. *Global Telecommunications Conference*, **2**, 841–845.

Boyer J, Falconer D.D, Yanikomeroglu H 2004 Multihop diversity in wireless relaying channels. *IEEE Trans. Commun.*, Oct. 2004.

Dousse O and Thiran P 2004 Connectivity vs. capacity in dense ad hoc networks. In *Proc. of 23rd Annual Joint Conf. of the IEEE Computer and Commun. Societies (Infocom)*, vol. 1.

Draper S and Wornell GW 2004 Side information aware coding strategies for sensor networks. *IEEE Journal on Selected Areas in Communication*.

Gamal HE and Aktas D 2003 Distributed space-time filtering for cooperative wireless networks. In *Proc. IEEE Global Telecomm. Conf. (Globecom 2003)*, vol. 4, pp. 1826–1830.

Gupta P and Kumar P 2000 The capacity of wireless networks. *IEEE Trans. on Information Theory*.

Gastpar M, Vetterli M 2005 On the capacity of large Gaussian relay networks. *IEEE Transactions on Information Theory*, **51**(3);765-779.

Hong Y-W, Scaglione A B 2003 Energy-efficient broadcasting with cooperative transmission in wireless sensory ad hoc networks. *Proc. of Allerton Conf. on Commun., Contr. and Comput. (ALLERTON)*, Oct. 2003.

Hong YW and Scaglione A On multiple access for correlated sources: A content-based group testing approach. *submitted to IEEE Transactions on Information Theory*.

Hua Y, Mei Y and Chang Y 2003 Wireless-antennas making wireless communications perform like wireline communications. *IEEE Topical Conference on Wireless Communication Technology*, pp. 47–73.

Jing Y and Hassibi B 2004 Wireless networks, diversity and space-time codes. In *Proc. of IEEE Information Theory Workshop*, pp. 463–468.

Jovicic A, Viswanath P and Kulkarni SR 2004 Upper bounds to transport capacity of wireless networks. *IEEE Transactions on Information Theory* **50**(11), 2555–2565.

Kramer G, Gastpar M, Gupta P 2005 Cooperative strategies and capacity theorems for relay networks. *IEEE Trans. Inform. Theory*, **51**(9).

Laneman JN and Wornell GW 2003 Distributed space-time-coded protocols for exploiting cooperative diversity in wireless networks. *IEEE Trans. Inform. Theory* **49**(10), 2415–2525.

Larsson EG and Stoica P 2003 *Space-Time Block Coding for Wireless Communications*. Cambridge University Press, Cambridge.

Luo J, Blum R.S., Cimini L.J., Greenstein L, and Haimovich A 2005 Link-failure probabilities for practical cooperative relay networks. In *Proc. of Vehicular Technology Conference*, vol. 3, 30 May–1 June 2005, pp. 1489–1493.

Maric I, Yates R D 2004 Cooperative multihop broadcast for wireless networks. *IEEE J. Select. Areas Commun.*, **22**(6).

Marsaglia G 1972 Choosing a point from the surface of a sphere. *Ann. Math. Stat.*

Mergen G and Tong L 2006 Type based estimation over multiaccess channels. *IEEE Transactions on Signal Processing* **54**(2), 613–626.

Muller ME 1959 A note on a method for generating points uniformly on n-dimensional spheres. *Comm. Assoc. Comp. Mach.*

Rappaport TS 2001 *Wireless Communications: Principles and Practice* second edition, Prentice Hall, New Jersey.

Ribeiro A, Cai X and Giannakis G B 2004 Symbol error probabilities for general cooperative links. *IEEE Trans. on Wireless Comm.*, **4**, 1264–1273.

Scaglione A and Y.-W.Hong 2003 Opportunistic large arrays: Cooperative transmission in wireless multihop adhoc networks to reach far distances. *IEEE Trans. Signal Processing* 51(8), 2082–2092.

Scaglione A, Sirkeci-Mergen B, Geirhofer S and Tong L 2006 Randomized distributed multi-antenna systems in multi-path channels. *Proc. of 14th European Signal Processing Conference*, Sep. 2-8 2006.

Sharp M, Scaglione A, Sirkeci-Mergen B, Sung T 2007 Randomized distributed multi- antenna systems in multi-path channels. *submitted to IEEE Transactions on Communications*, Jan. 2007.

Sirkeci-Mergen B and Scaglione A 2005a A continuum approach to dense wireless networks with cooperation. *Proc. of Annual Joint Conf. of the IEEE Computer and Commun. Societies (Infocom)*, 2005.

Sirkeci-Mergen B and Scaglione A 2005b Randomized distributed space-time coding for cooperative communication in self-organized networks. In *Proc. of IEEE Workshop on Signal Process. Advances in Wireless Commun. (SPAWC)*, pp. 500–504.

Sirkeci-Mergen B and Scaglione A 2006a Randomized space-time coding for distributed cooperative communication. In *Proc. of IEEE Proc. of IEEE International Conference on Communications*.

Sirkeci-Mergen B and Scaglione A 2006b Randomized space-time coding for distributed cooperative communication: Fractional diversity. In *Proc. of IEEE Inter. Conf. on Acoustics, Speech, and Signal Process. (ICASSP)*.

Sirkeci-Mergen B and Scaglione A 2007 Randomized space-time coding for distributed cooperative communication. *accepted to be published in IEEE Transactions on Signal Processing*.

Sirkeci-Mergen B, Scaglione A and Mergen G 2006 Asymptotic analysis of multi-stage cooperative broadcast in wireless networks. *Joint special issue of IEEE Transactions on Information Theory and IEEE/ACM Trans. On Networking*, **52**(6): 2531–2550.

Stefanov A and Erkip E 2004 Cooperative coding for wireless networks. *IEEE Transactions on Communications*.

Tarokh V, Seshadri N and Calderbank A 1998 Space-time codes for high data rate wireless communication: performance criterion and code construction. *IEEE Trans. Inform. Theory* **44**(2), 744–765.

Tse D and Viswanath P 2005 *Fundamentals of Wireless Communication*. Cambridge University Press, Cambridge.

Warneke B, Last M, Liebowitz B and Pister KSJ 2001 Smart dust: communicating with a cubic-millimeter computer. *IEEE Trans. on Computer*.

Wei S, Goeckel D and Valenti M 2004 Asynchronous cooperative diversity. In *Proc. of 2004 Conference on Information Sciences and Systems (CISS)*.

Yiu S, Schober R and Lampe L 2006 Distributed space-time block coding. *accepted for publication in IEEE Trans. on Communication*.

11

Application Dependent Shortest Path Routing in Ad-Hoc Sensor Networks

Saswat Misra, Lang Tong, and Anthony Ephremides

11.1 Introduction

Consider the generic communications network depicted in Figure 11.1. Such a network consists of nodes (sometimes known as switches or routers) and communication links (which may be wireline or wireless). Each node may generate (the same or different) data destined for some or all other nodes in the network. For example, consider the data generated at a node S intended for a node D in Figure 11.2. As shown in the figure, there are many possible paths on which the data can be *routed* from S to D. In general, routing seeks to determine the 'best' path(s) from each potential source to each potential destination subject to network constraints. Typical routing schemes may be chosen to address some combination of the following: (i) minimization of end-to-end routing delay, (ii) maximization of network throughput, (iii) avoidance of unreliable communications links, and (iv) minimization of communication costs. In this chapter, we will see that the definition of the 'best' path, and the specification of network constraints, depends strongly on the network's class, e.g., criteria that are sensible for wireline networks may not be so for wireless networks, and/or mobile ad-hoc sensor networks.

Suppose that we have a well-defined notion of what constitutes the 'best' route for a set of known network constraints for a given network. It is not yet clear how to determine the optimal routes. A conceptually straightforward strategy to determine optimal routes would

Wireless Sensor Networks: Signal Processing and Communications Perspectives   A. Swami, Q. Zhao, Y.-W. Hong and L. Tong
© 2007 John Wiley & Sons, Ltd

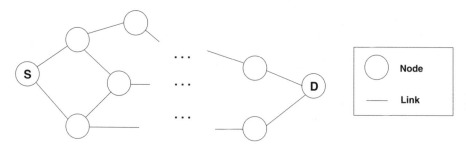

Figure 11.1 A generic communications network. Reproduced by permission of © 2007 IEEE.

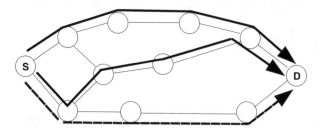

Figure 11.2 Three possible routing strategies from node S to D in a communications network.

be to test each possible route sequentially and select the one that provides best performance while satisfying the network constraints. Routes could then be hardcoded into nodes for each possible source-destination pair before deployment. However, such an approach is prohibitively complex, quickly becoming intractable as the network size increases. Further, implementation would be difficult or impossible in wireless or wireless sensor networks where the network topology is variable and unpredictable in advance, and inaccessible after deployment. *Shortest Path Routing* (SPR) is a classic routing technique that simplifies the determination of optimal routes. Central to SPR is the definition of a *link metric*, a number assigned to each link in the network, that quantifies its benefits and costs. Once an appropriate link metric has been defined, the optimal shortest path routes can be computed in polynomial time and in a distributed fashion.

In this chapter we study the evolution of shortest path routing, including relevant algorithms and protocols. Our aim is to show how each generation of network (wire-line, wireless, and sensor) has different routing requirements, and how SPR was adapted to meet those requirements. Our main contribution to the literature is the argument for new *application-dependent* strategies for SPR in ad-hoc wireless sensor networks, which we exemplify through a detailed example in Section 11.4.3. In Section 11.2, we review shortest path routing and describe its advantages and disadvantages for wireline commu-nications networks. In Section 11.3, we extend our discussion to wireless networks. We discuss the new challenges presented by such networks, and the SPR techniques developed to address them. In Section 11.4, we discuss the application of SPR to ad-hoc wireless sensor networks. First, we argue that routing for such networks should be designed in an application dependent manner. Then, we present an in-depth example of such routing to stimulate discussion and serve as an archetype for future research.

11.1.1 Major Classifications

We describe the key criteria used to differentiate networks and routing algorithms. These concepts will be referred to throughout the chapter.

The first criteria is the *switching mode*; we describe the cases of circuit switched and packet switched networks. Consider data generated at an arbitrary node S destined for another (arbitrary) node D as shown in Figure 11.1. *Circuit switching* is often used when data is transmitted in long (relative to the time required to establish the connection) steady streams and must be received in real-time at the destination. In circuit switched networks, a path from S to D is selected and reserved for the entire duration of the data transmission. This is the switching mode used in traditional voice telephone calls and corresponds to reservation-based constant bit rate (CBR) connections on the internet. By contrast, *packet switching* is often preferred when S generates bursty or intermittent traffic. In packet switching, the stream of data to be transmitted from S to D is decomposed into smaller blocks that are individually transmitted to D. The idea is that if the network consists of many source-destination pairs, each of which generates such bursty traffic, messages are transmitted more efficiently by interleaving blocks from different links as they traverse the network. Packet switching is the most common switching method in use today (e.g., in the Internet). Packet switching networks can be further classified as *virtual circuit* based, in which all packets travel the same route from S to D and arrive in sequential order, or *datagram* based, in which two packets from S can travel in different routes and appear out of sequence at D.

Routing algorithms can also be classified as *static* or *adaptive*. In static routing, routes are chosen based on the network topology and remain fixed over a long period of time, changing only due to events such as link failure. In *adaptive* routing, nodes monitor congestions levels and other network parameters and dynamically change routes in real-time.

Routing algorithms can also be classified as *centralized* or *distributed*. In centralized routing algorithms, a central controller determines routes and periodically updates routers with this information. In distributed routing algorithms, routers send messages to each other to determine routes based on some prescribed metric.

Finally, routing algorithms can be classified as *flat* or *hierarchical*. In flat algorithms, all nodes are functionally equal and run identical processes. In hierarchical algorithms, certain nodes are equipped with special functionality and form a 'backbone' network. Data generated at non-backbone nodes is first routed to a backbone node. From there, data travels the backbone network until it reaches the general area of the destination. At this point, data leave the backbone and travel through regular (non-backbone) nodes to the destination.

11.2 Fundamental SPR

We begin by reviewing SPR for wireline communications networks. We will see that many of the routing concepts encountered for wireline networks will serve as a basis for more advanced implementation in wireless and ad-hoc sensor networks. Standard references include (Bertsekas and Gallager 1992), (Kershenbaum 1993), and (Leon-Garcia and Widjaja 2000). Before proceeding, the reader may wish to consult Section 11.6 for a review of basic graph theory and related terminology.

11.2.1 Broadcast Routing

Before discussing SPR algorithms, we briefly describe broadcast routing. It is often the case that a message has to be *broadcast* from a common origin to all nodes in a network. For example, control and update messages may fit this paradigm. Broadcasting can also be viewed as a simple yet robust form of routing messages from *point-to-point* that may be useful when nodes lack topological information, or (as will be the case in Section 11.2) when links are unreliable. The price paid for this robustness is efficiency. We first study a basic form of broadcasting known as flooding, and then a more energy-efficient variant based on the idea of minimum weight spanning trees. In this section we assume that G is a undirected graph and, for simplicity, that G is connected.

Flooding

The simplest broadcast strategy is flooding. In flooding, the origin node (which may change with time, based on network events) sends its packetized information to its neighbors, i.e., those nodes to which it is connected via an arc. Neighbors then send the message to their neighbors, and so on. It is clear that the message will eventually reach all nodes in G. To reduce unnecessary transmissions and to ensure a finite stopping time two rules are used: (i) a node will not send a packet to a node from which it was received, and (ii) a node will send a packet to each of its neighbors at most once.

Let $L = |\mathcal{A}|$ denote the number of arcs in the network. By condition (i) a node will transmit a message out of all of its arcs except for arcs through which the message was received. Thus, each arc is utilized by some node at least once. From (ii), no link is used more than twice. Thus, the total number of transmissions in flooding is between L and $2L$. In a network, messages may be generated by many different nodes in rapid succession. It is assumed that each message is associated with a unique identifer to avoid the ambiguities when simultaneously routing several messages, and that nodes follow rules (i) and (ii) on a per message basis. An important feature of flooding is that the process is decentralized; it can be carried out if the local nodes follow rules (i) and (ii) independently, and without the use of a central controller.

Minimum weight spanning trees

A broadcasting scheme that strives for a better tradeoff between robustness and efficiency is based on construction of a minimum weight spanning tree (MWST). Suppose that to each arc in the network we assign a real valued link cost. A MWST finds the a set of routes from a source node S to all other nodes that travels the set of links with the minimum sum link cost, and therefore, that floods the network at the minimal total cost. It can be shown that a unique MWST exists if all arc weights are distinct (Bertsekas and Gallager 1992, p. 393), and we will make this assumption in the sequel. The generalization of the algorithms below to non-distinct arc lengths is straightforward, but uniqueness is not guaranteed. Let

$G = (\mathcal{N}, \mathcal{A})$ be a graph with node set $\mathcal{N}$ and arc set $\mathcal{A}$, and assign a weight $w_{ij} \in \mathbb{R}$ to each $(i, j) \in \mathcal{A}$.

Prim-Dijkstra algorithm A centralized method for constructing a MWST $G' = (\mathcal{N}', \mathcal{A}')$ from G that terminates in $|\mathcal{N}| - 1$ iterations is given by the Prim-Dijkstra algorithm. The idea is to construct G' iteratively using a greedy approach. We start by including in G' a single arbitrary node. Then at each step we add to G' the outgoing arc, whose terminating node is not already contained in G', that has the least weight. The algorithm is as follows.

1. Set $\mathcal{I} = 0$. Let $\mathcal{N}' = \{n\}$, where $n \in \mathcal{N}$ is arbitrarily chosen

2. If $\mathcal{I} = |\mathcal{N}| - 1$, stop ($G'$ is a MWST). Otherwise, add to G' the arc which has exactly one node contained in G' and which has minimum outgoing weight, i.e., let

$$(\overline{\gamma}, \overline{\delta}) = \arg \min_{\gamma \in \mathcal{N}', \, \delta \in \mathcal{N} - \mathcal{N}'} w_{\gamma\delta},$$

and let $\mathcal{N}' := \mathcal{N}' \cup \{\overline{\gamma}\}$ and $\mathcal{A}' := \mathcal{A}' \cup (\overline{\gamma}, \overline{\delta})$. Let $\mathcal{I} := \mathcal{I} + 1$, and return to step 2.

While the required number of transmissions for flooding G is between $|\mathcal{N}|$ and $2 |\mathcal{N}|$, the required number for flooding G' is only $|\mathcal{A}| = |\mathcal{N}' - 1| \le |\mathcal{N}|$, as each arc in the spanning tree is traversed exactly once using the flooding operation described in Section 11.2.1.

Distributed algorithms Distributed Algorithms also exist for constructing the MWST. Seminal works include (Spira 1977), (Humblet 1983), (Gallager et al. 1983), and the references therein.

11.2.2 Static Shortest Path Routing

Broadcasting is inefficient when a network is large. In this case SPR is a desirable alternative. We will describe SPR in the remainder of this chapter. We assume directed graphs. We assign to each directed arc (n_i, n_j) in G a real number $d_{i,j}$. By convention, we set $d_{i,j} = \infty$ if an arc does not exist. As in the case of broadcasting, the number $d_{i,j}$ represents the cost of using a particular arc in the network. Let $(n_1, \ldots, n_\ell)$ denote a directed walk. We define the length of the walk to be

$$d_{1,2} + d_{2,3} + \cdots + d_{\ell-1,\ell}.$$

Consider an arbitrary pair of nodes, n_i and n_j. The goal of shortest path routing is to find the minimum length path from n_i to n_j. We assume that G is connected, so that there exists at least one path from every source to every destination. To make the problem well defined and ensure finite stopping times for the algorithms below, we assume that every cycle in the network has a positive length, although certain arcs may have negative costs. Finally, it should be clear that the discussion below can be applied to undirected networks simply by setting $d_{i,j} = d_{j,i}$ for all i, j.

One challenge in implementing SPR is the determination of a link cost that is representative of the routing goals and constraints (see Section 11.4 for common metrics in ad-hoc sensor networks). After this, the protocols described in this section make implementation of SPR fairly straightforward. We consider both centralized and decentralized algorithms.

Centralized implementation

We review two fundamental centralized algorithms for determining shortest path routes below.

The Bellman-Ford algorithm   We describe the centralized Bellman-Ford algorithm (CBF) which finds the shortest paths from every origin node to a particular destination node. To determine all shortest paths in a network we must run the CBF algorithm $|\mathcal{N}|$ times, once for every possible destination node. It will be seen that the CBF algorithm terminates in at most $|\mathcal{N}|$ steps and has a worst-case complexity of $O(|\mathcal{N}|^3)$. One important feature of CBF is that it works when arc length are negative. When arc lengths are positive, Dijkstra's algorithm (see the next subsection) often solves the shortest path problem with fewer computations.

First, we show how to compute the shortest path lengths. Without loss of generality let the destination node be 1. Let D_i denote the length of the shortest walk from node i to node 1, and let D_i^h denote the shortest walk from a node i to node 1 that contains at most $h \geq 0$ arcs. The algorithm for computing D_i is given by the following steps

1. *Initial Conditions*. Set $D_1^h = 0$, $\forall h \in \{0, 1, \ldots\}$, $D_i^0 = \infty$, $\forall i \in \{2, \ldots, |\mathcal{N}|\}$, and set $h = 0$.

2. *Evaluate*.

$$D_i^{h+1} = \min_j \left[d_{i,j} + D_j^h \right], \forall i \neq 1, \tag{11.1}$$

   and let $h := h + 1$.

3. If $D_i^h = D_i^{h-1}$ $\forall i$, stop. Let $D_i = D_i^h$ $\forall i$. Otherwise goto step 2.

The worst case complexity is evaluated as follows: consider a fixed origin node i. The algorithm above terminates after h iterations, where $h \leq |\mathcal{N}|$ since the number of iterations cannot exceed the number of nodes in the network. For each iteration at most $|\mathcal{N}| - 1$ terms are compared to find the minimum. Finally, the process must be carried out for all $|\mathcal{N}|$ potential origin nodes, leading to a *worst-case* complexity of $O(|\mathcal{N}|^3)$.

Suppose that the CBF algorithm above has terminated and the shortest path lengths $\{D_i\}_i$ obtained. The corresponding shortest path routes can be found by observing that, upon termination,

$$D_1 = 0, \text{ and}$$

$$D_i = \min_j \left[d_{i,j} + D_j \right], \tag{11.2}$$

$\forall i \in \{2, \ldots, |\mathcal{N}|\}$. Now, select $|\mathcal{N}| - 1$ arcs from this network as follows: For each $i \neq 1$ select a single arc (i, j) corresponding to $d_{i,j}$ which achieves the minimum in (11.2). The resulting set of arcs is a subgraph G that contains $\mathcal{N} - 1$ arcs and that does not contain any cycles (a consequence of the positive cycle length assumption). Therefore, it is a spanning tree for G. The resulting spanning tree can be verified to specify the shortest path routes from every node to the destination node. The process of determining the shortest path lengths and corresponding routes is illustrated in Figure 11.3.

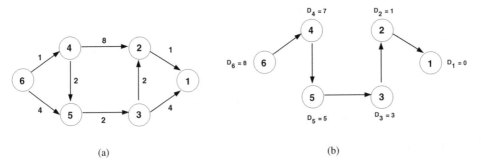

(a)   (b)

Figure 11.3 (a) The network $(\mathcal{N}, \mathcal{A})$ with $\mathcal{N} = \{1, \ldots, 6\}$ and $\mathcal{A} = \{$ (6, 5), (6, 4), (4, 5), (5, 3), (4, 2), (3, 2), (2, 1), (3, 1) $\}$. From this graph, the Bellman-Ford algorithm returns the ordered set of shortest path lengths, $(D_1, \ldots, D_6) = (0, 1, 3, 7, 5, 8)$. (b) The corresponding set of shortest path routes is constructed using the procedure described in the text, yielding this graph.

The CBF algorithm is a centralized approach to the implementation of SPR. This is because the algorithm is most readily implemented by a central controller (e.g., a particular node) that has complete topological information of the network. After computing $\{D_i\}_i$ and the corresponding shortest path routes, the central controller distributes this information to all network nodes, finalizing the implementation of SPR.

Dijkstra's algorithm Suppose that we wish to determine the shortest path routes for a network containing only positive links costs. A well-known algorithm that has a smaller worst-case complexity than CBF is given by Dijkstra's algorithm (DA). We describe a centralized version.

DA is implemented using an iterative algorithm that operates as follows. Suppose that the destination node is 1. Let $D_i, i \in \{1, \ldots, |\mathcal{N}|\}$, denote the shortest path length of the ith closest node to node 1 (in terms of shortest path lengths). We seek to determine $\{D_i\}$. In the first step of the algorithm, we determine the shortest path length of the node closest to 1. In the second step, we determine the shortest path length of the second closest node to 1. In general, in the ith step we determine the shortest path length of the ith closest node to 1. Clearly, the algorithm terminates after exactly $|\mathcal{N}| - 1$ iterations. DA hinges on the following straightforward observation: Suppose that we know $D_1, \ldots, D_K$ and the corresponding nodes, $\bar{n}_1, \ldots, \bar{n}_K$. Then the $(K + 1)$th closest node, $\bar{n}_{K+1}$, has a shortest path to node 1 consisting of either a single arc, directly connected to node 1, or else a route that only passes through nodes in $\{\bar{n}_1, \ldots, \bar{n}_K\}$. (To see this, assume that the shortest path of node $\bar{n}_{K+1}$ passes through a node $v \notin \{\bar{n}_1, \ldots, \bar{n}_K\}$. It follows immediately that v is closer to node 1 than $\bar{n}_{K+1}$ since links costs are positive, which is a contradiction.) Using this observation, node $\bar{n}_{K+1}$ and its shortest path distance D_{K+1} can be found efficiently. Since we know $\bar{n}_1 = 1$ and $D_1 = 0$ initially, we have a recursive algorithm for determining $\{D_i\}_{i \geq 1}$ which is given below.

Let P be a set of 'permanently labeled' nodes, i.e., those which for which we have determined the shortest path distance to node 1. Formally, the DA is:

1. Initialization. Set $P = \{1\}$, $D_1 = 0$, and $D_j = d_{j,1}$ for $j \neq 1$.

2. Find next closest node. Determine $i \notin P$ such that

$$D_i = \min_{j \notin P} D_j.$$

   Add node i to the set of permanently labeled nodes, i.e., $P := P \cup \{i\}$. If P contains all nodes, then stop; the algorithm is complete.

3. Updating of labels. For all $j \notin P$ set

$$D_j := \min_{i \in P} \left[D_j, d_{j,i} + D_i \right] \tag{11.3}$$

   Go to step 2.

Once the algorithm terminates, the $\{D_i\}$ are determined. The shortest path routes corresponding to $\{D_i\}$ can be found via a MWST as described in the last part of Section 11.2.2. To estimate the *worst case* complexity of the DA note that there are $|\mathcal{N}| - 1$ iterations, and that $|\mathcal{N}|$ comparisons are made in each iteration (due to the minimization in step (11.3)). Thus, the worst case complexity is $O(|\mathcal{N}|^2)$.

The DA algorithm is a centralized approach to SPR. This is because the algorithm is most readily implemented by a centralized controller that has complete topological information of the network. After the controller computes shortest path routes, it delivers this information to all network nodes.

Distributed implementations

In many networks there does not exist a central controller to implement the algorithms described above. In this case, distributed shortest path algorithms are needed in which individual nodes exchange messages among each other to determine link costs and optimal routes in a way that is efficient, stable, and robust to link disruptions. First, we describe two broad classes of distributed algorithms, known as Link State Routing and Distance Vector Routing. Then we present a detailed example of a Distance Vector Routing algorithm known as the Distributed Bellman Ford algorithm. Finally, we discuss the effect of link failures and routing loops in distributed algorithms.

Link State Routing (LSR) is a class of distributed shortest path algorithms in which each node is responsible for determining the shortest path route to every possible destination. To do this, each node maintains (estimates of) the full topological information of the network. To keep the network topology up to date, each node periodically broadcasts its entire topological map through the network. The shortest paths are then computed locally at each node using the centralized Dijkstra's algorithm described above. Widely used link state protocols in wireline networks include OSPF (Moy 1998) and IS-IS (Oran 1990).

Distance Vector Routing (DVR) is a class of distributed shortest path algorithms in which each node maintains a list of its (estimated) distance to every destination, and the next node in the path to reach the destination, i.e., (destination, next hop) pairs. DVR algorithms are typically implemented variants of the distributed Bellman-Ford algorithm, described below.

In LSR, nodes require large storage space to maintain full topological tables and CPU power to calculate shortest path routes. Further, large amounts of network traffic are generated by flooding large packets over all the network links. However, convergence is typically faster than for DVR making routing loops (created when different nodes have routing tables of varying degrees of staleness) less common and more short-lived. Importantly, LSR avoids the potentially serious problem known as 'counting to infinity' that occurs in many DVR algorithms (explained below).

The distributed Bellman Ford algorithm   As an example of DVR, we discuss the Distributed Bellman-Ford (DBF) algorithm. DBF is a distributed implementation of the CBF algorithm discussed in Section 11.2.2. Each node in DBF requires only communications with, and knowledge of the link cost of, the neighbors to which it is directly connected. In the following discussion, we assume that all cycles of the network G are of positive length, and that G is an undirected graph. For convenience, we consider the shortest path from all nodes to a single destination node, taken to be node 1 (in practice, the algorithm would be run once for each potential destination node).

Suppose first that the network topology does not change with time. Operation of DBF is summarized in Figure 11.4 and explained below. All nodes send (and receive) update messages at synchronized, regular time epochs, which we take to be $0, T, 2T, \ldots$. In every epoch, each node i generates a ([*next hop*], [*estimated cost*]) pair containing the length of the shortest known path length from node i to node 1 (that node i is currently aware of), as well as next node to which node i must forward data to remain on that path. This message is sent to all of node i's directly connected neighbors. It follows that each node i will also receive similar updates from its neighbors advertising their ([*next hop*], [*estimated cost*])

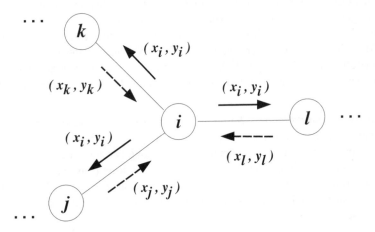

Figure 11.4 Illustration of DBF. Consider the node i in a network. At time epochs, $T, 2T, \ldots$, node i sends its ([*next hop*], [*estimated cost*]) pair (x_i, y_i) to each of the nodes to which it is directly connected; nodes j, k, and l. Similarly, node i receives ([*next hop*], [*estimated cost*]) pairs from each of its direct neighbors; also j, k, and l. From these received packets, node i updates its estimate of the total path length, and next hop to be taken, to reach the destination.

pairs to the destination. From these updates, node i can compute its revised estimated shortest path to node 1, as well as the next node to which it should forward data to remain on the path. This is done using (11.1) from CBF, but with the minimization taken over just those nodes from which it has received updates. This procedure must be repeated in parallel for all $|\mathcal{N}|$ nodes in the network. Therefore, each update message actually consists of a *vector* of $|\mathcal{N}|$ ([*next hop*], [*estimated cost*]) pairs (one for each potential destination node). It can be shown that the DBF algorithm converges to the correct shortest path routes in a finite amount of time. At the termination of DBF, each node in the network maintains a table of consisting of $|\mathcal{N}|$ entries, each of format ([*destination node*], [*next hop*], [*estimated cost*]). A packet following these tables from source to destination will travel the shortest path route.

There are two more facets of DBF that have made it a practically viable strategy for implementing SPR in real-world networks (e.g., DBF was the first routing algorithm used on the ARPANET and has been since used as the basis for algorithms found in many standards). First, it can be proven that DBF converges in finite time to the correct SPR solution even when nodes are free to send update messages in an *asynchronous* manner independently of each other. Therefore, it is not necessary to have update messages transmitted in well-coordinated time epochs, as described above. The price paid for asynchronous operation is the potential for slower convergence of the algorithm to the SPR solution (the convergence rate is highly dependent on the particular order in which the nodes send messages). Second, DBF can be shown to be fairly robust in the presence of a time-varying topology (which may include link failures and discoveries in addition to changing link costs). Specifically, it can be proven that a distributed, asynchronous version of the DBF algorithm presented above converges to the correct SPR solution, in finite time, as long as no new topological changes occur after some time t_0, and as long as the network remains connected.[1] In practice, this means that if the rate of topological change is much less than the rate at which DBF converges, then overall, the network will almost always route data along the correct SPR routes. For an expanded discussion of DBF, we refer the reader to (Steenstrup 1995, Chapter 3), which discusses several variants of DBF designed to ensure faster convergence time, and (Bertsekas and Gallager 1992), which provides a rigorous mathematical proof of the convergence of asynchronous DBF in a connected network that incurs link failures.

Link failures and routing loops   We describe routing loops, a phenomenon that occurs in the distributed algorithms discussed above. A routing loop is a network failure in which packets are routed in a cyclical erroneous pattern among certain nodes rather than arriving at the destination. Such loops may persist for a finite duration (this can occur in both DVR and LSR algorithms) or an infinite duration (this can occur in some DVR algorithms). The onset of such behavior is typically caused by the failure of a link. We will use DBF to illustrate these concepts.

Finite duration   Consider the network shown in Figure 11.5 with node 3 as the destination. The routing table at each node created by DBF is shown in Table 11.1, and explained as follows. After running for some time, DBF converges to the correct solution as shown

---

[1]If the network does not remain connected, then routing loops are possible, and convergence cannot be guaranteed. Routing loops are a general point of concern in distributed algorithms, and are discussed next.

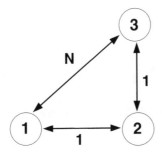

Figure 11.5 A network. It is assumed that the arc $(2, 3)$ fails at some time.

Table 11.1 Routing table for Figure 11.5 when the destination node is 3 and for $N > 2$. The notation (x, y) denotes that the estimated shortest path length to 3 is y and the next hop node is x.

Update	Node 1	Node 2	Node 3
Before link failure	$(2, 2)$	$(3, 1)$	$(-, 0)$
At link failure	$(2, 2)$	$(-, \infty)$	$(-, 0)$
1	$(2, 4)$	$(1, 3)$	$(-, 0)$
2	$(2, 6)$	$(1, 5)$	$(-, 0)$
3	$(2, 8)$	$(1, 7)$	$(-, 0)$
..	..	..	..
N_0	$(3, N)$	$(1, N{+}1)$	$(-, 0)$

in the first row of the table. Now, suppose that arc $(2, 3)$ fails. Consider the following sequence of events: node 2 detects failure and sets its path distance to the destination to ∞ (see the second line of Table 11.1). Next, node 2 receives a periodic update message from node 1 advertising a distance to the destination of 2 (units). Node 2, not realizing this path is based on outdated information in node 1's routing table, updates its own table to state that it can reach the destination in a distance of 3 through node 1 (reflected in the third row of Table 11.1). Node 2 now sends an update to 1 stating a distance to destination of 3. Since node 1 realizes that its shortest hop to the destination is through node 2, it updates its own table to reflect a revised distance to destination of 4. This process continues until, eventually at an update time N_0, the estimated path length exceeds N. At this point node 1 will correctly establish a route to the destination directly through node 3 and, after receiving an update message from node 1, node 2 will follow suit. Such a process can take a long time to converge (depending on the value of N). Meanwhile, packets generated at nodes 1 or 2 bounce between the two nodes. However, assuming no overflow, they will eventually arrive at their intended destination.

Infinite duration – the 'count to infinity' problem As an example of a loop of infinite duration consider the network of Figure 11.6. Initially, DBF is run and the routing table

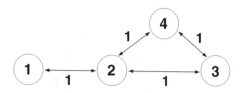

Figure 11.6 A network graph. It is assumed that the arc $(1, 2)$ fails at some time.

Table 11.2 Routing table for Figure 11.6 when the destination node is 3. The notation (x, y) denotes that the estimated shortest path length to 3 is y and the next hop node is x.

Update	Node 1	Node 2	Node 3	Node 4
Before link failure	$(-, 0)$	$(1, 1)$	$(2, 2)$	$(2, 2)$
At link failure	$(-, 0)$	$(-, \infty)$	$(2, 2)$	$(2, 2)$
1	$(-, 0)$	$(4, 3)$	$(2, 4)$	$(2, 4)$
2	$(-, 0)$	$(4, 5)$	$(2, 6)$	$(2, 6)$
3	$(-, 0)$	$(4, 7)$	$(2, 8)$	$(2, 8)$
4	$(-, 0)$	$(4, 9)$	$(2, 10)$	$(2, 10)$
..	..	..	..	..

for node 1 as the destination is given by the first row in Table 11.2. Now suppose that arc $(1, 2)$ fails. Consider the following sequence of events: (i) node 2 observes the failure and updates its routing table, producing the second row of Table 11.2. (ii) Node 4 advertises its (outdated) distance to destination of 2 to the network. This causes node 2 to update its routing table entry to $(4, 3)$. (iii) Node 2 advertises its distance to the network, this causes both nodes 3 and 4 to update their distances to the destination since both nodes use node 2 as the next hop. This produces the third row. Suppose that node 3 advertises its distance. No changes are made to the routing tables at nodes 1 or 2. Now, suppose that this sequence continues, node 4 advertises, followed by node 2 and then by node 3. The full routing table is given by Table 11.2. Packets generated at nodes 2,3, or 4 after the link failure bounce around the network indefinitely, but fail to reach their destination. This phenomenon is known as 'counting to infinity'.

Possible remedies   There are many possible remedies for looping in data networks. For example, modifications to Bellman-Ford that are typically implemented include *triggered updates* (in which any node which detects a link failure immediately sends this information to its neighbors), bounding the maximum allowed distance to the destination, split horizon and split horizon with poisoned reverse (Leon-Garcia and Widjaja 2000, p. 496). Such techniques are useful, but each can be foiled. Further, such schemes require far too much overhead in the form of control messages and loop detection algorithms to be used in networks where link failures are common (e.g., wireless networks).

11.2.3 Adaptive Shortest Path Routing

The static shortest path routing approach described so far lacks certain features of generally optimal routing schemes. It does not take into account the arrival rate of packets at various nodes or the capacity of various links in the network. As a result, it cannot adapt to congestion in network. An approach that partially addresses these issues is given by *adaptive shortest path routing* (ASPR).

The premise of ASPR is to measure the amount of traffic crossing each link in the network during a given time interval, and then to assign a link cost in the next time interval that reflects this measurement. More congested links in a given interval are relabeled with larger link costs in the next interval, while less congested links are relabeled with smaller costs. By increasing the link cost of a congested link, it is less likely to be used in the next interval. Similarly, increasing the link cost of a uncongested link makes it more likely to be used in the next interval. In a properly designed instance of ASPR, congestion is spread out evenly over the network, ensuring that average congestion (and delay) are kept to a minimum.

Although ASPR can be useful in some settings, it may be expected that such a strategy can cause oscillation in the assigned routes, as 'heavily congested' links are rotated around the network with each update. Indeed, ASPR can *increase* the congestion in the network, significantly in some cases! This happens when a cluster of close-by links in the network go unused in a given interval. Sensing this under-utilization, many routes may simultaneously use this set of (previously under-utilized) links in the next interval. Thus, ASPR will result in a strategy where many orutes contend for the same few links in each interval (although the set of links may rotate with time). This problem is especially true when all nodes perform updates at the same set of decision epochs.

To lessen the likelihood of unstable behavior two techniques can be used: (i) nodes can perform link updates out of sync with each other (in which case congestion adaption is done gradually), and/or (ii) a dampening (i.e., constant) factor can be added to link lengths. However, even when ASPR works well, it is missing elements of a truly robust protocol: (i) it fails to formally address the feedback effect of the current routes on congestion levels, and (ii) link capacity and the delay of a link as a function of its congestion level are not explicit parts of the formulation. A survey of ASPR routing issues and literature is given in (McDonald 1997). Theoretical justification for the near-optimality of ASPR routing in certain types of networks is given in (Gafni and Bertsekas 1987).

11.2.4 Other Approaches

We briefly describe two other popular classes of routing algorithms that do not fall within the shortest path framework. In *flow model routing* (Kleinrock 1964), (Bertsekas and Gallager 1992, Sections 5.4-5.7), models for link delay as a function of link congestion (traffic level) and capacity are used. The feedback effect of a particular route assignment on other routes is taken into account and packets from a given origin-destination pair may be split among several different paths in order to minimize the routing metric (which seeks to capture the average delay of packets in the network). Although such a characterization is valuable, solutions to the 'optimal' routing problem under this formulation are not as easily implementable as in SPR, especially for large networks.

In *deflection routing* (Steenstrup 1995, Chapter 9), the source node forwards a packet out a preferred link. However, if the link is congested or full, the packet is immediately rerouted (or deflected) to another preassigned link. The process continues as the packet is routed to the destination. Deflection routing is designed for scenarios where nodes have small or nonexistent buffers. Upon receiving a packet either forwards it or immediately deflects it to a free link (thus, the original name of 'hot potato' routing). This scheme tends to work well for networks with balanced and invariant topologies.

11.3   SPR for Mobile Wireless Networks

Routing for mobile wireless networks is a broad area. Overview papers and chapters can be found in (Akkaya and Younis 2005), (Al-Karaki and Kamal 2004), (Marina and Das 2005), and (Kowalik and Davis 2006). SPR seems well adapted to the wireless problem by providing both a mathematical basis for routing algorithms and an implementable structure. However, wireless networks present new challenges not accounted for in the development of Section 11.2. First, routing algorithms in wireless networks must be energy-efficient, and as a related issue, they should prolong the system lifetime. Second, routing algorithms must contain a medium access control (MAC) that limits the interference level at non-participatory nodes caused by transmissions at nearby nodes. This is particularly important when many nodes route unrelated data simultaneously in the network. Third, wireless networks may have a rapidly time-varying topology due to mobility and/or the time-varying nature of the communications channel. Thus, routing algorithms must be able to handle scenarios where links may fail and new links may be formed.

In parallel to Section 11.2, we first discuss broadcasting approaches before describing SPR techniques. For SPR, it will be seen that there are two fundamentally different views, known as proactive and reactive routing, on how to route messages in wireless networks.

11.3.1   Broadcast Methods

Pure flooding is an inappropriate routing strategy for large wireless networks due to the excessive interference levels that it generates, and due to the vast amounts of energy consumed by redundant and useless transmissions. We start by surveying refined flooding methods designed for the wireless environment. These methods seek to eliminate some of the unnecessary transmissions of pure flooding.

Suppose our goal is to reach all nodes connected to the source with a minimum sum cost (which, for simplicity, we take to be power). A seemingly reasonable approach would be to apply the MWST theory and algorithms exactly as discussed for wireline networks in Section 11.2.1. However, the MWST approach ignores the multiaccess nature of wireless communications. When wireless packets are transmitted, many nodes hear the transmission. Therefore, a transmission that requires more energy to connect the source to a particular node, say D, may also simultaneously connect several destination nodes to the source, and cannot be ruled out a priori, as is the case for MWST. In (Wieselthier et al. 2000), the authors propose the Broadcast Incremental Power Algorithm (BIP) as a MWST-inspired algorithm that captures this key feature of the wireless medium. BIP is an iterative algorithm. Suppose that some subset of the final broadcast tree has been determined. The main idea of BIP is to now add a single new node by defining link costs based on the *additional* power required to connect each possible new node to the broadcast tree. Such a definition of link costs is adaptive, changing after each iteration of the algorithm. However, standard MWST algorithms can be run at each iteration. Further analysis of BIP and variants can be found in (Wan et al. 2002; Wan et al. 2004), and the references therein.

A different, probabilistic, approach to efficient flooding is Gossiping (Haas et al. 2002). Gossiping operates as flooding except that when a node would normally relay a message in flooding, it relays it only with probability p in gossiping. Using arguments from percolation theory, it is shown that such a strategy can have a high probability of successfully connecting source and destination in large networks, and that such strategies, with some straightforward modifications, adapt well to smaller networks as well.

A still more efficient routing algorithm was derived using the theory of random walks in (Servetto and Barrenechea 2002). The proposed routing scheme is similar to that of gossiping except that instead of forwarding a message to all neighbors with probability p, a node forwards messages only to neighbors closer to the destination than itself according to some probability distribution on those neighbors (unlike gossiping, each message is forwarded to exactly one neighbor with probability 1). Compared to flooding (and gossiping) this method concentrates packet forwarding on short routes between source and destination, and secondly, it evenly balances network traffic on all nodes towards the destination. For a standard rectangular network topology closed form expressions can be derived for the optimal forwarding probabilities to evenly balance network traffic. It is shown that the random walk strategy results in a much better load distribution than gossiping. Also, such strategies perform well even when the topology is time varying.

Two more intelligent flooding strategies were presented in (Ni et al. 1999) and (Kozat et al. 2001). In (Ni et al. 1999), a delayed-flooding approach is proposed where, upon receipt of a flooded message, a node n_0 waits idly for a certain amount of time, T. If n_0 hears its neighbors rebroadcast the message more than N times in this interval, then it chooses *not* to rebroadcast. Clearly, the parameters T and N can be optimized for a given network topology. In (Kozat et al. 2001) a set of forwarding nodes are identified (at low overhead) that are used to flood messages. Clearly, these schemes result in fewer transmissions than pure flooding.

11.3.2 Shortest Path Routing

Shortest Path routing remains popular for wireless networks. However, implementations of SPR have to address the new features of wireless networks discussed above. A central question is: does a node need to store routing information to all possible destinations nodes, or only those likely to be used in the near future? Two broad classes of protocols have been designed to address this issue, as we discuss below.

In *proactive protocols*, each network node stores information on routes to all possible destinations at all times. The advantage of this approach is that when a routing request is needed, it can be initiated with minimal delay. The disadvantage is the overhead in the form of control messages that are required to establish and maintain all routes, even those that are used infrequently and/or subject to frequent failure. Extra control messages consume battery power and cause interference, and therefore reduce system lifetime.

In *reactive protocols*, routes are established on an as-needed basis (these protocols are also known as on-demand protocols). This approach requires less control information, particularly when the network topology is rapidly changing or when the data bursty or directed to only a small subset of the network nodes that are unknown in advance. The disadvantage of such an approach is that route requests can experience a large delay before being propagated from the source node, as a route to the destination has to first be determined.

Proactive routing

In proactive routing each node maintains routes between itself and all possible destinations in the network. Since this is the same approach as taken in the wireline case, many proactive algorithms are modifications of the distance vector and link state approaches described in Section 11.2.2. Below, we describe a well-known example of each type of protocol.

Distance vector protocols   Destination Sequenced Distance Vector (DSDV) (Perkins and Bhagwat 1994) was among the first DVR protocols for wireless networks. Based on the idea of DBF, the main feature of DSDV is that it avoids the count to infinity problem illustrated in Section 11.2.2, while requiring minimal internodal coordination or overhead (Perkins and Royer 1999). It can best be understood as a generalization of DBF. In DBF, each node keeps a triplet of information items for each destination, (D, DI, N), where D is the destination node, DI is the (estimated) shortest path distance to D, and N is the next hop node towards D. In DSDV, each node maintains a quadruple, (D, DI, N, SEQ) where D, DI, and N are as before, and SEQ is monotonically increasing sequence number issued by the *destination* node as part of its periodically issued status message. Whenever an intermediate node chooses to update its own routing table based on such a status message it also stores the original sequence number issued by the destination. Further, an intermediate node chooses to incorporate the update message only if at least one of the following conditions holds: (i) the intermediate node does not already have a table entry for the destination, (ii) the update message contains a newer sequence number than the currently stored information, or (iii) the sequence number is the same as in the table, but the update message contains a shorter route. The sequence number allows each intermediary node to know at what relative time an update message was issued by the destination. This is the mechanism by which the count to infinity problem is avoided. Other popular distance vector protocols for wireless networks include WIRP (Garcia-Luna-Aceves et al. 1997), ADV (Boppana and Konduru 2001), WRP (Murthy and Garcia-Luna-Aceves 1996), and PRNET (Jubin and Tornow 1987).

Link state protocols   For implementation in a wireless network, the size and frequency of link states updates (LSUs) must be reduced relative to that of traditional algorithms such as OSPF (see Section 11.2.2). Optimized Link State Routing (OLSR) (Clausen et al. 2003) is a modified version of OSPF that seeks to accomplish this goal. In OLSR a certain subset of nodes in the network are designated 'multi-point relays'. These nodes are responsible for forwarding link state updates, which contain information on only some of the network nodes. Clearly, the number and size of LSUs have been reduced. In this scheme each node has only partial topological information of the network. However, if correctly designed, this information is still sufficient for each node to recreate the entire topology graph. From this information, each node can determine shortest path routes via Dijkstra's algorithm. A mathematically framed comparison of the efficiency of OLSR to OSPF can be found in (Adjih et al. 2003).

Reactive (on-demand) routing

In reactive routing, routes are established on an as-needed basis. One of the earliest forms was proposed in (Corson and Ephremides 1995). Here, we discuss an archetypical protocol known as Dynamic Source Routing (DSR) (Johnson and Maltz 1996), (Perkins 2001, Chapter 5). DSR makes heavy use of two concepts: (i) source routing, in which the origin node knows the entire route to destination and embeds this information into the packet (in this way, intermediary nodes forward packets without having to make their own routing decisions), and (ii) route caches, in which each node maintains a list of routes to those destinations which it has 'discovered'. The protocol itself has two modes of operation *route discovery* and *route maintenance*, which we now describe.

Route discovery is used when a source node needs to forward a packet to a particular destination, but does not have a corresponding route stored in its cache. In this case the source node floods a query packet through the network. Intermediary nodes append their own identity to the query and forward it, *unless* they are either the destination node, or have a route for the destination stored in their cache. Nodes which do not forward the query send a reply message back to the source. (This is also done using source routing, which is possible since every packet traversing the network has embedded its route traveled so far; that is, the route for the reply packet is simply reversed.) When the reply arrives back the source, this newly 'discovered' route is stored in the source's cache and routing can commence.

Route maintenance is used when a previously discovered route is traversed and it is learned that there has been a link failure. In this case an error packet is generated and, through source routing, is sent back towards the source along the same route that it originally traversed. At each intermediate node along the route and at the source, the corresponding entry is *erased* from the cache (thus, the route is appropriately purged). In order to reach the destination the source must either use an alternate route also stored in its cache, or initiate a new route discovery.

One drawback of using source routing is that DSR suffers from poor scalability. As the network becomes larger, so does the size of the embedded route information in each packet. Secondly, DSR does not have a preemptive way of purging outdated routes. Because DSR relies on advertising mechanisms, incorrect information tends to propagate throughout the network. It was shown in (Perkins et al. 2001) that such overhead can have a significant impact on the performance of DSR. Several variants of DSR have been proposed to address these issues. Other well-known reactive protocols include Ad hoc On-demand Distance Vector (AODV) (Perkins and Royer 1999), the Temporally Ordered Routing Algorithm (TORA) (Park and Corson 1997), and Signal Stability Routing (SSR) (Dube et al. 1997).

11.3.3 Other Approaches

We briefly mention some other classes of routing algorithms that are widely cited for wireless networks, but that may not fit within the shortest path framework. The first approach is known as a *Hybrid Routing*. Routing algorithms in this class incorporate elements of both proactive and reactive protocols. A well-known example of such a protocol is the Zone Routing Protocol (ZRP) (Haas et al. 2002). A second approach is known as *Location Based Routing* (also as Geographic Routing), in which each node knows the physical location of other nodes in the nodes in the network (e.g., through Global Positioning System (GPS) data). When a node wants to send information to a given destination it simply forwards its information to the neighbor closest to the destination. The process continues until the message reaches the destination. Note that once locations for all nodes have been obtained, this schemes requires very little overhead, and it also has the potential to scale well. Examples include (Witt and Turau 2005) and (Basagni et al. 1998).

Backpressure routing and its variants are based on work originally proposed in (Tassiulas and Ephremides 1992). The idea is to choose the routing algorithm to maximize the stability region of the network (where stability is determined in terms of permissible arrival rates of data to network nodes). The optimal routing policy was found, and was seen to select a particular flow at the source node, and route it to the particular recipient node, in a way that maximizes the differential backlog. Although not originally proposed as an SPR algorithm, extensions have been made which incorporate SPR and other routing techniques (e.g., see (Neely et al. 2005)).

11.4 SPR for Ad-Hoc Sensor Networks

We now depart significantly from the types of networks considered thus far by focusing on ad-hoc wireless sensor networks. We will continue to use the shortest path methodology for such networks. First, we discuss the novel features of ad-hoc sensor networks and survey current SPR algorithms in Section 11.4.1. We show that the current approaches are largely application independent. In Section 11.4.2 we argue that, in contrast to previous networks, application dependent design of routing protocols is both appropriate and necessary. To address this issue, we present a detailed example of how application dependent SPR can be derived and implemented to improve routing performance in sensor networks in Section 11.4.3.

11.4.1 A Short Survey of Current Protocols

We start with a brief summary of routing protocols that were developed, or that are appropriate, for sensor networks. Like conventional wireless networks, ad-hoc wireless sensor networks are severely energy constrained, and experience an unpredictable and time-varying topology (due to low complexity, low power sensor nodes that may be prone to failure, and/or a mobile deployment environment). Not all of the works surveyed below are defined purely as shortest path algorithms. However, common to all is the choice of a link metric that measures the goodness of a given route, the minimization of which is assumed to provide improved application performance given these constraints. There is a plethora of works which are based upon the ideas presented here. We have omitted these for brevity.

It is important to realize that the routing algorithms discussed below are designed based on, at best, a vague notion of what constitutes 'good' application performance (less delay, more throughput, etc.). In no case is the true application performance metric (e.g., minimization of the bit error rate, minimization of estimation error, maximization of the detection performance, etc.) stated mathematically as a function of an arbitrarily chosen route. Thus, there is no convenient way to maximize application performance over all routes satisfying the relevant network constraints. We will refer to these approaches as application independent.

Two immediate strategies for routing in sensor networks are *minimum energy routing* (Rodoplu and Meng 1998), (Meng and Rodoplu 1998), and (Ettus 1998), in which the link metric is the energy required to send a wireless transmission over the link, and *minimum hop routing*, in which the link metric is constant for all links. However, neither approach is well suited to sensor networks. The first approach conserves energy, and therefore, is often thought to prolong network lifetime. However, such a scheme actually depletes energy in certain frequently used routes. This leads to premature link failure and network partitioning. The end result is reduced system lifetime compared to schemes which attempt to homogenize energy use throughout the network, or which take into account residual battery life. Similarly, minimum hop routing is often thought to minimize the end-to-end routing delay in a network. In reality, such an approach also tends to use the same subset of links to transmit the network's traffic. This leads to congested routes which in turn lead to relatively long routing delays. Again, strategies that homogenize traffic throughout the network will have better delay characteristics. Some more sophisticated approaches follow.

First, we describe approaches which address the system lifetime constraint. In (Shah and Rabaey 2002), the premature burnout of 'preferred routes' is addressed by determining the set of N routes in the network with the least energy usage, and then using each of these routes with a frequency (or probability) that is related to the residual lifetime of the nodes in the route. In (Barrett et al. 2003), a protocol is proposed that forwards each message probabilistically to a single neighbor, where the probability is chosen as a function of the number of shortest path routes to the destination on which the neighbor lies. Another approach to increasing lifetime is given in (Singh et al. 1998), where a link metric is proposed that is an arbitrary function of the energy expended by the transmitting node thus far (clearly related to the residual battery power in the node). A similar residual battery life metric is used in (Chang and Tassiulas 1999), and it is seen that an optimal approach evenly distributes power consumption throughout the network. Another residual battery metric was proposed in (Toh 2001). Specifically, *conditional max-min battery capacity* routing was proposed in which a shortest path route to the destination is selected subject to the constraint that all nodes in the path have a battery power above a certain threshold. In both (Michail and Ephremides 2000) and (Aslam et al. 2003), link metrics are proposed that balance choosing a route that entails minimum energy consumption with choosing a route that has sufficiently high residual energy.

Second, we cite approaches which address the unpredictable and time-varying topology of sensor networks. In (Gerharz et al. 2002) and (Gerharz et al. 2003), several link metrics are proposed which encourage the formation of routes comprised only of 'stable' links (those which are less likely to fluctuate in an out of service). In (De Couto *et al.* 2003), stable links are chosen by using shortest path routing with a link metric that measures the expected number of transmissions needed to traverse a link (this number generally exceeds one due to the probabilistic unreliability of a link). In (Lundgren et al. 2002), (Chin et al. 2002), and (Dube et al. 1997), the time-varying topology is addressed by omitting from consideration links whose reliability falls below a certain threshold. SPR can then be run on the remaining network to find routes (if such a network remains connected).

11.4.2   An Argument for Application Dependent Design

The layered architecture used in most data networks dictates that routing (at the network layer) be designed independently of the application performance (at application layer). Such an approach provides modularity at the expense of efficiency. This paradigm has been reasonably successful in wireline and some wireless networks. Modularity simplifies design and implementation via a divide and conquer approach. As a result, new applications can be developed independently of the medium access and routing strategies that are specified at lower layers, and new physical layer techniques can be implemented without changing upper layer implementations. Perhaps the most remarkable feature of the layered architecture is that it makes the network design scalable. A classic example is the Internet, which has grown from a handful of nodes in the ARPAnet to hundreds of millions of nodes today.

However, not all networks should be constrained to fit this paradigm. Ad-hoc sensor networks are designed for specific applications. In contrast to the Internet, i.e., a general-purpose traffic carrying network, sensor networks do not exist to serve individual nodes. Consider, for example, a sensor network deployed for target detection and tracking, environmental monitoring, or the detection of a specific chemical compound. In these applications, network performance should not be measured by general purpose metrics such as the data rate at the link level or by the throughput over the network. Conventional performance metrics such as throughput and delay do not necessarily translate to a performance measure suitable for the application. For application specific networks, and sensor networks in particular, performance should be measured instead by application defined metrics such as the miss detection and false alarm rates, the network lifetime for performing these tasks, and the energy efficiency of target detection, tracking, and estimation.

11.4.3   Application Dependent SPR: An Illustrative Example

The basic challenge in designing application specific routing using the SPR methodology is the determination of a link metric that captures the application performance (in such a way that minimization of the sum of link metrics is equivalent to maximization of application performance), while also accounting for network constraints. To show that such a link metric can indeed be derived in some cases, we provide an illustrative example in which we seek to maximize the detection performance of an ad hoc network subject to a constraint on energy consumption.

The development in this section assumes that the reader has a basic knowledge of binary hypothesis testing and detection theory (Poor 1994). The reader unfamiliar with these concepts may skip Section 11.4.3, making note of the application performance measure (11.7) and energy consumption expression (11.8), without loss of continuity. The ideas presented in this section are based on (Sung et al. 2006) and (Sung et al. 2007).

Shown in Figure 11.7 is a large network with geographically distributed sensors, each taking measurements of a certain phenomenon. We assume that there is a fusion center (or gateway node) that is responsible for collecting data from sensors and drawing inferences from that data. We are interested in the detection of a spatially correlated Gaussian random signal field. The two hypotheses are

$$
\begin{cases}
\mathcal{H}_0 & : \quad \text{independent and identically distributed (IID) Gaussian noise,} \\
\mathcal{H}_1 & : \quad \text{correlated Gaussian random field observed in IID Gaussian noise.}
\end{cases}
$$

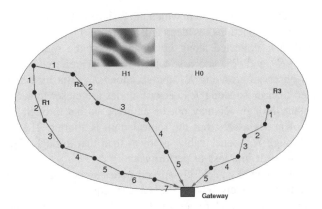

Figure 11.7 Routing for a detection application. Reprinted with permission from (Sung et al. 2007). © 2007 IEEE.

We assume that the sensors have already been placed in an arbitrary configuration (consistent with the notion of an ad-hoc network) and that the correlation structure of the Gaussian signal field is known. Later we will impose a specific structure on the correlation to facilitate aspects of the analysis.

Sensors have limited transmission range, and they have to deliver their data to the fusion center over certain routes. Obviously, the more data that is collected, the more accurate the inferences drawn about the phenomenon at the fusion center. What compounds the problem is the severe energy constraint on sensor communications. Each transmission will cost a certain amount of energy which will depend on the distance between the transmitter and the receiver and also on the amount of data that needs to be delivered. It is this tradeoff between performance and energy consumption that demands application dependent design.

There are numerous network-related problems associated with this application which are beyond the scope of this chapter. For example, the effect of transmission interference and the design of a carefully constructed MAC and PHY layer would be fundamental to a real-world implementation of the proposed design, but are ignored. Here, we are interested in determining the *route* over which data collection should be performed. The problem would not be interesting if, under hypothesis $\mathcal{H}_1$, the sensor measurements of the signal field were independent, i.e., the conditionally IID model. In this case, each sensor provides equally valuable information conditioned on the observations of previously sampled sensors, and the 'optimal' routing strategy would be to simply collect data from sensors closest to the fusion center, in order to save transmission energy. However, routing becomes a non-trivial issue when sensor measurements are not IID under $\mathcal{H}_1$. In this case, sensors further away from previously sampled sensors may have more valuable observations of the signal field (it may be that less correlated observations are more valuable). However, this data can only be collected at the expense of using more transmission energy. Hence, a tradeoff emerges between detection performance and energy consumption. Furthermore, the measurements of sensor nodes are imperfect, and it is necessary to consider this measurement inaccuracy and aggregate the measurements of multiple sensors for the final decision.

Consider a clock-driven application in which the fusion center issues regularly scheduled (possibly random) data collections, as depicted in Figure 11.7. For any fixed route, we will assume that data is collected along *all* nodes along the route. We highlight three potential routes $\mathcal{R}_1, \mathcal{R}_2$ and $\mathcal{R}_3$, and ask, which route is preferred for detection? There are eight potential observations along $\mathcal{R}_1$ and six along $\mathcal{R}_2$. But measurements along $\mathcal{R}_1$ are more correlated than those along $\mathcal{R}_2$ because nodes are closer to each other. Thus the 'information' content through $\mathcal{R}_1$ may not be as great as that through $\mathcal{R}_2$. Now $\mathcal{R}_3$ has the same number of nodes as $\mathcal{R}_2$, but the route length is shorter. The energy consumed in the collection through $\mathcal{R}_3$, conceivably, is lower than that through $\mathcal{R}_2$. But the limited coverage of $\mathcal{R}_3$ may result in a loss of performance.

Intuitive concepts alone, e.g., 'closely-spaced nodes sometimes provide less information' and 'collections over widely separated distances require more energy', will not carry us far in determining the optimal routes for the tradeoffs described above. To develop optimized application dependent routing, we will need an analytical characterization of performance which can lead to the right tradeoff between detection performance and energy consumption.

Chernoff routing, Schweppe's recursion, and Kalman aggregation

Unfortunately, there is no simple analytic expression that describes detection performance along a given route or the contribution of a given link to the overall detection performance. However, using some well-motivated approximations, we show that there does indeed exist a link metric in our case; one that quantifies the contribution of each link to detection performance in such a way that the sum of link metrics along a route is proportional to the detection performance of the route. Thus, routing can be greatly simplified using SPR techniques. Our tool is the use of detection performance bounds that are functions of network parameters such as the distance between a pair of nodes and signal parameters such as signal-to-noise ratio (SNR) and signal field correlation strength.

Chernoff routing

To find a suitable bound, we digress briefly into the theory of large deviations (Dembo and Zeitouni 1993). The Chernoff bound (Poor 1994) is a well known upper bound on the probability of detection error. Consider the simple binary hypotheses $\mathcal{H}_i : Y_k \sim p_i(y), i \in \{0, 1\}$, for $k = 1, 2, \ldots, n$. Define $Y \overset{\Delta}{=} [Y_1, \ldots Y_n]$ and let $P_i(Y)$ denote the probability distribution of Y under $\mathcal{H}_i, i \in \{0, 1\}$. The optimal detector for this test (under either the Bayesian or Neyman-Pearson formulation) compares the log-likelihood ratio

$$l(Y) \overset{\Delta}{=} \log \frac{P_1(Y)}{P_0(Y)} \underset{<\mathcal{H}_0}{\overset{\geq \mathcal{H}_1}{}} \tau$$

to an appropriate threshold value τ. The false alarm probability can be upper bounded by

$$\Pr(l(Y) > \tau | \mathcal{H}_0) < \exp\{-\Lambda(\tau)\}$$

where the so-called error exponent $\Lambda(\tau)$ is the Fenchel-Legendre transform of the cumulant generating function $\mu(s) \overset{\Delta}{=} \log \mathbb{E}\{e^{sl(Y)} | \mathcal{H}_0)\}$:

$$\Lambda(\tau) \overset{\Delta}{=} \sup_{s>0}\{s\tau - \mu(s)\}.$$

When we have IID measurements, the error exponent $\Lambda(\tau)$ and the log-likelihood ratio are additive, namely,

$$l(Y_1, \ldots, Y_n) = \sum_k l(Y_k),$$

and the corresponding Chernoff bound on the false-alarm probability has the form

$$\Pr(l(Y) > \tau | \mathcal{H}_0) < \exp\{-n\Lambda_1(\tau)\},$$

where $\Lambda_1(\tau)$ is the Fenchel-Legendre transform of $\log \mathbb{E}\{e^{s\,l(Y_1)} | \mathcal{H}_0)\}$. It is this additivity that makes it possible to obtain a link metric that captures the desired performance measure in our case. Additionally, the Chernoff bound is tight when n is large.

We have a similar lower bound on the false alarm probability that states

$$\Pr(l(Y_1, \ldots, Y_n) > \tau | \mathcal{H}_0) = \exp\{-n\Lambda_1(\tau) + o(n)\},$$

where $o(n)$ is such that $\lim_{n \to \infty} o(n)/n = 0$. Here, we interpret $\Lambda_1(\tau)$ as the decay rate of the false alarm probability. In fact, $\Lambda_1(\tau)$ can be shown to be the largest possible decay rate

$$\Lambda_1(\tau) = \lim_{n \to \infty} \frac{1}{n} \log \Pr(l(Y_1, \ldots, Y_n) > \tau | \mathcal{H}_0).$$

Under the Bayesian setup, the two types of detection-related error probabilities, the probability of false alarm and miss detection, are balanced by the priors of the two hypotheses. However, the largest decay rate for the average error probability, $P_e = \Pr(\mathcal{H}_0) \Pr(\text{Error} | \mathcal{H}_0) + \Pr(\mathcal{H}_1) \Pr(\text{Error} | \mathcal{H}_1)$, does not depend on prior probabilities, and is given by the Chernoff information defined as

$$C \triangleq \Lambda_1(0) = \sup_{s > 0} \{-\mu(s)\}. \tag{11.4}$$

This concludes our digression.

By Chernoff routing we mean routing where the Chernoff information is used as a route metric. For a fixed route, say $\mathcal{R}_1$ in Figure 11.7, we have a set of measurements $\{y_i\}$. Note, however, that y_i's are not IID, the Chernoff information for such a case is a function of the distribution of $l(y_1, \ldots, y_n)$, which, in turn, is a function of the route. Denoting the Chernoff information associated with a specific route $\mathcal{R}$ as $C(\mathcal{R})$, Chernoff routing aims to select a route that maximizes $C(\mathcal{R})$. Denoting by $E(\mathcal{R})$ the energy consumed when data are routed through route $\mathcal{R}$, we obtain an energy constrained form of Chernoff routing

$$\max_{\mathcal{R}} C(\mathcal{R}) \quad \text{subject to} \quad E(\mathcal{R}) \leq \mathcal{E}. \tag{11.5}$$

Link metric via innovations representation Although (11.5) captures the essence of optimal routing subject to an energy constraint, it does not provide an implementable routing protocol. To use SPR, we seek an additive link metric such that the accumulated value of the link costs on $\mathcal{R}$ is proportional to the value of $C(\mathcal{R})$. Unfortunately, the standard expression of the Chernoff information for the Gaussian hypotheses is given in terms of the eigenvalues of the covariance matrix of signal samples (Poor 1994), and does not allow the decomposition of the overall performance into a sum of the incremental performance gains at each link.

The key to obtaining an additive link metric, as proposed in (Sung et al. 2005), is the use of the innovations representation of the log-likelihood function (Schweppe 1965). To understand this important step, we note that the Chernoff information associated with y_i is not additive because the log-likelihood function under $\mathcal{H}_1$ is not additive. We seek innovations that are naturally independent. In the context of signal processing, this can often be achieved using recursive techniques. The idea of using the innovations representation to obtain the likelihood function recursively was first proposed by Schweppe (Schweppe 1965), and Schweppe's recursion leads to the decomposition of Chernoff information into an additive link metric.

For a fixed route $\mathcal{R}$, assuming a Gaussian signal along the route, it is shown in (Sung et al. 2005) that the Chernoff information $C(\mathcal{R})$ is approximately equal to the sum of the *logarithm of the innovations variance $R_{e,i}$* (normalized by the measurement noise variance) at each link, i.e.,

$$C(\mathcal{R}) \approx \sum_i C_i, \quad C_i = \frac{1}{2} \log \frac{R_{e,i}}{\sigma_w^2}, \tag{11.6}$$

at high SNR, where SNR is defined as the observational SNR at each sensor, σ_w^2 is the variance of measurement noise at each sensor, $\hat{y}_{i|i-1} \triangleq \mathbb{E}\{y_i|y_0, \ldots, y_{i-1}\}$ is the MMSE estimate of y_i given all upstream measurements, and $R_{e,i} \triangleq \mathbb{E}\{|y_i - \hat{y}_{i|i-1}|^2\}$ is the MMSE of the estimation process. When the random process is Markovian, the link metric is almost memoryless. This crucial property makes it possible to use shortest path routing.

Next, we need to connect (11.6) to physical parameters such as the SNR, node spacing, and field correlation. For the Gauss-Markov random field, we have the following approximation (Sung et al. 2005)

$$C_i \approx \frac{1}{2} \log\{\text{SNR} + 1 - (\text{SNR} - 1)e^{-2A\Delta_i}\}, \tag{11.7}$$

where $\Delta_i > 0$ is the link length, and where $A > 0$ describes the correlation strength and is the diffusion constant of the first order stochastic differential equation of the Gauss-Markov model. Note that as $A \to \infty$, the sensor observations approach statistical independence, and that as $A \to 0$, they become fully correlated.

A numerical evaluation of C_i as a function of link length Δ_i provides useful insights. Figure 11.8(a) shows the link metric as a function of link length Δ_i. For SNR ≥ 1, the metric is strictly increasing, strictly concave, bounded from above, and achieves a maximum value of $\frac{1}{2} \log(1 + \text{SNR})$. Thus, this value represents the maximum information that a link can provide, and it is attained if the two sensors at each end of the link have independent observations of the signal field (such may be the case if the sensor are spaced far enough apart, or if the field is sufficiently weak in correlation).

We can now bring energy consumption into the framework. The energy used by a particular node en route can represented by the sum of the processing energy $E_p \geq 0$ and transmission energy to the next node (e.g., see (Ephremides 2002)), i.e.,

$$E_i = E_p + E_{t,0}\Delta_i^{\nu}, \quad \nu \geq 2. \tag{11.8}$$

where $E_{t,0} \geq 0$ is a constant. Thus, the link *efficiency* can be defined as

$$\eta \triangleq \frac{C_i}{E_i}.$$

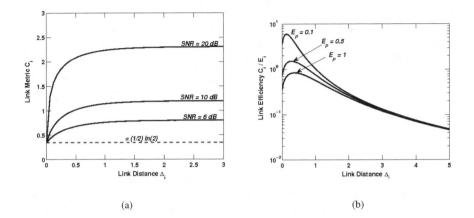

(a)   (b)

Figure 11.8 Detection-based link metric: (a) link metric C_i as a function of link length ($A = 1$) and (b) link efficiency as a function of length link ($A = 1$, SNR $= 10$ dB, $v = 2$, and $E_{t,0} = 1$). Reprinted with permission from (Sung et al. 2007). © 2007 IEEE.

Now, the tradeoff between having large C_i and low E_i becomes clear. Figure 11.8(b) shows the detection efficiency for several values of processing energy E_p at each sensor. The transmission energy at each link increases without bound as the link length increases. However, note that the link efficiency peaks before decreasing with increasing link length. Hence, we conclude that there is an optimal link length for optimal detection efficiency.

Definition of a link metric

We now define a link metric for shortest path routing which uses a modified version of C_i. We propose to balance the detection performance with the energy consumed through the following link metric

$$\gamma_{i,j} = \begin{cases} (E_i - \lambda C_i)_\epsilon^+ & \text{if nodes } i \text{ and } j \text{ are connected,} \\ \infty & \text{otherwise,} \end{cases}$$

where

$$(x)_\epsilon^+ \triangleq \begin{cases} x & x > 0, \\ \epsilon & x \leq 0, \end{cases}$$

and $\epsilon > 0$ is a constant. Note that the cases $\lambda \to \infty$ and $\lambda \to 0$ correspond to the minimum-hop and minimum-energy routing strategies, respectively.

We are now ready to provide numerical insights by considering a sensor network with 100 sensors placed on a circular field with radius one. Figure 11.9 shows the shortest-path route from each node to the fusion center, which is located at the center of the field and denoted by S, for each of the following routing strategies: minimum-hop routing, minimum-energy routing, and Chernoff-routing[2] (i.e., shortest path routing based on the metric (11.9)

---

[2]For simplicity, the Gauss-Markov model with diffusion constant A is used to describe the signal evolution along the route.

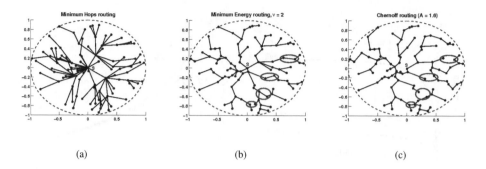

(a)   (b)   (c)

Figure 11.9 The shortest-path route from each node to the fusion center S for three different routing strategies: (a) Minimum-hop routing, (b) Minimum-energy routing, and (c) Chernoff-routing. Parameters are $A = 1.6$, $\lambda = 0.01$, SNR = 15 dB, $\nu = 2$, $N = 100$, $E_{t,0} = 1$, and $E_p = 0$. Reprinted with permission from (Sung et al. 2007). © 2007 IEEE.

with a nontrivial value of λ, i.e., $0 < \lambda < \infty$). The differences in the route topology for these schemes is evident. While minimum-hop routing results in a few, large, well-directed hops to the fusion center, minimum-energy and Chernoff routing take smaller and more scattered steps. This is because the transmission energy is a convex increasing function of link length. The nodes which lead to major topological differences between the two latter strategies are circled in the figure. As expected, Chernoff routing produces routes which deviate from those of minimum-energy routing, and for which the detection performance is presumably better.

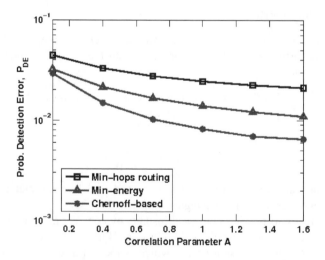

Figure 11.10 Probability of detection error P_{DE} averaged over all routes for the network topology of Figure 11.9 under three different routing strategies ($\Pr(\mathcal{H}_0) = 0.75$ and all other parameters are the same as in Figure 11.9). Reprinted with permission from (Sung et al. 2007) © 2007 IEEE.

Figure 11.10 is a plot of the probability of detection error P_{DE} for the topology shown in the previous figure averaged over all N routes (i.e., $P_{DE} = \frac{1}{N} \sum_{n=1}^{N} P_e(\mathcal{R}_n)$ where $\mathcal{R}_n$ is the optimal route from node n to the fusion center, determined separately for each routing scheme). The probability that the signal field is absent is $\Pr(\mathcal{H}_0) = 0.75$. Here, we used a more realistic signal correlation for the detection, i.e., the actual correlation between two sensors is a function of their Euclidean distance. That is, while the Chernoff-routes are assigned assuming the Gauss-Markov model, the P_{DE} is determined assuming the more realistic model. Note that Chernoff-routing provides about a 40-percent reduction in the P_{DE} compared to minimum energy routing.

Figure 11.11 shows the average routing characteristics when SNR $= 15$ dB. Here, for each value of the network size, the performance is averaged over realizations of the

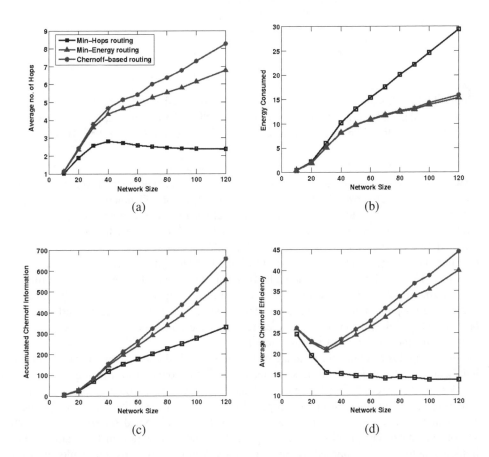

Figure 11.11 Performance analysis of the three routing strategies in terms of the: (a) average no. of hops, (b) total energy consumed, (c) accumulated Chernoff information, and (d) average detection efficiency, and (d) average error probability (SNR $= 15$ dB, $\nu = 2$, $\lambda = 0.01$, $A = 0.5$). Reprinted with permission from (Sung et al. 2007). © 2007 IEEE.

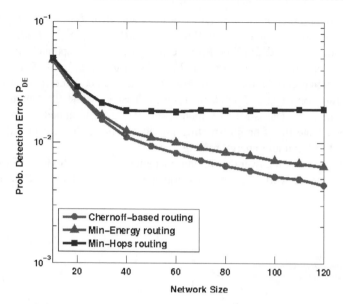

Figure 11.12 Probability of detection error P_{DE} averaged over all routes and network topology under three different routing strategies (SNR = 15 dB, $\nu = 2$, $\lambda = 0.01$, $A = 4$, $\Pr(\mathcal{H}_0) = 0.75$). Reprinted with permission from (Sung et al. 2007). © 2007 IEEE.

network topology, (i.e., the sensor locations) to extract the fundamental network behavior. Figure 11.11(a) shows the average number of hops from all potential sources to the fusion center. As expected, the minimum-hop routing gives the smallest number of hops while Chernoff routing provides the largest. Figure 11.11(b) shows the average energy required by each scheme. It is seen that Chernoff routing requires almost the same as the minimum-energy routing, providing the largest accumulated Chernoff information as shown in Figure 11.11(c). In Figure 11.11(d) it is seen that Chernoff routing results in the maximum detection efficiency as expected. Finally, Figure 11.12 shows the average detection error probability as a function of the network size. Note that the network size can be reduced significantly in the same area for the same error rate when we use Chernoff routing over the conventional routing methods.

In closing this example, we reiterate that we have not addressed several network-related issues, such as the effect of transmission interference, the potential for frequency reuse and the design of a carefully constructed MAC and PHY layer, that would be fundamental to a real-world implementation of the proposed design. Also of practical and perhaps theoretical interest is the effect of quantization on both the real-valued observations at each sensor and the scalar quantities used for prediction and propagation in the Kalman filter implementation of the proposed routing algorithm. Although some partial results exist for general models, they have not yet been applied to the specific model used in the current work. Such issues present opportunities for further research.

11.5   Conclusion

In this chapter we have studied the evolution of shortest path routing (SPR) in data networks, and argued that the next step in this evolution is the design of application dependent routing schemes for ad-hoc wireless sensor networks.

In Section 11.2, we studied shortest path routing for wireline networks, including common algorithms and centralized and distributed implementations. We studied the strengths and weaknesses of SPR in this context. In Section 11.3, we outlined the new issues in wireless networks, and the SPR algorithms developed to address these. In Section 11.4, we provided the main contribution of this chapter. We discussed SPR for ad-hoc wireless sensor networks. First, we conducted a survey of routing protocols applicable to these networks. It was seen that the current protocols are designed based on general metrics such as energy consumption and delay rather than on explicit application performance. We then argued that, since ad-hoc sensor networks are application driven, their routing protocols should be application *dependent*. We provided a detailed example of how such an application dependent, distributed, SPR routing scheme can be constructed, and showed through simulation that such an approach can indeed provide improved application performance.

11.6   A Short Review of Basic Graph Theory

Data networks are often modeled and analyzed using graph theory. In this chapter we review basic graph theoretic concepts. A graph $G = (\mathcal{N}, \mathcal{A})$ is defined by a finite non-empty set of *nodes* $\mathcal{N}$ and a set of node pairs, or *arcs*, denoted $\mathcal{A}$. For example, the graph $(\mathcal{N}, \mathcal{A}) = (\{1, 2, 3, 4, 5\}, \{(1, 2), (2, 3), (1, 3), (4, 5)\})$ is depicted in Figure 11.13(a). We require that if $(a_1, a_2) \in \mathcal{A}$, then $a_1 \neq a_2$. G specifies the connectivity of nodes, but not their physical locations. Thus, Figure 11.13(a) represents an arbitrarily chosen arrangement. We discuss undirected and directed graphs separately.

11.6.1   Undirected Graphs

In an undirected graph, the elements of $\mathcal{A}$ are unordered. For example, since the graph in Figure 11.13(a) is undirected, it equivalently represents the networks $(\mathcal{N}, \mathcal{A}) = (\{1, 2, 3, 4,$

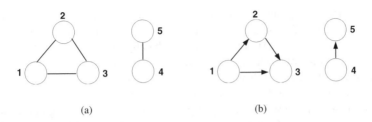

(a)   (b)

Figure 11.13 The graph $G = (\mathcal{N}, \mathcal{A}) = (\{1, 2, 3, 4, 5\}, \{(1, 2), (2, 3), (1, 3), (4, 5)\})$ when (a) G is undirected, (b) G is directed.

5}, {(2, 1), (2, 3), (3, 1), (4, 5)}) and $(\mathcal{N}, \mathcal{A}) = (\{1, 2, 3, 4, 5\},$ {(2, 1), (3, 2), (3, 1), (5, 4)}). There may be at most one arc between any pair of nodes. The terminology of undirected graphs is given as follows: (i) A *walk* in G is a sequence of nodes $(n_1, \ldots, n_\ell)$ such that $(n_k, n_{k+1}) \in \mathcal{A}, \forall k \in \{1, \ldots, \ell - 1\}$, (i-a) a *path* is a walk with no repeated nodes, (i-b) a *cycle* is a walk with $n_1 = n_\ell, \ell \geq 3$, and no repeated nodes other than $n_1 = n_\ell$, (ii) G is *connected* if for each node i there is a path $(n_1 = i, \ldots, n_\ell = j)$ to each node j $(j \neq i)$, (iii) $G' = (\mathcal{N}', \mathcal{A}')$ is a *subgraph* of $G = (\mathcal{N}, \mathcal{A})$ if G' is a graph, $\mathcal{N}' \subseteq \mathcal{N}$, and $\mathcal{A}' \subseteq \mathcal{A}$, (iv) a *tree* is a connected graph that contains no cycles, and (iv-a) a *spanning tree* of a graph G is a subgraph of G that is a tree and that includes all the nodes of G.

Spanning trees

We have the following proposition (Bertsekas and Gallager 1992).

Proposition. Let $G = (\mathcal{N}, \mathcal{A})$ be a connected graph. Then G contains a spanning tree. Furthermore, the following algorithm constructs a spanning tree $G' = (\mathcal{N}', \mathcal{A}')$ from G:

1. Let $\mathcal{N}' = \{n\}$ where $n \in \mathcal{N}$ is chosen arbitrarily. Let $\mathcal{A}' = \emptyset$.

2. If $\mathcal{N}' = \mathcal{N}$, then stop ($G' = (\mathcal{N}', \mathcal{A}')$ is a spanning tree). Otherwise, go to step 3.

3. Let $(i, j) \in \mathcal{A}$ be an arc with $i \in \mathcal{N}', j \in \mathcal{N} - \mathcal{N}'$. Update $\mathcal{N}'$ and $\mathcal{A}'$ by

$$\mathcal{N}' := \mathcal{N}' \cup \{j\}$$
$$\mathcal{A}' := \mathcal{A}' \cup \{(i, j)\}$$

and go to step 2.

It is often desirable to work with the spanning tree of a network's graph rather than with the complete graph, G, when designing a routing strategy. Doing so removes the redundancy and possible ambiguity caused by cycles while at the same time maintaining the connectivity of G. We note that the spanning tree of G is not in general unique. In Section 11.2.1 we discuss the minimum weight spanning tree of G, which is unique (under mild conditions).

11.6.2 Directed Graphs

In a directed graph, or digraph, arcs are ordered and called directed arcs. For example, the digraphs $(\mathcal{N}, \mathcal{A}) = (\{1, 2, 3, 4, 5\}, \{(2, 1), (2, 3), (3, 1), (4, 5)\})$ and $(\mathcal{N}, \mathcal{A}) = (\{1, 2, 3, 4, 5\}, \{(2, 1), (3, 2), (3, 1), (5, 4)\})$ are distinct. Graphically, a directed arc is drawn with an arrow at the end point of each arc, as shown in Figure 11.13(b). We require that there exists at most one directed arc from a node n_i to a node n_j $(i \neq j)$, and thus, that there exist at most two directed arcs between n_i and n_j.

Much of the terminology for undirected graphs applies to directed graphs. Specifically, for a digraph $G = (\mathcal{N}, \mathcal{A})$ form the associated undirected graph $G^U = (\mathcal{N}^U, \mathcal{A}^U)$, where $\mathcal{N}^U = \mathcal{N}$ and $(i, j) \in \mathcal{A}^U$ if $(i, j) \in \mathcal{A}$, $(j, i) \in \mathcal{A}$, or both. We say that $(n_1, \ldots, n_\ell)$ is a walk, path, or cycle in G if it is walk, path, or cycle in G^U. We say that G is connected if G^U is connected. Additionally, we introduce the following terminology: $(n_1, \ldots, n_\ell)$ is a directed walk in G if (n_k, n_{k+1}) is a directed arc, $\forall k \in \{1, \ldots, \ell - 1\}$. A directed path is directed walk with no repeated nodes, and a directed cycle is a directed walk with $n_1 = n_\ell$ for $\ell \geq 3$ and no other repeated nodes. We say that G is strongly connected if for each pair of nodes i and j there is a directed path from i to j.

Acknowledgements

This work is supported in part by the Army Research Office under Grant ARO-W911NF-06-1-0346. The U. S. Government is authorized to reproduce and distribute reprints for Government purposes notwithstanding any copyright notation thereon.

Bibliography

Adjih C, Baccelli E, and Jacquet P 2003 Link state routing in wireless ad-hoc networks. In *Proc. IEEE Military Communications Conference (MILCOM 2003)*. Boston, MA, USA, pp. 1274–1279.

Akkaya K and Younis M 2005 A survey of routing protocols in wireless sensor networks. *Elsevier Ad Hoc Network Journal*, **3/3**, 325–349.

Al-Karaki JN and Kamal AE 2004 Routing techniques in wireless sensor networks: A survey. *IEEE Trans. Wireless Commun.*, **11**(6), 6–28.

Aslam J, Li Q, and Rus D 2003 Three power-aware routing algorithms for sensor networks. *Wireless Communications and Mobile Computing*, **2**(3),187–208.

Barrett L, Eidenbenz SJ, and Kroc L 2003 Parametric probabilistic sensor network routing. In *Proc. of the WSNA'03*, San Diego, CA, USA.

Basagni S, Chlamtac I, Syrotiuk VR, and Woodward BA 1998 A distance routing effect algorithm for mobility (DREAM). In *Proc. of ACM/IEEE Int. Conf. on Mobile Computing and Networking*, pp. 76–84.

Bertsekas D and Gallager R 1992 *Data Networks*, 1st edn. McGraw-Hill, Marderhead.

Boppana RV and Konduru S 2001 An adaptive distance vector routing algorithm for mobile, ad hoc networks. In *Proc. IEEE Infocom (INFOCOM 2001)*. pp. 1753–1762.

Chang J and Tassiulas L 1999 Routing for maximum system lifetime in wireless ad-hoc networks, In *Proc. of 37th Annual Allerton Conference on Communication, Control, and Computing*, Monticello, IL, USA, Sept. 1999.

Chin KW, Judge J, Williams A, and Kermode R 2002 Implementation experience with manet routing protocols. *ACM SIGCOMM Computer Communications Review*, **32**(5), 49–59.

Clausen T et al. 2003 Optimized Link State Rotuing Protocol, http://www.ietf.org/internet-drafts/draft-ietf-manet-olsr-11.txt.

Corson MS and Ephremides A 1995 A distributed routing algorithm for mobile wireless networks. *ACM-Baltzer Journal of Wireless Networks*, **1**(1), 61–81.

De Couto DSJ, Aguayo D, Bicket J, and Morris R 2003 A high-throughput path metric for multihop wireless routing. In *Proc. MOBICOM 2003*, San Diego, CA, USA, pp. 14–19.

Dembo A and Zeitouni O 1993 *Large Deviations Techniques and Applications*. Jones and Bartlett.

Dube R, Rais C, Wang K, and Tripathi S 1997 Signal stability based adaptive routing (SSA) for ad hoc mobile networks. *IEEE Pers. Commun.*, **4**(1), 36–45.

Ephremides A 2002 Energy concerns in wireless networks. *IEEE Trans. Wireless Commun.*, **9**(4) 48–59.

Ettus M 1998 System capacity, latency, and power consumption in multihop-routed SS-CDMA wireless networks. In *Proc. IEEE Radio and Wireless Conference (RAWCON 1998)*, Colorado Springs, CO, USA, pp. 55–58.

Gafni EM and Bertsekas DP 1987 Asymptotic optimality of shortest path routing algorithms. *IEEE Trans. Inf. Theory*, **33**(1), 83–90.

Gallager RG, Humblet PA, and Spira PM 1983 A distributed algorithm for minimum weight spanning trees. *ACM Trans. Program. Lung. Sysr.*, **5**, 66–77.

Garcia-Luna-Aceves JJ, Fullmer CL, Madruga E, Beyter D, and Frivold T 1997 Wireless internet gateways (WINGS). In *Proc. IEEE Military Communications Conference (MILCOM 1997)*. Monterey, CA, USA, pp. 1271–1276.

Gerharz M, de Waal C, Frank M, and Martini P 2002 Link stability in mobile wireless ad hoc networks. In *Proc. IEEE Conference on Local Computer Networks (LCN 2002)*. Tampa, FL, 2002.

Gerharz M, de Waal C, Martini P, and James P 2003 Strategies for finding stable paths in mobile wireless ad hoc networks. In *Proc. IEEE Conference on Local Computer Networks (LCN 2003)*, Kongiswinter, Germany, Oct. 2003, pp. 130–139.

Haas ZJ, Halpern JY, and Li L 2002 Gossip-based ad hoc routing. In *Proc. IEEE Infocom (INFOCOM 2002)*, New York.

Humblet PA 1983 A distributed algorithm for minimum weight directed spanning trees. *IEEE Trans. Commun.*, **31**(6), 756–762.

Johnson DB and Maltz DA 1996 *Mobile Computing*, 1st edn (edited). Kluwer, Dordrecht.

Jubin J. and Tornow J 1987 The DARPA packet radio network protocols. *Proc. IEEE*, **75**(1), 21–32, Jan. 1987.

Kershenbaum A 1993 *Telecommunications Network Design Algorithms*, 1st edn. McGraw-Hill, Maiderhead.

Kleinrock L 1964 *Communication Nets*. Dover.

Kowalik K and Davis M 2006 Why are there so many routing protocols for wireless mesh networks?. In *Proc. of IEE Irish Signals and Systems Conference 2006 (ISSC 2006)*, Dublin, Ireland, June 2006.

Kozat UC, Kondylis G, Ryu B, and Marina MK 2001 Virtual dynamic backbone for mobile ad hoc networks. In *Proc. of IEEE Int. Conf. on Communications (ICC 2001)*, pp. 250–255.

Leon-Garcia A and Widjaja I 2000 *Communication Networks: Fundamental Concepts and Key Architectures*, 1st edn. McGraw-Hill, Maiderhead.

Lundgren H, Nordstrom E, Tschudin C 2002 Coping with communication grey zones in IEEE 802.11b based ad hoc networks. In *Proc. IEEE Int. Symp. on a World of Wireless, Mobile and Multimedia Networks (WoWMoM 2002)*.

Marina MK and Das SR 2005 *Routing in Mobile Ad Hoc Networks*, 1st edn (edited). Springer, New york.

McDonald BA 1997 Survey of adaptive shortest-path routing in dynamic packet-switched networks, available at citeseer.ist.psu.edu/mcdonald97survey.html, April 24, 1997.

Meng TH and Rodoplu V 1998 Distributed network protocols for wireless communication. In *Proc. IEEE Int. Symposium on Circuits and Systems (ISCAS 1998)*, Monterey, CA, USA, June 1998.

Michail A and Ephremides A 2000 Energy efficient routing for connection-oriented traffic in ad-hoc wireless networks. In *Proc. IEEE Personal, Indoor and Mobile Radio Communications (PIMRC 2000)*, pp. 762–766.

Moy J 1998 *Anatomy of an Internet Routing Protocol*. Addison-Wesley, Reading, MA.

Murthy S and Garcia-Luna-Aceves JJ 1996 An efficient routing protocol for wireless networks. *Mobile Networks and Applications*, **1**(2), 183–197.

Neely MJ, Modiano E, and Rohrs CE 2005 Dynamic power allocation and routing for time varying wireless networks. *IEEE J. Sel. Areas Commun., Special Issue on Wireless Ad-Hoc Networks*, **23**(1), 89–103.

Ni SY, Tseng YC, Chen YS, and Sheu JP 1999 The broadcast storm problem in a mobile ad hoc networt., In *Proc. MOBICOM 1999*, pp. 151–162.

Oran D 1990 OSI IS-IS Intra-domain Routing Protocol. RFC 1142, http://ietf.org/rfc/rfc1142.txt, 1990.

Park VD and Corson MS 1997 A highly adaptive distriubted routing algorithm for mobile wireless networks. In *Proc. IEEE Infocom (INFOCOM 1997)*.

Perkins CE 2001 *Ad-Hoc Networking*, 1st edn. Addison-Wesley, Reading, MA.

Perkins CE and Bhagwat P 1994 Highly dynamic destination sequenced distance-vector routing (DSDV) for mobile computers. In *Proc. of SIGCOMM 1994*, pp. 234–244.

Perkins CE and Royer EM 1999 Ad hoc on-demand distance vector routing. In *Proc. of IEEE Workshop on Mobile Computing Systems and Applications (WMCSA 1999)*, pp. 90–100.

Perkins CE, Royer EM, Das SR, and Marina MK 2001 Performance comparison of two on-demand routing protocols for ad hoc networks. *IEEE Pers. Commun.*, pp. 16–28, 2001.

Poor HV 1994 *An Introduction to Signal Detection and Estimation*, 2nd edn. Springer, New York.

Rodoplu V and Meng TH 1998 Minimum energy mobile wireless networks. In *Proc. of the IEEE Int. Conf. on Communications (ICC 98)*, Atlanta, GA, USA, June 1998.

Schweppe FC 1965 Evaluation of likelihood functions for Gaussian signals. *IEEE Trans. Inf. Theory*, vol. IT-1, pp. 61–70.

Servetto SD and Barrenechea G 2002 Constrained random walks on random graphs: routing algorithms for large scale wireless sensor networks. In *Proc. of WSNA 2002*.

Shah RC and Rabaey J 2002 Energy aware routing for low energy ad hoc sensor networks. In *Proc. IEEE Wireless Communications and Networking Conference (WCNC 2002)*, Orlando, FL, USA.

Singh S, Woo M, Raghavendra CS 1998 Power-aware routing in moible ad hoc networks. In *Proc. MOBICOM 1998*, Dallas, TX, USA.

Spira PM 1977 Communication complexity of distributed minimum spanning tree algorithms. In *Proc. 2nd Berkeley Conf. Distributed Data Manag. Comput. Networks*, June, 1977.

Steenstrup M 1995 *Routing in Communications Networks*, 1st edn. Prentice Hall, New Jersey.

Sung Y, Misra S, Tong L, and Ephremides A 2006 Signal processing for application-specific ad hoc networks. *IEEE Trans. Signal Process., Special Issue on Signal Processing for Wireless Ad hoc Communication Networks*, 23(5), 74–83.

Sung Y, Misra S, Tong L, and Ephremides A 2007 Cooperative routing for distributed detection in large sensor networks. *IIEEE J. Sel. Areas Commun., Special Issue on Cooperative Communications and Networking*, 25(2), 471–483.

Sung Y, Tong L, and Ephremides A 2005 A new metric for routing in multi-hop wireless sensor networks for detection of correlated random fields. In *Proc. IEEE Military Communications Conference (MILCOM 2005)*. Atlantic City, NJ, USA.

Tassiulas L and Ephremides A 1992 Stability properties of constrained queueing systems and scheduling for maximum throughput in multihop radio networks. *IEEE Trans. Autom. Control*, 37(12), 1936–1949.

Toh CK 2001 Maximum battery life routing to support ubiquitous mobile computing in wireless ad hoc networks. *IEEE Commun. Mag.*, pp. 138–147, June 2001.

Wan PJ, Calinescu G, Li XY, and Frieder O 2002 Minimum-energy broadcasting in static ad hoc wireless networks. *Wireless Networks*, 8(6), 607–617.

Wan PJ, Calinescu G, and Yi CW 2004 Minimum-power multicast routing in static ad hoc wireless networks. *IEEE/ACM Trans. Netw.*, 12(3), 507–514.

Wieselthier JE, Nguyen GD, and Ephremides A 2000 On the construction of energy-efficient broadcast and multicast trees in wireless networks. In *Proc. IEEE Infocom (INFOCOM 2000)*, pp. 585–594.

Witt M and Turau V 2005 BGR: blind geographic routing for sensor networks. In *Proc. of the Third Workshop on Intelligent Solutions in Embedded Systems (WISES 2005)*. Hamburg, Germany.

12

Data-Centric and Cooperative MAC Protocols for Sensor Networks

Yao-Win Hong and Pramod K. Varshney

12.1 Introduction

Medium Access Control (MAC), which is conventionally viewed as part of the Data Link Layer in the OSI model, coordinates the use of a shared transmission medium in multiuser systems and ensures reliable communication over interference-free channels for each user. Due to the limited bandwidth, the fast-varying channels and the energy-costly transmissions in wireless systems, it is particularly important to derive MAC protocols that efficiently utilize the channel resources, e.g. bandwidth and energy. However, most MAC protocols proposed in the literature (Bertsekas and Gallager 1991; Gummalla and Limb 2000) assume that the users are independent of each other and that they are competing for the use of the common transmission channel. System attributes such as throughput, delay, fairness and reliability are often used as the objective functions when designing conventional MAC protocols. Unfortunately, these Quality-of-Service (QoS) attributes do not accurately describe the performance of sensor applications and may lead to inefficient solutions for sensor networks.

Due to the strict resource constraints of sensor networks, two properties of the system are often exploited for designing sensor network MAC protocols: (1) *the application-dependent objectives* and (3) *the cooperative nature of the distributed sensors*. Conventionally, wireless MAC protocols do not exploit the cooperative and application-dependent properties and

Wireless Sensor Networks: Signal Processing and Communications Perspectives   A. Swami, Q. Zhao, Y.-W. Hong and L. Tong
© 2007 John Wiley & Sons, Ltd

focus on generic designs that are suitable for a wide variety of applications. While this is useful in most applications, the attempt to serve a wide variety of applications causes a loss in the efficiency of resource utilization. In fact, it is often necessary to compromise the throughput, fairness or delay performances in order to save energy and bandwidth.

The key to designing efficient sensor MAC protocols is to recognize the fact that sensors are cooperating instead of competing and that they are actually highly dependent users as opposed to being independent in conventional systems. Therefore, the energy and bandwidth resources can be allocated to optimize the application-dependent performance measures, eg. the probability of detection errors of the estimation distortion, without fairness constraints. Hence, new sensor network MAC protocols must be designed, taking into consideration the application-specificity through cross-layered approaches and the advantages of cooperative/distributed signal processing. In this chapter, we introduce three classes of sensor MAC protocols that exploit these properties: (1) the energy-efficient MAC protocols; (2) the data-centric MAC protocols; and (3) the cooperative MAC protocols. The emphasis will be on the third class of strategies, where the cooperation is used to improve both the throughput of the user-oriented system, e.g. computer networks, and the performance of data-centric systems that consists of correlated users, e.g. sensor networks.

In the first part of this chapter, we give a short survey of several popular sensor MAC protocols that optimize the system parameters of conventional schemes, such as Carrier Sensing Multiple Access (CSMA) or Time Division Multiple Access (TDMA), to achieve energy-efficient communication in sensor networks. These strategies take into consideration the power-saving mode of sensor devices and design adaptive sleep-wake patterns to avoid the major sources of energy-waste, such as idle-listening, overhearing, protocol overhead, and collision/interference. The performance in terms of throughput, delay or fairness are compromised in exchange for energy efficiency. These strategies provide a generic design for a variety of sensor network applications since the functionalities are confined within the conventional layered architecture. However, the advantages of cooperation and application-specificity were not exploited.

To derive application-dependent protocols, it is necessary to consider during the design the statistical properties of the sensors' data and the performance measures corresponding to the application. Capitalizing on the high correlation between the sensors' data, several cross-layered MAC protocols have been proposed to reduce the redundant transmissions made by the local sensors, and, thus, the total energy and bandwidth consumption is minimized. Three main approaches have been taken in the literature: (1) data aggregation; (2) distributed source coding; and (3) spatial sampling. The savings are realized at the cost of additional computations at the local sensors. However, these approaches have been considered promising due to the rapid advances of the sensor hardware technology as opposed to the slow increase of the energy-density in batteries.

While many sensor MAC protocols have been designed to facilitate the signal processing performed in the application-layer, the cooperative signal processing techniques that can be used to enhance the networking and communication efficiency are often neglected. The main focus of this chapter is to introduce the cooperative signal processing techniques that can be used to improve the efficiency of resource utilization and the application-specific performances. When the sensors are treated as independent users, MAC protocols based on cooperative relaying have been proposed to improve the throughput of the system. When the system capitalizes on the high dependency between sensors, cooperative signal processing can also be used to aggregate the sensors data and, thus, improve the MAC performance. Two specific applications are used to illustrate these advantages: the data gathering application and the decentralized detection and estimation application. In the first application, the sink or data gathering node obtains an estimate of the sensors' data under a certain distortion constraint while, in the second application, the central processor obtains the optimal decision or estimate of a common event based on the information provided by the local sensors. With cooperation, we are able to closely integrate the functionalities of different layers and achieve a better performance while utilizing less bandwidth or energy resources.

In the following, we first give a short introduction to two basic MAC protocols and extend these two approaches throughout this chapter.

12.2 Traditional Medium Access Control Protocols: Random Access and Deterministic Scheduling

MAC protocols have been studied extensively for many years, ranging from wire-line telephony networks to the internet to wireless ad hoc networks. The efficiency of the MAC design is crucial for sensor networks due to the scarce energy and bandwidth resources of these systems. Based on the assumption that the users in the system are transmitting independent data and that they are competing for the use of the transmission channel, MAC protocols are designed to allocate separate interference-free channels to each user. This is typically achieved through either *random access* or *deterministic scheduling*. In this section, we shall introduce briefly two specific protocols as an example for each of these two approaches: the Carrier Sense Multiple Access (CSMA) protocol and the Time Division Multiple Access (TDMA) protocol. In fact, most recent MAC protocols, including those tailored for sensor network applications, were proposed based on these two strategies.

12.2.1 Carrier Sense Multiple Access (CSMA)

CSMA is one of the most popular choices for random access networks due to its simple and effective design. In random access systems, users access the network based only on its local information, e.g. the state of its queue or the state of the channel that it senses, which is relatively random for other users in the network. As proposed in (Kleinrock and Tobagi 1975), in CSMA, each user, that has a message to transmit, first senses the channel to see whether or not there is an on-going transmission from other users before it sends its own message. This is done in an attempt to avoid collision with other users. However, CSMA does not completely avoid collisions since two users may transmit simultaneously if they are not able to sense each other's transmissions, which may be caused by the large propagation delay or signal attenuation between the users. When a collision occurs at the destination, each transmitting user waits for a random backoff time before it makes an attempt to transmit the message again. CSMA can be proposed with various backoff models and is shown to achieve an efficiency of 50% ~ 80% depending on the specifics of the protocol.

CSMA serves as the basis for many wireless MAC protocols. One widely adopted protocol is the Multiple Access with Collision Avoidance (MACA) protocol (Karn 1990) that introduces a three-way handshake between the transmitter and receiver to solve

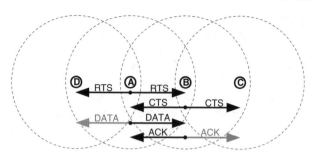

Figure 12.1 Illustration of the hidden terminal problem and the three-way handshake.

the well-known hidden terminal problem present in conventional CSMA. Specifically, the hidden terminal problem, as illustrated in Figure 12.1, occurs when two users, say user A and user C, intend to transmit to a common user, e.g. user B. Since users A and C are located outside the transmission range of each other, they are not able to sense each other's transmissions and, thus, cannot avoid collision through carrier sensing. In this case, we say that user C is a hidden terminal of user A and, vice versa. The three-way handshake proposed in MACA resolves this problem by having each user transmit a Request-To-Send (RTS) message whenever it has a packet to send, indicating the destination and the length of the intended data transmission. If the destination user successfully receives the RTS message and has not completed a handshake previously with other nodes, it will then respond with a Clear-To-Send (CTS) packet indicating that it is ready for reception. The source then sends a DATA packet to the destination once the CTS is received. This process is also illustrated in Figure 12.1. The key to resolving the hidden terminal problem is to have all users that overhear the RTS or CTS packets remain silent for a duration corresponding to the length of the DATA packet. The MACA Wireless (MACAW) protocol (Bharghavan et al. 2004) further defines an ACK message in response to the DATA packet to take into account the reliability of the wireless channel. In fact, CSMA serves as the basis of the Distributed Coordination Function (DCF) in the contention period of the IEEE 802.11 standard, and also the Contention Access Period (CAP) in the IEEE 802.15.4 standard (Callaway et al. 2002; Yedavalli and Krishnamachari 2006).

12.2.2 Time-Division Multiple Access (TDMA)

As opposed to CSMA, TDMA provides each user with interference-free transmission channels through deterministic scheduling. Specifically, TDMA divides the use of the channel into fixed time slots and schedules the transmission of the active users among these time slots based on the users' demands and the total resources available. TDMA requires strict synchronization among users and a centralized control to coordinate the use of the channels. Benefitting from the extra coordination, it is easier for TDMA to achieve the users' QoS demands, e.g. the rate, delay or bit-error-rate (BER) requirements, while consuming less resources. Even with the complexity of computing the optimal channel allocation and the increase of control messages, it is often worth-while for delay-constrained or energy-constrained applications. In addition, the coordination also allows TDMA to achieve better throughput under heavy traffic loads. The IEEE 802.11 standard also employs deterministic

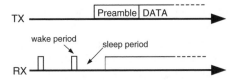

Figure 12.2 Illustration of the preamble sampling method.

scheduling in the contention-free period through the defined Point Coordination Function (PCF) to take account of applications with strict QoS constraints. This is also used in the period of guaranteed time slots (GTS) in the IEEE 802.15.4 standard.

Both the CSMA and the TDMA protocols have been used as a basis for many sensor network MAC protocols. Several of these methods are introduced in the following section with the emphasis on energy efficiency.

12.3 Energy-Efficient MAC Protocols for Sensor Networks

One of the primary challenges of designing sensor network MAC protocols is the limited battery-lifetime of sensor devices. The large-scale deployment of sensors, possibly in hostile environments, may prohibit the use of human maintenance and manual replacement of batteries. Therefore, most existing sensor network MAC protocols focus on eliminating the sources of energy-waste inherent in conventional protocols. However, the improvement comes at the cost of reduced fairness, throughput and increased delay, which are fortunately less relevant in sensor network applications. In the following, we introduce a few of these protocols while a comprehensive study can be found in (Demirkol et al. 2006; Langendoen and Halkes 2005).

The major sources of energy-waste in conventional MAC protocols, as identified in (Langendoen and Halkes 2005; Ye et al. 2004), are idle-listening, overhearing, collisions, protocol overheads, and over-emitting. The effect of these aspects differs according to the application. For example, under low traffic rates, sensors consume most of their energy in idle-listening since the transmission occurs sporadically. However, under high traffic rates, the collisions and protocol overheads cause a significant increase in energy consumption. We classify the sensor network MAC protocols as either random access or deterministic scheduling. In fact, most of these strategies inherit the basic structures of the CSMA and TDMA protocols introduced previously and impose an intelligent sleep-wake policy to reduce the energy consumption of idle users. A tradeoff between energy efficiency and adaptivity is observed between the two classes of strategies. Specifically, while ideal coordination eliminates many sources of energy waste in deterministic scheduling protocols, flexibility and adaptivity are lost when compared to random access protocols.

The random access approach

Taking the random access approach, El-Hoiydi (2002) and Hill and Culler (2002) proposed sensor MAC protocols based on the well-known ALOHA and CSMA protocols, respectively. The proposed strategies reduce the energy waste, due to idle-listening and overhearing, by having the users enter a sleep state when they becomes idle, i.e. when they are neither transmitting nor receiving. To maintain communication, each user must wake up periodically to listen for a packet intended for itself. If, in fact, such a packet is heard during the wake-up period, the sensor will remain awake until the data packet is received. To guarantee that a packet is always heard, each transmitter must emit a preamble at the beginning of each data packet that is sufficiently long to cover at least one sleep-wake period, as shown in Figure 12.2. This method, i.e. the preamble sampling method, was implemented as part of the TinyOS in Mica sensors (Hill and Culler 2002). Under low traffic load, the increased energy consumption for transmitting a long preamble is overcome by the reduced idle-listening and overhearing. The WiseMAC protocol (El-Hoiydi and Decotignie 2004) improves the performance of the preamble sampling scheme under high traffic loads by reducing the length of the preambles. This is achieved by allowing sensors to exchange their local sleep-wake cycles and by aligning each sensors' transmissions with the wake-up period of the corresponding receiver so that long preambles are no longer necessary. However, this is achieved at the cost of additional coordination.

The deterministic scheduling approach

Although random access achieves good flexibility and low latency for applications with low traffic loads, deterministic scheduling is actually the most effective way of eliminating the sources of energy waste. In fact, with perfect scheduling, only one transmitter-receiver pair would be active during each transmission period, therefore, reducing collision and eliminating idle-listening and overhearing. However, deterministic TDMA scheduling[1] requires a large overhead in order to maintain accurate synchronization between sensors and to exchange local information, such as the network topology and the communication pattern. Moreover, the latency increases linearly with the total number of sensors sharing the channel since TDMA assigns a separate time-slot to each transmitting sensor. This can be improved by applying spatial channel reuse in TDMA-based sensor network MAC protocols. For example, (Arisha et al. 2002) and (Wu and Biswas 2005) applied TDMA in cluster-based sensor networks where the channel (which is divided in either time, frequency or code) is reused in different clusters. Furthermore, with the knowledge of the communication pattern, the sensors on the same packet delivery path can be scheduled to transmit in order while having the sensors wake-up accordingly to match the transmit and receive periods. As proposed in (Kulkarni and Arumugam 2004), this method reduces significantly the delay but relies on the application-specific knowledge of the communication pattern and loses the flexibility of random access systems.

The balanced approach

In sensor networks, it is equally important for the MAC designs to be adaptable to dynamic changes of the environment as well as to achieve energy efficiency and to prolong the network lifetime. To strike a balance between these two desired properties, several works consider the case where sensors generate their sleep-wake patterns in a distributed fashion,

---

[1]Even though other channelization methods such as CDMA and FDMA are equally effective, most current sensor devices utilize only a single frequency channel and cannot afford the complexities of a CDMA transceiver.

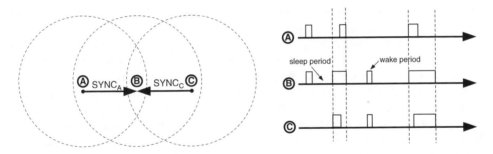

Figure 12.3 Illustration of SMAC.

so that adaptability is maintained, and exchange the local patterns with neighboring sensors to coordinate the transmissions and to avoid the energy waste caused by idle-listening, overhearing and over-emitting.

S-MAC (Ye et al. 2004) was proposed based on this approach, where each sensor generates a local sleep-wake schedule and broadcasts the information to neighboring sensors through the exchange of SYNC packets. If a sensor receives a sleep-wake schedule from other sensors before it broadcasts its own, it will operate under the received schedule instead of the one generated locally. As a result, the network will form virtual clusters that contain sensors running a common sleep-wake schedule. To further guarantee connection between neighboring clusters, sensors that receive multiple schedules must adopt a wake-up period that is equal to the union of all schedules. For example, as shown in Figure 12.3, user *B* receives the schedule from *A*, *C* and adopts a wake-up schedule that is the union of both schedules. In S-MAC, the energy consumption is reduced by having sensors operate under low duty cycles while the communication is maintained with synchronized wake-up periods between neighboring sensors. The drawback is that the sleep-wake schedule must be determined beforehand and cannot be adjusted adaptively for different traffic loads or communication patterns.

To overcome this disadvantage, T-MAC (van Dam and Langendoen 2003) uses a time-out mechanism to terminate the wake-up period and to mark the beginning of a sleep period. Specifically, T-MAC allows a sensor to enter the sleep state if no activation event occurs over a certain amount of time. The activation event may be the reception of data on the radio, the sensing of a collision or the overhearing of RTS/CTS packets from its neighbors etc. In this case, each sensor's wake-up period is automatically adjusted to match the communication pattern. Hence, the protocol achieves better throughput and delay performances when compared to fixed scheduling policies that do not match accurately to the dynamic changes of the communication pattern.

The common problem of MAC protocols that adopt sleep-wake policies is the large latency that occurs when the receiver of a certain packet enters the sleep state before the packet is transmitted. In this case, the transmitter must wait until the next wake-up period before this packet can be sent. This is avoided when the sensors have full knowledge of the communication pattern and are able to run a staggered wake-up pattern that matches the sensor's receive period to the transmission period of its upstream node. In fact, D-MAC (Lu et al. 2004) utilizes such a staggered sleep-wake cycle for data gathering communication patterns where packets are sent from the distributed sensors to a single sink node. The latency is significantly reduced for data gathering applications but may incur large delays for applications that generate different communication patterns.

The protocols introduced in this section demonstrate improvements in energy efficiency by making simple modifications to the conventional CSMA or TDMA protocols. The advantage of these schemes is that they are readily implementable due to their similarity with conventional protocols and that they can be adopted in different applications since they are largely confined within the layered architecture. When the knowledge of the network topology, the traffic load or the communication pattern is further exploited to optimize the MAC, a significant improvement in energy efficiency, as well as throughput and delay, can be achieved but a loss in performance is experienced if the strategies are applied to non-matching applications. Nonetheless, for sensor networks that are strictly constrained in resources, application-specific designs may be more important than generic designs that are applicable for a large class of applications.

12.4 Data-Centric MAC Protocols for Sensor Networks

The advantage of application-specific designs can be exploited further by taking into consideration the statistical knowledge of the sensors' data. Namely, due to the high dependency between the sensors, the traffic that goes through the network may be highly redundant and may not contribute equally to the system objective. In this case, in order to save energy and bandwidth, one should either combine the sensors' data in a compressed form through in-network processing or simply gather information only from the minimum number of sensors that are sufficient to achieve the goal of the application. This leads to the class of data-centric MAC protocols. Three approaches have been used to reduce the redundancy in the sensors' transmissions: data aggregation, distributed compression and spatial sampling of correlated data. Although the data aggregation and distributed compression schemes are done mostly in the application layer, the performance depends largely on the specific MAC protocol and motivates the study of cross-layered protocols.

12.4.1 Data Aggregation

Data aggregation (Heinzelman et al. 1999; Intanagonwiwat et al. 2003; Rajagopalan and Varshney 2006) is a general concept that aims at eliminating the redundancy of the messages (and, thereby, reducing the number of transmissions and energy consumption) by jointly processing the received messages and the local data at each sensor enroute to its destination. The aggregation is generally performed sequentially and depends on the specific application at hand. For example, in distributed detection systems (Varshney 1996; Viswanathan and Varshney 1997), the aggregation at each sensor is achieved by performing local decision fusion on the received messages and the local observations; in data gathering applications where the goal is to obtain an estimate of the entire sensor field, joint compression of the received messages and the local data (Scaglione and Servetto 2002) is used as the data aggregation method; in decentralized optimization problems (Rabbat and Nowak 2005), the aggregated data is generated by the incremental optimization of some cost function.

Most data aggregation methods do not depend on the MAC and can be implemented on top of most existing protocols (Ditzel and Langendoen 2005; Krishana-machari et al. 2002). However, the choice of MAC protocols may affect the aggregation performance in terms of the compression rate, the energy efficiency or the overall application performance. Specifically, in distributed detection applications, the detection performance can be improved if the energy and bandwidth resources are allocated with respect to the reliability of the sensors' observations (Lei et al. 2007), e.g. sensors with better reliability should be allocated a larger bandwidth or transmission power. In the data gathering application where the joint compression scheme (Scaglione and Servetto 2002) is used to achieve aggregation, aggregation or compression should occur between highly correlated sensors in order to reduce the total transmission energy. When data from less correlated sensors are aggregated first, more energy must be expended to forward the message to the next sensor due to the low compression rate. Therefore, the order of transmission that is determined by both the MAC and the routing protocols is crucial to achieve good aggregation efficiency.

12.4.2 Distributed Source Coding

Distributed source coding (DSC) for sensor networks, as described in (Pradhan et al. 2002; Xiong et al. 2004), is based on the Slepian-Wolf (Slepian and Wolf 1973) theory which shows that, by knowing only the joint statistics of the data at each sensor, the sensors are able to compress their local data without the specific knowledge of the other sensors' data and achieves a compression rate equal to that of the centralized compression. The Slepian-Wolf encoding scheme is illustrated in Figure 12.4(a). Here, two users transmit with rates R_1 and R_2 to the destination where the messages are jointly decoded. Slepian-Wolf theory proves that the users can perform lossless source coding with rates R_1 and R_2 that fall within the shaded region of Figure 12.4(b). In particular, the DSC achieves the maximal sum rate $R_1 + R_2 = H(X, Y)$, which is the optimal compression obtained in the centralized case. With DSC, the sensors are able to transmit their data in parallel to the sink node as opposed to the serial transmission path required for efficient data aggregation. However, the DSC scheme requires the messages to be jointly decoded at the receiver, which makes the system prone to errors and induces a large latency as in data aggregation. As a result, the performance of DSC is greatly affected by the MAC protocol, which is illustrated in the following example and described in detail in (Tsai et al. 2007).

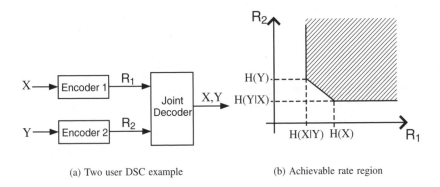

(a) Two user DSC example   (b) Achievable rate region

Figure 12.4 Slepian-Wolf theory.

Consider a network of N sensors that generate data $X_1, X_2, \ldots, X_N$, respectively, and let the sensors encode with the rates $R_1 = H(X_1)$, $R_2 = H(X_2|X_1)$, $R_3 = H(X_3|X_2, X_1), \ldots, R_N = H(X_N|X_{N-1}, \ldots, X_1)$. Due to the specific encoding order, the data of sensor i can be decoded only if the data from sensors 1 to $i-1$ are successfully received. Therefore, the conventional MAC protocols that maximize the packet throughput are not sufficient to ensure a high percentage of decoded packets, i.e. the decoded throughput. To improve the decoded throughput or to reduce the decoding delay, the sensors encoded with less dependency with other sensors should be granted with a higher probability of transmission. For example, in a slotted ALOHA system, the transmission probability assignment should be such that $p_1 \geq p_2 \geq \cdots \geq p_N$, where p_i is the probability that user i transmits in a time slot. We compare three policies – Policy I: $p_i = \frac{1}{N}$, for all i; Policy II: $p_i = a - \frac{a-b}{N-1}(i-1)$, where $(a+b)/2 = 1/N$; and Policy III: $p_i = \frac{2^\alpha - 1}{2^{\alpha i}}$, where α is a policy parameter. Note that the probabilities are assigned such that $\sum_i p_i \approx 1$. For $N = 21$ and the probability assignments plotted in Figure 12.5(a), we show, in Figure 12.5(b), the throughput of the system with respect to the index of the time slots. The solid lines represent the conventional throughput under each policy, in terms of the number of received packets (not necessarily decodable), while the dashed lines represent the decoded throughput. We can see that, as the transmission probabilities becomes more biased towards the users with smaller indices, the decoded throughput is increased at the early stage of the process while the throughput of the packet is decreased. However, later on in the process, the decoded throughput increases rapidly for the policies with high throughput since most data are collected at this point. To see the effect on energy consumption, let us consider the binary Markov data model where each sensor has a binary data with probability 0.5 and the consecutive sensors have a Markov relation with transition probability equal to 0.05. Suppose that the energy consumed to transmit each bit is equal and normalized to 1. We can then obtain the average decoded throughput per

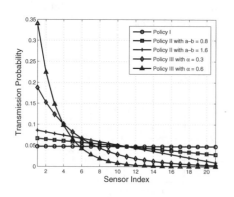

(a) Transmission Probabilities

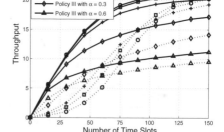

(b) Throughput versus the number of time slots

Figure 12.5 The performance of DSC under different random access policies where the solid line is the throughput and the dashed line is the decoded throughput.

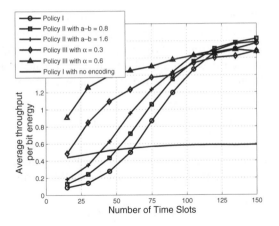

Figure 12.6 Throughput per energy bit.

bit-energy as

$$\frac{\text{Decoded Throughput}}{H(X_i|X_{i-1}, \ldots, X_1) \times (\text{\# of transmissions})}$$

for sensor i. It is shown in Figure 12.6 that the biased transmission probabilities yield a high decoded throughput per energy bit while the case without DSC loses significantly in terms of energy consumption. Interestingly, one can also trade off the compression efficiency to increase the decoded throughput by having sensors group into clusters and encode their messages based only on the sensors within its cluster. A discussion of reliability and efficiency tradeoff can be found in (Marco and Neuhoff 2004).

12.4.3 Spatial Sampling of a Correlated Sensor Field

For sensor networks that are densely deployed in a spatially correlated field, sensors in the vicinity of each other may contain highly redundant information and the resources should not be expended to transmit the data from all the sensors. The spatial sampling technique exploits this fact and leads to sensor MAC protocols that enable only a subset of sensors to access the channel, much like the sampling procedure in digital signal processing.

A sensor MAC protocol based on spatial sampling was proposed in (Vuran and Akyildiz 2006) for a data gathering application described as follows. Consider a network of N sensors, denoted by $\mathcal{S} = \{1, 2, \ldots, N\}$, that take noisy measurements of a common source Ξ. The data measured by sensor i is modeled by

$$X_i = \Xi_i + W_i$$

where the W_i's are *i.i.d.* Gaussian with zero mean and variance σ_W^2. Suppose that the measurements are spatially correlated where Ξ_i and Ξ_j are modeled as jointly Gaussian random variables with zero mean and correlation coefficient $\rho_{\Xi_i, \Xi_j} = e^{(-d_{i,j}/\theta_1)^{\theta_2}}$, for $\theta_1 > 0$ and $\theta_2 = (0, 2]$, which decreases monotonically with the distance between i and j, denoted by

$d_{i,j}$. Similarly, Ξ is also jointly Gaussian with Ξ_i with variance σ_Ξ^2 and the correlation coefficient is $\rho_{\Xi,\Xi_i} = e^{(-d_{0,i}/\theta_1)^{\theta_2}}$ where $d_{0,i}$ is the distance between the source and sensor i.

Let $\mathcal{S}_M \subset \mathcal{S}$ be the subset of sensors whose data is to be sent to the sink node and let $|\mathcal{S}_M| = M$. Suppose that each sensor in the subset transmits a scaled version of their local measurement to the data fusion center. The message sent by sensor i is then modeled by

$$Y_i = \sqrt{\frac{P_E}{\sigma_\Xi^2 + \sigma_W^2}} X_i, \qquad \text{for all } i \in \mathcal{S}_M,$$

where P_E is the power constraint at each sensor. The sink computes the estimate of Ξ by taking the average of the minimum mean square error (MMSE) estimates of Y_i's, i.e.

$$\hat{\Xi}(\mathcal{S}_M) = \frac{1}{M} \sum_{i \in \mathcal{S}_M} Z_i,$$

where

$$Z_i = \frac{\mathbf{E}[\Xi_i Y_i]}{\mathbf{E}[Y_i^2]} Y_i = \frac{\sigma_\Xi^2}{\sigma_\Xi^2 + \sigma_W^2}(\Xi_i + W_i).$$

The mean square distortion of the estimate is then given as follows:

$$D(\mathcal{S}_M) = \mathbf{E}\left[\left(\Xi - \hat{\Xi}(\mathcal{S}_M)\right)^2\right]$$

$$= \sigma_\Xi^2 - \frac{\sigma_\Xi^4}{M(\sigma_\Xi^2 + \sigma_W^2)}\left(2\sum_{i \in \mathcal{S}_M} \rho_{\Xi_i,\Xi} - 1\right) + \frac{\sigma_\Xi^6}{M^2(\sigma_\Xi^2 + \sigma_W^2)^2}\sum_{i \in \mathcal{S}_M}\sum_{j \neq i} \rho_{\Xi_i,\Xi_j}. \quad (12.1)$$

If there exists a tolerable distortion $D^* > D(\mathcal{S})$, then it is sufficient to gather information only from a subset of sensors $\mathcal{S}_M \subset \mathcal{S}$ such that $D(\mathcal{S}_M) < D^*$.

This example illustrates the effectiveness of spatial sampling. Specifically, as shown in (12.1), there are two conflicting factors that determine the sensors' selection: first of all, sensors close to the event Ξ should be chosen since these sensors have high correlation with the event Ξ and reduces the distortion through the parameter $\rho_{\Xi_i,\Xi}$ in the second term; secondly, sensors in close vicinity of each other should not be chosen simultaneously since the distortion increases due to the effect on ρ_{Ξ_i,Ξ_j} in the third term. Hence, from a MAC perspective, sensors closer to the source should be granted the priority of using the channel, but using the limited resources to transmit all the data in the vicinity of the source will increase the distortion due to the close location (and, thus high correlation) between sensors.

The Correlation-based Collaborative Medium Access Control (CC-MAC) protocol was proposed in (Vuran and Akyildiz 2006) based on this argument. During the initial phase of CC-MAC, the sink node computes a 'correlation radius' based on the statistics of the sensor field and the distortion constraint of the system. Then, broadcasts this information to the sensors. Similar to CSMA, CC-MAC adopts the RTS/CTS/DATA/ACK handshake every time it has a message to transmit. To eliminate the redundant transmissions from highly correlated sensors, each sensor listens to the channel for RTS packets and checks if these packets belong to sensors within its correlation radius. If in fact an RTS belongs to one of these neighbors, the sensor will enter a sleep state to avoid idle-listening, overhearing and transmission redundancy. This method effectively reduces the energy and bandwidth consumption but relies strongly on knowledge of the statistical model of the sensor field.

The delay in communications may also have a large impact on the application performance especially when the sensor field is changing rapidly over time. However, this aspect is often neglected in the studies of sensor MAC protocols. In (Cristescu and Vetterli 2005), a spatial sampling scheme is considered for the gathering of a time-varying sensor field. The author quantifies the distortion caused by both the MAC delay and the number of sensor samples that are gathered at the sink node. Although an increase of sensor samples will improve the distortion, there is an optimal point beyond which the MAC delay caused by channel contention will dominate the performance. Therefore, it is important to derive MAC protocols that meet the distortion constraint at each instant in time with the careful selection of sensors and an appropriate packet drop rate, which is subject to future research.

12.5 Cooperative MAC Protocol for Independent Sources

As shown in the previous sections, the signal processing performed in the sensor network applications can be facilitated with cross-layered designs between the application and the MAC layers. However, instead of making slight modifications to existing MAC protocols, new signal processing techniques can be used to improve the efficiency of sensor network MAC protocols both in terms of conventional QoS attributes, such as throughput and delay, and also in terms of the application performance. In fact, capitalizing on the collaborative nature of sensor networks, we show that these improvements can be obtained through cooperative communications or collaborative signal processing between sensors.

Cooperative communications

Cooperative communications, as proposed for user-oriented systems in (Laneman and Wornell 2003; Laneman et al. 2004; Sendonaris et al. 2003), allows distributed users to share and to coordinate the use of their resources in a wireless environment. To achieve this goal, users must communicate their local information to their respective partners and help relay the messages received from their cooperating partners. As a result, users that momentarily experience a deep fade in their link towards the destination can utilize quality channels provided by their partners to transmit their data. This is the *spatial diversity gain* achieved with cooperative communications, which is similar to that of multiple antenna systems.

The cooperative system is best illustrated with a canonical three node example, as shown in Figure 12.7, where the two users, user 1 and user 2, are transmitting cooperatively their messages to the destination. Without cooperation, each user's transmission will go through an independent fading path as shown in the figure, in which case the transmission will fail with high probability if the signal-to-noise (SNR) ratio of its own path falls below a certain value. However, if the users are able to cooperate by transmitting each other's messages, the transmission will fail only when both channels experience deep fading simultaneously, i.e. the duration between the dashed vertical lines in Figure 12.7. Specifically, if each channel enters a deep fade with probability p, the probability of an unsuccessful transmission in the cooperative system will be equal to p^2 instead of the probability p in a non-cooperative system.

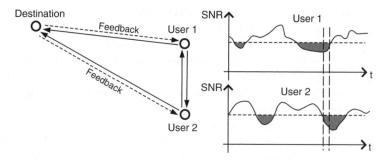

Figure 12.7 Two user cooperative communications system.

A cooperative communication scheme typically consists of two phases: *the coordination phase* and *the cooperative transmission phase*. In contrast to multiple antenna systems, coordination is necessary for cooperative communications since the users are distributed in space, but it is often the cause of inefficiency in many cases. Typical ways of achieving coordination are through either direct transmission between sensors (solid lines in Figure 12.7) or feedback from the destination (dashed lines). Depending on the messages received in the coordination phase, the users (sensors) encode their messages to enhance the reception at the receiver. A few popular cooperative schemes are the selective relaying (SR) scheme, the amplify-and-forward (AF) scheme and the decode-and-forward (DF) scheme. A survey of cooperative communication strategies can be found in (Hasna and Alouini 2004; Hong et al. 2007). Most of these cooperation schemes can be extended to a network with an arbitrary number of users. In sensor network applications, cooperation can be employed under different network topologies, such as: the cluster-based topology (Bandyopadhyay and Coyle 2003), where sensors in the same cluster cooperatively transmit to the cluster-head; the star topology, where sensors transmit through direct links to a central access point (Heinzelman et al. 2002; Venkitasubramaniam et al. 2004); or the multi-hop topology, where the destination is simply an intermediate relay in a multi-hop route.

Two features of the cooperative system allow us to improve upon conventional MAC protocols. On the one hand, users that experience bad channels due to their distant location or deep fading may utilize other users to relay their messages. One the other hand, the destination is able to combine the signals received at different time instants or from different users to enhance the reception performance. In non-cooperative systems, redundant transmissions at different time instants occur only when the previous transmissions fail and the failed transmissions are always discarded even though they actually contain partial information of the transmitted messages. With cooperation, transmissions that are usually considered as failures will be combined to enhance the reception. In this case, multiple failures in the conventional sense may add up to be a successful transmission in the cooperative system. Even with the cost of increased overhead due to coordination, there is an overall advantage in terms of outage probability and throughput (Laneman et al. 2004; Liu et al. 2005).

IEEE 802.11 legacy cooperative MAC

In (Liu et al. 2005), a MAC protocol was proposed based on cooperative transmissions and provides backward compatibility with current IEEE 802.11 legacy systems. Specifically, the cooperative MAC inherits the three-way handshake procedure adopted in the DCF of IEEE 802.11. In the proposed MAC protocol, each user maintains a table of all possible cooperating partners (or *helpers*), denoted by the set $\mathcal{C}$, by recording three fields: the helpers' MAC address, the achievable transmission rate between the source and the helper (R_{sh}) and the rate between the helper and the destination (R_{hd}). The rate R_{sh} is estimated by measuring the relative channel conditions of the transmissions when the source overhears the RTS/CTS message from the partner during previous transmissions. The rate R_{hd} and the MAC address of the helper are contained in the RTS/CTS messages that are emitted by the partners and are recorded from previous handshake procedures. When a user has a message to transmit, it will check within its table and select the user with the best rate as its partner. Suppose that each data packet contains L bits, the user will select the partner $h \in \mathcal{C}$ if and only if $L/R_{sh} + L/R_{hd} < L/R_{sk} + L/R_{kd}$ for all $k \in \mathcal{C}$ and $L/R_{sh} + L/R_{hd} < L/R_{sd}$, where R_{sd} is the rate of the direct transmission from the source to the destination. As a result, cooperation is used only when it outperforms direct transmission and the helper that yields the least amount of transmission time will be chosen as the potential cooperating partner.

As shown in Figure 12.8, once the partner is chosen, the source broadcasts an RTS message containing the partner's MAC address along with the rates R_{sh} and R_{hd}. If the RTS is successfully received and that the rates R_{sh} and R_{hd} are in fact achievable, the helper will reply with an HTS message, whose format is similar to that of the CTS message. After receiving the HTS at the destination, it will then reply with a CTS message that reserves the channel for the amount of time needed for the relay transmission. If the HTS was not received within a certain time-out period, the destination will instead emit a CTS message reserving the channel for the duration needed for direct transmission. The proposed strategy allows us to achieve spatial diversity with simple modifications to the IEEE 802.11 legacy system.

The cooperative method used in the above MAC protocol is a variant of the selective relaying scheme and does not utilize the advantage of signal combining at the receiver.

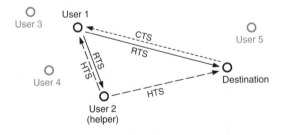

Figure 12.8 Cooperative MAC based on the IEEE 802.11 system.

In (N et al. 2005), a cooperative MAC protocol was proposed based on the DF relaying scheme. This work considers a random access system where all users have direct links to each other and the access point, hence, there is no need for the exchange of RTS/CTS messages. Similar to the DCF of the IEEE 802.11, each user that has a message to send will transmit whenever they sense the channel idle. If the message was not successfully received (and, thus, not acknowledged) by the access point, the users that overhear the transmission will backlog the packet and retransmit the packet in place of the source user after a random backoff time. The retransmission from cooperating terminals will continue until an acknowledgement is received from the access point or until a time-out period ends, in which case the backlogged packets will be dropped. The partially decoded packets are queued at the destination until a combined detection is able to correctly decode the packet (or until the time-out expires).

Collision resolution based on cooperation

Cooperative transmission is also used to resolve collisions in random access networks. Suppose we consider a slotted random access network where the users transmit in each time slot with independent probabilities. If the destination employs only a single-user receiver, multiple transmissions in the same time slot will result in a collision and no message can be extracted from the corrupted signal. When this occurs in conventional non-cooperative systems, the corrupted messages will be discarded by the receiver and the failed users will retransmit the same message in later time slots. However, there is a loss in efficiency since the information embedded in the discarded packets were not exploited in the reception.

In (Lin and Petropulu 2005), a collision resolution method was proposed for slotted random access networks by using cooperative transmissions and optimal combining at the receiver. When a collision occurs in a certain time slot, the network will enter a *cooperative transmission epoch* (CTE) for a duration of $\hat{K} - 1$ time slots. During each of these time slots, a user is selected to transmit based on a predetermined order. If the selected user is one of the colliding users, it will retransmit its own message in the time slot; otherwise, the user serves as a relay and retransmits the mixture of signals that was overheard during the collision.

Let $\mathbf{X}_i[n] = [X_{i,1}[n], \ldots, X_{i,J}[n]]$ be the J-symbol message transmitted by user i at time n. Suppose that K users, denoted by $S[n] = \{i_1, \ldots, i_K\}$, transmits during the n-th time slot and each user not in the set $S[n]$ receives a mixture of signals modeled by

$$\mathbf{Y}_r[n] = \sum_{i \in S[n]} h_{ir}[n]\mathbf{X}_i[n] + \mathbf{w}_r[n], \quad r \notin S[n]$$

where $h_{ir}[n]$ is the channel gain from user i to r during the n-th time slot and $\mathbf{w}_r[n]$ is the 1-by-J additive white Gaussian noise vector. Similarly, the access point (AP), denoted by d, will receive the signal

$$\mathbf{Z}[n] = \sum_{i \in S[n]} h_{id}[n]\mathbf{X}_i[n] + \mathbf{w}_d[n].$$

Based on the received signal, the AP will detect the occurrence of a collision and estimate the number of users $\hat{K}$ contributing to the mixed signal. If a collision is detected, the access point will notify the users of the beginning of the CTE and a set of users $\mathcal{R}[n] =$

$\{r_1, \ldots, r_{\hat{K}-1}\}$ will be chosen, in a distributed fashion, as the ones transmitting in the time slots $n + 1$ to $n + \hat{K} - 1$. If the selected user r_i does not belong to the set $\mathcal{S}[n]$, it will relay a scaled version of the mixed signals, similar to that of the amplify-and-forward cooperation scheme. Therefore, the signals received by the AP during the time slots of the CTE are given as

$$\mathbf{Z}[n+k] = \begin{cases} h_{rd}[n+k]\mathbf{X}_r[n] + \mathbf{w}_d[n+k], & r \in \mathcal{R}[n] \bigcap \mathcal{S}[n] \\ h_{rd}[n+k]\alpha[n+k]\mathbf{Y}_r[n] + \mathbf{w}_d[n+k], & r \in \mathcal{R}[n] \setminus \mathcal{S}[n], \end{cases}$$

for $k = 1, \ldots, \hat{K} - 1$, where $\alpha[n + k]$ is the scaling used to meet the power constraints of each user. In this case, the signal received over the time slots n to $n + \hat{K} - 1$ can be written in the form of a MIMO signal, i.e. we have

$$\mathbf{Z} = \mathbf{HX} + \mathbf{W}$$

where $\mathbf{Z} = [\mathbf{z}^T[n], \ldots, \mathbf{z}^T[n + \hat{K} - 1]]^T$, $\mathbf{X} = [\mathbf{X}_{i_1}^T[n], \ldots, \mathbf{X}_{i_K}^T[n]]^T$ and $\mathbf{W} = [\mathbf{w}^T[n], \ldots, \mathbf{w}^T[n + \hat{K} - 1]]^T$. For $\hat{K}$ sufficiently greater than K, the data can be reliably estimated with the maximum likelihood detector, i.e. $\hat{\mathbf{X}} = \arg\min_{\mathbf{X}} \|\mathbf{Z} - \mathbf{HX}\|_F$ (the Forbenius norm), or the zero-forcing receiver, i.e. $\hat{\mathbf{X}} = \mathbf{H}^{\dagger}\mathbf{Z}$ where † represents the pseudo-inverse.

Ideally, it is sufficient to have $\hat{K} \approx K$ and, thus, there is no loss of throughput even when collision occurs. However, in practice, the conditions of the matrix $\mathbf{H}$ may not be well enough to solve for $\mathbf{X}$ with only K slots of the CTE. Also, the overhead required to obtain the estimate of K and the error in the estimation will both cause a loss in throughput. Nonetheless, the overall throughput still exceeds that of random access protocols without cooperation. A similar collision resolution method can be obtained without the use of cooperation by simply having users retransmit their own packets during the CTE, as proposed in (Tsatsanis et al. 2000). However, there is a loss in performance due to the lack of spatial diversity in this scheme.

Two advantages are exploited in the cooperative MAC protocols: the spatial diversity that is used to provide each user with a more reliable transmission path and the spatial multiplexing gain that is used to separate the sources embedded in the mixed signal. However, in sensor network applications, the sensors' data are often highly correlated and it is often unnecessary to decode separately the messages from each sensor. Instead, we can extract only the data that is relevant for detection or estimation. This is explored in the remainder of this chapter.

12.6 Cooperative MAC Protocol for Correlated Sensors

Conventionally, MAC is used to distinguish, at the receiver, the different messages received from multiple independent sources. However, in sensor networks, the messages transmitted by the users often represent only a small number of sources or events, such as object tracking or target detection, or it may represent a sensor field that is highly correlated in space, such as temperature or humidity measurements. Therefore, the amount of informative data generated by the sensor field would not be proportional to the number of sensors. Consequently, the sensor network MAC need only to extract the data relevant for computation instead of collecting all the data packets transmitted by the sensors, which are inherently redundant. Although strategies such as DSC or spatial sampling can be used to achieve this task, they do not benefit from the energy efficiency and improved reliability of cooperative systems. In this section, we describe a class of Data-Centric Cooperative MAC (DC-MAC) protocols and show the effectiveness of this class of strategies in eliminating the sensors' redundant transmissions using cooperative signal processing techniques. This method is illustrated in two applications: the data retrieval problem and the distributed detection or estimation problems.

12.6.1   Data Retrieval from Correlated Sensors

Consider a data retrieval problem where a data gathering node is to obtain a reliable estimate of each sensor's observation using the minimum number of transmissions or time slots. When using a spatial sampling scheme, the data gathering node collects data only from a number of sensors whose data are sufficient to meet the distortion constraint. However, the efficiency degrades dramatically under strict distortion requirements since the data from a large fraction of sensors must be collected. In this case, the redundant transmissions will increase dramatically if distributed source coding or multiple description coding schemes (Cover and Thomas 1991) are not employed. This disadvantage is overcome with DC-MAC.

The key intuition of the DC-MAC is to have multiple highly correlated sensors share the same transmission channel instead of assigning a separate channel to each individual user. In the extreme case where all sensors observe the same data, it should be sufficient to assign a single channel for the transmission of this data and have all sensors transmit cooperatively in the same time slot. In practical scenarios, we can choose a subset of sensors that has a high probability of containing the same data to transmit cooperatively in each time slot. In fact, this occurs among sensors that are closely located in a spatially correlated sensor field. The concept is similar to that of group testing (Dorfman 1943; Sobel and Groll 1959) in blood testing applications where multiple blood samples are pooled together and tested simultaneously to see if one of the samples is infected with the disease.

As a proof of concept, we first show the performance of DC-MAC for a binary correlated model and, then, generalize the concept to include different data models and cooperation strategies.

Example: One-dimensional Binary Markov Field

Consider a network of N sensors, denoted by $\mathcal{S} = \{1, 2, \ldots, N\}$. Let $\mathbf{X} = [X_1, \ldots, X_N]$ be the set of data to be collected from the sensors, where $X_i \in \{0, 1\}$ represents the data observed at sensor i, e.g. the binary quantization of a continuous random field. For the simplicity of illustration, we shall assume that the sensors are located in a one-dimensional area $[0, D]$, as shown in Figure 12.9(a) and that the sensors are taking binary samples of a continuous Markov random field. In this case, the data vector $\mathbf{X}$ forms a first-order shift invariant Markov chain with two possible states 0 and 1, i.e.

$$P(\mathbf{X}) = P(X_1) \prod_{i=2}^{n} P(X_i | X_{i-1}) \tag{12.2}$$

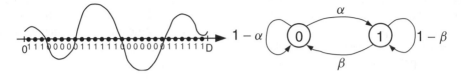

(a) The line network   (b) The two state Markov chain

Figure 12.9 Illustration of the binary Markov data model.

with the transition probabilities $\alpha = P(X_i = 1|X_{i-1} = 0)$ and $\beta = P(X_i = 0|X_{i-1} = 1)$, as illustrated in the state diagram shown in Figure 12.9(b).

Assume that the probability of X_i, for all i, is equal to the steady state distribution where

$$p \triangleq \Pr(X_i = 1) = \frac{\alpha}{\alpha + \beta} \tag{12.3}$$

and, similarly, $q \triangleq \Pr(X_i = 0) = \frac{\beta}{\alpha+\beta} = 1 - p$. The correlation coefficient is defined as

$$\rho \triangleq \frac{\mathrm{Cov}(X_i, X_{i+1})}{\sigma_{X_i}\sigma_{X_{i+1}}} = 1 - (\alpha + \beta) = 1 - \overline{\rho} \tag{12.4}$$

where $\sigma_{X_i}^2 = \mathrm{E}[X_i - E(X_i)]^2 = p(1 - p)$ and $\mathrm{Cov}(X_i, X_{i+1}) = p(1 - p)[1 - (\alpha + \beta)]$. We note that each value of (p, ρ) uniquely specifies a pair of transition probabilities (α, β). In this example, we consider the case where ρ takes on values within the interval $[0, 1]$. When $\rho = 0$, the model reduces to the *i.i.d.* Bernoulli probability model, which is the model adopted in the group testing literature (Dorfman 1943; Sobel and Groll 1959). Even though the sensors are independent in this case, the total number of transmission channels can still be reduced due to the low aggregate entropy when p is close to 0 or 1.

DC-MAC: A query-and-response data retrieval strategy

To efficiently retrieve the data from the sensors, the receiver polls a group of sensors to transmit in each time slot depending on the statistics of the sensors' data. Specifically, we pick a group of sensors that is likely to contain the same quantized measurement. For example, when the sensors in a group $\mathcal{G} \subset \mathcal{S}$ are likely to contain the bit $b = 0$, the receiver polls the sensors in $\mathcal{G}$ along with a query asking whether or not the sensors actually contain the bit 0. This query is denoted by $Q = (\mathcal{G}, b) \in 2^{\mathcal{S}} \times \{0, 1\}$ (where $b = 0$ in this specific example). As a response to this query, sensor i transmits a pulse in the assigned time slot only if $i \in \mathcal{G}$ and $X_i \neq b$ (in protest of the wrong guess imposed by the data gathering node); otherwise, the sensor remains silent. Let us split the group $\mathcal{G}$ into two distinct subgroups $\mathcal{G}^{(0)} \triangleq \{i \in \mathcal{G} : X_i = 0\}$ and $\mathcal{G}^{(1)} \triangleq \{i \in \mathcal{G} : X_i = 1\}$. Thus, we have $\mathcal{G} = \mathcal{G}^{(0)} \cup \mathcal{G}^{(1)}$ and the signal arriving at the receiver can be denoted by

$$r(t) = \sum_{i \in \mathcal{G} \setminus \mathcal{G}^{(b)}} h_i(t) * p(t) + n(t) \tag{12.5}$$

where $p(t)$ is the signal emitted when the query was not satisfied, $h_i(t)$ is the channel response between sensor i and the data gathering node and $n(t)$ is the additive white Gaussian noise process. Let us consider a simple receiver structure[2] where the receiver only detects for the existence of a pulse. In this case, the receiver makes a decision among the two hypotheses:

$$
\begin{aligned}
H_0 : \quad & r(t) = n(t) \\
H_1 : \quad & r(t) = \sum_{i \in \mathcal{G} \backslash \mathcal{G}^{(b)}} h_i(t) * p(t) + n(t) \qquad \text{for } |\mathcal{G} \backslash \mathcal{G}^{(b)}| > 0.
\end{aligned}
\tag{12.6}
$$

In this example, we assume that the noise is negligible and focus on analyzing the compression capability of the proposed algorithm. In the absence of noise, when H_0 occurs, the receiver knows that all the sensors contain the bit b and, therefore, has resolved the set using only one time-slot. However, when H_1 occurs, the receiver knows that there exists at least one sensor in the group that does not possess the bit b but no information is given on the specific identity or even the total number of these sensors. In this case, smaller subsets of the group $\mathcal{G}$ must be polled in the subsequent time slots in order to identify the sensors possessing the opposite message. By appropriately choosing the queries in each time slot, one can eventually resolve the entire set of sensors' data. In fact, our goal is to obtain a lossless estimate of $\mathbf{X}$ at the data gathering node using the minimum number of time slots.

Considering the noise-free version of (12.5), the data gathering node receives a binary symbol $Z[m] = 1$ in the m-th transmission time-slot if H_1 occurs, and receives $Z[m] = 0$ if H_0 occurs. The sequence of outcomes combined with the sequence of queries $\{Q[m]\}$ allows the receiver to reconstruct the vector $\mathbf{X}$. Suppose that L is the total number of queries needed to obtain a lossless reconstruction of the vector $\mathbf{X}$. Then, the vector $\mathbf{Z}[1 : L] = [Z[1], Z[2], \ldots, Z[L]]$ serves as a lossless data representation of the vector $\mathbf{X}$ and the expected number of queries $\mathbf{E}[L]$ is lower bounded by the entropy of $\mathbf{X}$ (Cover and Thomas 1991), i.e.

$$
H(\mathbf{X}) \le \mathbf{E}[L] \le N.
\tag{12.7}
$$

The upper bound is achieved when we assign an individual time-slot to each sensor. To minimize the total number of transmissions, the sequence of queries $\{Q[m]\}$ must be optimally designed based on the available information at the data gathering node, i.e. the previously received channel outputs $\mathbf{Z}[1 : m - 1]$ and the statistical distribution of the data $P_{\mathbf{X}}$. Although there is no tractable method for obtaining the optimal set of queries, the advantage of DC-MAC can be shown through the analysis of known suboptimal algorithms.

Two methods have been proposed in (Hong and Scaglione 2004b): (1) *the optimized recursive algorithm* and (2) *the tree splitting algorithm*. These methods follow the same approach as those used to analyze the performance of group testing problems, as given in (Berger et al. 1984; Capetanakis 1979). The first scheme allows us to illustrate the effectiveness of this strategy with finite number of sensors while the second strategy provides an analytical study on the scaling of the performances as N increases.

Strategy I: Optimized recursive algorithm

The intuition of the proposed scheme is to query groups of sensors that are likely to contain the same data bit. In the binary Markov model, it is desirable in most cases to

---

[2]Note that more complicated receivers can be used, such as to estimate the number of pulses instead of simply detecting for the existence, and the data can be resolved more rapidly. However, this simple receiver is sufficient for the performance discussions given in the following.

choose groups of contiguous sensors since they are most likely to contain the same data. This spatial dependency is consistent with most sensor network applications. Capitalizing on this observation, we derive the optimal strategy that queries the sensors in the order of their indices and always takes groups of contiguous sensors. In this case, a node must be included in the current query if it contains the smallest index among the unresolved sensors. Our search for the optimal group $\mathcal{G}[m]$ (for the m-th query) is reduced from the set of size 2^N (i.e. the power set of $\mathcal{S}$) to a set of size N by taking groups of contiguous sensors in the order of their index. Note that, in addition to the optimal group, the optimal question must also be determined.

Let $L_{rec}(\mathbf{X}_i^j)$ be the minimum number of queries needed to resolve the data $\mathbf{X}_i^j = [X_i, X_{i+1} \ldots, X_j]$ using the optimized recursive strategy. For the sensor network $\mathcal{S}$, our goal is to find the expected value of $L_{rec}(\mathbf{X}_1^N)$. To initialize the querying process, we start by allocating a transmission slot to the first sensor in the group, i.e. sensor 1, and ask either one of the two questions $b = 0$ or $b = 1$. Due to the correlation between adjacent sensors, the realization of the first node will help us determine the best question to ask in the subsequent queries. Following the approach given in (Berger et al. 1984), we can obtain the expected number of queries under the optimized recursive scheme as

$$\mathbf{E}[L_{rec}(\mathbf{X}_1^N)] = 1 + p \cdot \mathbf{E}[L_{rec}(\mathbf{X}_2^N)|X_1 = 1] + (1 - p) \cdot \mathbf{E}[L_{rec}(\mathbf{X}_2^N)|X_1 = 0]$$

$$= 1 + p \cdot G_1(N - 1) + (1 - p) \cdot G_0(N - 1) \tag{12.8}$$

where

$$G_a(m) \triangleq \mathbf{E}[L_{rec}(\mathbf{X}_{i+1}^{i+m})|X_i = a].$$

The second equality follows from the spatial homogeneity of the Markov Chain which yields the fact that

$$\mathbf{E}[L_{rec}(\mathbf{X}_1^m)|X_0 = a] = \mathbf{E}[L_{rec}(\mathbf{X}_{i+1}^{i+m})|X_i = a], \quad \text{for all positive integer } i.$$

To solve for $G_a(m)$, we define the functions

$$F_a(m, x, b) \triangleq \mathbf{E}[L_{rec}(\mathbf{X}_{i+1}^{i+m})|X_i = a, \exists j \in \{i+1, \ldots, i+x\} \text{ s.t. } X_j = \overline{b}]$$

$$J_a(m, x, b, y) \triangleq \mathbf{E}\left[L_{rec}(\mathbf{X}_{i+1}^{i+m}) \left| \begin{array}{l} X_i = a, \exists j \in \{i+1, \ldots, i+x\} \text{ s.t. } X_j = \overline{b}, \\ \exists r \in \{i+1, \ldots, i+y\} \text{ s.t. } X_r = b \end{array} \right. \right],$$

for $m \geq x, r$, where $\overline{b}$ is the complement of b. Let us define the events

$$E_{xb} = \{X_j = b, \text{ for } j = i+1, \ldots, i+x\},$$

$$E_1 = \{X_i = a, \exists j \in \{i+1, \ldots, i+x\} \text{ s.t. } X_j = \overline{b}\}$$

$$E_2 = \{X_{i-1} = a, \exists j \in \{i+1, \ldots, i+x\} \text{ s.t. } X_j = \overline{b}, \exists r \in \{i+1, \ldots, i+y\} \text{ s.t. } X_r = b\}.$$

Then, the functions in (12.8) can be evaluated with the following set of recursive equations:

$$G_a(m) = 1 + \min_{\substack{b \in \{0,1\} \\ 1 \leq x \leq n}} P(E_{xb}|X_i = a)G_b(m - x) + [1 - P(E_{xb}|X_i = a)] F_a(m, x, b)$$

$$F_a(m, x, b) = 1 +$$

$$\min_{c \in \{0,1\}} \begin{cases} \min_{1 \leq y < x} P(E_{yc}|E_1)F_c(m - y, x - y, b) + \left[1 - P(E_{yc}|E_1)\right] F_a(m, y, b), \\ \qquad\qquad\qquad\qquad\qquad\qquad\qquad\qquad\qquad\qquad \text{for } c = b \\ \min_{1 \leq y \leq m} P(E_{yc}|E_1)G_c(m - y) + \left[1 - P(E_{yc}|E_1)\right] J_a(m, x, b, y), \\ \qquad\qquad\qquad\qquad\qquad\qquad\qquad\qquad\qquad\qquad \text{for } c = \overline{b}. \end{cases}$$

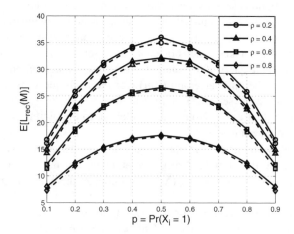

Figure 12.10 For $N = 36$ and $\rho = 0.2, 0.4, 0.6, 0.8$, we show the performance of the optimized recursive algorithm (solid line) and the entropy lower bound of (12.7) (dashed line).

$$J_a(m, x, b, y) = 1 +$$

$$\min_{d \in \{0,1\}} \begin{cases} \min_{1 \leq z < x} P(E_{zd}|E_2)F_d(m - z, x - z, b) + \left[1 - P(E_{zd}|E_2)\right] J_a(m, z, b, y), \\ \hspace{7cm} \text{for } d = b, \\ \min_{1 \leq z < y} P(E_{zd}|E_2)F_{\overline{b}}(m - z, y - z, \overline{b}) + \left[1 - P(E_{zd}|E_2)\right] J_a(m, z, \overline{b}, x), \\ \hspace{7cm} \text{for } d = \overline{b}. \end{cases}$$

In Figure 12.10, we show the performance of the optimized recursive algorithm for a network of $N = 36$ sensors and for various values of ρ. The dashed lines represent the entropy lower bound of (12.7) for each case of ρ and the solid lines represent the performance of the proposed algorithm. We observe that the proposed querying strategy closely approximates the optimal performance (i.e. the entropy lower bound that can be achieved asymptotically with Huffman coding). More importantly, we can see that the expected number of time-slots vary with the entropy of the data as opposed to using a fixed number of time slots that are proportional to number of sensors, as with user-oriented MAC protocols such as conventional CSMA or TDMA schemes. The advantage of DC-MAC is most promising when p is close to 0 or 1 and when ρ is close to 1, i.e. high spatial correlation.

Strategy II: Tree splitting algorithm

With Strategy I, we have shown that the proposed cooperative MAC protocol yields comparable performances to the optimal compression scheme for a finite number of sensors. However, the recursive formulation of (12.8) does not show explicitly the effect of N, p and ρ on the performance. In this section, we propose a tree splitting algorithm to determine the group of sensors for each query and capitalize on the simple structure of the algorithm to attain the asymptotic performance of the cooperative MAC protocol.

To simplify our analysis, we consider a suboptimal scheme where each group of sensors is queried twice whenever it is selected, e.g. for some $n \geq 1$, let the two queries $Q[2n - 1] = (\mathcal{G}[n], 0)$ and $Q[2n] = (\mathcal{G}[n], 1)$ be imposed on the same group $\mathcal{G}[n]$. This approach yields a pair of outputs $(Z[2n - 1], Z[2n])$ and provides us with the ternary information on $\mathcal{G}[n]$:

0: $(Z[2n - 1], Z[2n]) = (0, 1)$
1: $(Z[2n - 1], Z[2n]) = (1, 0)$
e: $(Z[2n - 1], Z[2n]) = (1, 1)$

where **0** indicates the fact that all sensors in $\mathcal{G}[n]$ contain the bit 0 and, vice versa, for **1**. When **e** occurs, the receiver identifies the fact that both 0 and 1 are contained in the group of sensors but it is not able to identify the exact sensors that have the bit 1 and those that have the bit 0. This is referred to as the *erasure* case. When **e** is received, a subset of the original group of sensors must be taken in the subsequent queries until the group is resolved. By constraining the strategy such that both questions are asked using consecutive queries, what remains to be determined is the sensors chosen in each group. The approach is suboptimal but we shall show that it is sufficient to observe the scaling behavior of the DC-MAC.

Consider the case where $N = 2^M$ for some positive integer M. The binary tree splitting protocol initially splits the group of sensors into two distinct subgroups of equal size and each of these subgroups are queried twice in consecutive time slots. If the outcome of the query results in an erasure, the original group is divided again into two subgroups of the same size where each subgroup is tested separately in subsequent time slots. Otherwise, the outcome is **0** or **1** and the receiver goes on to query a new group of sensors.

For example, as shown in Figure 12.11, we consider a network of 16 nodes where each vertex $\mathcal{G}_{ij}$ denotes a group that consists of all the sensors within its subtree. In the proposed algorithm, the sequence of queries starts from the subgroups of $\mathcal{G}_{00}$, i.e. $\mathcal{G}_{10}$ and $\mathcal{G}_{11}$, and continues splitting and querying the smaller subgroups each time the larger group cannot be resolved through a single query. If the query on $\mathcal{G}_{ij}$ results in either **0** or **1**, the system continues to query the group $\mathcal{G}_{i,j+1}$ since it is the smallest group that is not yet resolved. However, if the test results in an erasure, the vertex $\mathcal{G}_{ij}$ branches into two subgroups where the group $\mathcal{G}_{i+1,2j}$ is queried next. For the data vector as shown at the bottom of the tree in Figure 12.11, the sequence of tests are done in the order of the following groups: $\mathcal{G}[1] = \mathcal{G}_{10}$, $\mathcal{G}[2] = \mathcal{G}_{20}$, $\mathcal{G}[3] = \mathcal{G}_{30}$, $\mathcal{G}[4] = \mathcal{G}_{40}$, $\mathcal{G}[5] = \mathcal{G}_{31}$, $\mathcal{G}[6] = \mathcal{G}_{21}$, $\mathcal{G}[7] = \mathcal{G}_{11}$, $\mathcal{G}[8] = \mathcal{G}_{22}$, $\mathcal{G}[9] = \mathcal{G}_{23}$, $\mathcal{G}[10] = \mathcal{G}_{36}$ and $\mathcal{G}[11] = \mathcal{G}_{37}$. We note that, after the query on the group $\mathcal{G}_{40}$, the query on $\mathcal{G}_{41}$ is skipped since we already know that $\mathcal{G}_{40}$ and $\mathcal{G}_{41}$ contain different bits from the erasure resulting from the query on $\mathcal{G}_{30}$.

Considering the Markov model in (12.2), we can compute the minimum number of time-slots needed to resolve the sensors' data using the binary tree splitting algorithm.

Theorem 12.6.1 (Hong and Scaglione (2004b)) *Consider a network of $N = 2^M$ sensors and the binary data* **X** *modeled by the two-state Markov Chain with the parameters (p, ρ).*

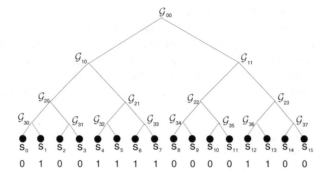

Figure 12.11   Example of the realization of a sensor field with the binary sequence 0100111100001100.

With the $\{0, 1, e\}$ *information, the binary tree splitting algorithm yields an expected number of time-slots equal to*

$$\mathbf{E}[L_{tree}(N)] = 2 + \sum_{i=1}^{M-2} 2^{i+1}\psi(M-i, p, \rho) + 2^{M-1}\psi(1, p, \rho). \tag{12.9}$$

where $\psi(M-i, p, \rho) = 1 - p \cdot [1 - (1-p)(1-\rho)]^{2^{M-i}-1} - q \cdot [1 - p(1-\rho)]^{2^{M-i}-1}$.

When the correlation is high, i.e. ρ close to 1, the result in (12.9) can be approximated as

$$\mathbf{E}[L_{tree}(N)] \cong 2 + pq(1-\rho)[4N \cdot \log_2 N - 9N + 8]. \tag{12.10}$$

This result shows that the expected number of queries scales with the parameters p and ρ which determines the statistics of the sensor field. Specifically, $\mathbf{E}[L_{tree}]$ decreases as p or q approaches 0 and also for ρ approaching 1. This is consistent with the behavior observed in Figure 12.10. However, since we claim no optimality in this scheme, poor performances may occur when the sensors have small correlation. Nonetheless, the performance depends on the data statistics rather than having a fixed transmission cost with respect to the number of users.

In the binary tree splitting algorithm, we initiate the process by splitting the network into 2 subgroups and then proceed with the binary splitting within each group. However, when the correlation is low, the large initial groups will almost certainly result in an erasure. Therefore, it would be desirable to partition the network initially into 2^K subgroups, where $K \geq 1$, and choose the optimal K for each values of (p, ρ). Thus, we have a 2^K-ary tree splitting scheme, as shown in Figure 12.12, where each partition proceeds in a similar way as the binary tree splitting algorithm described in Figure 12.11. Following the same approach as that shown in Capetanakis (1979), we can find the optimal K that minimizes the expected number of tests $\mathbf{E}[L]$ for each value of (p, ρ). We note that the optimal K, as a function of p and ρ, decreases monotonically with respect to ρ and it is symmetric around $p = 1/2$ (Hong and Scaglione 2006). In the following, we derive the optimal K for two cases: (i) the case where there is a fixed value of $p = 0.5$; and (ii) the case where ρ is close to 1.

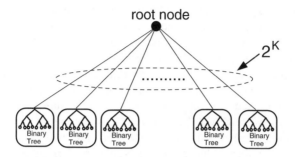

Figure 12.12 The illustration of the tree under the 2^K-ary tree splitting algorithm.

Theorem 12.6.2 (Hong and Scaglione (2004b)) **Case I:** *For fixed $p = 0.5$, it is optimal to split the root node immediately into 2^K branches where*

$$1 \le K = \left\lceil M - \log\left(\frac{1}{1 - \log(1 + \rho)}\right) \right\rceil \le M. \qquad (12.11)$$

Case II: *For ρ close to 1,*

$$1 \le K \approx \left\lceil M - \log\left(1 + \frac{1}{4p(1 - p)(1 - \rho)}\right) \right\rceil \le M. \qquad (12.12)$$

From (12.11), we can see that the optimal splitting of the network for $(p, \rho) = (0.5, 0)$ is equivalent to querying each individual sensor separately, since $K = M$ in this case. Under the Markov data model, the entropy of the data $\mathbf{X}$ scales with N for fixed values of p and ρ since $H(\mathbf{X}) = h(p) + (N - 1) \cdot H(X_1|X_0)$. However, if optimized splitting is not applied, the average number of queries for the binary tree splitting algorithm increases as $O(N \log N)$ according to (12.10). Interestingly, from (12.10) and Theorem 12.6.2, we can instead achieve an increase of only $O(N)$ when the optimal splitting is performed. Therefore, it is crucial to adopt the optimal splitting in the tree algorithm.

Discussions on the optimal DC-MAC for the binary Markov case

Let L_{opt} be the number of time-slots used by the optimal querying strategy under the binary Markov model in (12.2). Let L^*_{tree} be the number of time-slots used by the optimized tree splitting algorithm shown in Figure 12.12. The performance of the tree algorithm serves as an upper bound to the optimal strategy, whose complexity is exponential in the number of sensors (Du and Hwang 1993). Hence, we have

$$H(\mathbf{X}) \le \mathbf{E}[L_{opt}] \le \mathbf{E}[2L^*_{tree}]. \qquad (12.13)$$

The multiplication of 2 on the right side of the inequality is due to the two queries that we impose on each chosen group, which does not apply to the optimal scheme. From the bounds given in (12.13), we derive the achievable scaling performances of the expected number of time slots for the optimal DC-MAC strategy. We consider two cases: (i) the case where the number of sensors increases while the density remains constant; and (ii) the case where the density increases linearly with the number of sensors.

As described in the system model, the binary data at each sensor can be viewed as the binary quantization of a spatially continuous Markov random process, which results in a process similar to the random telegraph process (Papoulis and Pillai 2001). Therefore, in the first scenario, the correlation coefficient ρ remains constant as the size of the network increases. In the second case, the distance between sensors decreases as the number of sensors increases and, thus, increases the correlation between sensors. Suppose that the sensors are placed in a fixed interval $[0, D]$, as shown in Figure 12.9(a) and the distance between sensors is approximately D/N. It is easy to show that the correlation coefficient between adjacent sensors satisfies $1 - \rho = c'/N$. For these two cases, we can show that the optimal DC-MAC protocol achieves the best scaling performances in the sense that the scaling with respect to N is the same as that of the entropy, for the cases indicated in the following theorem.

Theorem 12.6.3 (Hong and Scaglione (2006)) *Let $E[L_{opt}]$ be the expected number of queries necessary for the optimal DC-MAC strategy for the binary Markov data model. Then, the following properties hold:*

1) for fixed (p, ρ) such that $(1 - \rho) \ll 1$,

$$E[L_{opt}] = O(N) = O(H(\mathbf{X})); \tag{12.14}$$

2) for fixed p and $\overline{\rho} = c'/N$ for some $c' > 0$,

$$E[L_{opt}] = O(\log(N)) = O(H(\mathbf{X})). \tag{12.15}$$

For the binary source example, we have shown that data-centric cooperative MAC protocols can significantly reduce the total number of time-slots necessary for the sensor to convey its information to a data gathering node. The key feature of this strategy is to impose a guess or query on $\mathbf{X}$, given the knowledge obtained from previous transmissions, that will yield the most amount of information from the sensors response. This method simultaneously compresses the sensors' information while scheduling the transmission of each sensor. However, we note that the class of DC-MAC protocols are not restricted to binary sources. A trivial extension to m-ary sources is given in (Hong and Scaglione 2004a) where each query is to ask m questions corresponding to the symbols in the m-ary alphabet. In fact, DC-MAC can be generalized for different data statistics and for different receiver structures, e.g. a receiver that gives an estimate of the number of sensors transmitting in each time-slot. A generalized formulation is given in the following and a heuristic algorithm is given for the selection of queries.

12.6.2   Generalized Data-Centric Cooperative MAC

Consider a set of N sensors, denoted by $\mathcal{S} = \{1, 2, \ldots, N\}$ and let $\mathbf{X} = [X_1, X_2, \ldots, X_N]$ be the set of random variables, defined on the probability space $(\Omega, \mathcal{B}, P_{\mathbf{X}})$, which represents a snapshot of measurements obtained by the sensors at a particular time instant. In general, the data at sensor i, i.e. X_i, may belong to an arbitrary alphabet $\mathcal{X}_i$, which can be either finite or infinite. The goal is to obtain a reliable estimate of the vector $\mathbf{X}$ at the data gathering node with the minimum number of transmission time-slots.

As described in the binary example, DC-MAC is based on a query-and-response system which can be modeled as a multiple-access channel with feedback, as shown in

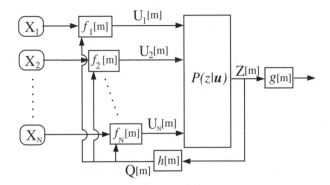

Figure 12.13 Cooperative MAC with Feedback Communications.

Figure 12.13. The feedback from the data gathering node is a way to achieve coordination among cooperating users, as described in Section 12.5. In this model, each sensor transmits through a direct link to the data gathering node and the desired information is extracted from the sequence of channel outputs that are transmitted in response to the queries. Each channel output Z is the result of the simultaneous transmission from multiple sensors and the feedback from the receiver characterizes the query that is sent by the data gathering node.

Let $U_i[m] \in \mathcal{U}_i$, for $m \geq 1$, be the symbol transmitted by sensor i during the m-th time-slot and let $Z[m] \in \mathcal{Z}$ be the corresponding channel output, which is determined by the conditional probability function $p(Z[m]|\mathbf{U}[m])$ where $\mathbf{U}[m] = [U_1[m], \ldots, U_N[m]]$. Assume that the channel is invariant to the time index m, i.e. $p(Z|\mathbf{U}) \triangleq p(Z[m]|\mathbf{U}[m])$ for all m, and assume that the statistics of $\mathbf{X}$ and the channel $p(Z|\mathbf{U})$ are known at both the transmitter and the receiver. The sequence of channel outputs are used to form a query or feedback symbol through the function $h[m] : \mathcal{Z}^m \rightarrow \mathcal{Q}$, where $\mathcal{Q}$ is the set of possible queries, i.e. $Q[m] = h[m](Z[1], \ldots, Z[m])$. In the binary example, we defined the set of queries to be $\mathcal{Q} = 2^S \times \{0, 1\}$ and utilized these queries to notify the sensors implicitly about their partners' data and to coordinate the sensors' transmissions. When the query is sent, each sensor performs a symbol-by-symbol encoding (Gastpar and Vetterli 2003) and emits the symbol $U_i[m]$, which depends on both the local data X_i and the query $Q[m]$. Thus, we define the encoding function $f_i[m] : \mathcal{X}_i \times \mathcal{Q} \rightarrow \mathcal{U}_i$. Depending on the nature of the problem, the functions $f_i[m]$, for all i and m, are subject to certain constraints, e.g., in practice, it is natural to impose a power constraint on the sensors (or, more specifically, on the variance of $U_i[m]$). With the information obtained through m channel outputs $Z[1], Z[2], \ldots, Z[m]$, the receiver computes an estimate of the observation with the decoding function $g[m] : \mathcal{Z}^m \rightarrow \prod_{i=1}^N \mathcal{X}_i$. The estimate obtained after the m-th transmission is denoted by $\hat{\mathbf{X}}[m]$ and the initial estimate obtained before any transmission occurs is denoted by $\hat{\mathbf{X}}[0]$.

Let L be the random variable representing the total number of channel accesses used to retrieve the data $\mathbf{X}$. Given the distortion measure $d(\mathbf{X}, \hat{\mathbf{X}}[L])$ and the constraint D, our goal is to minimize the expected number of time-slots, i.e. $\mathbf{E}[L]$, such that the estimate achieves the distortion $\mathbf{E}[d(\mathbf{X}, \hat{\mathbf{X}}[L])] \leq D$ (through the design on $f_i[m]$ and $g[m]$). In the previous example, we considered the lossless reconstruction of $\mathbf{X}$ where D is set to 0. We note that, in general, the symbol-by-symbol encoding considered above may not reach the distortion constraint most efficiently. However, it is well known that the joint source and channel coding over correlated sources (Cover et al. 1980) and the feedback structure (Cover and Leung 1981) adopted in DC-MAC can improve the capacity of the multiple access channel. This is the form of cooperative advantage that we look to exploit with DC-MAC.

A drawback of this scheme is the complexity involved in computing the optimal sequence of queries since it requires optimization over all possible groups of sensors and over all time-slots. In fact, this is the reason for the two suboptimal strategies proposed in the binary Markov problem. A standard approach to reducing the complexity is to reduce the size of the search over a reasonably large set or to reduce the problem to a step-by-step optimization where the optimization is performed separately for each time-slot. For example, in the binary Markov case, we restrict the search over sets of consecutive sensors, which reduces the problem to require only polynomial complexity and experiences little loss in performance. In the following, we provide a heuristic search of the queries that allows us to reduce the problem to a step-by-step optimization and the algorithm is equally applicable for all cases.

Heuristic algorithm based on the mutual information criterion

Suppose that the data gathering node is able to compute the best estimate of $\mathbf{X}$, at any instant in time. Then, we need only to determine the query function $h[m]$ and the encoding functions $f_i[m]$, for all i and m. In order to achieve the distortion D with the minimum number of time slots, it is reasonable to choose the functions that provide the data gathering node with the most amount of information about $\mathbf{X}$. Therefore, we propose to use the following mutual information as the design criterion of these functions.

Definition 12.6.4 (Mutual Information Criterion) *During the m-th time slot and given the channel outputs* $\mathbf{Z}_1^{m-1} = \mathbf{z}_1^{m-1}$, *the functions* $h[m]$ *and* $\{f_i[m], \forall i\}$ *are chosen as*

$$\{h[m], f_1[m], \ldots, f_N[m]\} = \underset{h[m], f_1[m], \ldots, f_N[m]}{\arg\max} I\left(\mathbf{X}; Z[m] \middle| f_i[m](X_i, h[m](\mathbf{z}_1^{m-1})), \forall i\right)$$

If the channel is deterministic (i.e. it is noiseless as assumed in the binary example), the criterion is reduced to the following

$$\{h[m], f_1[m], \ldots, f_N[m]\} = \underset{h[m], f_1[m], \ldots, f_N[m]}{\arg\max} H\left(Z[m] \middle| f_i[m](X_i, h[m](\mathbf{z}_1^{m-1})), \forall i\right).$$

For a channel output that is binary, the entropy is maximized if

$$\Pr\left(Z[m] \middle| f_i[m](X_i, h[m](\mathbf{z}_1^{m-1})), \forall i\right) \approx \frac{1}{2}.$$

In the binary example given previously, the specific DC-MAC strategy is restricted to having $h[m] = (\mathcal{G}, b)$,

$$f_i[m] = \begin{cases} 1 & \text{if } i \in \mathcal{G} \text{ and } X_i \neq b \\ 0 & \text{otherwise.} \end{cases}$$

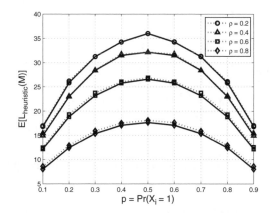

Figure 12.14 The performance of the heuristic algorithm versus the optimized recursive algorithm for $N = 36$. The solid lines represent the performance of the optimized recursive algorithm and the dotted lines represent that of the heuristic algorithm.

and the channel is deterministic with output $z = 0$ if $\mathbf{u} = \mathbf{0}$ and $z = 1$, otherwise. The optimization is then over the groups $\mathcal{G}[m]$ and the queried question $b[m]$. The solution given by the heuristic algorithm is then equal to

$$Q[m] \triangleq (\mathcal{G}[m], b[m]) = \arg \min_{\mathcal{G}, b} \left| \Pr\left(X_i = b, \forall i \in \mathcal{G} \,\Big|\, \mathbf{Z}_1^{m-1} = \mathbf{z}_1^{m-1}\right) - \frac{1}{2} \right|.$$

As shown for the case of $N = 36$ in Figure 12.14, we observe only a slight loss in performance for the heuristic algorithm when compared to the optimized recursive scheme.

The mutual information criterion determines the transmission of each sensor based on a step-by-step optimization of the functions $\{h[m], f_1[m], \ldots, f_N[m]\}$. This method has been used to optimize tree structured algorithms such as in group testing problems (Chen et al. 1987) and in decision tree construction problems (Hartmann et al. 1982). The method is justified in the sense that it reduces a general upper bound derived for $\mathbf{E}[L]$ [see (Hartmann et al. 1982; Hong et al. 2005a)]. The strategy does not achieve the optimal performance in general, but the design criterion can be applied for different data models, channel models and distortion measures. While this criterion determines the query selection during each time-slot, a stopping rule must be defined to terminate the process at the desired distortion.

Definition 12.6.5 *The total number of queries to achieve the distortion D is defined as*

$$L = \inf\left\{\ell : \mathbf{E}[d(\mathbf{X}, \hat{\mathbf{X}}[\ell])|Z[1], \ldots, Z[\ell]] \le D\right\}. \tag{12.16}$$

where $\hat{\mathbf{X}}[\ell]$ is the optimal estimate of $\mathbf{X}$ given the channel outputs $Z[1], \ldots, Z[\ell]$.

When the channel output alphabet $\mathcal{Z}$ is finite, we observe that the query-and-response structure of the cooperative MAC protocol can be illustrated with a tree diagram, as shown in Figure 12.15, where each node in the tree represents a query and each branch extending

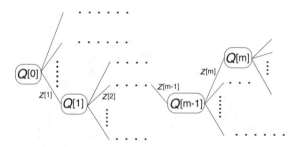

Figure 12.15 Special Case: Tree structured illustration of the querying process.

from the node corresponds to a realization of the channel outputs. The tree terminates at the point where the distortion constraint is satisfied as defined above. Therefore, the leaves of the tree represent different estimates of $\mathbf{X}$ corresponding to different sequences of channel outputs $\mathbf{Z}_1^L$ or, equivalently, the different paths of the tree. Since the problem can be described as a tree algorithm, several techniques proposed in the literature for general tree structured problems can be used to improve our strategy. Specifically, we can utilize the optimal tree pruning algorithm proposed in (Chou et al. 1989) to derive a more efficient querying tree. First of all, we construct a tree based on the proposed heuristic algorithm but overshoot in performance, i.e. we terminate each branch in the tree at a distortion D' that is much less than the constraint D. Then, we apply the optimal tree pruning algorithm to eliminate the additional extensions in the tree such that the average distortion satisfies $\mathbf{E}[d(\mathbf{X}, \hat{\mathbf{X}}[L])] \leq D$. Although there may exist a path down the tree that terminates at a distortion greater than D, the overall distortion is satisfied with less expected time slots. Details of the tree pruning for DC-MAC and examples of different applications are given in (Hong et al. 2005a).

The advantages of DC-MAC can be twofold: (1) it improves the throughput and delay by exploiting the high correlation among densely deployed sensors; and (2) it achieves good energy efficiency due to the cooperative transmissions. The first advantage is clear from the previous discussions while the second advantage is inherent in all cooperative communications [see (Hong et al. 2005b) for details]. The underlying concept of the DC-MAC is to assign the same transmission channel to sensors that are likely to transmit the same information. In this section, we have illustrated the effectiveness of this method in data gathering applications. Even more interesting, this also applies to detection and estimation applications as well.

12.6.3 MAC for Distributed Detection and Estimation

A classical example of highly correlated sources can be found in distributed detection and estimation applications. In these systems, the sensors observe and generate data according to a common source or event. These local observations are then communicated to the data gathering node where a final decision or estimate is made. The performance depends on the power constraints at each sensor and the specific channel model. Since these sensors' observations correspond to the same event, it is likely that they

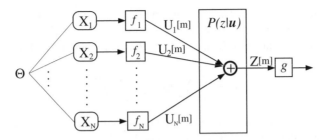

Figure 12.16 Distributed detection and estimation problems.

will send the same message to the data gathering node. Therefore, we should not allocate a separate channel to each sensor, but should assign only a common channel for sensors that are transmitting identical messages. This is the key intuition of the DC-MAC and has been applied independently to distributed detection and estimation problems in (Liu and Sayeed 2004) and (Mergen and Tong 2006), where it is referred to as the Type-Based Multiple Access (TBMA) protocol.

In (Liu and Sayeed 2004) and (Mergen and Tong 2006), the authors considered a sensor network model as shown in Figure 12.16. In this setup, each sensor, say sensor i, observes an *i.i.d.* vector of data $X_i[1], \ldots, X_i[M]$ with respect to the event $\theta \in \Theta$, where Θ is the set of all possible values of the event and $X_i[k] \in \mathcal{X}_i$ is the observation made at time k. Assume that the data is also *i.i.d.* over different sensors. Thus, let $P_{X;\theta}$ be the distribution of $X_i[k]$ for all k and i. Note that the parameter space is discrete in detection problems, e.g. $\Theta = \{\theta_0, \theta_1\}$ for binary hypotheses testing, while it may be continuous in estimation problems, e.g. $\Theta \subset \mathbf{R}$.

In the centralized scenario where the data gathering node has noise-free access to the observations, the optimal detector for binary hypotheses testing is obtained as follows:

$$\hat{\theta}(\mathbf{x}[1], \ldots, \mathbf{x}[M]) = \begin{cases} 1, & \gamma \geq 0 \\ 0, & \gamma < 0 \end{cases} \tag{12.17}$$

where $\Theta = \{0, 1\}$, $\mathbf{x}[k] = [x_1[k], \ldots, x_N[k]]^T$ and

$$\gamma = \frac{1}{NM} \sum_{i=1}^{N} \sum_{k=1}^{M} \log \frac{P_{X;1}(x_i[k])}{P_{X;0}(x_i[k])}.$$

γ is called the normalized *log-likelihood ratio* (LLR). For all reasonable detection strategies, the average error probability P_e decreases as N, M increase. As shown in (Cover and Thomas 1991), the P_e for this optimal centralized scheme decreases with an error exponent equal to

$$E_c = \lim_{M,N \to \infty} \sup -\frac{1}{NM} \log P_e = CI(P_{X;0}, P_{X;1}) \tag{12.18}$$

where

$$CI(P_{X;0}, P_{X;1}) = -\min_{0 \leq s \leq 1} \log \mathbf{E}_{P_{X;0}} \left[\frac{P_{X;1}}{P_{X;0}} \right]^s \geq 0$$

is called the Chernoff information between $P_{X;0}$ and $P_{X;1}$.

In the decentralized case, the local observations are reported to the data gathering node through the multiple access channel as shown in Figure 12.16. When the sensors have knowledge of the statistics $P_{X;0}$ and $P_{X;1}$, they can compute and transmit their local likelihood ratios

$$\gamma_i = \frac{1}{M} \sum_{k=1}^{M} \log \frac{P_{X;1}(x_i[k])}{P_{X;0}(x_i[k])}.$$

When these values are transmitted cooperatively and coherently over the multiple access channel, the data gathering node receives the combined signal

$$Z = \sum_{i=1}^{N} \rho \gamma_i + w \qquad (12.19)$$

where ρ is the scaling used to satisfy the sum power constraints and w is the additive white Gaussian noise. The receiver makes the decision $\hat{\theta} = 1$ if $Z > 0$ and $\hat{\theta} = 0$, otherwise. It is shown in (Liu and Sayeed 2004) that the decision based only on one transmission of the kind in (12.19) achieves the error exponent of the optimal centralized detector.

When the sensors have no knowledge of the data statistics, the LLR cannot be computed and the previous method would not be applicable. However, by using the TBMA, one can still achieve the error exponent of the optimal P_e for the centralized detector. Suppose that the observations belong to a finite alphabet $\mathcal{X} = \{a_1, \ldots, a_{|\mathcal{X}|}\}$, e.g. the sensors observe quantized measurements. In TBMA, each sensor first computes the 'type', i.e. the empirical distribution, based on the sequence of data $X_i[1], \ldots, X_i[M]$. The type computed at sensor i is represented by

$$\mathbf{T}_i(X_i[1], \ldots, X_i[M]) = [T_{i,a_1}, \ldots, T_{i,a_{|\mathcal{X}|}}] = \left[\frac{\sum_k \mathbf{1}_{X_i[k]=a_1}}{M}, \ldots, \frac{\sum_k \mathbf{1}_{X_i[k]=a_{|\mathcal{X}|}}}{M} \right],$$

$$(12.20)$$

where

$$\mathbf{1}_{X_i[k]=a_j} = \begin{cases} 1, & \text{if } X_i[k] = a_j \\ 0 & \text{otherwise.} \end{cases}$$

The local type information is transmitted by the sensors cooperatively using a total of $|\mathcal{X}|$ time slots. Specifically, we assign a separate time slot to each element in the alphabet instead of to each sensor, similar to that with DC-MAC. During the m-th time slot, the sensors transmit the type T_{i,a_m} and the signal arriving at the receiver is

$$Z[m] = \frac{\rho}{N} \sum_{i=1}^{N} T_{i,a_m} + w[m] \quad \text{for } m = 1, \ldots, |\mathcal{X}|. \qquad (12.21)$$

Based on the $|\mathcal{X}|$ channel outputs, the receiver computes the log-likelihood ratio

$$\tilde{\gamma} = \sum_{m=1}^{|\mathcal{X}|} Z[m] \log \frac{P_{X,1}(a_m)}{P_{X,0}(a_m)}$$

and makes the decision $\hat{\theta} = 1$ if $\tilde{\gamma} \geq 1$ and $\hat{\theta} = 0$, otherwise. This detector also achieves an error exponent equal to that of the centralized detector as the number of sensors increases.

More importantly, the cost of transmitting the sensors data is only equal to the number of possible observation values, i.e. $|\mathcal{X}|$, instead of increasing linearly with the number of users.

The TBMA was proposed independently for the distributed estimation problem in (Mergen and Tong 2006) where it considers a parameter space $\Theta \subset \mathbf{R}$ and adopts the MMSE $\mathrm{E}\{(\hat{\theta} - \theta)^2\}$ as the performance measure. It is shown that the TBMA with coherent transmissions as in (12.21) yields an asymptotically efficient estimator in the sense that

$$\sqrt{n}(\hat{\theta} - \theta)^2 \to \mathcal{N}\left(0, \frac{1}{I(\theta)}\right)$$

where $I(\theta) = \sum_{j=1}^{|\mathcal{X}|} \frac{(dP_{X;\theta}(a_j))^2}{P_{X;\theta}(a_j)}$ is the Fisher information (Poor 1994). However, if the sensors do not transmit coherently and the signal appearing at the receiver is

$$Z[m] = \frac{\rho}{N} \sum_{i=1}^{N} h_i T_{i,a_m} + w[m] \quad \text{for } m = 1, \ldots, |\mathcal{X}|, \tag{12.22}$$

where h_i's are random fading gains with non-zero mean, then the ML variant estimator proposed in (Mergen and Tong 2006) satisfies

$$\sqrt{n}(\hat{\theta} - \theta)^2 \to \mathcal{N}\left(0, \frac{1 + \sigma_h^2}{I(\theta)}\right)$$

where $\sigma_h^2 = \mathrm{Var}(\Re(h_i/\mathrm{E}[h]))$. A slight loss in performance is observed due to the random fading gains experienced by the sensors. Note that, if each sensor makes only one observation, i.e. $M = 1$, the type vector $\mathbf{T}_i$ will simply be a canonical vector with a 1 at only one position and 0 everywhere else. In other words, the sensor i transmits only in the time slot that corresponds to its local data. This is consistent with the strategy used in the binary example and the extension to M-ary variables in (Hong and Scaglione 2004a).

Interestingly, not only does the data-centric cooperative transmission reduce the total amount of bandwidth resources, but the resulting detectors or estimators also outperform the case where each sensor transmits individually in separate channels. This is the loss of separating source and channel coding in the multiple access environment (Cover et al. 1980). The topic of source-channel communication in sensor networks is discussed in (Gastpar and Vetterli 2003, 2005) where they showed that the uncoded transmission for a finite network of Gaussian sources outperforms any kind of approach based on the separation principle between source and channel coding. In fact, separation may involve an exponential penalty in terms of communication resources when the number of sensors increases. Other work on detection and estimation with communication constraints can be found in (Chamberland and Veeravalli 2003; Jayaweera 2005; Longo et al. 1990; Xiao and Luo 2006).

12.7 Conclusion

In this chapter, we surveyed different classes of sensor network MAC protocols with the emphasis on those adopting cooperative transmissions. Most of the early sensor network MAC protocols were attained by optimizing the system parameters of conventional schemes, such as CSMA or TDMA, with the addition of sleep-wake mechanisms to improve the energy efficiency of the system. With the help of cross-layered policies, the sensor network MAC protocols can be made more efficient by incorporating knowledge of the application, such as the possible communication pattern, the traffic load or the network topology. The dependence of the MAC on the application layer is more explicit in the class of data-centric protocols. In these schemes, the MAC is derived with respect to the statistics of the data or is optimized with respect to the performance measure of the application. These protocols exploit the collaborative nature of the sensors in the application layer to reduce the redundancy of the transmitted messages. To achieve this goal, advanced signal processing techniques, such as the various data aggregation methods, distributed source coding or spatial sampling, must be performed in the application layer.

Although these methods are easier to achieve in practice due to their similarity with conventional MAC, they do not fully exploit the cooperative advantage of sensor networks. By utilizing cooperative signal processing at the sensors, we showed that cooperative MAC improves upon the conventional MAC in three perspectives: (1) the QoS attributes such as throughput and latency can be improved due to the spatial multiplexing gain; (2) the reliability and energy efficiency are obtained from the spatial diversity gains; (3) the bandwidth utilization is reduced since the simultaneous transmissions effectively compresses the sensors' redundant messages. However, the work proposed in the literature provides more of a theoretical study rather than a practical scheme. Future work in this direction is necessary, especially with the data-centric MAC protocols. Specifically, the robustness of the DC-MAC must be studied and error correction mechanisms must be derived to overcome the costly error propagation that may occur due to its sequential data gathering structure, a preliminary study is given in (Hong and Scaglione 2005). Similar problems also occur in DSC and data aggregation, but are dealt with by compromising the performance with clustering approaches. The extensions of DC-MAC to random access networks are also desirable since TDMA requires strict coordination between sensors, which are often hard to achieve without depleting the network resources.

In all sensor network MAC protocols, there is a trade off in reliability, flexibility, energy efficiency and the application performance. Therefore, numerous sensor network MAC protocols have been proposed with none of them having a dominant performance over all the others. The performance, however, is highly dependent on the application at hand and the users should choose carefully the sensor network MAC protocols adopted in their system. With the limited resources being the major constraint in sensor networks, it is often necessary to perform advanced signal processing within the sensors to avoid unnecessary expenditure of the resources.

Bibliography

Arisha KA, Youssef MA and Younis MF 2002 Energy-aware TDMA-based MAC for sensor networks. In *Proceedings of IEEE Workshop on Integrated Management of Power Aware Communications, Computing and Networking (IMPACCT)*, New York.

Bandyopadhyay S and Coyle EJ 2003 An energy efficient hierarchical clustering algorithm for wireless sensor networks. In *Proceedings of the IEEE Twenty-Second Annual Joint Conference of the IEEE Computer and Communications Societies (INFOCOM)*, vol. 3, pp. 1713–1723.

Berger T, Mehravari N, Towsley D and Wolf J 1984 Random multiple-access communication and group testing. *IEEE Transactions on Communications* **32**(7), 769–779.

Bertsekas D and Gallager R 1991 *Data Networks* 2nd edn. Prentice Hall, New Jersey.

Bharghavan V, Demers A, Shenker S and Zhang L 2004 Macaw: a media access protocol for wireless LANs. In *Proceedings of the conference on communications architectures, protocols and applications*, pp. 210–225.

Callaway E, Gorday P, Hester L, Gutierrez JA, Naeve M, Heile B and Bahl V 2002 Home networking with IEEE 802.15.4: A developing standard for low-rate wireless personal area networks. *IEEE Communications Magazine* pp. 70–77.

Capetanakis J 1979 Generalized TDMA: The multi-accessing tree protocol. *IEEE Transactions on Communications* **27**(10), 1476–1484.

Chamberland JF and Veeravalli VV 2003 Decentralized detection in sensor networks. *IEEE Transactions on Signal Processing* **51**(2), 407–416.

Chen P, Hsu L and Sobel M 1987 Entropy-based optimal group testing procedures. *Probability in the Engineering and Informational Sciences* **1**, 497–509.

Chou PA, Lookabaugh T and Gray RM 1989 Optimal pruning with applications to tree-structured source coding and modeling. *IEEE Transactions on Information Theory* **35**(2), 299–315.

Cover T and Leung C 1981 An achievable rate region for the multiple-access channel with feedback. *IEEE Transactions on Information Theory* **27**(3), 292–298.

Cover TM and Thomas JA 1991 *Elements of Information Theory*. Wiley-Interscience, New York.

Cover T, Gamal AE and Salehi M 1980 Multiple access channels with arbitrarily correlated sources. *IEEE Transactions on Information Theory* **26**(6), 648–657.

Cristescu R and Vetterli M 2005 On the optimal density for real-time data gathering of spatio-temporal processes in sensor Networks *Proc. of IEEE and ACM International Symposium on Information Processing in Sensor Networks (IPSN)*, Los Angeles, CA.

Demirkol I, Ersoy C and Alagöz F 2006 MAC protocols for wireless sensor networks: a survey. *to appear in IEEE Communications Magazine.*

Ditzel M and Langendoen K 2005 D3: Data-centric data dissemination in wireless sensor networks. In *Proceedings of European Conference on Wireless Technology.*

Dorfman R 1943 The detection of defective members of large population. *The Annals of Mathematical Statistics* **14**(4), 436–440.

Du DZ and Hwang FK 1993 *Combinatorial Group Testing and Applications* vol. 12 of *Series on Applied Mathematics* 2 edn. World Scientific Pub Co Inc., New York.

El-Hoiydi A 2002 Aloha with preamble sampling for sporadic traffic in ad hoc wireless sensor networks. In *Proceedings on IEEE International Conference on Communications (ICC)*, pp. 3418–3423.

El-Hoiydi A and Decotignie JD 2004 Wisemac: an ultra low power MAC protocol for the downlink of infrastructure wireless sensor networks. In *Proceedings on IEEE International Symposium on Computers and Communications (ISCC)*, pp. 244–251.

Gastpar M and Vetterli M 2003 Source-channel communication in sensor networks. In *Proc. of IEEE/ACM International Symposium on Information Processing in Sensor Networks (IPSN).*

Gastpar M and Vetterli M 2005 Power, spatio-temporal bandwidth, and distortion in large sensor networks. *IEEE Journal on Selected Areas in Communications* **23**(4), 745–754.

Gummalla ACV and Limb JO 2000 Wireless medium access control protocols. *IEEE Communications Surveys and Tutorials* **3**(2), 2–17.

Hartmann CRP, Varshney PK, Mehrotra KG and Gerberich CL 1982 Application of information theory to the construction of efficient decision trees. *IEEE Transactions on Information Theory* **28**(4), 565–577.

Hasna MO and Alouini MS 2004 Optimal power allocation for relayed transmissions over Rayleigh-fading channels. *IEEE Transactions on Wireless Communications* **3**(6), 1999–2004.

Heinzelman WR, Chandrakasan AP and Balakrishnan H 2002 An application-specific protocol architecture for wireless microsensor networks. *IEEE Transactions on Wireless Communications* **1**(4), 660–670.

Heinzelman WR, Kulik J and Balakrishnan H 1999 Adaptive protocols for information dissemination in wireless sensor networks. In *Proceedings of the 5th Annual ACM/IEEE International Conference on Mobile Computing and Networking (MobiCom)*, pp. 174–185.

Hill JL and Culler DE 2002 Mica: a wireless platform for deeply embedded networks. *IEEE Micro* **22**(6), 12–24.

Hong YW and Scaglione A 2004a Content-based multiple access: Combining source and multiple access coding for sensor networks. In *Proc. of IEEE International Workshop on Multimedia Signal Processing*, Siena, Italy.

Hong YW and Scaglione A 2004b On multiple access for correlated sources: A content-based group testing approach. In *Proc. of IEEE Information Theory Workshop*, San Antonio, TX.

Hong YW and Scaglione A 2005 A scalable communication architecture for the sensor broadcast problem. In *Proc. of SPAWC*, New York.

Hong YW and Scaglione A 2006 Group testing for binary markov sources: Data-driven group queries for cooperative sensor networks. *submitted to IEEE Transactions on Information Theory*.

Hong YW, Huang WJ, Chiu FH and Kuo CCJ 2007 Cooperative communications in resource-constrained wireless networks. *to appear in IEEE Signal Processing Magazine*.

Hong YW, Scaglione A and Varshney P 2005a An information-theoretic approach for multiple access transmissions in sensor networks *in preparation*.

Hong YW, Scaglione A, Manohar R and Sirkeci-Mergen B 2005b Dense sensor networks that are also energy efficient: When 'more' is 'less'. In *Proc. of IEEE Military Communications Conference (MILCOM)*, pp. 1–7.

Intanagonwiwat C, Govindan R, Estrin D, Heidemann J and Silva F 2003 Directed diffusion for wireless sensor networking. *IEEE/ACM Transactions on Networking* **11**(1), 2–16.

Jayaweera S 2005 Large system decentralized detection performance under communication constraints. **9**(9), 769–771.

Karn P 1990 Maca - a new channel access method for packet radio. In *Proceedings of ARRL/CRRL Amateur Radio 9th Computer Networking Conference*, pp. 134–140.

Kleinrock L and Tobagi FA 1975 Packet switching in radio channels: Part i–carrier sense multiple-access modes and their throughput-delay characteristics. *IEEE Transactions on Communications* **23**(12), 1400–1416.

Krishanamachari B, Estrin D and Wicker S 2002 The impact of data aggregation in wireless sensor networks. In *Proc. on International Workshop of Distributed Event Based Systems (DEBS)*, Vienna, Austria.

Kulkarni SS and Arumugam M 2004 TDMA services for sensor networks. In *Proc. of the 24th International Conference on Distributed Computing Systems Workshops (ICDCSW)*.

Laneman J and Wornell G 2003 Distributed space-time-coded protocols for exploiting cooperative diversity in wireless networks. *IEEE Transactions on Information Theory* **49**(10), 2415–2425.

Laneman J, Tse D and Wornell G 2004 Cooperative diversity in wireless networks: Efficient protocols and outage behavior. *IEEE Transactions on Information Theory* **50**(12), 3062–3080.

Langendoen K and Halkes G 2005 Energy-efficient medium access control. In *Embedded Systems Handbook*. CRC Press.

Lei KU, Hong YW and Chi CY 2007 Distributed estimation in wireless sensor networks with channel-aware slotted aloha. In *Proceedings of the IEEE Signal Processing Advances in Wireless Communications Workshop (SPAWC)*.

Lin R and Petropulu A 2005 A new wireless network medium access protocol based on cooperation. *IEEE Transactions on Signal Processing* **53**(12), 4675–4684.

Liu K and Sayeed AM 2004 Optimal distributed detection strategies for wireless sensor networks. In *Proc. on the 42nd Annual Allerton Conference on Communications, Control and Computing*, Monticello, IL.

Liu P, Tao Z and Panwar S 2005 A cooperative MAC protocol for wireless local area networks. In *Proc. on IEEE International Conference on Communications (ICC)*.

Longo M, Lookabaugh TD and Gray RM 1990 Quantization for decentralized hypothesis testing under communication constraints. *IEEE Transactions on Information Theory* **36**(2), 241–255.

Lu G, Krishnamachari B and Raghavendra CS 2004 An adaptive energy-efficient and low-latency MAC for data gathering in wireless sensor networks. In *Proceedings of IEEE International Parallel and Distributed Processing Symposium (IPDPS)*.

Marco D and Neuhoff DL 2004 Reliability vs. efficiency in distributed source coding for field-gathering sensor networks. In *Proc. on the Third International Symposium on Information Processing in Sensor Networks (IPSN)*.

Mergen G and Tong L 2006 Type based estimation over multiaccess channels. *IEEE Transactions on Signal Processing* **54**(2), 613–626.

N SS, Chou CT and Ghosh M 2005 Cooperative communication MAC (CMAC) - a new MAC protocol for next generation wireless LANs. In *Proceedings of IEEE International Conference on Wireless Networks, Communications and Mobile Computing (WiCOM 2005)*, vol. 1, pp. 1–6.

Papoulis A and Pillai SU 2001 *Probability, Random Variables and Stochastic Processes*, 4th edn. McGraw-Hill, New York.

Poor HV 1994 *An Introduction to Signal Detection and Estimation*. Springer, New York.

Pradhan S, Kusuma J and Ramchandran K 2002 Distributed compression in a dense microsensor network. *IEEE Signal Processing Magazine* **19**(2), 51–60.

Rabbat MG and Nowak RD 2005 Quantized incremental algorithms for distributed optimization. *IEEE Journal on Selected Areas in Communications* **23**(4), 798–808.

Rajagopalan R and Varshney PK 2006 Data-aggregation techniques in sensor networks: a survey. *IEEE Communications Surveys and Tutorials* **8**(4), 48–63.

Scaglione A and Servetto S 2002 On the interdependence of routing and data compression in multi-hop sensor networks. In *Proceedings of the 8th Annual International Conference on Mobile Computing and Networking*, Atlanta, GA.

Sendonaris A, Erkip E and Aazhang B 2003 User Cooperation Diversity–Part I: System Description and User Cooperation Diversity–Part II: Implementation Aspects and Performance Analysis. *IEEE Transactions on Communications*.

Slepian D and Wolf J 1973 Noiseless coding of correlated information sources. *IEEE Transactions on Information Theory* **19**(4), 471–480.

Sobel M and Groll PA 1959 Group testing to eliminate efficiently all defectives in a binomial sample. *The Bell System Technical Journal* **38**, 1179–1253.

Tsai YR, Hong YW, Liao YY and Yang KJ 2007 The efficiency and delay of distributed source coding in random access sensor networks. In *Proceedings of the IEEE Wireless Communications and Networking Conference (WCNC)*.

Tsatsanis MK, Zhang R and Banerjee S 2000 Network-assisted diversity for random access wireless networks. *IEEE Transactions on Signal Processing* **48**(3), 702–711.

van Dam T and Langendoen K 2003 An adaptive energy-efficient MAC protocol for wireless sensor networks *The First ACM Conference on Embedded Networked Sensor Systems (SenSys 2003)*, Los Angeles, CA.

Varshney PK 1996 *Distributed Detection and Data Fusion*. Springer, New York.

Venkitasubramaniam P, Adireddy S and Tong L 2004 Sensor networks with mobile access: optimal random access and coding. *IEEE Journal on Selected Areas in Communications* **22**(6), 1058–1068.

Viswanathan R and Varshney PK 1997 Distributed detection with multiple sensors: Part I - fundamentals. In *Proceedings of the IEEE*, vol. 85, pp. 54–63.

Vuran MC and Akyildiz IF 2006 Spatial correlation-based collaborative medium access control in wireless sensor networks. *IEEE/ACM Transactions on Networking*.

Wu T and Biswas S 2005 A self-reorganizing slot allocation protocol for multi-cluster sensor networks. *Proceedings of the Fourth International Symposium on Information Processing in Sensor Networks (IPSN)*, pp. 309–316.

Xiao JJ and Luo ZQT 2006 Universal decentralized detection in a bandwidth-constrained sensor network. *IEEE Transactions on Signal Processing* **53**(8), 2617–2624.

Xiong Z, Liveris AD and Cheng S 2004 Distributed source coding for sensor networks. *IEEE Signal Processing Magazine* **21**(5), 80–94.

Ye W, Heidemann J and Estrin D 2004 Medium access control with coordinated adaptive sleeping for wireless sensor networks. *IEEE/ACM Transactions on Networking* **12**(3), 493–506.

Yedavalli K and Krishnamachari B 2006 Enhancement of the ieee 802.15.4 MAC protocol for scalable data collection in dense sensor networks. Technical Report CENG-2006-14, USC Computer Engineering.

13

Game Theoretic Activation and Transmission Scheduling in Unattended Ground Sensor Networks: A Correlated Equilibrium Approach

Vikram Krishnamurthy, Michael Maskery, and Minh Hanh Ngo

13.1 Introduction

In this chapter, we control the dynamical behavior of an unattended ground sensor network (UGSN) to acquire information about intruders. An UGSN comprises of a large number of inexpensive sensors that contribute to area surveillance to detect and track slowly moving intruders (targets). Intruders may be either moving on foot or in ground vehicles. The following fundamental tradeoff between the cost of acquiring data (e.g., sensor battery usage for sensing and data transmission) and the usefulness of the data is typical in such sensor networks: Each sensor can measure, with limited accuracy, the range (distance) and bearing (angle with respect to a reference direction) of nearby targets and then transmit these measurements to a local hub node for data fusion. More sensor readings, and more aggressive transmission of measurement data, will result in better target awareness, but also

Wireless Sensor Networks: Signal Processing and Communications Perspectives   A. Swami, Q. Zhao, Y.-W. Hong and L. Tong
© 2007 John Wiley & Sons, Ltd

greater consumption of limited battery resources. The aim of this chapter is to optimally (in a sense to be made more specific below) trade off target awareness, data transmission and energy consumption using a two-time scale, hierarchical approach.

The rest of this introductory section is organized as follows. To fix ideas, we first outline our methodology for designing sensor activation and transmission scheduling algorithms in Sec. 13.1.1. Then to provide more perspective, in Sec. 13.1.2, we outline the fundamental tools and related literature.

13.1.1   UGSN Sensor Activation and Transmission Scheduling Methodology

We will show in this chapter that the sensor activation and transmission scheduling problem naturally decomposes into two cross-coupled decentralized algorithms that operate on two different time scales. They are as follows:

- Decentralized Sensor Activation as the Correlated Equilibrium of a Non-cooperative Game: At the slow time-scale (typically in the order of seconds), we use a game-theoretic adaptive learning strategy to activate sensors according to their proximity to targets of interest. The sensors are interpreted as players in a non-cooperative game. We present a decentralized learning algorithm which each sensor deploys. The algorithm determines at each time instant whether the sensor should be activated or remain asleep. When each sensor plays according to this decentralized algorithm, the number of active sensors in the UGSN converges to a *correlated equilibrium*. The concept of correlated equilibria in game theory was introduced in Aumann (1974, 1987) and is more general than the widely used Nash equilibrium. Moreover, correlated equilibria are easier to characterize and more natural to decentralized adaptive algorithms such as considered here.[1] The resulting sensor activation control powers down all but a limited number of sensors that have the best readings of the target.

- Markov Decision-based Threshold Transmission Controller: If a sensor is activated by the above game theoretic learning algorithm, then it measures bearing and range signals of nearby targets and transmits data to a local hub node over a wireless multi-access channel shared by several other sensors. The transmission of packets takes place at a fast time-scale (typically in the order of milliseconds). The wireless channel quality is adversely affected by the number of other sensors transmitting data, and can be modelled as a finite state Markov chain. At each time instant, based on the channel quality, each sensor has to decide whether to transmit data and waste battery power, or wait and increase delay. We formulate the sensor transmission scheduling problem as a Markov decision process with a penalty terminal cost. The key result shown is that the optimal transmission policy has a threshold structure. This threshold structure is proved using the concept of supermodularity.

---

[1] Aumann was awarded the 2005 Nobel Prize in economics. The following extract is from the Nobel Prize press release in October 2005: 'Aumann also introduced a new equilibrium concept, correlated equilibrium, which is weaker than Nash equilibrium, the solution concept developed by John Nash, an economics laureate in 1994. Correlated equilibrium can explain why it may be advantageous for negotiating parties to allow an impartial mediator to speak to the parties either jointly or separately, and in some instances give them different information.'

13.1.2 Fundamental Tools and Literature

The above methodological description of activation and transmission control in an UGSN mentioned three fundamental concepts, which are of broad interest in statistical signal processing, stochastic control and wireless communications. To give more perspective, we briefly describe these three tools and related literature.

1. Non-cooperative games and Correlated Equilibria: In Sec. 13.3.1, we describe static non-cooperative games and correlated equilibria. Here we briefly outline recent work in the use of non-cooperative games in sensor networks and motivate our approach. Several problems related to communications in sensor networks have been addressed from a game theoretic perspective to date, and Goldsmith and Wicker (2002) and MacKenzie and Wicker (2001) provide a good overview of the area. For specific areas, Kannan et al. (2003) and Rogers et al. (2005) treat multihop routing, MacKenzie and Wicker (2003) investigate random access over a common channel, transmission power control is treated by Xing and Chandarmouli (2004), and topology control by Li and Hou (2006) and Borbash and Jennings (2002). The common approach in these papers is to define a utility function that each system component selfishly maximizes, and then analyze system performance at the Nash equilibrium point of the resulting game. What has not been studied is games in which sensors must decide how to sense their environment, as opposed to how to transmit their information. In addition, research so far has used only the Nash equilibrium concept; no attention is paid to correlated equilibria. We attempt to fill these gaps in our game theoretic sensor activation approach described in detail in Sec. 13.3. A related area is the study of correlated equilibria in stochastic, dynamic games, which we consider in applications to missile deflection in Maskery and Krishnamurthy (2007).

2. Stochastic approximation algorithms: We present a decentralized stochastic approximation algorithm that each sensor deploys to adapt its probability of being activated. When each sensor deploys this stochastic approximation algorithm, it will be shown that the overall behaviour of the UGSN converges to set of correlated equilibria (in terms of the sensor activation probabilities). Stochastic approximation/optimization algorithms are widely used in electrical engineering to recursively estimate the optimum of a function or its root, Kushner and Yin (2003) and Benveniste et al. (1990) give excellent expositions of this area. The well known least mean squares (LMS) adaptive filtering algorithm is a simple example of a stochastic approximation algorithm with a quadratic objective function. Stochastic approximation algorithms have been applied to reinforcement learning for stochastic control problems in Bertsekas and Tsitsiklis (1996), learning equilibria in games such as in Section 13.3.2, as well as optimization and parametric identification problems (e.g., recursive maximum likelihood and recursive expectation maximization algorithms, see Krishnamurthy and Yin (2002)).

In tracking applications, the step size of a stochastic approximation algorithm is chosen as a small constant. For such constant step size algorithms, one typically proves weak convergence of the iterates generated by the stochastic approximation algorithm. Weak convergence is a generalization of convergence in distribution to a function space. The weak convergence analysis of stochastic approximation algorithms with Markovian noise has been pioneered by Kushner and co-workers, see Kushner and Yin (2003) and references therein. It was demonstrated in the 1970s that the limiting behaviour of a stochastic approximation algorithm can be modelled as a deterministic ordinary-differential-equation (ODE). This is the basis of the so-called ODE method for convergence analysis of stochastic approximation algorithms. Actually, for the correlated equilibrium learning algorithm presented in this chapter, the limiting behaviour is captured by a differential inclusion Benaim et al. (2005), rather than a differential equation. We briefly outline this approach in Section 13.3.3.

3. Optimality of Threshold Policies for Markov Decision Processes: Once a sensor is activated by the above game theoretic learning algorithm, it gathers data and transmits this data to a local hub node. In Sec. 13.4 we formulate the problem of optimal transmission scheduling for each sensor in the UGSN as a finite state, finite action MDP over a finite horizon with a terminal cost. Bertsekas (1995); Kumar and Varaiya (1986); Puterman (1994); Ross (1983) are comprehensive references for MDPs. Translated to the transmission scheduling problem in an UGSN, the objective is for each sensor to exploit channel state information (CSI) to minimizes the expected sum of transmission and data loss costs.

The globally optimal policy for a MDP can be obtained via stochastic dynamic programming[2] which leads to a functional equation called Bellman's equation. However, dynamic programming suffers from the curse of dimensionality. In this chapter we investigate the use of supermodularity to obtain structural results for the optimal scheduling policy. In particular, given the state x of the MDP at any time k, we give sufficient conditions for which the optimal policy is threshold of the form

$$u_k^*(x) = \begin{cases} \text{Action 1} & \text{If state } x < s_k \\ \text{Action 2} & \text{Otherwise.} \end{cases} \qquad (13.1)$$

Here s_k denotes the threshold state at time k. Thus if a MDP has a threshold policy, one only needs to compute the threshold (s_k in the equation above) to implement the optimal policy. This serves as the main motivation for proving structural results for MDP in general. In the sensor transmission scheduling problem, we prove that the optimal transmission policy is threshold in the residual transmission time and the buffer occupancy.

The main idea involved in proving that the optimal policy is threshold is quite straightforward: Bellman's equation yields that the optimal action at time k is

$$u_k^*(x) = \arg\min_a Q_k(x, a)$$

where $\min_a Q_k(x, a)$ is the value function (optimal cost to go) at stage k (a precise definition is given in Sec. 13.4). It is obvious that for a two-action MDP if $u_k^*(x)$ is an increasing function of x, then the optimal policy is threshold, i.e., it is of the form (13.1). Therefore to show the existence of an optimal threshold policy, one only needs to verify that $u_k^*(x) = \arg\max_a Q_k(x, a)$ is increasing in x. The natural question then is: What conditions on the function $Q_k(x, a)$ guarantee that the $\arg\max Q_k(x, a)$ is increasing? *Supermodularity* which, roughly speaking, is a generalization of convexity to a lattice, is a sufficient condition.

We refer the reader to the text of Heyman and Sobel (1984), for a graduate level course treatment of supermodularity and its use in proving monotone optimal policies for MDPs. The use of supermodularity in MDPs and games was championed by Topkis (1978). A comprehensive exposition of the topic supermodularity in decision problems (e.g., MDPs)

---

[2]The MDPs we consider in this chapter are finite horizon and do not have global constraints. For infinite horizon average cost MDPs with global constraints, the optimal (randomized) policy can be computed as the solution of a linear programming problem, or alternatively the Lagrangian dynamic programming approach presented in Altman (1999) can be used.

and noncooperative and cooperative games is given in Topkis (1998). An introductory tutorial on supermodularity with applications in decision problems and games can also be found in Amir (2003).

The supermodularity proof presented in Sec. 13.4 requires a substantial relaxation of the sufficient conditions in the textbook treatment of Heyman and Sobel (1984). Indeed, the results we present can be viewed as a generalization of the classical results in Derman et al. (1976) and Ross (1983). In Derman et al. (1976) and Ross (1983), the authors consider the problem of constructing a finite number of components subject to penalty costs and use supermodularity to prove the monotonicity of the optimal resource allocation policy. The optimal sensor transmission scheduling problem in this chapter can be abstracted as the problem of constructing a finite number of components (transmitting a batch of packets) within a finite number of time slots in *Markovian systems* subject to penalty costs. The proofs of the monotone results in Derman et al. (1976) and Ross (1983) cannot be generalized for the case of a Markovian system state (which in our case is required to model the Markovian wireless channel). In this chapter, we provide proofs that work straightforwardly for both cases, i.e. the case of a Markovian system state and the classical case in Derman et al. (1976) and Ross (1983) where there is no system state (or equivalently, the system state is a constant). The threshold structure in the residual transmission time and the buffer occupancy substantially reduces the computational complexity required to implement the optimal transmission policy. In addition, the proofs we present in this chapter give an easily accessible tutorial example into the concepts and application of supermodularity.

Besides the literature above, for an exposition of target tracking, detection and recognition, see Bar-Shalom et al. (2001). In addition, work that also uses a MDP approach for energy-efficient channel-aware transmission control includes Johnston and Krishnamurthy (2005), which formulates the problem of designing a channel-aware ARQ protocol to balance between average energy consumption and throughput over infinite time for a two-state Gilbert-Elliot channel model as a partially observed MDP (POMDP). Wang and Mandayam (2005) analyze the problem of transmission scheduling in Markovian fading channels for a power-delay tradeoff under energy constraints with the use of a channel aware ARQ protocol. Arulselvan and Berry (2005) exploit CSI for optimal transmission control for i.i.d channels. Finally, Krishnamurthy and Djonin (2007) derive threshold policies via supermodularity with respect to the monotone likelihood ratio for POMDPs in the context of dynamic sensor scheduling.

The rest of the chapter is organized as follows. In Section 13.2 we describe the unattended ground sensor network scenario and trade offs. Section 13.3 treats the activation control problem as a game, and outlines a decentralized algorithm for sensor activation. In Section 13.4, we describe supermodularity in Markov decision processes and show how it leads to a simple threshold policy for optimal transmission scheduling for the sensors. Simulation results are presented in Section 13.5 and conclusions follow in Section 13.6.

13.2 Unattended Ground Sensor Network: Capabilities and Objectives

In order to maximize energy efficiency in unattended ground sensor networks, it is important to understand the capabilities and limitations of the network. In this section we present details on the integrated activation control and transmission control in an UGSN. This provides the context for the analysis in this chapter.

13.2.1   Practicalities: Sensor Network Model and Architecture

We consider a network of L energy constrained sensors labelled $l = 1, 2, \ldots, L$, uniformly distributed in a plane. The sensors are capable of short-range CDMA (code division multiple access) data transmission, target detection and localization through vibration (e.g. acoustic) or electromagnetic (e.g. visual) components, and monitoring of RF activity of other nearby sensors through a specially tuned RF power detector. Computationally, sensors are able to collect and process target readings, make local decisions for activation control, and decide when to transmit data according to pre-computed transmission scheduling policies.

An important feature of this chapter is that communication is unidirectional; sensors are only required to transmit information. This feature greatly simplifies sensor network operation, since the complex task of communicating from hubs to multiple sensors is avoided.

Along with the sensor nodes, a smaller number of more powerful hub nodes are deployed. These hubs, which can be either static or mobile (e.g., mini unmanned aerial vehicles (UAVs) can be used to collect information from sensors), form a backbone which receives, processes and reroutes sensor data to an end user. The hubs are highly functional, with reliable communication and efficient routing protocols for transmitting information error-free along a multi-hop path to an end user. However, in this chapter we concern ourselves only with sensor behaviour; route optimization, topology control, and optimal transmission among hubs is beyond our scope. Indeed, the purpose of the hubs is to abstract away these issues. The topology of the unattended ground sensor network, along with a block diagram indicating sensor capabilities, is shown in Figure 13.1.

Below we provide nominal specifications for various sensor functions, including their power requirements and frequencies. These specifications are derived from various sources, including the Crossbow Inc. Mica mote sensor, and Akyildiz et al. (2002) and Merrill et al. (2003). A more functional (and consequently more energy-intensive) sensor package is available in the form of the military REMBASS-II unattended ground sensor network, as outlined by Marandola et al. (2002). As an alternative, (Maskery and Krishnamurthy 2007) calculate the power requirements for Zigbee-enabled sensors in a game-theoretic activation setting, accounting for power variation due to crowded channel conditions.

Sensing components

Each sensor is equipped with passive detectors, which indicate the direction and intensity of signals emanating from a target (e.g. motor vibrations), subject to measurement noise. Assuming signal intensity falls off as an inverse square of the range, we have for sensor l at time t,

$$\text{Proximity}(l, t) = \min \left\{ \sum_{m=1}^{M} \frac{A_m(t)}{(\Delta(l, z_m(t)))^2} + \eta_l(t), K_I \right\}, \qquad (13.2)$$

where there are M targets, with A_m represents the intensity of Target $m = 1, 2, \ldots, M$ at one metre, $\Delta(l, z^m)$ the Euclidian distance from Sensor l to Target m, K_I represents the sensor saturation level, and η_I is white Gaussian noise.

Sensors also measure the local RF activity of other sensors through an RF power sensor. Merrill et al. (2003) argue that RF power falls off as an inverse fourth power law. Hence we have

$$\text{Traffic}(l, t) = \min \left\{ \sum_{k=1}^{L} \frac{B_k(t)}{(\Delta(l, k))^4} + \eta_T(t), K_T \right\}, \tag{13.3}$$

where B_k represents the RF intensity of Sensor k at one metre, $\Delta(l, k)$ is the distance from Sensor l to Sensor k, K_T represent saturation values for the RF sensor, and η_T is white Gaussian noise.

Processor and sensor

When active, raw sensor measurements are taken at the rate of 200 kHz and processed locally. In particular, for every batch of 100 measurements, i.e. at the rate of 2 kHz, some bearing information (e.g., direction of arrival) is computed and encoded into one packet. Hence, each active sensor produces bearing data at the rate of 2000 packets per second. There is a finite-size buffer to store packets that are waiting to be transmitted. Packet transmission decisions are by pre-computed channel-aware transmission scheduling policies. Activation control decisions are made at a slow rate of 20 Hz. Processing is assumed to consume 16.5 mW, as in the Mica mote, and its duty cycle is taken to be 100% in active mode, and only 5% in sleep mode. Both target and RF sensing have a similar duty cycle, but requires 10 mW for sampling, amplifying and A/D conversion.

RF transceiver

When active, the RF Transmitter consumes 36 mW of power, while the receiver consumes 4.5 mW. Since transceiver startup time is nonnegligible due to the lock time of the phase-lock loop, it is assumed that the transmitter has 100% duty cycle in active mode. The transceiver is completely shut down in sleep mode. We do not require the receiver at all in our implementation, although it may be useful for higher level configuration purposes not considered here.

Power supply

Merrill et al. (2003) specify that Lithium battery technology contains approximately 340 WH of power per 1000 cm³. Assuming a 8 cm³ battery pack, this yields 2.72 WH of power. From the above specifications, active sensors consume 77 mW, while sleeping sensors consume only 1.83 mW. A sensor is therefore capable of 35 hours of active operation, or 61 days of standby operation.

13.2.2 Energy-Efficient Sensor Activation and Transmission Control

The task of each sensor is to efficiently perform target detection and localization, which involves two key components: (1) sensors must be fully active only when needed, and (2) they must operate as efficiently as possible when active. We quantify these trade-offs in this section, and investigate how to achieve them in Sections 13.3 and 13.4, respectively.

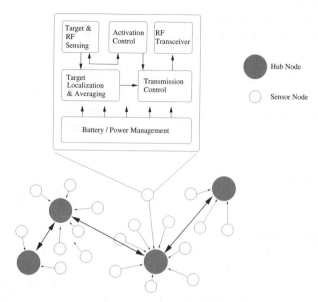

Figure 13.1 Sensor network layout and sensor architecture. Sensors transmit data to a backbone of larger hubs. The two-time scale sensor activation and transmission control algorithm is given in Section 13.4. The activation control algorithm is established separately in Section 13.3.

Activation control specifies when sensors should be actively transmitting data and when they should power down to save energy. This should result in activation of the most well-positioned sensors (closest to target) at any time, but should limit the number of sensors on to avoid redundancy.

We consider a decentralized activation control solution for two reasons. First, it provides us with a simple and scalable algorithm for sensor activation since there is no central controller or infrastructure for activation. Second, it is more energy efficient since sleeping sensors do not have to periodically turn on a receiver to obtain instructions. We assume that each sensor activates or sleeps to maximize a utility given by:

$$u^l(t) = \begin{cases} \alpha_I \text{Proximity}(l, t) - \alpha_T \text{Traffic}(l, t) - \text{Energy Cost(active)}, & \text{if active,} \\ -\text{Energy Cost(asleep)}, & \text{if asleep.} \end{cases} \quad (13.4)$$

Here α_I and α_T are system parameters, and the energy costs can be taken as 3.625 and 0.0912 mJ for active and sleep mode, respectively, which are the energy costs per period (at 20 Hz) as specified in Section 13.2.1. We use the Proximity and Traffic functions of (13.2) and (13.3).

If a mobile hub is used, a sensor may adapt (α_I, α_T) slowly in time as follows. If the hub is present, then (α_I, α_T) carry their normal values. If the hub is not currently near the cluster, then α_I may be decreased and/or α_T may be increased. This will reduce the incentive for a sensor to activate when the hub is not present, since one may desire fewer measurements to be taken (and stored locally) in this case.

In contrast to our approach, consider a more traditional framework for sensor activation. Groups of sensors would typically agree on a monitoring schedule, based on expected activity levels. When activity is detected by a sensor, it would wake up other nearby sensors to extract sufficient information, and also wake up sensors along expected paths of the target, in the case of mobile target tracking. Note that to implement such a scheme requires dedicated sensor-to-sensor communication, which requires a highly complex multiple access communication architecture. The power requirements of such communication is also high, and can only be lowered by reducing the responsiveness of the network, e.g. by scheduling synchronization periods every T seconds to reduce the radio duty cycle to roughly $1/T$. We propose the game theoretic approach as one method of circumventing these considerations.

For our scenario, a sensor in active mode produces data at the rate of 2000 packets per second or 100 packets per each activation decision period (50 ms). In Section 13.4 we formulate the problem of optimal scheduling of the transmission of a finite number of packets in a Markovian fading channel to trade off energy efficiency and throughput as a finite-horizon Markov Decision Process with a terminal penalty cost. The objective is to minimize an expected total cost, which is the sum of accumulated transmission costs and a penalty cost on the amount of data lost. The optimality of threshold transmission policy is proved in Section 13.4 using the concept of supermodularity. The threshold structure substantially reduces the complexity required to implement the optimal transmission policy.

The interaction between the activation control and transmission scheduling layers is shown in Figure 13.2. The activation control layer affects the transmission scheduling layer in two ways. First, transmission only occurs when the sensor is turned on. Second, when nearby sensors activate, the transmission scheduling layer sees a noisier channel, and this change in the channel state will be reflected in the transmission decisions of the sensor. In particular, the activation decisions $\mathbf{X}_n^{-l}$ of sensors near l (see (13.9)) affects the SINR seen by the transmission scheduling layer through (13.29) and hence affects the

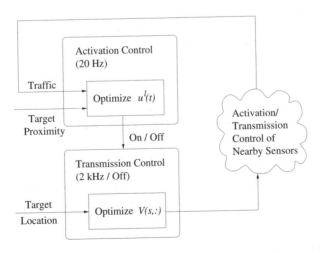

Figure 13.2 Interaction between the Activation Control and Transmission Scheduling Layers. Transmission scheduling indirectly affects activation control through neighbouring sensors.

transmission decisions, which are determined by a precomputed optimal channel-aware transmission policy.

Conversely, the transmission scheduling layer affects the activation control layer by feeding back to the sensor activation layer the total transmission energy and data loss costs during each activation period. A higher cost fedback from the transmission scheduling layer will cause a sensor l to associate a lower reward with activating (see (13.4)), and hence activate less often by the algorithm of Section 13.3.2. Lastly, the activation control of sensors nearby sensor l also affects the activation control of l because transmissions occur in the active mode.

13.3   Sensor Activation as the Correlated Equilibrium of a Noncooperative Game

In this section we describe the applicability of game theory to sensor network design, with a focus on adaptive decentralized implementation. We then describe the regret tracking algorithm as it pertains to the sensor activation game described above. The main result is Algorithm 1, which provides a completely decentralized rule for sensor activation that converges to a correlated equilibrium.

13.3.1   From Nash to Correlated Equilibrium – An Overview

Modern noncooperative game theory originated with John von Neumann in 1928 in the context of economics, but is increasingly finding applications in computer science and engineering. In particular, systems with a large number of independent components are well modeled by game theory when these components act independently. This is particularly attractive for sensor networks, since such decentralization promises to reduce or eliminate costly overhead associated with coordinating sensor activity.

Briefly, game theory treats the problem of optimal, interactive decision making among multiple players. For a game with L players, the problem of each player $l = 1, 2, \ldots, L$ is to select action s^l from a set $\mathbf{S}^l$ (with size S^l), to maximize a given utility function $u^l(s^1, s^2, \ldots, s^L)$. Since each player only controls one of L variables, the problem requires careful consideration of the actions of other players.

The central concept in noncooperative game theory is an equilibrium, which identifies stable operating points of the system under certain conditions, such as common knowledge of rationality. The most important such equilibrium is due to Nash (1951), defined as follows:

Definition 13.3.1 *For each player l, define a strategy π^l to be a random variable on $\mathbf{S}^l$, with distribution $f^l(s) = Pr(\pi^l = s)$, and define a strategy profile π to be a random variable on the product space $\mathbb{S} = \mathbf{S}^1 \times \mathbf{S}^2 \times \ldots \times \mathbf{S}^L$, with distribution $f(s) = \prod_{k=1}^{L} f^k(s^k)$. We may write any strategy π as (π^l, π^{-l}) for any l, where π^{-l} is the strategy profile of all players but l. The expected utility to l resulting from π is*

$$u^l(\pi) = \sum_{s \in \mathbb{S}} u^l(s) f(s). \tag{13.5}$$

π *is a Nash equilibrium if each* π^l *is an optimal response to the collection* π^{-l} *of strategies of other players. That is,*

$$u^l(\pi^l, \pi^{-l}) \geq u^l(\sigma^l, \pi^{-l}), \qquad (13.6)$$

for all $l = 1, 2, \ldots, L$ *and all possible alternative strategies* σ^l.

The notation (σ^l, π^{-l}) means that l uses strategy σ^l instead of π^l. If σ^l has distribution g^l, then the profile (σ^l, π^{-l}) has distribution $f(s) = g^l(s^l) \cdot \prod_{k \neq l} f^k(s^k)$.

In general there can be many Nash equilibria, which are either pure (a single action is chosen with probability one) or mixed. Significant research has been done into investigating game structures with certain types of equilibria, e.g. supermodular games by Topkis (1998), potential games by Monderer and Shapley (1996), etc.

An important generalization of the Nash equilibrium, proposed by Aumann (1974, 1987) is the *correlated equilibrium*. Note that in Definition 13.3.1, f is restricted to the space of product distributions on $\mathbb{S}$. The *correlated equilibrium* condition simply removes this restriction, so that players may correlate their random action choices:

Definition 13.3.2 *Define a correlated strategy* π *on* $\mathbb{S}$ *with distribution* $f(s) = Pr(\pi = s)$. π *is a correlated equilibrium if*

$$u^l(\pi^l, \pi^{-l}) \geq u^l(\sigma^l, \pi^{-l}), \qquad (13.7)$$

for all $l = 1, 2, \ldots, L$ *and all possible alternative strategies* σ^l *that are a function of* π^l.

Strategy π provides each player l with an action 'recommendation' a^l. Based on this, and knowing π, the player can calculate an *a posteriori* probability distribution for the actions of other players, and hence an expected utility for each action. The equilibrium condition states that there is no deviation rule (represented by a function σ^l of π^l), that would award l a better utility in expected terms. Substituting (13.5) into (13.7), we obtain a commonly used equivalent condition:

$$\sum_{s^{-l} \in \mathbb{S}^{-l}} f(j, s^{-l})[u^l(k, s^{-l}) - u^l(j, s^{-l})] \leq 0, \qquad (13.8)$$

for all $l = 1, 2, \ldots, L$ and all $j, k \in \mathbb{S}^l$. That is, for any recommendation j to l, there is no profitable deviation k.

The correlated equilibrium concept permits coordination between players, and its potential for improved performance over a Nash equilibrium is shown by Aumann (1987). The set of correlated equilibria of a game is also structurally simpler; Nau et al. (2004) show it is a convex set, whereas the set of Nash equilibria are fixed points, coincident with some of the extrema of this set. Since the set of correlated equilibria is convex, fairness between players can also be addressed in this domain. Finally, decentralized, online adaptive procedures (see below) naturally converge to the set of correlated equilibria, whereas the same is not true for Nash equilibria (the so-called law of conservation of coordination of Hart and Mas-Colell (2003)).

For engineering approaches, the role of game theory is to analyze the behaviour of decentralized adaptive behaviour in a repeated game. These procedures, outlined below, allow system components to adapt to a changing environment without explicit coordination. Players take a sequence of actions, receive utilities, and adjust their actions over time to optimize their performance. It is important in this case that the players' utilities are based only on local information; if players based their utilities on information from a central source, it is a clue that the decentralized algorithm being considered could easily be replaced by a more efficient centralized scheme.

We outline a few adaptive schemes below. In what follows, let $n = 0, 1, 2, \ldots$ be discrete time, let $\mathbf{X}_n^l$ denote the action of Player l at time n,g and let $\mathbf{X}_n^{-l}$ denote the joint actions of all player but l at time n.

Best response   The simplest scheme, 'best response', simply takes

$$\mathbf{X}_{n+1}^l = \arg\max_{x \in \mathbf{S}^l}\{u^l(x, \mathbf{X}_n^{-l})\},$$

that is, l reacts optimally, assuming the other players will repeat their previous actions. Since it fails to account for simultaneous adaptation of multiple players, this approach converges only in special cases, such as two-player zero-sum games, supermodular games, potential games, and certain types of submodular games.

Fictitious play   This is an improvement over best response, wherein each player calculates a best response assuming the historical distribution of play is a good predictor of future actions. That is,

$$\mathbf{X}_{n+1}^l = \operatorname{argmax}_{x \in \mathbf{S}^l}\{u^l(x, \hat{\boldsymbol{\pi}}_n^{-l})\},$$

where $\hat{\boldsymbol{\pi}}_n^{-l}$ is the empirical joint distribution of play up to time n. Fictitious play and its variants are extensively studied by Fudenberg and Levine (1999), and enjoy good convergence properties in practice, although convergence to Nash equilibrium is known to be false in general. One drawback is the need to explicitly observe and model the behaviour of all opponents, which may not be appropriate for sensor networks.

Regret-based algorithms   Alternatively, one can consider a class of algorithms due to Hart (2005); Hart and Mas-Colell (2000, 2001a,b). These algorithms replace explicit opponent modelling with an implicit 'regret matrix', $\theta^l(n)$ which tracks, for every pair of actions $i, j \in S^l$, the difference in utility if l had taken action j in the past everywhere he took action i. Given $\mathbf{X}_n^l = i$, the probability of $\mathbf{X}_{n+1}^l = j$ is proportional to $\theta_n^l(i, j)$, the regret from i to j. Learning proceeds by exploring actions and switching to actions that are perceived as 'better' according to this regret measure.

We consider this last class of algorithms in the rest of this chapter. The main advantages of regret-based procedures are ease of implementation, and provable convergence to the set of correlated equilibria. Maintenance of $\theta^l(n)$ requires minimal computation and no explicit awareness of other players. The main disadvantage is that players are required to know $u^l(k, \mathbf{X}_n^{-l})$ for all possible $k \in \mathbf{S}^l$ at each n. This requirement is removed in modified regret matching in Hart and Mas-Colell (2000), and through Step One in Algorithm 1 below, which estimates the values where needed.

13.3.2   Adaptive Sensor Activation through Regret Tracking

In this section, we apply a regret-based algorithm to solve the decentralized sensor activation problem. We use a modified version of the regret matching procedure of Hart and

Mas-Colell (2000), which is presented as a distributed stochastic approximation algorithm. This formulation allows us to specify an adaptive variant of the original algorithm, called 'Regret Tracking', which uses a constant step size to dynamically adapt to time-varying game conditions, thus allowing sensors to function in a changing environment. The algorithm is provably convergent to the set of correlated equilibria.

At each decision period (at a rate of 20 Hz, see Section 13.2.1), sensors take an action, either to activate and transmit information or to sleep. Define the sequence of joint sensor actions $\{\mathbf{X}_n : n = 1, 2, \ldots\}$ as a vector-valued, finite state stochastic process in $\mathbb{S}$, such that

$$\mathbf{X}_n = \begin{bmatrix} \mathbf{X}_n^1 & \mathbf{X}_n^2 & \ldots & \mathbf{X}_n^L \end{bmatrix}'. \tag{13.9}$$

The regret-based algorithm works by averaging the potential costs associated with past joint actions $\{\mathbf{X}_n : n = 1, 2, \ldots\}$ and using them to choose future actions. The potential costs to Sensor l under joint action $\mathbf{X}$ are collected in a $\mathbf{S}^l \times \mathbf{S}^l$ matrix $\mathbf{H}^l$, with entries

$$\mathbf{H}_{jk}^l(\mathbf{X}) = I\{\mathbf{X}^l = j\}\left(u^l(k, \mathbf{X}^{-l}) - u^l(j, \mathbf{X}^{-l})\right). \tag{13.10}$$

For convenience we collect these into an aggregate matrix,

$$\mathbf{H}(\mathbf{X}) = \begin{bmatrix} (\mathbf{H}^1(\mathbf{X}))' & (\mathbf{H}^2(\mathbf{X}))' & \ldots & (\mathbf{H}^L(\mathbf{X}))' \end{bmatrix}', \tag{13.11}$$

where A' denotes the transpose of A.

The potential cost requires sensors to know the utility for being active (Action 2) even when they are asleep (Action 1). Since this is not strictly possible, we approximate the utility by $\hat{u}^l(2, \cdot)$ using a sample and hold method. On an average of every K iterations, a sleeping sensor activates its sensor inputs (but not its radio) long enough to estimate the utility for being active. See Step one of Algorithm 1 below. The value K is preprogrammed; larger values mean less power consumption but also less accuracy. In practice, and for our two action scenario, we therefore use the approximation:

$$\mathbf{H}^l(\mathbf{X}) = \begin{bmatrix} 0 & \hat{u}^l(2, \mathbf{X}^{-l}) - u^l(1, \mathbf{X}^{-l}) \\ u^l(1, \mathbf{X}^{-l}) - u^l(2, \mathbf{X}^{-l}) & 0 \end{bmatrix}. \tag{13.12}$$

The matrices $\{\mathbf{H}(\mathbf{X}_1), \mathbf{H}(\mathbf{X}_2), \ldots \mathbf{H}(\mathbf{X}_n)\}$ are averaged to yield the sensor regrets $\{\boldsymbol{\theta}_n : n = 1, 2, \ldots\}$, where $\boldsymbol{\theta}_n$ is an aggregate $L S^l \times S^l$ regret matrix,

$$\boldsymbol{\theta}_n = \begin{bmatrix} (\boldsymbol{\theta}_n^1)' & (\boldsymbol{\theta}_n^2)' & \ldots & (\boldsymbol{\theta}_n^L)' \end{bmatrix}'. \tag{13.13}$$

$\boldsymbol{\theta}_{ij}^l$ tracks the regret from i to j for Player l. The regret matrix $\boldsymbol{\theta}_n$ defines a probability distribution according to which $\mathbf{X}_{n+1}$ is chosen, with higher regret-values leading to higher probabilities.

The regret-based procedure is summarized Algorithm 1, which is carried out independently by each sensor. There are two versions of this algorithm. For regret matching, define a decreasing sequence $\{\varepsilon_n\}$ satisfying $\sum_n \varepsilon_n = \infty$, e.g. $\varepsilon_n = 1/(n + 1)$. For regret tracking, take $\{\varepsilon_n\}$ to be a constant $\varepsilon > 0$. That is, we allow for either a decreasing or constant step size algorithm.

Algorithm 1 *The regret-based algorithm for sensor activation has parameters $(u^l, \mu, \{\varepsilon_n : n = 1, 2, \ldots\}, K, \boldsymbol{\theta}_0^l, \mathbf{X}_0^l)$, where u^l are the player costs, μ is a function of the costs as in (13.16), $\{\varepsilon_n\}$ is the SA step size defined above, K is the average interval between monitoring instances for sleeping sensors, and $\boldsymbol{\theta}_0^l$, $\mathbf{X}_0^l$ are arbitrary initial regrets and actions.*

1. *Initialization: Set $n = 0$, initialize regret $\boldsymbol{\theta}_0^l$ and take action $\mathbf{X}_0^l$.*

2. *Repeat for $n = 1, 2, \ldots$:*

   (a) **Cost Estimate**

      i. If $\mathbf{X}_n^l = 1$ (the sensor is asleep),

         A. Generate random $\kappa \sim \text{Uniform}(0, 1)$.

         B. If $\kappa < 1/K$, activate target and RF sensing capabilities and update $u_{last} = u^l(2, \mathbf{X}_n^{-l})$.

         C. Set $\hat{u}^l(2, \mathbf{X}^{-l}) = u_{last}$.

      ii. If $\mathbf{X}_n^l = 2$ (the sensor is awake), set $u_{last} = u^l(2, \mathbf{X}_n^{-l})$.

   (b) **Regret Value Update** Update $\boldsymbol{\theta}_n$ using the stochastic approximation (SA):

   $$\boldsymbol{\theta}_{n+1}^l = \boldsymbol{\theta}_n^l + \varepsilon_n(\mathbf{H}^l(\mathbf{X}_{n+1}) - \boldsymbol{\theta}_n^l), \tag{13.14}$$

   (c) **Action Update** Choose $\mathbf{X}_{n+1}^l = k$ with probability

   $$Pr(\mathbf{X}_{n+1}^l = k | \mathbf{X}_n^l = j, \boldsymbol{\theta}_n^l = \theta) = \begin{cases} \max\{\theta_{jk}^l, 0\}/\mu, & k \neq j, \\ 1 - \sum_{i \neq j} \max\{\theta_{ji}^l, 0\}/\mu, & k = j. \end{cases} \tag{13.15}$$

The value μ is a normalization constant, which can be taken as

$$\mu > (S^l - 1)(u_{max}^l - u_{min}^l), \tag{13.16}$$

over all l, which is obtainable from the parameters of the game.

We are most interested in the constant step size algorithm, since it allows sensors to adapt their activation strategies due to changes in their environment. For example, when target positions change over time or sensors fail.

In our approach, $S^l = 2$, with sleep action $x^l = 1$, and wake action $x^l = 2$. Recall the player utility as given by (13.4), (13.2) and (13.3). This can be rewritten as

$$u^l(x, z) = (x^l - 1)\Big(\alpha_I \min\Big\{\sum_{m=1}^{M} \frac{A_m}{(\Delta(l, z^m))^2} + \eta_I, K_I\Big\}$$

$$- \alpha_T \min\Big\{\sum_{k=1}^{L} \frac{B_k(x^k - 1)}{(\Delta(l, k))^4} + \eta_T, K_T\Big\} - K_E\Big). \tag{13.17}$$

Note that in addition to the sensor state $\mathbf{X}_n$, we also have the target position Z_n, which denotes the vector of target positions at time n, and (α_I, α_T), which may be allowed to vary if the hub is mobile, see the discussion after (13.4). However, for slowly moving targets (and hubs) these may be regarded as slowly varying parameters. Noting that games are equivalent under addition of a constant to each utility, we have taken the sleep cost to be zero, and K_E to be 3.53 mJ, the difference between active and sleep costs in Section 13.2.2. This yields

$$\mu > \max\{\alpha_I K_I - K_E, \alpha_T K_T + K_E\}. \tag{13.18}$$

The game is submodular. It has many pure Nash equilibria and hence many correlated equilibria (any convex combination of Nash equilibria is a correlated equilibrium). Best response is not guaranteed to converge since it is possible for sensors to synchronize their actions. Moreover, best response requires that a sleeping sensor still knows the actions of its neighbors, a condition that cannot be met by definition.

13.3.3 Convergence Analysis of Regret-based Algorithms

In this section, we analyze some of the convergence results for Algorithm 1. Due to (13.15), it is clear from Algorithm 1 that $\{\mathbf{X}_n : n = 1, 2, \ldots\}$ is a Markov chain with transition matrix $\mathbf{A}(\boldsymbol{\theta}_n)$ at time n. This is the key to convergence analysis. Formally, it helps to define a smooth version of the transition matrix of the resulting Markov chain as follows.

Definition 13.3.3 *Define $\rho(\cdot)$ to be a twice continuously differentiable matrix-valued function such that for any $\delta_\rho > 0$, $\rho(\theta^l) = \theta^l$ for $\theta^l > 0$ and θ^l is outside $N(0, \delta_\rho)$ a δ_ρ-neighborhood of 0 and $\rho(\theta^l) = 0$ for $\theta^l < 0$ and θ^l is outside $N(0, \delta_\rho)$. For any small δ, we define the entries of $\mathbf{A}(\boldsymbol{\theta})$ by*

$$a_{jk}(\boldsymbol{\theta}) = Pr(\mathbf{X}_{n+1} = k | \mathbf{X}_n = j, \boldsymbol{\theta}_n = \boldsymbol{\theta}) = \prod_{l=1}^{L} a^l_{j^l k^l}, \tag{13.19}$$

where

$$a^l_{jk}(\boldsymbol{\theta}) = Pr(\mathbf{X}^l_{n+1} = k | \mathbf{X}^l_n = j, \boldsymbol{\theta}_n = \boldsymbol{\theta})$$

$$= (1 - \delta)e'_j \left(\frac{1}{\mu} \rho(\theta^l)(\mathbf{I} - \mathbf{1}e'_j) + \mathbf{I} \right) e_k + \frac{\delta}{S^l}, \tag{13.20}$$

where $\mathbf{I}$ is the identity matrix, $\mathbf{1}$ is a vector of ones, and $\mathbf{e}_j$ is unit basis vector. The matrix $\mathbf{A}(\boldsymbol{\theta})$ is the transition matrix of the Markov chain $\{\mathbf{X}_n : n = 1, 2, \ldots\}$ in the regret-based algorithms. It is multilinear (L−linear) in $\boldsymbol{\theta}$. In addition, each process $\{\mathbf{X}^l_n : n = 1, 2, \ldots\}$ constitutes an independent Markov chain with transition matrix $\mathbf{A}^l(\boldsymbol{\theta})$, with entries defined as in (13.20).

Note that $\rho(\cdot)$ is a smoothed version of $\max\{\cdot, 0\}$, and that δ is a small probability of choosing an action from a uniform distribution. These two components are added to facilitate our analysis; to recover the original algorithm, one would set δ and δ_ρ to zero.

Using asymptotics to characterize the stationary distribution of the Markov chain, and following a Liapunov-type stability argument, one can prove the following theorem, as in Hart and Mas-Colell (2001b).

Theorem 13.3.4 *Under multiplayer regret matching, the joint empirical distribution of play converges to the set of correlated equilibria.*

Convergence for regret tracking, and for slowly varying Z_n follows in a similar fashion, although the results are correspondingly weaker. On the other hand, regret tracking allows the players to behave adaptively, whereas regret matching does not respond well to changes in the game structure unless the procedure is restarted periodically.

More powerful stochastic approximation techniques, such as differential inclusions and the ODE method used by Benaim et al. (2005); Kushner and Yin (2003), offer a promising alternative to proving these convergence results. These methods allow one to define a deterministic continuous time dynamical system based on a stochastic approximation that captures the average behaviour of the original system.

The ODE method defines an interpolated process $\boldsymbol{\theta}^{\varepsilon}(t)$ from the stochastic approxima-tion, such that $\boldsymbol{\theta}^{\varepsilon}(t) = \boldsymbol{\theta}_n$ for $t \in [\varepsilon n, \varepsilon(n + 1))$. As $n \to \infty$ and $\varepsilon \to 0$, with εn moderate, averaging theory can be used to show that the limiting behaviour of the ODE is equivalent to the limiting behaviour of the stochastic approximation. This is true almost surely for decreasing step size ε_n and in probability for constant step size ε_n.

For more complicated systems, the stochastic approximation algorithm must be studied through the more general equation

$$\dot{\boldsymbol{\theta}}^l(t) \in F(\boldsymbol{\theta}(t)).$$

This differential inclusion approach can handle more general situations, including cases where the dynamics are not uniquely known.

The key advantage of these two methods is that random behaviour is 'averaged out', so that deterministic analysis can be carried out. For example, Liapunov's direct method can be applied to the differential systems to establish stability of the stochastic approximation algorithms.

Formally, we may apply the ODE method to obtain a differential equation describing the two regret algorithms:

$$\frac{d}{dt}\theta(t) = \overline{\mathbf{H}}(\theta(t)) - \theta(t), \tag{13.21}$$

$$\frac{d}{dt}\widehat{\pi}(t) = \pi(\theta(t)) - \widehat{\pi}(t), \tag{13.22}$$

where

$$\overline{\mathbf{H}}(\theta) = \sum_{x_i \in S} \mathbf{H}(x_i)\pi_i(\theta). \tag{13.23}$$

For $\delta = 0$, $\mathbf{A}(\theta)$ may possibly have multiple stationary distributions. The differential inclusion method must be used in this case:

$$\frac{d}{dt}\theta(t) \in \mathbf{F}(\theta(t)), \tag{13.24}$$

$$\frac{d}{dt}\widehat{\pi}(t) \in \mathbf{G}(\widehat{\pi}(t), \theta(t)), \tag{13.25}$$

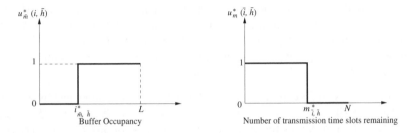

Figure 13.3 The optimal transmission policy is of the form (13.41) and (13.38), i.e. monotonically increasing and threshold in the buffer occupancy and monotonically decreasing and threshold in the residual transmission time.

where

$$\mathbf{F}(\theta) = \left\{ \sum_{x_i \in S} \mathbf{H}(x_i)\boldsymbol{\pi}_i(\theta) - \theta : \boldsymbol{\pi}(\theta) = \boldsymbol{\pi}(\theta)\mathbf{A}(\theta) \right\}, \tag{13.26}$$

$$\mathbf{G}(\widehat{\boldsymbol{\pi}}, \theta) = \left\{ \boldsymbol{\pi}(\theta) - \widehat{\boldsymbol{\pi}} : \boldsymbol{\pi}(\theta) = \boldsymbol{\pi}(\theta)\mathbf{A}(\theta) \right\}. \tag{13.27}$$

While we have stated the results formally, it can be shown that $\rho(\boldsymbol{\theta})^2$ is a Liapunov function for both systems whenever $\boldsymbol{\theta} > 0$ in some component. This can be used to show that regret values tend to zero as play continues, and hence is an alternate way of showing convergence to the set of correlated equilibria. Thus, the ODE methods provide a promising framework for stability analysis for the decentralized stochastic approximation described here.

13.4 Energy-Efficient Transmission Scheduling in UGSN – A Markov Decision Process Approach

When a sensor is activated (by Algorithm 1), it measures bearing and range of a target and forwards data to a hub node over a wireless link. In this section we formulate the problem of optimal channel-aware transmission scheduling subject to a hard delay constraint as a Markov Decision Process (MDP) and illustrate the use of the supermodularity concept to prove monotonicity of the optimal transmission policy.

Assume time is divided into slots of equal size so that at each time slot, one packet can be transmitted. For generality, we assume that the channel between every sensor and the hub node is correlated fading. The channel is correlated fading for the following scenarios:

- The hub node is static (or mobile), and the sensors communicate with the local hub node using the Direct Sequence Code Division Multiple Access (DS-CDMA) scheme with random (i.e. non-orthogonal) signature sequences. Then the channel state of every sensor is influenced by interference from other sensors. As sensors have random awake/ sleep time and unsynchronized, different transmission policies, the channel state of every sensor is correlated over time.

- The hub node is mobile, for example, a mini unmanned aerial vehicle (UAV) that cruises (e.g., at a height in the range of 100 m above the ground and a velocity of 40km/h) through different clusters of sensors to collect information. In this case, the channel between a ground sensor and the hub node is correlated fading regardless of the multiple access and modulation schemes.

In a discrete time system, a correlated fading channel is typically modelled by a finite state Markov chain (FSMC). In Guan and Turner (1999), Wang and Moayeri (1995), Zhang and Kassam (1999), the procedures for obtaining FSMC models for different physical channels are given.

At the beginning of each transmission time slot, a sensor knows its quantized channel state perfectly and decides whether to transmit a packet. A sensor can estimate its channel state by measuring the signal strength of a beacon broadcast by the local base station, and possibly the RF activity of other sensors. Intuitively, the motivation for transmitting a packet is to get data through, and the motivation for possibly delaying a transmission is not to waste energy, when the channel is in a bad state.

The optimization problem is formulated as a finite horizon MDP, with the objective being to minimize the expected total cost, which is the sum of transmission costs and the terminal data loss penalty cost. In other words, we use a MDP approach to optimize channel-aware transmission policies to conserve energy, subject to a hard delay constraint that will be described later. Given the MDP formulation, the optimal transmission policy is proved to be monotone, and hence threshold, in the buffer occupancy and the residual transmission time using the concept of supermodularity (see Figure 13.3).

The plan of this section is as follows. We will first give a brief introduction to Markov decision processes and the concept of supermodularity. The second part of the section is dedicated to formulating the optimal transmission scheduling problem as a MDP and proving the threshold structure of the optimal transmission policy. At the end of the section, we briefly describe how the optimal transmission scheduling problem can be formulated as an infinite horizon, average cost, countable state constrained MDP and how the supermodularity concept can still be applied to obtain similar structural results.

13.4.1 Outline of Markov Decision Processes and Supermodularity

Markov decision processes

Formally, a *Markov decision process* (MDP) consists of five tuple (S, T, A, R, P), where S is the system state space, T denotes the decision epochs, A is the action sets, R is the rewards or costs associated with action-state pairs, and P is a state transition probability matrix. In this chapter, for simplicity we consider a finite horizon MDP, i.e. $T = 0, 1, \ldots, M$ for some finite M, with the objective being to minimize the total expected cost. (We refer the reader to Ngo and Krishnamurthy (2007a,b) for an extension of the results in this section to constrained infinite horizon average cost MDPs). A policy μ for such a finite horizon MDP consists of M decision rules $u_m(\cdot) : S \rightarrow A$, one for each decision epoch. The value of each policy μ is given by

$$V(s, \mu) = \mathbb{E}\left\{ \sum_{m=0}^{m=M} r(s_m, a_m) \big| s_0 = s, \mu \right\},$$
(13.28)

where $r(s, a)$ denotes the instantaneous reward for action a and state s. The optimality criterion in solving a MDP is then to find a policy to minimize the total expected cost or maximize the total expected reward above. An optimal policy for a MDP can be found by solving a stochastic dynamic programming called Bellman's equation which is based on the Principle of Optimality developed by Bellman in 1950's. In Section 13.4.2, the problem of optimal transmission scheduling is formulated as a MDP and the Bellman's equation is (13.34).

Supermodularity in MDPs

In Def. 13.4.1 below we define a supermodular function and its connection to complementarity in a simple context, where the objective function is a real-valued function of real variables. The topic of supermodular optimization is studied in great depth in Topkis (1998). Therein, the theory relating to lattices, supermodularity, complementarity and monotone comparative statics is established.

Definition 13.4.1 *A function* $F(x, y) : X \times Y \to \mathbb{R}$ *is supermodular in* (x, y) *if*

$$F(x_1, y_1) + F(x_2, y_2) \geq F(x_1, y_2) + F(x_2, y_1) \; \forall x_1, x_2 \in X, y_1, y_2 \in Y, x_1 > x_2, y_1 > y_2.$$

If the inequality is reversed, the function $F(\cdot, \cdot)$ *is called submodular.*

Supermodularity is a sufficient condition for optimality of monotone policies. Specifically, the main punch line is the following result regarding the arg max of a supermodular function:

Result 1 *(Topkis (1998)) If* $F(x, y)$ *defined as in Def. 13.4.1 is supermodular (submodular) in* (x, y) *then* $y(x) = \arg\max_y F(x, y)$ *is non-decreasing (non-increasing) in* x.

We will establish supermodularity of the state action cost function that arises in Bellman's equation and then use the above result to prove that the optimal policy (which is given by the arg max of the state action cost function) is monotone and hence threshold.

13.4.2 Optimal Channel-Aware Transmission Scheduling as a Markov Decision Process

When a sensor is in the active mode, it measures bearing and range of a target and forwards the information to a mobile hub node via a wireless link. In particular, an active sensor samples the target noise at a fast rate (e.g., 200 kHz) and processes the measurements locally to produce data packets that contain useful information at a lower rate, e.g. 2000 packets per second. When a sensor is in the active mode, it must remain active for at least one activation/deactivation decision period, during which a large number of packets are produced. The problem that we consider here is to exploit CSI to transmit the packets efficiently subject to a hard delay constraint.

Time is slotted so that at each time slot one packet can be transmitted. Assume a quantized channel state space. As the channel is correlated fading due to mobility of the hub node (or alternatively, due to the random interference from other active sensors), we model the evolution of the channel state by a FSMC. At the beginning of each transmission time slot, a sensor estimates its channel state by measuring the signal strength of a beacon broadcast by the local base station, and possibly the RF activity of other sensors.In what follows, we assume that at the beginning of each time slot a sensor knows its quantized channel state perfectly and decides whether to transmit a packet.

Assume that every batch of L packets must be transmitted in $M \geq L$ transmission time slots. In other words, we assume there is a buffer of size L to store packets and there is a mechanism to reserve some bandwidth for delayed transmissions, e.g., an ARQ protocol, to implement the transmission of L packets within $M \geq L$ transmission time slots. The values of L and M can be designed to meet some system performance requirements. When a transmission is attempted, a positive transmission cost, which depends on the Markovian channel state, occurs. When a transmission is not attempted, the packet is stored in the buffer and can be transmitted later. At the end of all M transmission time slots, if not all L packets are transmitted, a penalty cost on the number of untransmitted packets will occur.

The optimal transmission scheduling problem is to find a transmission policy (Def. 13.4.2) to minimize the expected total cost, which is the sum of the transmission costs and the data loss penalty cost. The problem is formulated as a MDP and optimality of threshold transmission policies is proved using the concept of supermodularity. The threshold structure of the optimal transmission policies is outlined in Figure 13.3. In Section 13.5 we provide numerical examples that illustrate the trade-off between the energy consumption and data throughput that is achieved by the monotone optimal channel-aware transmission policy.

Finite state Markov chain transmission channel model

For generality we use a FSMC channel model. Due to the random wake/sleep time of other sensors and their transmit activities the communication channel state of a sensor is correlated in time and can be modeled by a Finite State Markov Chain (FSMC). For example, consider a network where sensors use CDMA with random code sequences for communications. Assume additive white Gaussian noise (AWGN) of variance $\sigma^2 = 1$, the SINR of a sensor with signal to noise ratio (SNR) γ is given approximately by

$$h = \frac{\gamma}{1 + \sum_{j \in \mathcal{A}} \gamma_j / S}, \qquad (13.29)$$

where γ_j is the SNR of sensor j, $\mathcal{A}$ contains the indexes of all other active sensors.

It is easy to see that h evolves as a Markov chain if when the number of sensors that are transmitting data is a Markov chain. In fact, if the number of transmitting sensor is a Markov chain with a known transition probability matrix $\mathcal{P}$, and the probability distribution functions of the SNRs are known, then the transition probabilities of h can be computed straightforwardly for any given quantization of the channel state (SINR) space. Furthermore, it is reasonable to assume that the number of transmitting sensors is a Markov chain as each sensor has a random (exponentially distributed) sleeping time and when a sensor wakes, it deploys a transmission policy that maps the Markovian channel state (SINR), the buffer occupancy state and the residual transmission time into an action of whether to transmit a packet.

If the hub node is mobile, e.g., a mini UAV, the communication channel between a sensor and the hub node is correlated fading regardless of the multiple access scheme and can be modelled by a FSMC. The derivation of a FSMC channel model includes partitioning the channel state space and computing the transition probability matrix. The details are covered by Guan and Turner (1999); Wang and Moayeri (1995); Zhang and Kassam (1999).

Now we assume that SINR represents the channel state and that the channel state domain is divided into K non-overlapping ranges corresponding to K channel states. Denote the channel state space by $\mathcal{H} = \{\Gamma_1, \Gamma_2, \ldots, \Gamma_K\}$, where Γ_i is a better channel state than Γ_j for all $i > j$. Denote the channel state at time slot m by h_m, then $h_m \in \mathcal{H}$ and h_m evolves as a Markov chain according to a transition probability matrix $\mathcal{P} = (p_{ij} : i, j = 1, 2, \ldots, K)$, where $p_{ij} = \mathbb{P}(h_{m+1} = \Gamma_j | h_m = \Gamma_i)$ for all $m = 1, 2, \ldots$.

Optimal channel-aware transmission scheduling problem statement

Assume every batch of L packets needs to be transmitted within M transmission time slots while the channel state evolves according to a FSMC as described above. Let the time index $m = 0, 1, \ldots, M$ denote the number of *remaining* transmission time slots, i.e. the residual transmission time. At each time slot, the system state is defined by the CSI $h \in \mathcal{H}$, and the number of packets that are not yet transmitted (i.e. the number of packets in the buffer; the buffer occupancy state) $i \in \mathcal{I}$, where $\mathcal{I} = \{1, 2, \ldots, L\}$ is the buffer state space. In summary, at residual time m, the system state can be denoted by $s_m = [i_m, h_m] \in \mathcal{S}$, where $\mathcal{S} = \mathcal{I} \times \mathcal{H}$ is the system state space.

Denote the action set by $\mathcal{A} = \{0, 1\}$, where 0 and 1 stand for the actions of not transmitting and transmitting respectively. In a time slot, if action $a \in \mathcal{A}$ is selected, the sensor has to pay an instantaneous transmission cost

$$g(\cdot, \cdot) : \mathcal{H} \times \mathcal{A} \to \mathbb{R}, \tag{13.30}$$

where $g(h, a)$ is decreasing in h, increasing in a and $g(h, 0) = 0$. A transmission policy consists of M decision rules the decision epochs $m = 1, 2, \ldots, M$. A decision rule at time m is a function mapping states to actions $u_m : \mathcal{S} \to \mathcal{A}$. Then a transmission policy can be written as $\mu = (u_1(\cdot, \cdot), \ldots, u_m(\cdot, \cdot))$. A formal definition of a transmission policy is given below. We also denote the space of admissible transmission policies defined by Def. 13.4.2 by $\mathcal{M}$.

Definition 13.4.2 *A transmission policy is a function mapping the residual transmission time, the buffer state and the channel state information into an action:*

$$\mu : \{1, \ldots, M\} \times \mathcal{S} \to \mathcal{A}. \tag{13.31}$$

At the end of all M time slots, i.e. when $m = 0$, a penalty cost incurs on the untransmitted packets. A data loss cost function can be modified to meet specific requirements of the application but must be nondecreasing in the number of untransmitted packets, i.e. the terminal buffer occupancy state. We assume the general model where the penalty cost function maps the terminal buffer state to a cost: $C(\cdot) : \mathcal{I} \to \mathbb{R}$, where $C(i + 1) \geq C(i)\ \forall i \in \mathcal{I}$ and $C(0) = 0$.

The expected total cost is given for a transmission policy $\mu = (u_m(\cdot, \cdot) : m = 1, 2, \ldots, M)$ and an initial state $s \in \mathcal{S}$ by

$$V(s, \mu) = \mathbb{E}\left\{ \sum_{m=1}^{m=M} c(s_m, u_m(\cdot, \cdot)) + C(i_0) \Big| s_M = s, \mu \right\}, \tag{13.32}$$

where $c(s_m, u_m(\cdot, \cdot)) = g(h_m, u_m(h_m, i_m))$. The MDP problem is to find a transmission policy $\mu^* = (u_m^*(\cdot, \cdot) : m = 1, 2, \ldots, M)$ that minimizes the expected total cost given above for all initial state $s \in \mathcal{S}$:

$$V(s, \mu^*) = V(s) \overset{\triangle}{=} \inf_{\mu \in \mathcal{M}} V(s, \mu). \tag{13.33}$$

The MDP formulated above is a generalization of the classic terminal cost MDP considered in Derman et al. (1976); Ross (1983), where the monotone structure of the optimal policy is proved for the case of a constant channel state. The methodology deployed in Derman et al. (1976) and Ross (1983) can only work for a constant or i.i.d. channel state. The monotone structure proof in this chapter, however, works for the FSMC channel model, and hence for the i.i.d. or constant channel model. In addition, in Ngo and Krishnamurthy (2006), we consider a more general resource allocation problem, where the success transmission probabilities depend on the channel state, and transmitted packets that are not successfully received can be retransmitted. In that case there needs be an error-free feedback channel so that the outcome of each transmission is known at the transmitter. In this chapter, the success probability is assumed to be equal to 1 for the action $a = 1$. The motivation for this simplification is to obtain a simple optimal channel-aware transmission policy that exploits CSI to conserve energy and that does not require a feedback channel.

13.4.3   Optimality of Threshold Transmission Policies

The optimal transmission policy can be denoted by $\mu^* = (u_m^*(\cdot, \cdot) : n = 1, 2, \ldots, M)$, where the decision rules $u_m^*(\cdot, \cdot) : n = 1, 2, \ldots, M$ are the solution of the following stochastic dynamic programming recursion called Bellman's equation (Puterman (1994)):

$$V_m(i, h) = \min_{a \in \mathcal{A}} Q_m(i, h, a), \tag{13.34}$$

$$u_m^*(i, h) = \arg\min_{a \in \mathcal{A}} Q_m(i, h, a), \tag{13.35}$$

where

$$Q_m(i, h, a) = \left\{ g(h, a) + \sum_{t \in \mathcal{H}} p_{ht} V_{m-1}(i - a, t) \right\}, \tag{13.36}$$

and $V_m(0, h) = 0$, $V_0(i, h) = C(i)$. We refer to the $V_m(\cdot, \cdot)$ defined by (13.34) as the *value function* and $Q_m(\cdot, \cdot, \cdot)$ defined by (13.36) as the *state action cost function* respectively.[3]

The optimality of threshold transmission policies is proved using supermodularity in two steps, see Heyman and Sobel (1984) and Puterman (1994),

---

[3] In reinforcement learning of MDPs, the state action cost function is also called the Q function, see Bertsekas and Tsitsiklis (1996).

1. Step 1: Prove monotonicity of the value function in the state using mathematical induction. For an arbitrary finite state, finite action MDP, as shown in Heyman and Sobel (1984), a sufficient condition for this to hold is that for any fixed action, each row of the transition probability matrix is first order stochastically dominated by the next row. It turns out that in our case due to the block diagonal structure of the transition probability matrix for the system state $s = [i, h]$, this property straightforwardly holds.

2. Step 2: Prove Supermodularity of the state action cost function: We show that the state action cost function $Q_m(i, h, a)$ defined by (13.36) is supermodular/submodular (see Def. 13.4.1) using mathematical induction. The threshold structure of the optimal transmission policy then follows from Result 1.

In Heyman and Sobel (1984), assume that the value function is monotonically nondecreasing in the action, Step 2 (supermodularity) of the state action cost function is established by imposing a condition on the matrix obtained by taking the difference between the two transition probability matrices for the two different actions. The condition is that each row of this matrix has a tail sum that is dominated by the next row's tail sum. In the optimal transmission scheduling problem that we consider, we have an augmented system state $s = [i, h]$ and this condition is not satisfied. Hence, we deploy a different approach in our proof of Theorem 13.4.4, that corresponds to Step 2. We show that the value function is monotone and convex (in the buffer state) and submodular (in the buffer state and the residual transmission time). From submodularity of the value function, we prove supermodularity of the state action cost function in the action and the residual transmission time and submodularity of the state action cost function in the action and the buffer state.

We start with establishing Step 1 (monotonicity of the value function).

Lemma 13.4.3 *The value function $V_m(i, h)$ defined by (13.34) is increasing in the number of remaining packets, i.e. buffer state i and decreasing in the residual transmission time m.*

Proof. See the appendix

We now establish Step 2 (supermodularity).

Theorem 13.4.4 *If $C(\cdot)$ is an increasing function (of the number of untransmitted packets, i.e. the terminal buffer state) then the state action cost function $Q_m(i, h, a)$ is supermodular in (m, a), i.e.*

$$Q_m(i, h, 1) - Q_m(i, h, 0) \le Q_{m+1}(i, h, 1) - Q_{m+1}(i, h, 0). \qquad (13.37)$$

As a result, the optimal transmission policy is threshold in the residual transmission time, i.e.

$$u_m^*(i, h) = \begin{cases} 1 & \text{if } m < m_{i,h}^* \\ 0 & \text{otherwise,} \end{cases} \qquad (13.38)$$

where $m_{i,h}^*$ is the optimal residual transmission time threshold for buffer state i and channel state h.

Furthermore, if the penalty cost $C(\cdot)$ is an increasing function and satisfies (13.39)

$$C(i + 2) - C(i + 1) \ge C(i + 1) - C(i) \; \forall i \ge 0 \qquad (13.39)$$

then the state action cost function $Q_m(i, h, a)$ *is submodular in* (i, a), *i.e.*

$$Q_m(i, h, 1) - Q_m(i, h, 0) \geq Q_m(i + 1, h, 1) - Q_m(i + 1, h, 0). \tag{13.40}$$

As a result, the optimal transmission policy is threshold in the buffer state, i.e.

$$u_m^*(i, h) = \begin{cases} 1 & \text{if the buffer state } i > i_{m,h}^* \\ 0 & \text{otherwise,} \end{cases} \tag{13.41}$$

where $i_{m,h}^*$ *is the optimal buffer state threshold for residual transmission time m and channel state information h. Furthermore,* $i_{m,h}^*$ *is increasing in m.*

   Proof. See the appendix.

Summary of sensor activation and transmission scheduling algorithm for UGSN

For the reader's convenience, we now summarize the entire two-time scale algorithm developed in this chapter. The algorithm is schematically depicted in Figure 13.1. The optimal transmission thresholds $i_{m,h}^*$ in (13.41) are computed offline (as described below) and then stored in a look-up table.

Algorithm 2 *Set the sensor activation time index* $n = 0$, *initialize* θ_0^l *arbitrarily and take arbitrary initial action* X_0^l *as in Algorithm 1 in Section 13.3.2. At each subsequent time index n, repeat*

   1. **Sensor Activation:** *By Algorithm 1 in Section 13.3.2*

   2. **While in active mode**

   Initialize the residual transmission time index $m = M$, *Buffer state* $i = L$

   For $m = M, M - 1, \ldots, 1$

   Estimate channel state h by measuring RF activity of other sensors

   Check current buffer state i

   If $i \geq i_{m,h}^*$, *where* $i_{m,h}^*$ *is the optimal transmission threshold in (13.41)* *then* *TRANSMIT,* $i = i - 1$

   End For

   Reset $k = M$

Significance of threshold structural result

The threshold results proved in Theorem 13.4.4 can be used to reduce the computational complexity required for solving the dynamic programming problem and implementing the optimal policy. Heyman and Sobel (1984) give several algorithms (e.g., value iteration, policy iteration) that exploit monotone results to efficiently compute the optimal policies. For a finite horizon MDPs, threshold structural results can be exploited in forward/backward induction. In order to illustrate the significance of the threshold results, let us consider an example where we need to transmit $L = 10$ packets in $M = 15$ time slots in a Markovian fading channel with $K = 2$ channel states. The total number of transmission policies is $2^{MLK} = 2^{300}$. In comparison, the number of transmission policies that are threshold both in the number of transmission time slots remaining and the buffer state is $ML^K = 150^2$, i.e. the space of admissible optimal policies is reduced significantly.

The threshold results also reduce the memory required to store to optimal transmission policies. Consider an example with a larger state space. Assume that we need to transmit $L = 100$ packets within $M = 125$ time slots in a Markovian fading channel with $K = 8$ channel states. Without structural results, the memory required to store the optimal trans-mission policy is $MLK * 1 = 100$ kbits. By the result of Theorem 13.4.4, for each channel state h, we only have to store the increasing threshold $i^*_{m,h}$. It is easy to design a scheme for storing only $i_{0,h}$ and the incremental values, i.e. $i^*_{m+1,h} - i^*_{m,h}$. In this case, the memory required to store to optimal transmission policies is approximately $MK * \Delta = \Delta$ kbits, where Δ is the number of bits required to store the incremental values. For $L = 100$, Δ should be at most 2 bits, hence the memory required for storage of the optimal transmission policy is 2 kbits.

Hence, assuming the channel state transition matrix is known and the transmission cost and the terminal penalty cost are properly selected, the optimal transmission scheduling policy can be precomputed with much less computational complexity (see Heyman and Sobel (1984), Puterman (1994)), and stored with minimal memory as described above.

Monotonicity of average cost constrained optimal transmission policy

The delay-critical optimal transmission scheduling problem can also be formulated as an average cost constrained MDP. Ngo and Krishnamurthy (2007a), and (2007b) consider the transmission scheduling/rate control problem for the same system model with a buffer of infinite capacity and an ARQ protocol for retransmission. We formulate the optimization problem as an infinite horizon, average cost, countable state MDP with a constraint on the average delay cost.

The most common procedure to derive the structure of the optimal policy for an average cost countable state constrained MDP involves the following steps:

1. Determine conditions for recurrence of the Markov chains, which is essential to show the existence of a stationary optimal policy.

2. Use the Lagrange multiplier method to convert the constrained MDP into a parame-terized unconstrained MDP.

3. Analyse the structure of the Lagrange unconstrained average cost optimal policy. Typ-ically, an average cost optimal policy can be viewed as a limit point of a sequence of discounted cost optimal policies. In addition, proving structural results for discounted cost MDPs is normally more straightforward due to convergence of the value itera-tion algorithm. In other words, by relating average cost and discounted cost optimal

policies, structure of the Lagrange unconstrained average cost optimal policies can be derived using the usual approach, e.g., using the supermodularity concept.

4. Typically, it then follows that the constrained optimal policy is a randomized mixture of two Lagrange average cost optimal policies with the structure that has been proved in the previous step.

In Ngo and Krishnamurthy (2007a) and (2007b), we follows the above steps and prove that the constrained optimal transmission/resource allocation policy is a randomized mixture of two stationary, deterministic transmission policies that are monotonically increasing in the buffer occupancy state. That is the stationary average cost constrained optimal policy is $\mu^* = (u^*(\cdot))^\infty$ with

$$u^*(\cdot) = q u_1^*(\cdot) + (1 - q) u_2^*(\cdot), \tag{13.42}$$

where $q \in [0, 1]$ is the randomize factor and $u_1^*(\cdot)$ and $u_2^*(\cdot)$ are deterministic and monotone in the buffer state, i.e. of a form similar to (13.1) if there are only two actions allowed.

The first advantage of the infinite horizon, average cost countable state constrained MDP formulation is that the optimal policy is stationary, hence the implementation of the optimal policy is simplified. Furthermore, for an infinite horizon MDP, online estimation of the optimal policy is possible via algorithms such as Q-learning. Therefore, monotone structural results are particularly useful since they substantially improve the efficiency of the reinforcement learning algorithms. In Ngo and Krishnamurthy (2007a), the monotone structure is also exploited not only in a real-time Q-learning-based algorithm but also to derive a gradient-based *monotone policy search* algorithm. For the latter, a deterministic, monotone transmission policy is represented by a parameterized smooth function with the use of a mollifier, and the discrete monotone policy search problem is converted into estimating the optimal parameters, i.e. a continuous optimization problem, which can be numerically solved via available gradient-based algorithms.

13.5 Numerical Results

In this section we present simulation results illustrating the algorithms in Sections 13.3.2 and 13.4 for representative scenarios. We present the results for each section separately, which best demonstrates the efficiency of the algorithms.

13.5.1 UGSN Sensor Activation Algorithm

The sensor activation algorithm (Algorithm 1) specifies that sensors should activate only when they are close enough to a target, and far enough from other active sensors. In this section we discuss the performance of the regret tracking algorithm for the following scenario:

- The UGSN comprises of $L = 500$ sensors deployed at random over a 100×100 meter grid.

- Each sensor l radiates RF transmission power B_l μW at one metre, according to a Gaussian distribution with mean 10 and variance 0.5.

- For simplicity, we consider $M = 1$ target, which radiates an acoustic signature with intensity A_m at one metre, according to a Gaussian distribution with mean 0.01 and variance 0.0005. The algorithm works similarly for $M > 1$ targets.

- The activation cost K_E is taken as 3.53 mJ, as in (13.17).

- The acoustic saturation value is sensor $K_I = 0.0004$, which is equivalent to the acoustic intensity of a single target at a range of five metres.

- The RF saturation value is $K_T = 0.123$ μW, which is equivalent to the RF intensity of a single neighbour at a range of three metres.

We also specify the reward weights α_I, α_T, according to the following design criteria. If no other sensors are active, a sensor should receive positive utility if a target is less than ten metres away, and negative utility otherwise. Since utility decreases with distance, we therefore specify:

$$u^l(2e_l, 10) = \alpha_I \frac{0.01}{10^2} - 3.53 = 0 \Rightarrow \alpha_I = 35300,$$

where e_l denotes a unit vector in the lth component in L–space. Likewise, we specify that a sensor should receive positive utility if there is one other active sensor at a range greater than six metres, when the target is closer than six metres. That is,

$$u^l(2e_l + 2e_k, 6) = \alpha_I \frac{0.01}{6^2} - \alpha_T \frac{0.01}{6^4} - 3.53 = 0 \Rightarrow \alpha_T = 813312.$$

The above parametrization of the utility is a design choice, intended to result in a certain level of sensor participation. If other parameters are chosen, the sensor activation pattern will be correspondingly different, however, convergence to the (different) set of correlated equilibria will still hold, and convergence times will be approximately unchanged.

The sensor activation protocol was simulated using `Matlab`, and the outcomes from several runs were used to obtain the results below. In Figure 13.4, we show the sensor response to a jump change in the system. With $\varepsilon_n = 0.3$ set in (13.14) in Algorithm 1, it takes an average of 16 iterations (approx. one second at 20 Hz) for sensors to react to the sudden appearance of a target. After another three seconds the sensors have reconfigured to an equilibrium position. When the target disappears, it takes an average of 22 iterations (one second) for all nearby sensors to return to sleep mode. For a few runs, a small number of sensors remain active for approximately 100 iterations (five seconds).

We also show the sample convergence time for a stationary target in Figure 13.5.

Next, we analyze the tracking performance for a moving target, with an average velocity of six metres per second. Specifically, we simulate a target that updates its position $(x(n), y(n))$ 20 times per second, such that

$$x(n + 1) = x(n) + v_x(n) + w_x(n), \tag{13.43}$$

$$y(n + 1) = y(n) + w_y(n), \tag{13.44}$$

where $w_x(n)$ and $w_y(n)$ are independent, uniform random variables with mean 0 and variance 1/12. We also specify hard boundaries at the edge of the sensor field; if x_n or y_n

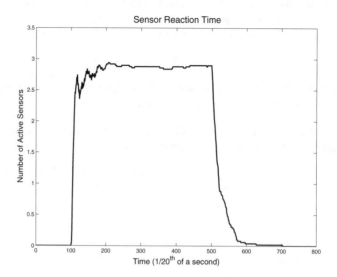

Figure 13.4 Sensor activation response to sudden appearance/disappearance of a target (500 sensors in a 100 m^2 field). At $n = 100$, iterations a target appears and an average of three sensors activate. At $n = 500$, the target disappears and sensors sleep. We show average behaviour over 50 simulation runs.

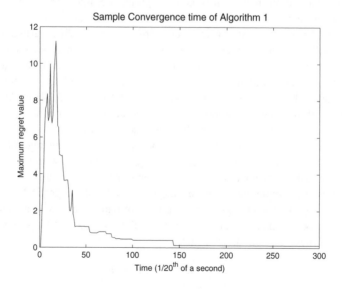

Figure 13.5 Time for convergence of regret values to zero. The data was obtained from 10 independent simulations of Algorithm 1 with a stationary target appearing at time zero.

are outside the range (0, 100), they are truncated to lie within this range. We simulate the system over 40 seconds (800 iterations). For $n <= 266$ we set $v_x(n) = 0.3$, which causes the target to move from left to right, eventually sticking to the right-hand edge of the sensor field at about $n = 220$ iterations. We then change the target direction, setting $v_x(n) = -0.3$ for $266 < n < 533$. For $n >= 534$, $v_x(n) = 0.3$ again.

The base station receives all the sensor measurements and estimates the instantaneous target positions. (We assume here that the target dynamics are not taken into account in the estimates; improved estimates could be obtained if the dynamics are known.) Assume that variance of the target position estimates is inversely proportional to the total acoustic energy received by the active sensors, such that if the received energy is 0.0012, (equivalent to three readings at five metres), then the variance is 0.12 m². That is,

$$\sigma_{\tilde{x}}^2(n) = \sigma_{\tilde{y}}^2(n) = 1.44 \times 10^{-4} \left(\sum_{l=1}^{L} \min \left\{ \sum_{m=1}^{M} \frac{A_m}{(\Delta(l, z^m))^2} + \eta_I, K_I \right\} \right)^{-1}. \qquad (13.45)$$

The measurement variance, averaged over 50 simulation runs, is plotted in Figure 13.6.

The activation algorithm tracks the target well, with only two or three sensors close to the target activating at any time. Due to the requirement for sensor separation, sensors tend to activate on all sides of the target, thereby obtaining diverse measurements. Simulation also shows that the regret values tend to zero, as expected.

Lastly, we perform a simulation to illustrate how active sensors cluster around the target. A target moves slowly from bottom right to top left of the sensor field of 500

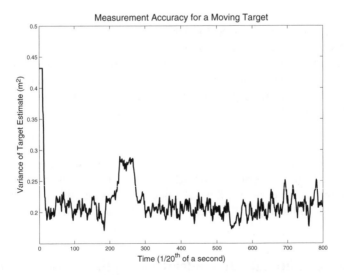

Figure 13.6 Measurement accuracy of target at the base station for 500 sensors in a 100 m² field. An average of three sensors are active during most of the simulation; the peak occurs when the target is at the edge of the sensor field, where on average only two sensors are active.

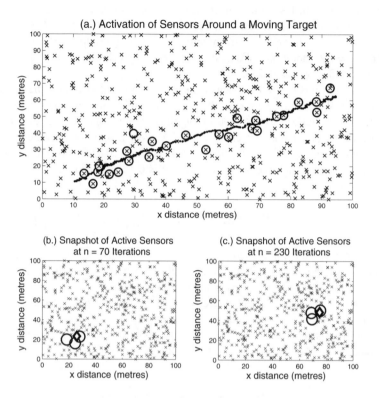

Figure 13.7 (a.) Activation history for sensors tracking a single moving target in the unattended ground sensor network (UGSN). The scenario contains 500 sensors in a 100 m^2 field. The target trajectory is shown by a solid line, the activated sensors are shown as circles. Figures (b.), (c.) depict snapshots of the same target trajectory and activated sensors at iterations 70 and 230. The target positions are shown as diamonds. The active sensors in the UGSN surround the target. The x's indicate the position of inactive sensors.

sensors along the trajectory in Figure 13.7. The figure also shows the history of all sensors that were activated under Algorithm 1 to track the target. The results show that only sensors close to the target are ever active. Several snapshots illustrate the exact configuration of active sensors around the target at a given time. The sensors surround the target, indicating good spatial diversity.

13.5.2 Energy Throughput Tradeoff via Optimal Transmission Scheduling

When a sensor in the UGSN is activated by the correlated equilibrium learning algorithm (Algorithm 1), it has to collect and then forward data to a local hub node for data fusion in a timely manner while conserving energy. The structure of the optimal transmission policy has been analyzed in Section 13.4.2, here we provide a numerical example to illustrate these structural results and analyse the energy-throughput tradeoff that can be achieved.

In particular, we assume the delay constraint is that every batch of 70 packets has to be transmitted within 100 transmission time slots, i.e. 30% of the bandwidth is reserved for delayed transmissions. Assume SNR presents channel state and the underlying channel is Rayleigh fading. Assume a carrier frequency $f_c = 1.9GHz$, the packet transmission time $T = 0.5$ ms (as the packet transmission rate is 2000 packets per second). Assume the local hub node is moving at the velocity of 20 km/h, which leads to a Doppler frequency of 35.1852 Hz. Using the SNR partitioning method in Zhang and Kassam (1999), the channel state can be modelled by a five state Markov chain, i.e. $\mathcal{H} = (\Gamma_1, \Gamma_2, \ldots, \Gamma_5)$, where $\Gamma_1, \ldots, \Gamma_5$ correspond to the SNR ranges $(-\infty, 0.7)$, $[0.7, 6)$, $[6, 9)$, $[9, 12)$, $[12, \infty)$ dB, respectively. The corresponding transition probabilities are given approximately by

$$p_{kk} = 0.8, \; p_{k,k-1} = p_{k,k+1} = 0.1 \text{ for } k = 2, 3, 4,$$

$$p_{11} = 0.8, \; p_{12} = 0.2, \; p_{54} = 0.1, \; p_{55} = 0.9.$$

Assume that the transmission energy cost function is given by

$$g(u, h) = 2I(u = 1)/\sqrt{h}, \tag{13.46}$$

and the data loss cost function is given by

$$C(i) = \lambda i^2, \tag{13.47}$$

for some $\lambda \in [0.05, 1]$. A higher value of λ emphasizes the importance of data throughput. It should be noted that the transmission cost and data loss penalty cost functions can be of any form other than the functions we consider here. In fact, designing proper transmission and data loss cost functions is an important problem, which is beyond the scope of this chapter.

For the above numerical example, the optimal transmission scheduling policies are computed using monotone-policy forward induction (see Puterman (1994)) and plotted in Figure 13.8. It can be seen from the figure that the optimal policies indeed have the threshold structure as proved in Theorem 13.4.4. In particular, when there are 60 transmission time slots remaining and the channel state is $h = \Gamma_4$, the optimal action is to transmit if and only if there are more than 46 packets in the buffer (i.e. less than 24 packets have been transmitted).

In addition, Figure 13.9 depicts the tradeoff between energy efficiency and throughput that can be achieved by a sensor via the optimal threshold transmission policy. It can be seen that for 30% of the bandwidth reserved for delayed transmissions, the energy saving is very significant. In particular, in comparison to a (non channel-aware) policy of always transmitting, the optimal channel aware transmission policy offers a throughput rate of approximately 98% for about 60% of the total transmission energy cost. It should be noted that, the optimal transmission scheduling policy requires that some bandwidth is reserved for opportunistic transmission. This is not a stringent requirement since the rate at which packets are produced is relatively low in comparison to standard wireless data transmission rates.

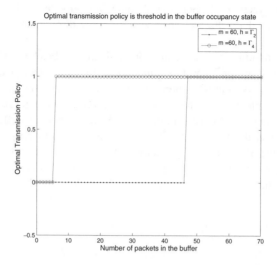

Figure 13.8 At any give time slot, for any given channel state h, the optimal transmission policy is threshold in the buffer state, i.e. $u_m^*(i, h)$ is of the form (13.41), where the optimal buffer state threshold $i_{m,h}^*$ is increasing in m.

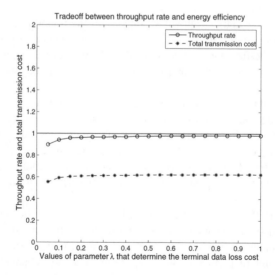

Figure 13.9 A throughput rate of 98% is achieved while the total energy cost is reduced to 60% assuming that 30% of the bandwidth is reserved for delayed transmission.

13.6 Conclusion

Energy-efficient design for unattended ground sensor networks (UGSN) requires a highly integrated approach. Increasing the autonomy and intelligence of sensors is critical to reducing costly communication overhead associated with coordinating sensors, but this must be done carefully. We have presented here computationally efficient algorithms for sensor activation and transmission scheduling. The tools used such as correlated equilibria of non-cooperative games, stochastic approximation, stochastic dynamic programming, and supermodularity, are useful in the design and analysis of sensor networks.

We conclude with a discussion of three useful extensions of the methods in this chapter.

1. Stochastic Sensor Scheduling: In this chapter we have considered sensor activation and communication in an UGSN. A third important aspect of an UGSN, which we have not considered, is sensor signal processing and scheduling, i.e., how each sensor unit in the UGSN locally senses and processes data. Typically, each individual sensor unit in an UGSN has a suite of sensors available for measurement of target type and target coordinates. Due to communication and battery costs, each sensor unit can only deploy one or a few of these sensors at each time instant. Usually more accurate sensors are more expensive to deploy. The stochastic *sensor scheduling* problem deals with how these sensors should be dynamically scheduled to optimize a cost function comprising of the Bayesian estimate of the target and the usage cost of the sensors. Such problems are partially observed stochastic control problems as considered in Baras and Bensoussan (1989). We refer the reader to the work of Krishnamurthy (2002) for the partially observed Markov decision process (POMDP) sensor scheduling problem. More recently, Krishnamurthy and Djonin (2007) show, using supermodularity on the information state space with respect to the monotone likelihood ratio ordering, that under reasonable conditions, the optimal sensor scheduling policy is monotone.

2. Analysis of stochastic approximation algorithms: The constant step size learning algorithm (Algorithm 1 in Sec. 13.3.2) presented in this chapter converges weakly to the set of correlated equilibria of a non-cooperative game. As mentioned in Sec. 13.3, the algorithm can be used to track a slowly time varying correlated equilibrium set caused due to a slowly moving target. Moreover, the limiting behaviour of the algorithm is captured by a differential inclusion. Suppose we were to assume that the target moves according to a slow Markov chain with transition probability matrix $I + \epsilon Q$ (where $\epsilon > 0$ is a small parameter and Q is a generator matrix with each row summing to zero). With this assumption on the target, how can one analyze the tracking performance of the learning algorithm (Algorithm 1) with step size ϵ? Note that the adaptation speed (step size ϵ) of the algorithm matches the speed at which the correlated equilibrium set changes (transition matrix $(I + \epsilon Q)$). In our recent work Yin et al. (2004) and Yin and Krishnamurthy (2005), we have shown that the limiting behaviour of the stochastic approximation algorithm for tracking a parameter evolving according to a Markov chain is captured by a Markovian switched ordinary differential equation. This result was somewhat remarkable, since typically the limiting process of a stochastic approximation algorithm is a deterministic ordinary differential equation. We conjecture that the limiting (see Benveniste et al. (1990)) behaviour of Algorithm 1 is captured by a Markovian switched differential inclusion. This analysis requires use of yet another extremely powerful tool in stochastic analysis namely, the so called 'martingale problem' of Strook and Varadhan, see Ethier and Kurtz (1986) and Kushner (1984) for comprehensive treatments of this area.

   3. *Constrained Markov Decision Processes. Monotone Policies and Reinforcement Learning Algorithms*: Due to time-critical nature of the data being monitored by the UGSN, the sensor transmission scheduling problem in Sec. 13.4 was formulated as a finite horizon MDP with a terminal penalty cost. Another possibility is to formulate the problem as an infinite horizon average cost MDP with delay constraints. It is well known that for such constrained MDPs, the optimal policy can be randomized function of the state. A natural question then is whether one can demonstrate some form of monotonicity of the optimal randomized policy. Since the optimal policy is randomized it cannot be directly obtained via stochastic dynamic programming – or put another way, stochastic dynamic programming cannot directly deal with global constraints.[4] Altman (1999) and coworkers have introduced a powerful methodology based on Lagrangian dynamic programming which comprises of a two level optimization: At the inner step one solves a dynamic programming problem with a fixed Lagrange multiplier to obtain a pure policy. At the outer step one optimizes the Lagrange multipliers – these Lagrange multipliers determine the randomization coefficients for switching between the pure policies. The supermodularity methods in this chapter can be applied to the inner step to show that the pure policy is threshold. Then as a result of the outer optimization step, the optimal policy is a randomized mixture of threshold policies. We refer the reader to Djonin and Krishnamurthy (2006) for a detailed exposition and also reinforcement learning algorithms that exploit this monotone structure. In Abad and Krishnamurthy (2003), a gradient estimation approach was used to devise reinforcement learning algorithms for such constrained MDPs.

13.7 Appendix

13.7.1 List of Symbols

Section 13.3, Sensor Activation

Variable	Interpretation
$l \in L$	Label of sensor; set of all sensors.
$u^l(\cdot)$	Utility of sensor l given action.
X_n; X_n^l; X_n^{-l}	Joint action at time n; action of l, action of all sensors but l.
$\mathbf{H}^l(X)$	Instantaneous regret matrix.
θ^l	Average regret matrix.
$\Delta(l, k)$	Distance from l to position of k (target or sensor)
A_m, B_l	Signal intensity of target m; sensor l.
η_I, η_T	Gaussian noise in measurement of target; sensor signal.
K_I, K_T, K_E	Physical sensor parameters.
α_I, α_T	Programmable utility parameters.

---

[4] The optimal randomized policy can be obtained as the solution of a linear programming problem.

Section 13.4, Transmission Scheduling

Variable	Interpretation
m	Residual transmission time
$\mathcal{M}$	Total number of transmission time slots
$s, \mathcal{S}$	System state; system state space
$h, \mathcal{H}$	Channel state; channel state space
$i, \mathcal{L}$	Buffer occupancy state; buffer occupancy state space
$\mathcal{A}$	Action space
$g(\cdot, \cdot)$	Transmission cost
$C(\cdot)$	Terminal penalty cost
$u(\cdot)$	Decision rule
$\mu = (u_m(\cdot) : m = 1, 2, \ldots, M)$	Transmission policy
$V(\cdot, \cdot)$	Value function
$Q(\cdot, \cdot, \cdot)$	State action cost function (or Q-function)
μ^*	Optimal transmission policy
$\mathbf{I}(\cdot)$	Indicator function

13.7.2 Proof of Lemma 13.4.3

$V_0(i, h)$ is increasing in i since $C(.)$ is increasing. From the definition of $V_m(i, h)$ given by (13.34) and (13.36), the monotonicity of $V_m(i, h)$ in i follows immediately by induction.

It is clear that $V_1(i, h) \leq V_0(i, h)$ since $V_1(i, h) \leq Q_1(i, h, 0) = \sum_{t \in \mathcal{H}} p_{ht} V_0(i, t) = C(i) = V_0(i, h)$. The monotonicity of $V_m(i, h)$ in m then follows straightforwardly from the definition of $V_m(i, h)$ given by (13.34), (13.36).

13.7.3 Proof of Theorem 13.4.4

First note that

$$Q_m(i, h, 1) - Q_m(i, h, 0) = g(h, 1) + \sum_{t \in \mathcal{H}} p_{ht} \left(V_{m-1}(i - 1, t) - V_{m-1}(i, t) \right). \quad (13.48)$$

Then it is clear that $Q_m(i, h, a)$ is supermodular in (m, a) if and only if $V_m(i, h)$ is submodular in (m, i). Similarly, $Q_m(i, h, a)$ is submodular in (i, a) if and only if $V_m(i, h)$ has increasing differences in the buffer state i. The proof consists of two parts as below.

Part 1: $V_m(i, h)$ is submodular in (i, m) and hence (13.37) and (13.38) hold

We prove by mathematical induction that

$$V_m(i + 1, h) - V_m(i, h) \geq V_{m+1}(i + 1, h) - V_{m+1}(i, h). \quad (13.49)$$

First, (13.49) holds for $m + i = 0$ since $V_m(i, h)$ is nonincreasing in m. Assume that (13.49) holds for $n + i = k$. We will prove that it holds for $m + i = k + 1$. Let $V_{m+1}(i +$

$1, h) = Q_{m+1}(i+1, h, a_{11})$, $V_{m+1}(i, h) = Q_{m+1}(i, h, a_{10}), V_m(i+1, h) = Q_m(i+1, h,$
$a_{01})$, $V_m(i, h) = Q_m(i, h, a_{00})$ for some $a_{00}, a_{01}, a_{10}, a_{11}$. We have to prove that

$$Q_{m+1}(i+1, h, a_{11}) - Q_{m+1}(i, h, a_{10}) - Q_m(i+1, h, a_{01}) + Q_m(i, h, a_{00}) \leq 0$$

$$\Leftrightarrow Q_{m+1}(i+1, h, a_{11}) - Q_m(i+1, h, a_{01}) - Q_{m+1}(i, h, a_{10}) + Q_m(i, h, a_{00}) \leq 0$$

$$\Leftrightarrow \underbrace{Q_{m+1}(i+1, h, a_{11}) - Q_{m+1}(i+1, h, a_{01})}_{\leq 0 \ \text{(By optimality)}} + \underbrace{Q_{m+1}(i+1, h, a_{01}) - Q_m(i+1, h, a_{01})}_{A}$$

$$- \underbrace{(Q_{m+1}(i, h, a_{10}) - Q_m(i, h, a_{10}))}_{B} + \underbrace{(-Q_m(i, h, a_{10}) + Q_m(i, h, a_{00}))}_{\leq 0 \ \text{(By optimality)}} \leq 0. \quad (13.50)$$

By induction hypothesis we have

$$A = \sum_{t \in \mathcal{H}} p_{ht} \left(I(a_{01} = 1) \left[V_m(i, t) - V_{m-1}(i, t) \right] + I(a_{01} = 0) \right.$$

$$\left. \left[V_m(i+1, t) - V_{m-1}(i+1, t) \right] \right)$$

$$\leq \sum_{t \in \mathcal{H}} p_{ht} \left[V_m(i, t) - V_{m-1}(i, t) \right].$$

Similarly, $B \geq \sum_{t \in \mathcal{H}} p_{ht} \left[V_m(i, t) - V_{m-1}(i, t) \right]$. Hence, $B \geq A$.

Therefore, $V_m(i, h)$ satisfies (13.49), which implies that $Q_m(i, h, a)$ is supermodular in (m, a) as in (13.37). It then follows that the optimal transmission policy $u_m(\cdot, \cdot)$ given by (13.35) satisfies (13.38).

Part 2: $V_m(i, h)$ has increasing differences in i hence (13.40) and (13.41) hold

Here we prove by mathematical induction that

$$V_m(i+2, h) - V_m(i+1, h) \geq V_m(i+1, h) - V_m(i, h). \quad (13.51)$$

First, (13.51) holds for $m = 0$ due to (13.39). Assume (13.51) holds for $m = k$. We will prove that it holds for $m = k + 1$. Let $V_{k+1}(i+2, h) = Q_{k+1}(i+2, h, a_2)$, $V_{k+1}(i+1, h) = Q_{k+1}(i+1, h, a_1)$, $V_{k+1}(i, h) = Q_{k+1}(i, h, a_0)$ for some a_0, a_1, a_2. We then have to prove that

$$Q_{k+1}(i+2, h, a_2) - Q_{k+1}(i+1, h, a_1) - Q_{k+1}(i+1, h, a_1) + Q_{k+1}(i, h, a_0) \geq 0$$

$$\Leftrightarrow \underbrace{Q_{k+1}(i+2, h, a_2) - Q_{k+1}(i+1, h, a_2)}_{A} + \underbrace{Q_{k+1}(i+1, h, a_2) - Q_{k+1}(i+1, h, a_1)}_{\geq 0 \ \text{(By optimality)}}$$

$$\underbrace{-Q_{k+1}(i+1, h, a_1) + Q_{k+1}(i+1, h, a_0)}_{\geq 0 \ \text{(By optimality)}} - \underbrace{(Q_{k+1}(i+1, h, a_0) - Q_{k+1}(i, h, a_0))}_{B} \geq 0.$$

$$(13.52)$$

In addition, it follows from the induction hypothesis that

$$A = \sum_{t \in \mathcal{H}} p_{ht} \left[I(a_2 = 1)(V_k(i+1,t) - V_k(i,t)) + I(a_2 = 0)(V_k(i+2,t) - V_k(i+1,t)) \right]$$

$$\geq \sum_{t \in \mathcal{H}} p_{ht}(V_k(i+1,t) - V_k(i,t)). \tag{13.53}$$

Similarly, $B \leq \sum_{t \in \mathcal{H}} p_{ht}(V_k(i+1,t) - V_k(i,t))$. Hence, $A - B \geq 0$.

Therefore, $V_m(i,h)$ satisfies (13.51), which implies that $Q_m(i,h,a)$ given by (13.48) is submodular in (i,a). Hence the optimal transmission policy $u_m(\cdot,\cdot)$ given by (13.35) satisfies (13.41).

Bibliography

Abad FV and Krishnamurthy V 2003 Constrained stochastic approximation algorithms for adaptive control of constrained Markov decision processes. *42nd IEEE Conference on Decision and Control*, pp. 2823–2828.

Akyildiz IF, Su W, Sankarasubramaniam Y and Cayirci E 2002 Wireless sensor networks: A survey. *Computer Networks* **38**(4), 393–422.

Altman E 1999 *Constrained Markov Decision Processes*. Chapman and Hall, London.

Amir R 2003 Supermodularity and complementarity in economics: an elementary survey. *Southern Econonomic Journal* **71**(3), 636–660.

Arulselvan N and Berry R 2005 Energy-throughput optimization for wireless ARQ protocols. In *Proc. of ICASSP*, pp. 18–23.

Aumann R 1974 Subjectivity and correlation in randomized strategies. *Journal of Mathematical Economics* **1**, 67–96.

Aumann RJ 1987 Correlated equilibrium as an expression of Bayesian rationality. *Econometrica* **55**(1), 1–18.

Baras J and Bensoussan A 1989 Optimal sensor scheduling in nonlinear filtering of diffusion processes. *SIAM Journal Control and Optimization* **27**(4), 786–813.

Bar-Shalom Y, Li XR and Kirubarajan T 2001 *Estimation with Applications to Tracking and Navigation: Theory, Algorithms, and Software*. Wiley, New York.

Benaim M, Hofbauer J and Sorin S 2005 Stochastic approximations and differential inclusions. *SIAM J. Control Optim.* **44**(1), 328–348.

Benveniste A, Metivier M and Priouret P 1990 *Adaptive Algorithms and Stochastic Approximations*, vol. 22 of *Applications of Mathematics*. Springer-Verlag, New York.

Bertsekas D 1995 *Dynamic Programming and Optimal Control* vol. 1 and 2. Athena Scientific, Belmont, MA.

Bertsekas D and Tsitsiklis J 1996 *Neuro-Dynamic Programming*. Athena Scientific, Belmont, MA.

Borbash S and Jennings E 2002 Distributed topology algorithm for multihop wireless networks. *Proc. Int'l Neural Networks* pp. 355–360.

Derman C, Lieberman G and Ross S 1976 Optimal system allocations with penalty cost. *Management Science* **23**(4), 399–403.

Djonin D and Krishnamurthy V 2007 Structural results on optimal transmission scheduling: A constrained Markov decision process approach pp.75–98 in *Wireless Communications*. Editors: P. Agarwal, et.al., IMA-Proceedings Springer-Verlag, New York.

Djonin D and Krishnamurthy, V. Q-Learning Algorithms for Constrained Markov Decision Processes with Randomized Monotone Policies: Applications in Transmission Control, IEEE Transactions Signal Processing, Vol.55, No.5, pp.2170–2181, 2007.

Djonin D and Krishnamurthy, V. V-BLAST Power and Rate Control under Delay Constraints in Markovian Fading Channels - Optimality of Monotonic Policies, IEEE Transactions Signal Processing, 2008 (to appear).

Ethier S and Kurtz T 1986 *Markov Processes:Characterization and Convergence*. Wiley, Chichester.

Fudenberg D and Levine D 1999 *The Theory of Learning in Games*. MIT Press, Cambridge, MA.

Goldsmith A and Wicker S 2002 Design challenges for energy-constrained ad hoc wireless networks. *IEEE Wireless Comm.* **9**(4), 8–27.

Guan Y and Turner L 1999 Generalised FSMC model for radio channels with correlated fading *IEE Proceedings Communications*, **146**, pp. 133–137.

Hart S 2005 Adaptive heuristics. *Econometrica* **72**, 1401–1430.

Hart S and Mas-Colell A 2000 A simple adaptive procedure leading to correlated equilibrium. *Econometrica* **68**(5), 1127–1150.

Hart S and Mas-Colell A 2001a A general class of adaptive strategies. *Journal of Economic Theory* pp. 26–54.

Hart S and Mas-Colell A 2001b A reinforcement procedure leading to correlated equilibrium, in Economic Essays, Springer, New York, pp. 181–200.

Hart S and Mas-Colell A 2003 Uncoupled dynamics do not lead to nash equilibrium. *American Economic Review* **93**(5), 1830–1836.

Heyman D and Sobel M 1984 *Stochastic Models in Operations Research*, vol. 2. McGraw-Hill, Maidenhead.

Johnston L and Krishnamurthy V 2005 Opportunistic file transfer over a fading channel: a POMDP search theory formulation with optimal threshold policies. *IEEE Trans. Wireless Commun.* **5**(2), 394–405.

Kannan R, Sarangi S, Iyengar SS and Ray L 2003 Sensor-centric quality of routing in sensor networks. *IEEE Infocom.*

Krishnamurthy V 2002 Algorithms for optimal scheduling and management of hidden Markov model sensors. *IEEE Trans. Signal Proc.* **50**(6), 1382–1397.

Krishnamurthy V and Djonin D 2007 Structured threshold policies for dynamic sensor scheduling – a partially observed Markov decision process approach. *IEEE Trans. Signal Proc.*, (accepted).

Krishnamurthy V and Yin G 2002 Recursive algorithms for estimation of hidden Markov models and autoregressive models with Markov regime. *IEEE Trans. Inform. Theory* **48**(2), 458–476.

Kumar P and Varaiya P 1986 *Stochastic Systems – Estimation, Identification and Adaptive Control*. Prentice-Hall, New Jersey.

Kushner H 1984 *Approximation and Weak Convergence Methods for Random Processes, with Applications to Stochastic Systems Theory*. MIT Press, Cambridge, MA.

Kushner H and Yin G 2003 *Stochastic Approximation and Recursive Algorithms and Applications* , second edn. Springer-Verlag, New York.

Li N and Hou J 2006 Localized fault-tolerant topology control in wireless ad hoc networks. *IEEE Transactions on Parallel and Distributed Computing* **17**(4), 307–320.

MacKenzie A and Wicker S 2001 Game theory and the design of self-configuring, adaptive wireless networks. *IEEE Communications Magazine* pp. 126–131.

MacKenzie A and Wicker S 2003 Stability of multipacket slotted aloha with selfish users and perfect information. *Proceedings of IEEE INFOCOM*.

Marandola H, Mollo J and Walter P 2002 Self-organized routing for wireless micro-sensor networks. *Proceedings of SPIE* **4743**, 99–107.

Maskery M and Krishnamurthy V 2007 Decentralized activation in a ZigBee-enabled unattended ground sensor network: a correlated equilibrium game theoretic analysis. *Proceedings of IEEE ICC*.

Maskery M, Krishnamurthy V and Yin G 2006 Stochastic approximation based tracking of correlated equilibria for game-theoretic reconfigurable sensor network deployment. *Proceedings of IEEE CDC*.

Maskery M and Krishnamurthy V 2006 Decentralized management of sensors in a multi-attribute environment under weak network congestion. *Proceedings of IEEE International Conference on Acoustics, Speech, and Signal Processing*.

Maskery M and Krishnamurthy V 2007 Decentralized algorithms for netcentric force protection against anti-ship missiles. *IEEE Transactions on Aerospace and Electronics Systems*.

Maskery M and Krishnamurthy V 2007 Network enabled missile deflection: games and correlated equilibrium. *IEEE Transactions on Aerospace and Electronics Systems*.

Merrill W, Newberg F, Sohrabi K, Kaiser W and Pottie G 2003 Collaborative networking requirements for unattended ground sensor systems. In *Proc. IEEE Aerospace Conference*.

Monderer D and Shapley LS 1996 Potential games. *Games and Economic Behavior* **14**(1), 124–143.

Nash J 1951 Non-cooperative games. *Annals of Mathematics* **54**(2), 286–295.

Nau R, Canovas SG and Hansen P 2004 On the geometry of Nash equilibria and correlated equilibria. *International Journal of Game Theory* **32**(4), 443–453.

Ngo MH and Krishnamurthy V 2006 Optimality of threshold policies for channel-aware ARQ video transmission in Markovian fading channels. *IEEE/ACM Trans. Netw.* Submitted.

Ngo MH and Krishnamurthy V 2007a Monotonicity of constrained optimal transmission policies in correlated fading channels with ARQ. *IEEE Transactions on Mobile Computing*. Submitted.

Ngo MH and Krishnamurthy V 2007b On optimality of monotone channel-aware transmission policies: a constrained Markov decision process approach *Proc. of ICASSP*, Hawai'i, USA.

Puterman M 1994 *Markov Decision Processes*. Wiley, Chichester.

Rogers A, David E and Jennings N 2005 REMBASS-II: the status and evolution of the Army's unattended ground sensor system. *IEEE Transactions on Systems, Man, and Cybernetics – Part A* **35**(3), 349–359.

Ross S 1983 *Introduction to Stochastic Dynamic Programming*. Academic Press, San Diego, CA.

Topkis DM 1978 Minimizing a submodular function on a lattice. *Operations Research* **26**, 305–321.

Topkis DM 1998 *Supermodularity and Complementarity*, vol. 2. Princeton University Press, Princeton, NJ.

Wang H and Mandayam N 2005 Opportunistic file transfer over a fading channel under energy and delay constraints. *IEEE Trans. Commun.* **53**(4), 632–644.

Wang HS and Moayeri N 1995 Finite-state Markov channel – A useful model for radio communications channels. **44**(1), 163–171.

Xing Y and Chandarmouli R 2004 Distributed discrete power control for bursty transmissions over wireless data networks. In *Proc. CISS?*

Yin G and Krishnamurthy V 2005 Least mean square algorithms with Markov regime switching limit. *IEEE Trans. Auto. Control* **50**(5), 577–593.

Yin G, Krishnamurthy V and Ion C 2004 Regime switching stochastic approximation algorithms with application to adaptive discrete stochastic optimization. *SIAM Journal on Optimization* **14**(4), 117–121.

Zhang A and Kassam SA 1999 Finite-state Markov model for Rayleigh fading channels. *IEEE Trans. Commun.* **47**, 1688–1692.

Index

σ-distance, 151

Achievable D-R region, 179–80
Activation control, sensor, 350, 355,
    357–8
  example scenario, 374–8
  transmission scheduling,
    interaction, 357
Acyclic networks, 59
ALOHA, 315, 320
Analog-amplitude data, 167
A priori information, 166
Architecture, sensor network, 354
Asymptotics, 150, 159, 162, 177

Bandwidth constraints, 149, 152, 157,
  162
Bayesian networks, 216
belief propagation evaluation, 84
Bellman-Ford algorithm, 282, 285
Berger-Tung achievable region, 180
Best response dynamic, 360
Binary observations, 150, 151, 153, 160,
  166
  dependent, 154
  vector, 160
Block computation, 51
Block coordinate descent, 172, 182
Broadcast network, 60
Broadcast routing, 280, 290

Capacity
  MAC with independent sources,
    24–25
  MAC with independent sources and
    feedback, 25–28

MAC with dependent sources,
    28–30
under received-power constraint,
    31
CAP, *see Contention Access Period*
Carrier Sense Multiple Access (CSMA),
  313–14, 322
CDMA, 365, 368
Centralized codes, 256–7
  centralized Alamouti, 264, 272
  centralized scheme, 265
Channel codes, 173
Channel state information, *see* CSI
Chernoff-based routing, 298
Circuit switching, 279
Clairvoyant estimator, 150, 151
Clear-To-Send, 314, 322,
    325
clique, 217
Collector node, 44
Collision-free protocols, 52
Collision Resolution, 326–7
Collocated networks, 49
Communication complexity model, 46,
  48
compatibility function, 217
Computational throughput, 51
Conditional graph entropy, 64
Consistent estimator, 159
Constructive interference, 139
Contention Access Period (CAP),
    314
control policy, 236
Convex optimization, 180, 181
convolutional codes, 89

Wireless Sensor Networks: Signal Processing and Communications Perspectives A. Swami, Q. Zhao, Y.-W. Hong and L. Tong
© 2007 John Wiley & Sons, Ltd